The Routledge Handbook of Critical Resource Geography

This Handbook provides an essential guide to the study of resources and their role in socio-environmental change. With original contributions from more than 60 authors with expertise in a wide range of resource types and world regions, it offers a toolkit of conceptual and methodological approaches for documenting, analyzing, and reimagining resources and the worlds with which they are entangled.

The volume has an introduction and four thematic sections. The introductory chapter outlines key trajectories for thinking critically with and about resources. Chapters in Section I, "(Un)knowing resources," offer distinct epistemological entry points and approaches for studying resources. Chapters in Section II, "(Un)knowing resource systems," examine the components and logics of the capitalist systems through which resources are made, circulated, consumed, and disposed of, while chapters in Section III, "Doing critical resource geography: Methods, advocacy, and teaching," focus on the practices of critical resource scholarship, exploring the opportunities and challenges of carrying out engaged forms of research and pedagogy. Chapters in Section IV, "Resource-making/world-making," use case studies to illustrate how things are made into resources and how these processes of resource-making transform socio-environmental life.

This vibrant and diverse critical resource scholarship provides an indispensable reference point for researchers, students, and practitioners interested in understanding how resources matter to the world and to the systems, conflicts, and debates that make and remake it.

Matthew Himley is an Associate Professor of Geography at Illinois State University. He is a nature-society geographer with interests in the political ecology and political economy of resource industries, especially in the Andean region of South America. His recent research focuses on the historical role of science in mineral extraction and state formation in Peru.

Elizabeth Havice is an Associate Professor of Geography at the University of North Carolina at Chapel Hill. She uses the lens of governance to explore distributional outcomes in marine spaces, food systems, and global value chains. She is a cofounder of the Digital Oceans Governance Lab that explores intersections of data technologies and oceans governance.

Gabriela Valdivia is a Professor of Geography at the University of North Carolina at Chapel Hill. She is a feminist political ecologist examining the relationship between resources and socio-environmental inequities. Gabriela is an author of the digital project Crude Entanglements, which explores the affective dimensions of oil production, and a coauthor of *Oil, Revolution, and Indigenous Citizenship in Ecuadorian Amazonia*.

The Routledge Handbook of Critical Resource Geography

Edited by *Matthew Himley, Elizabeth Havice, and Gabriela Valdivia*

LONDON AND NEW YORK

First published 2022
by Routledge
2 Park Square, Milton Park, Abingdon, Oxon OX14 4RN

and by Routledge
605 Third Avenue, New York, NY 10158

Routledge is an imprint of the Taylor & Francis Group, an informa business

British Library Cataloguing-in-Publication Data
A catalogue record for this book is available from the British Library

Library of Congress Cataloging-in-Publication Data
Names: Himley, Matthew, editor. | Havice, Elizabeth, editor. | Valdivia, Gabriela, 1974- editor.
Title: The Routledge handbook of critical resource geography / edited by Matthew Himley, Elizabeth Havice, and Gabriela Valdivia.
Description: Abingdon, Oxon ; New York, NY : Routledge, 2021. | Series: Routledge international handbooks | Includes bibliographical references and index.
Identifiers: LCCN 2020057616 (print) | LCCN 2020057617 (ebook) | ISBN 9781138358805 (hardback) | ISBN 9780429434136 (ebook)
Subjects: LCSH: Environmental geography--Handbooks, manuals, etc. | Natural resources--Handbooks, manuals, etc.
Classification: LCC G143 .R68 2021 (print) | LCC G143 (ebook) | DDC 333.709--dc23
LC record available at https://lccn.loc.gov/2020057616
LC ebook record available at https://lccn.loc.gov/2020057617

ISBN: 978-1-138-35880-5 (hbk)
ISBN: 978-1-032-02311-3 (pbk)
ISBN: 978-0-429-43413-6 (ebk)

Typeset in Bembo
by KnowledgeWorks Global Ltd.

Contents

Contents

Figures

Figures

Tables

Boxes

Contributors

About the editors

Elizabeth Havice is an Associate Professor of Geography at the University of North Carolina at Chapel Hill. She uses the lens of governance to explore distributional outcomes in marine spaces, food systems, and global value chains. She is a cofounder of the Digital Oceans Governance Lab that explores intersections of data technologies and oceans governance.

Matthew Himley is an Associate Professor of Geography at Illinois State University. He is a nature-society geographer with interests in the political ecology and political economy of resource industries, especially in the Andean region of South America. His recent research focuses on the historical role of science in mineral extraction and state formation in Peru.

Gabriela Valdivia is a Professor of Geography at the University of North Carolina at Chapel Hill. She is a feminist political ecologist examining the relationship between resources and socio-environmental inequities. Gabriela is an author of the digital project Crude Entanglements, which explores the affective dimensions of oil production, and a coauthor of *Oil, Revolution, and Indigenous Citizenship in Ecuadorian Amazonia*.

About the authors

Leslie Acton is an Assistant Professor in the Division of Coastal Sciences, School of Ocean Science and Engineering, at the University of Southern Mississippi. Her research focuses on oceans governance and human–ocean relationships across space, scale, and time.

Helene Ahlborg works as an Associate Professor at Chalmers University of Technology, Department of Technology Management and Economics, Gothenburg, Sweden. Her current work focuses on local and global transitions toward more diverse and decentralized energy systems. Her main themes are power, politics, energy access, and use in East Africa.

Karen Bakker is a Professor and Canada Research Chair in Political Ecology at the Department of Geography, University of British Columbia. Much of her earlier research centers on the political ecology and political economy of freshwater governance, and her more recent work focuses on the implications of digital technologies for environmental governance.

Rahul Basu is a member of The Future We Need, a global movement to make the intergenerational equity principle foundational for civilization. He is also a Research Director at Goa Foundation, an environmental nonprofit in India.

Anthony Bebbington is International Director of Natural Resources and Climate Change at the Ford Foundation, and Higgins Professor of Environment and Society in the Graduate School of Geography, Clark University (on leave). He is also a Director of Oxfam-America, Professorial Fellow at the University of Manchester, and a Research Associate with Rimisp-Centro Latinoamericano para el Desarrollo Rural.

Eloisa Berman-Arévalo is an Assistant Professor at the Universidad del Norte in Colombia. Her research explores the everyday politics of agro-extractivism in the Colombian Caribbean with a focus on conflicts over land and territory, gender, and ethno-racial relations. She engages these topics through an ethnographic lens informed by feminist methodologies and praxis.

Christine Biermann is an Assistant Professor of Geography and Environmental Studies at University of Colorado Colorado Springs. Her research addresses forest dynamics and biodiversity conservation.

Emily Billo is an Associate Professor of Environmental Studies in the Center for Geographies of Justice at Goucher College in Baltimore, MD. A current project develops a feminist analysis of the state through gendered resistance to mining in Intag, Ecuador.

Patrick Bond teaches at the University of the Western Cape, after two prior decades at Wits University in Johannesburg and the University of KwaZulu-Natal in Durban. His doctorate was supervised by David Harvey (Johns Hopkins 1993) and his books include *Elite Transition*, *The Politics of Climate Justice*, and *Looting Africa*.

Gavin Bridge is a Professor of Economic Geography at Durham University and has research expertise in the political economy of natural resources. His research centers on the spatial and temporal dynamics of extractive industries such as oil, gas, and mining.

Joe Bryan is an Associate Professor of Geography at the University of Colorado Boulder. He specializes in participatory mapping, human rights, and comparative politics of indigeneity in the Americas. He is the coauthor with Denis Wood of *Weaponizing Maps* (Guilford Press, 2015) and the coeditor with Bjorn Sletto of *Radical Cartographies* (University of Texas Press, 2020).

Beatriz Bustos-Gallardo is an Associate Professor in the Department of Geography, University of Chile. Her research focus is on the geographies of commodity production in rural landscapes and, in particular, on the salmon and mining economies of Chile.

Lisa M. Campbell is the Rachel Carson Professor of Marine Affairs and Policy in the Nicholas School of Environment at Duke University. She is broadly interested in how new technologies underwrite contemporary interest in oceans conservation and development.

Liam Campling teaches political economy at Queen Mary University of London. He is an editor of *Journal of Agrarian Change,* and the coauthor of *Capitalism and the Sea* (Verso, 2021) as well as *Free Trade Agreements and Global Labour Governance* (Routledge, 2021).

Ana Estefanía Carballo is a Research Fellow in mining and society at the School of Geography, University of Melbourne. Her research interests focus on the interplay between indigenous epistemologies and human-environment development ethics in Latin America. She is a founding editor at *Alternautas* and often collaborates with civil society organizations.

Wim Carton is an Assistant Professor at the Lund University Centre for Sustainability Studies (LUCSUS). His work focuses on the political ecology of climate change mitigation and carbon removal.

Jennifer J. Casolo is a critical human geographer who has worked and/or researched alongside women, indigenous peoples, and rights defenders in Central America for 35 years. Currently, her research-action focuses on the co-strengthening of *Othered* knowledges and praxis in the region that challenge colonial and capitalist designs on humans and natures.

Paul S. Ciccantell is a Professor of Sociology at Western Michigan University. His research agenda examines long-term change in the capitalist world-economy and the roles of raw materials and transport industries in extractive regions.

Christopher Courtheyn is an Assistant Professor in the School of Public Service at Boise State University. He holds a PhD in Geography from the University of North Carolina at Chapel Hill and BA in Latin American Studies from the University of California, Berkeley.

Andrew Curley is an Assistant Professor in the School of Geography, Development & Environment at the University of Arizona and a member of the Navajo Nation. His work focuses on Indigenous geography, land, resources, energy, water, tribal sovereignty, and infrastructure.

Elvin Delgado is an Associate Professor of Geography at Central Washington University. His research explores the political ecology and political economy of fossil fuels in Venezuela and Argentina and the socio-environmental impacts associated with the development of large-scale solar projects in rural communities across Washington state.

Karin Edstedt is a Master's student in Environmental Studies at the Norwegian University of Life Sciences (NMBU), Ås, Norway. She previously wrote her bachelor thesis on the Kachung Forest Project in Uganda.

Nicole Fabricant is an Associate Professor of Anthropology at Towson University in Maryland. Her teaching and research focus on the cultural politics of resource wars in the United States and Latin America. She has written several books on Bolivian social movements and is working on a book for University of California Press about environmental justice organizing in Baltimore.

Maria Fannin is Professor of Human Geography at the University of Bristol, United Kingdom. Her interests include the biopolitics of pregnancy and birth. Her recent work focuses on the human placenta and its role in scientific and cultural understandings of maternal-fetal relations.

Ashley Fent is a postdoctoral fellow at Vassar College. She holds a PhD degree in geography from the University of California, Los Angeles, and an MA in anthropology from Columbia University. Her research focuses on resource controversies in West Africa.

Kathryn Furlong is an Associate Professor of Geography and Canada Research Chair in Water and Urbanization at the Université de Montréal. Her current work focuses on water supply, infrastructure, debt and financialization from a political ecology perspective.

Noella J. Gray is an Associate Professor in the Department of Geography, Environment, and Geomatics at the University of Guelph. Her recent work focuses on knowledge conflicts in global environmental governance, governance of marine protected areas, and ocean territorialization.

Gillian Gregory is a Research Fellow in mining governance in the School of Geography at the University of Melbourne. She is also a visiting researcher at the Centro Interdisciplinario de Estudios sobre el Desarrollo at the Universidad de los Andes in Colombia as well as at the Instituto de Políticas Públicas at the Universidad Católica del Norte in Chile.

Rebecca Gruby is an Associate Professor in the Department of Human Dimensions of Natural Resources at Colorado State University. Her research focuses on contemporary transformations in ocean conservation and governance.

Conor Harrison is an Associate Professor in the Department of Geography with a joint appointment in the Environment and Sustainability Program in the School of Earth, Ocean, and Environment. He is a 2019 recipient of the Michael J. Mungo Undergraduate Teaching Award at the University of South Carolina.

Matthew T. Huber is an Associate Professor of Geography at Syracuse University. He is the author of *Lifeblood: Oil, Freedom and the Forces of Capital* (University of Minnesota Press, 2013). He is currently working on a book on class and climate politics for Verso Books.

Adrienne Johnson is an Assistant Professor in the Environmental Studies Program at the University of San Francisco. Her research interests include the politics of environmental governance, sustainability certification, and feminist approaches to research in extractive resource sectors.

Kärg Kama is an Assistant Professor in Human Geography at the University of Birmingham, United Kingdom. Her research combines critical resource geography with insights from science and technology studies and economic anthropology, with a focus on unconventional fossil fuel resources.

Ahsan Kamal is a Lecturer at Quaid-e-Azam University, Islamabad. He works on land, water, and ecology with riverine, desert, and urban activists in Pakistan. He completed his PhD in Sociology from the University of North Carolina at Chapel Hill and earned an MA in South Asian Studies and International Affairs from Columbia University, New York.

Vijay Kolinjivadi is a Postdoctoral Researcher at the Institute of Development Policy, University of Antwerp. His research has focused on household behavior, community preferences, and relations with downstream actors around ecosystem service policies. His research has involved engagement with farmers and other land-users in Nepal, Kyrgyzstan, and eastern Canada.

Stuart N. Lane is a Professor of Geomorphology at the University of Lausanne where he marries an interest in Alpine environments (ice, water, sediment, and ecosystems) with the challenges posed to a society of a rapidly changing climate.

Rebecca Lave is a Professor and Chair of Geography at Indiana University. Her research combines fluvial geomorphology, political economy, and STS.

Elizabeth Lunstrum is an Associate Professor at Boise State University's School of Public Service. Her research examines the political ecology of conservation—including green militarization, conservation-induced displacement, and illegal wildlife trade (IWT)—along with the political ecology of international borders. Her work focuses on Southern Africa and North America.

Francis Massé is a Lecturer in the Department of Geography and Environmental Sciences, Northumbria University. His research uses political ecology to examine conservation and wildlife crime, including commercial poaching, the illegal wildlife trade, and efforts to address them from the local to global scale.

Will McKeithen (they/them) is an independent scholar and educator who received their PhD in Geography from the University of Washington. Their research explores the uneven distribution of health, wealth, and care, particularly as they affect North American communities targeted by White supremacy, neoliberalism, heteropatriarchy, and mass incarceration. Their work can be found at www.willmckeithen.com.

Sharlene Mollett is an Associate Professor in the Department of Human Geography, Centre for Critical Development Studies, and Department of Geography and Planning. She is a Distinguished Professor in Feminist Cultural Geography, Nature, and Society at the University of Toronto.

Skye Naslund is an Instructor in the Department of Geography at the University of Washington where she also earned her PhD in 2019. Her research examines the changing nature of human-parasite relationships, the popularization of alternative health practices in the United States, and the communication of health (mis)information.

Heidi J. Nast is a Professor of International Studies and Geography at DePaul University. Her interests lie in the geopolitical economy of sex and sexual difference. Her current work focuses on African relational ontologies (the "maternal") prior to the Arab and transatlantic slave trades and how technologies and currency forms made these manifest; and how ontologies of private property and possession mortified and abstracted the "maternal," splitting it into "mother" and "child."

Andrea Joslyn Nightingale is a Professor of Geography at the University of Oslo and part-time Senior Researcher at the Swedish University of Agricultural Sciences where she previously held the Chair in Rural Development. Her research interests in Nepal, Scotland, and Kenya span political violence, public authority, and state formation; climate change; feminist work on emotion and subjectivity; development, transformation, and collective action. Her recent book is *Environment and Sustainability in a Globalizing World* (Routledge, 2019).

Gustavo de L. T. Oliveira obtained his PhD in geography from the University of California, Berkeley, and is now an Assistant Professor of Global and International Studies at the University of California, Irvine. He is a coeditor of *Soy, Globalization, and Environmental Politics in South America* (Routledge, 2018).

Nancy Lee Peluso is a Professor of political ecology in the Department of Environmental Science, Policy, and Management at the University of California, Berkeley. She draws inspiration from Indonesian people and places, where she has conducted ethnographic-historical research in resource geographies of forests, agrarian transformations, small-scale mining, and migration for more than three decades.

Tom Perreault is DellPlain Professor of Latin American Geography at Syracuse University (Syracuse, New York). His research concerns the political ecologies of resource governance, rural livelihoods, agrarian transformation, and rural peoples' social movements in the Andean region.

Elspeth Probyn is a Professor of Gender and Cultural Studies at the University of Sydney. She is the author of *Sexing the Self: Gendered Positions in Cultural Studies*; *Outside Belongings*; *Carnal Appetites*; *Blush: Faces on Shame*; and *Eating the Ocean*. Her latest book is the coedited (with Kate Johnston and Nancy Lee) *Sustaining Seas: Oceanic Space and the Politics of Care*.

Danielle M. Purifoy is an Assistant Professor in the Department of Geography at University of North Carolina at Chapel Hill. Her research traces the racial politics of development and environmental inequity in Black communities.

Kolson Schlosser is an Associate Professor of Instruction in the Department of Geography and Urban Studies at Temple University. He teaches primarily in the department's environmental studies program and publishes in the area of environmental geohumanities.

Elizabeth Shapiro-Garza is an Associate Professor of the Practice of Environmental Policy and Management at the Nicholas School of the Environment, Duke University. She has worked and conducted research with rural communities in Latin America and the United States for over 25 years.

Kathryn Snediker is a Research and Instruction Librarian at the University of South Carolina's Thomas Cooper Library. She has been the library's liaison to the Department of Geography for more than eight years, teaching research techniques and information literacy skills.

Alejandra Uribe-Albornoz is a PhD student in geography at Université de Montréal. Her research examines processes of ecosystem definition and their consequences on livelihoods and resource access.

Gert Van Hecken is an Assistant Professor in International Cooperation and Development at the Institute of Development Policy, University of Antwerp. For over 15 years, he has lived and worked in Nicaragua, both as a researcher on social-environmental dynamics in rural communities and as a representative for a development NGO.

Martine Verdy is a Postdoctoral Researcher in geography at Université de Montréal. Her research explores the territorial dispute between the provinces of Quebec and Newfoundland over the Churchill Falls hydroelectric development in Labrador, Canada.

Tim Werner is a Research Fellow at the University of Melbourne, School of Geography. With a background in environmental engineering and geographical science, his research employs mixed methods from economic geology, industrial ecology, and remote sensing to analyze various aspects of mining.

Catherine Windey is a PhD candidate and teaching assistant at the Institute of Development Policy, University of Antwerp. She conducts research on local socio-spatial experiences, land use knowledge, and practices in the Democratic Republic of Congo. She also worked in Cameroon on community-based forestry.

Anna Zalik is an Associate Professor at York University, Canada. Her scholarship centers on the political ecology, economy, and historical sociology of industrial extraction. Zalik's recent work concerns Mexican energy sector restructuring, financial risk in Canadian hydrocarbons, and the development of an extractive regime for the international seabed.

Annah Zhu is an Assistant Professor of environmental globalization in the Environmental Policy Group at Wageningen University in the Netherlands. She received her PhD in Society and Environment from the University of California, Berkeley. Her work analyzes "Global China" and the environment from a political ecology perspective.

Sarah Bess Jones Zigler is an ethnographer who studies how the worlds we live and perform are the worlds we strive to conserve. She advocates for the decolonization of conservation to conserve many natures.

Preface

Handbook-making

Handbooks are products of the careful curation of a diverse collection of contributions, but we rarely hear about how these collections come together—that is, about the processes of selecting, organizing, and looking after the items that end up in a collection. This Preface represents our effort to raise the curtain on our editorial process and the conditions under which the Handbook was created, which we see as part of our commitment to ongoing reflection on the question of how, why, and for whom the academic production of knowledge about resources matters (see Valdivia, Himley, and Havice, Chapter 1 in this volume). It offers an overview of the intellectual-political method that unfolded incrementally and collaboratively during the production of the Handbook.

Our initial motivation for putting together this Handbook was to take stock of the increased attention to resource-related processes in geographical (and, more broadly, social-scientific) scholarship. Noting that this work was characterized by the application of a variety of critical-theoretical approaches to resource dynamics, we aimed to probe if and how these multiplying strains of scholarly activity came together into a subfield—"critical resource geography"—and, if so, around which kinds of questions, methods, and forms of praxis it might cohere. Our initial review of the literature suggested that a core objective of this scholarship was to destabilize dominant understandings of resources by detailing the power-laden as well as historically and geographically contingent practices though which resources come to be and circulate through social-environmental life. And yet, as an arena of scholarly inquiry, "critical resource geography" seemed emergent: a field-in-the-making, encompassing work by scholars who perhaps identified more squarely with other subdisciplines in geography—cultural geography, economic geography, feminist geography, political ecology, political geography, among others—or other disciplines altogether and thus brought a wide range of perspectives to questions of resources. From the initial stages of this editorial project, then, we knew that our task would *not* be—as is often the case with volumes like this—to provide a state-of-the-art assessment of an already existing, well-bound, and well-defined body of scholarship. Rather, we developed an approach to the project that was more about probing boundaries than reporting on them; more about working to identify the analytical, methodological, and normative concerns that stitch together critical resource geography as a field of inquiry (to the extent that these common concerns exist), rather than describing or making them accessible to new audiences. And we imagined the process of addressing these issues to be an iterative one—something we would undertake in conversation with contributing authors.

We envisioned a volume consisting of various kinds of chapters. Some would be genealogical, focusing on the history of resource geography and the multiple and diverse intellectual traditions that scholars have brought to it. Others would center on particular theoretical or methodological approaches, with an emphasis on exploring the advantages and limitations of these for informing broader debates about the role of resources in socio-environmental change. We also wanted to provide a venue for reflection on the practices of critical resource scholarship. The focus here would be on "engaged" forms of teaching and research, including those carried out via sustained interactions with economic, political, and community actors beyond academia—for instance, as part of public, participatory, and/or collaborative scholarship. Finally, we envisioned chapters showcasing, through empirical research, the value of critical resource geography's theoretical and methodological tools. These chapters would be organized around the conceptual framing device of "resource-making/world-making," a heuristic that came to take on broad significance in this project (for elaboration, see Valdivia, Himley, and Havice, Chapter 1 in this volume). In terms of format, we were keen to take advantage of the relative openness that the Handbook model offered, and we imagined extending to contributing authors the opportunity to construct their chapters in inventive, nontraditional formats—something, we were delighted to see, several of our authors indeed ended up doing. Across these diverse chapters, our hope was that the volume would contribute an expansive approach to the study of resources and resource systems, one capable of making (more) visible the connections between the (un) making of resources and the (un)making of worlds.

In line with our approach of exploring the boundaries of critical resource geography—and aware of the role that Handbook editors play as "gatekeepers" in the production of academic knowledge (Schurr, Müller, and Imhof 2020)—our author-invitation strategy sought to cultivate a multiple-perspective exchange of ideas on the nature and potential futures of this body of scholarly work. We contacted potential authors who were at different career stages, from different disciplinary and sub-disciplinary "homes," and at institutions in a variety of countries. We encouraged prospective authors to invite coauthors as a way of bringing a broader and more diverse set of voices into the conversation. Several of our original invitees declined to contribute a chapter, including for reasons that brought the political economy of academic work and publishing directly to bear on the content and trajectory of this project. Several prospective authors declined simply due to being overcommitted—an all-too-common characteristic of work in the neoliberal university (Mountz et al. 2015). A pre-tenure scholar declined on the basis of advice from their department chair, who had said that a chapter in an edited volume like this would be of limited value for tenure and promotion. In another case, a potential author chose not to participate in part because of the model of knowledge production and circulation itself. Specifically, they were concerned by how commercial publishers benefit from established routes through which scholars accumulate academic prestige and by using paywalls that limit public access to knowledge—a concern about for-profit academic publishing that is shared by many and persists even with the emergence of creative and experimental ways of publishing scholarship (see Batterbury 2017; Kallio and Hyvärinen 2017; Kallio and Metzger 2018). In a number of cases, declined invitations from senior scholars came with alternative suggestions that led us to early-career scholars of whom we would not have been aware, and whose contributions have enriched the volume greatly. Several authors responded to our invitations with questions: what *is* critical resource geography? What makes it distinct and worthy of its own handbook in a crowded field of handbooks? Why this project and why now? What do you mean when you say *critical*? These questions prompted exchanges that helped to hone the project and to develop our ideas about what resource geography is, and what it might be.

Our editorial process involved deep engagement with contributors, with one of us serving as lead editor for each of the chapters. For each manuscript, the lead editor and one other editor provided comments on the first draft; for subsequent drafts, the second editor "switched" to the one of us who had not read the first draft. We chose this strategy so that authors would benefit from our collective energies and expertise while also ensuring that all three of the volume's coeditors developed up-close knowledge of each of the chapters. To foster conversation among authors, we organized a series of three panel sessions at the 2019 annual meeting of the American Association of Geographers held in Washington, DC, in which 16 of the volume's authors participated. This was a valuable opportunity to think across chapters and to identify cross-cutting themes at an early stage in the editorial process.

We took a collaborative, co-creative approach to write this Preface and the Handbook's introductory chapter. We outlined these pieces during a multi-day editors' meeting in North Carolina in early 2020. Soon after this, when we had final or near-final versions of all the volume's chapters, we began holding weekly virtual meetings to exchange ideas, to discuss the themes and literatures with which to engage, and to experiment with approaches to collaborative writing. Week after week, we drafted text, suggested revisions to each other's contributions, refined the volume's organizational structure, and worked through generative tensions in our approaches and perspectives in a process that embodied the "restless thinking" that we describe in Chapter 1.

Publication authorship and productivity are typically seen as key measures of the impact, relevance, and importance of scholarly work and function as "benchmarks" in academic hiring and promotion decisions. As a technology of accounting, authorship order is also significant, including because it can signal whose voices are represented, and can reproduce oppressive practices (Kobayashi, Lawson, and Sanders 2014; Mattingly and Falconer-Al-Hindi 1995; Mott and Cockayne 2017). In high-ranking geography journals, for instance, women are underrepresented in authorship positions that equate to respect and merit, and women more frequently write coauthored papers for which it may be difficult to clearly attribute credit for work (e.g., Rigg, McCarragher, and Krmenec 2012). For us, because we each have tenure, the stakes in deciding authorship order may have been lower than if we were in less-secure academic positions; but they still gave us pause. We settled on a strategy that involved cycling through our names on the different elements of the project and contextualizing our collaboration in this Preface. Yet, we remain aware that prevailing norms regarding authorship order and attribution—coupled with the strangeness and ambiguity of "unbounding" these norms—can act as a constraint on innovation in research and engagement in critical resource geography and other fields (see also Ahlborg and Nightingale, Chapter 2 this volume). We are heartened that it is now common to see authors add "equal-collaboration" statements to their writings or elect to list authors in reverse-alphabetical order. We also note that many natural-science journals now require a statement summarizing the contributions of each author to data collection, analysis, and writing. To attend directly to authorship questions like these is to highlight "the labor process, assumptions, possibilities, and risks" of coauthorship—undertaken among academics or in collaboration with nonacademic authors—as a tool for making nuanced critical interventions (Nagar 2013, 1).

Finally, over the last year, we concluded this project in the midst of the COVID-19 pandemic as well as a surge of anti-racist social mobilizations in numerous parts of the world. We acknowledge these world-transforming events here in part because they have shaped our thinking about resources, including by (further) drawing into sharp relief the centrality of resources and resource systems to the contemporary world. We see various ways in which thinking critically with resources sheds light on the emerging impacts of and responses to COVID-19, as well as the linkages between the pandemic and the systems of oppression that today's justice movements aim to dismantle. Several governments, for instance, have used the economic fallout of the pandemic as a justification to accelerate private investment in land appropriation and resource extraction (logging, mining, oil extraction, etc.), the negative socioecological effects of which tend to fall heaviest on marginalized groups (Davenport and Friedman 2020; El Comercio 2020; Torres and Branford 2020; Vila Benites and Bebbington 2020). More broadly, the making of and trade in resources has long been central to the systems of coloniality, patriarchy, and racial capitalism that have delineated patterns of vulnerability to COVID-19 and that are key for understanding the pandemic's syndemic characteristics. In particular, it is increasingly clear that people already living under the "noxious social conditions" of capitalism (Adams 2020; Herrick 2020) have been especially vulnerable to the virus and to the pandemic's socioeconomic effects, such as historically disenfranchised peoples suffering from long-term disinvestment in public health and basic services (Angelo 2020; Bagley 2020; Johns 2020; Kestler-D'Amours 2020; Saffron 2020); the descendants of original peoples who have experienced repressive state actions under COVID-19 emergency declarations (IWGIA 2020); and migrant workers whose mobility and livelihood opportunities are limited by stay-at-home and social distancing orders (Roy 2020a, 2020b) or who are at the frontlines of COVID-19 outbreaks as "essential workers" in food industries (Groves and Tareen 2020; Havice, Marschke, and Vandergeest 2020; Jabour 2020).

Our own personal and professional lives also have been impacted by the pandemic. During the last year of the project, we worked remotely from home while caring for and supporting the remote or home-based educations of our young children, whose own daily lives have been radically changed, including due to the closure of schools, the cancellation of out-of-school activities, and the inability to interact in-person with many of their friends and loved ones. Many of our contributing authors experienced similar situations, caring for isolated loved ones, and made their ways through these pandemic conditions with restricted or anemic support systems. During this time, we also learned that many authors had carved out time to support students, friends, and allies whose status had been jeopardized by sudden travel restrictions or changes to student visa rules, and others had committed to mutual aid and fundraising efforts to expand caregiving to those most vulnerable to COVID-19. Professionally, even if the research agendas of critical resource geographers and allied scholars are not directly tied to justice-oriented social mobilizations or COVID-19, the events of the past year have profoundly transformed everyday spaces of research, teaching, and advocacy. COVID-19-related mobility restrictions, for instance, have affected access and connectivity in research and advocacy (cf. Finn et al. 2020), while the pandemic's deepening of existing inequalities has reshaped the worlds about which critical resource scholars produce knowledge.

These events have shaped our thinking about this collection and, of course, impacted how we worked together and with contributing authors to complete the project as we faced instability and disruption at every turn. Working through this moment of rupture added to the urgency of advancing a critical resource geography that explores how resources come to be and how their histories, geographies, and the social and political-economic systems through which they circulate are intertwined with the worlds we inhabit *and* with the worlds that might be.

We continue to be inspired by the contributors who put trust in the idea of a Handbook of critical resource geography. We thank them for engaging seriously with our questions and suggestions. Their intellectual generosity and curiosity fill the pages that follow. We urge readers to draw upon the innovative insights and reflections in the contents of this Handbook—as we have—to navigate the present and envision future possibilities for resource geography and the world.

Elizabeth Havice, Gabriela Valdivia, Matthew Himley

References

Adams, Vincanne. 2020. "Disasters and capitalism…and COVID-19." *Somatosphere*, March 26. http://somatosphere.net/2020/disaster-capitalism-covid19.html/.

Angelo, Mauricio. 2020. "Lack of clean water leaves Brazil indigenous reserve exposed to coronavirus." *Reuters*, April 21. https://www.reuters.com/article/us-health-coronavirus-brazil-indigenous/lack-of-clean-water-leaves-brazil-indigenous-reserve-exposed-to-coronavirus-idUSKCN2232H1.

Bagley, Katherine. 2020. "Connecting the dots between environmental injustice and the coronavirus." *YaleEnvironment360, Yale School of the Environment*, May 7. https://e360.yale.edu/features/connecting-the-dots-between-environmental-injustice-and-the-coronavirus.

Batterbury, Simon. 2017. "Socially Just Publishing: Implications for Geographers and Their Journals." *Fennia-International Journal of Geography* 195 (2): 175–181.

Davenport, Coral, and Lisa Friedman. 2020. "Trump, citing pandemic, moves to weaken two key environmental protections." *The New York Times*, June 4. https://www.nytimes.com/2020/06/04/climate/trump-environment-coronavirus.html.

El Comercio. 2020. "Editorial: La veta intacta." *El Comercio*, July 14. https://elcomercio.pe/opinion/editorial/editorial-la-veta-intacta-mineria-noticia/?ref=ecr.

Finn, John C., Eugenio Arima, Martha Bell, Jessica Budds, Jörn Seemann, Gabriela Valdivia, Diana Tung, Eric Carter, and Alan Marcus. 2020. "Editorial: Mobility, Connectivity, and the Implications of Covid-19 for Latin American Geography." *Journal of Latin American Geography* 19 (4): 6–8.

Groves, Stephen, and Sophia Tareen. 2020. "US meatpacking industry relies on immigrant workers. But a labor shortage looms." *The Los Angeles Times*, May 26. https://www.latimes.com/food/story/2020-05-26/meatpacking-industry-immigrant-undocumented-workers.

Havice, Elizabeth, Melissa Marschke, and Peter Vandergeest. 2020. "Industrial Seafood Systems in the Immobilizing COVID-19 Moment." *Agriculture and Human Values* 37: 655–656.

Herrick, Clare. 2020. "Syndemics of COVID-19 and 'pre-existing conditions.'" *Somatosphere*, March 30. http://somatosphere.net/2020/syndemics-of-covid-19-and-pre-existing-conditions.html/.

IWGIA. 2020. "Safeguarding the health, wellbeing and livelihoods of indigenous peoples across the world in face of COVID-19." *Statement. The Group of Friends of Indigenous Peoples*. https://iwgia.org/en/news-alerts/news-covid-19/3573-statement-group-of-friends-ips.html.

Jabour, Anya. 2020. "Immigrant workers have borne the brunt of Covid-19 outbreaks at meatpacking plants." *The Washington Post*, May 22. https://www.washingtonpost.com/outlook/2020/05/22/immigrant-workers-have-born-brunt-covid-19-outbreaks-meatpacking-plants/.

Johns, Wahleah. 2020. "A Life on and Off the Navajo Nation: The reservation has one of the country's highest rates of infection." *The New York Times*, May 13. https://www.nytimes.com/2020/05/13/opinion/navajo-nation-coronavirus.html.

Kallio, Kirsi Pauliina, and Pieta Hyvärinen. 2017. "A Question of Time—or Academic Subjectivity?" *Fennia-International Journal of Geography* 195 (2): 121–124.

Kallio, Kirsi Pauliina, and Jonathan Metzger. 2018. "'Alternative' Journal Publishing and the Economy of Academic Prestige." *Fennia-International Journal of Geography* 196 (1): 1–3.

Kestler-D'Amours, Jillian. 2020. "Indigenous 'at much greater risk' amid coronavirus pandemic." *Al Jazeera*, March 20. https://www.aljazeera.com/indepth/features/indigenous-greater-risk-coronavirus-pandemic-200320185556308.html.

Kobayashi, Audrey, Victoria Lawson, and Rickie Sanders. 2014. "A Commentary on the Whitening of the Public University: The Context for Diversifying Geography." *The Professional Geographer* 66 (2): 230–235.

Mattingly, Doreen, and Karen Falconer-Al-Hindi. 1995. "Should Women Count? A Context for Debate." *The Professional Geographer* 47 (4): 427–435.

Mott, Carrie, and Daniel Cockayne. 2017. "Citation Matters: Mobilizing the Politics of Citation Toward a Practice of 'Conscientious Engagement." *Gender, Place & Culture* 24 (7): 954–973.

Mountz, Alison, Anne Bonds, Becky Mansfield, Jenna Loyd, Jennifer Hyndman, Margaret Walton-Roberts, Ranu Basu, et al. 2015. "For Slow Scholarship: A Feminist Politics of Resistance through Collective

Action in the Neoliberal University." *ACME: An International Journal for Critical Geographies* 14 (4): 1235–1259.

Nagar, Richa. 2013. "Storytelling and Co-Authorship in Feminist Alliance Work: Reflections from a Journey." *Gender, Place & Culture* 20 (1): 1–18.

Rigg, Lesley S., Shannon McCarragher, and Andrew Krmenec. 2012. "Authorship, Collaboration, and Gender: Fifteen Years of Publication Productivity in Selected Geography Journals." *The Professional Geographer* 64 (4): 491–502.

Roy, Arundhati. 2020a. "The pandemic is a portal." *Financial Times*, April 3. https://www.ft.com/content/10d8f5e8-74eb-11ea-95fe-fcd274e920ca.

Roy, Arundhati. 2020b. "After the lockdown, we need a reckoning." *Financial Times*, May 23. https://www.ft.com/content/442546c6-9c10-11ea-adb1-529f96d8a00b.

Saffron, Jesse. 2020. "Covid-19 shines light on Navajo water contamination." *Environmental Factor, National Institute of Environmental Health Sciences*, June. https://factor.niehs.nih.gov/2020/6/feature/1-feature-navajo-contamination/index.htm.

Schurr, Carolin, Martin Müller, and Nadja Imhof. 2020. "Who Makes Geographical Knowledge? The Gender of Geography's Gatekeepers." *The Professional Geographer* 72 (3): 317–331. doi: 10.1080/00330124.2020.1744169.

Torres, Mauricio, and Sue Branford. 2020. "Brazil opens 38,000 square miles of indigenous lands to outsiders." *Mongabay: News and Inspiration from Nature's Frontline*, May 8. https://news.mongabay.com/2020/05/brazil-opens-38000-square-miles-of-indigenous-lands-to-outsiders/.

Vila Benites, Gisselle, and Anthony Bebbington. 2020. "Political Settlements and the Governance of Covid-19: Mining, Risk, and Territorial Control in Peru." *Journal of Latin American Geography* 19 (3): 215–223. doi: 10.1353/lag.0.0135.

Acknowledgments

Over the course of the three years it took to complete this project, we learned that putting together a Handbook requires love, grit, perseverance, and a multitude.

We thank Andrew Mould at Routledge, who planted the seed for this project with the idea to compile a volume that would examine "current debates, controversies, and questions around resources." We are also grateful to Egle Zigaite at Routledge for shepherding us through the manuscript preparation and production process.

Editorial work is collective and collaborative. We are indebted to Trey Murphy for his detail-oriented, enthusiastic, and well-organized editorial assistance. We are also grateful to Ana Kinsley, who collaborated with Trey. Together they worked with surgical precision to format and copyedit chapters prior to submission to Routledge. We also thank the summer internship and mentoring program at the North Carolina School of Math and Science, which made possible Ana's participation in the Handbook's editorial process. Caryn O'Connell of Watertight Texts (www.watertight-texts.net) also played a key role in the manuscript's preparation. We are thankful for her eagle-eyed, thoughtful editing as well as for her excitement about the content at hand, and we thank Dr. John Florin for directing us to Caryn.

We thank Dr. Nancy Lee Peluso for permission to use one of her beautiful photographs on the cover of this Handbook. Nancy, your photography is exquisite. Dr. Peluso took this photo during fieldwork in West Kalimantan, Indonesia in 2014–2015. The photo captures a moment in the panning process when an expert panner consolidates the flakes of gold caught in the sluice carpets of a small-scale mining operation. The panner shown here is in the final step in the process, and his labor will produce a small, consolidated gold ball. This and other photos of miners are displayed in Dr. Peluso's exhibit entitled *Indonesia: Spectacles of Small-Scale Gold Mining*, which is part of the national project *Extraction: Art on the Edge of the Abyss* organized by publisher Peter Koch and Sam Pelts and sponsored by the University of California Berkeley's Environmental Design Library. For information about other *Extraction* exhibits, see https://www.extractionart.org/home.

We are grateful to Dr. Lisa Campbell for careful and thoughtful comments that helped us to push the introductory chapter of this volume to conclusion; thank you for taking the time for this when there was simply no time available in late-2020 pandemic conditions. We also thank our students at UNC Chapel Hill in GEOG 435 Global Environmental Justice, GEOG 803 Ontological Politics, and GEOG 130 Development and Inequality, who read draft versions of select chapters. Your feedback helped us keep the need for clear communication at the forefront of this project.

Putting together a Handbook costs money and takes time. The editors thank, for their generous financial support, the Department of Geography at the University of North Carolina at Chapel Hill; the University of North Carolina at Chapel Hill Office of Research Development;

the Department of Geography, Geology, and the Environment at Illinois State University; and the Illinois State University College of Arts and Sciences. Elizabeth Havice also thanks the Institute for the Arts and Humanities whose Faculty Fellowship Program afforded her time for work on this Handbook.

Our deepest gratitude goes to the 60 contributing authors for their faith and determination and for the rich and stimulating conversations we had with them as we poured over and commented on chapter drafts. Many authors also advanced the collective conversation by engaging in a three-part session on critical resource geography that we organized at the 2019 annual meeting of the American Association of Geographers, held in Washington DC. We thank the authors for continuing to work on and discuss with us the world of resources. We have learned so much from your intellectual labor and generosity. You have enriched our scholarship, and we are certain that your chapters will enrich the scholarship of Handbook readers too.

Last, but certainly not least, we thank our families—Reecia, Jona, and Ellis; James and Orion; Chris, Eva, and Noah. Their intellectual support, love, and time have made this project possible.

Critical resource geography

An introduction

Gabriela Valdivia, Matthew Himley, and Elizabeth Havice

Introduction

This Handbook is about the state of knowledge of one of geography's most cherished objects of study: resources. Resources can encompass a broad range of things, including, but not limited to, physical entities that are regularly disentangled from their existing relations and incorporated as parts or fragments within other sets of relations, in order to fulfill a promise. Resources, for example, are often thought of as means to an end, instruments to realize a goal or state, such as a life free of suffering or a "higher" level of socioeconomic development. Think of hydrocarbons extracted from the underground and refined to generate energy, or guano harvested from island ecosystems to fertilize depleted soils, or tuna captured from the oceans to meet food market demands. But resources, and the promises that they are expected to fulfill, are not simple or straightforward. Resources require systems of "resource-making" (Kama 2020; Li 2014; Richardson and Weszkalnys 2014), each with its own infrastructures, logics, temporalities, and valuation systems. Removing something from its existing relations in order to incorporate it as a resource into a new set of relations requires thought and action, all based on architectures of valuation through which some things and relations are rationalized as more valuable than others. This, in turn, raises questions about who is making these value judgments, in what context these valuations make sense and become dominant, and how systems of resource-making affect different constituencies in varied and uneven ways.

The systems through which resources are made and circulated have compounding effects, shaping the world and how people experience and know it. The idea for this Handbook emerged in the context of an expansion of "critical" resource-centered scholarship examining the relationship between resource systems and the uneven worlds they create (see, for example, Bakker and Bridge 2006; Bridge 2009; Furlong and Norman 2015; Huber 2018; Lawhon and Murphy 2012; Kama 2020; Robbins 2002). Broadly, this research coheres around three key elements: an approach that positions the resource itself as the analytical starting point; an emphasis on the interrelated materiality and spatiality of resources and resource systems; and a concern for unequal power relations, distributive outcomes, and the ethical dimensions of these systems. Thematically, the focus is often on the capitalist production, distribution, and consumption of "established" resources—that is, entities whose identities as resources are relatively well-consolidated (e.g., copper, oil, tuna)—as well as the emergence of new socio-spatial "frontiers" for these, as seen, for instance, in the march of hydrocarbon and mineral-mining operations offshore and into deeper waters. Scholars also examine how an increasing array of things-in-the-world are abstracted, monetized, and incorporated into social life as "novel" resources or in resource-like ways: things such as human tissues, wildlife, parasites, and ecosystem services. Across this body of scholarship, researchers are attentive to the contestations and crises—from climate change to species extinction to toxic contamination—that are generated by dominant (i.e., capitalist) modes of resource production, consumption, management, and disposal. And scholars are increasingly interested in resource futures (e.g., green transitions and degrowth) as well as worlds that exist (or might exist) without or against the notion of resources.

This editor's introduction and the chapters that follow in this Handbook examine various dimensions of the promises of resources and their worlds. Collectively, these contributions revisit geographical thought and bridge disciplinary divides to prompt (other) ways for thinking with and about resources. Two fundamental ideas about critical resource geography have emerged from the contents of the chapters that follow and have come to underpin this Handbook project.

The first is that *the critical analysis of resource systems and their historical and contemporary geographies is integral for understanding the state of the world.* Resources have long been central components of the political-economic systems of coloniality, patriarchy, and racial capitalism that have made and continue to remake the contemporary world. For example, the hunt for and trade in valued materials, like furs, gold, and silver, drove European colonialism from its earliest days (Galeano 1971; Wolf 1982; Yusoff 2018). In the Americas, the centuries-long oppression of enslaved peoples in large-scale commodity production (e.g., cotton, sugar, tobacco) was central to the rise of the integrated nineteenth-century Atlantic economy and its extension to the rest of the world (Inikori 2020; Mintz 1986), while unfree labor of multiple kinds built the infrastructures (e.g., railroads) that made this integration possible. Throughout, Indigenous societies were relocated, exploited, or eliminated to secure land and resources for settler colonialism and imperial expansion (Dunbar-Ortiz 2014; Lowe 2015; Wolfe 2006). These histories of colonialism, slavery, environmental degradation, and imperial trade in resources deepened uneven development and structural forms of inequality (Du Bois 2009; Johnson 2018; Pulido 2018) *and* seeded collective memories of loss, harm, and disenfranchisement at the center of transnational struggles to end systemic oppression (Erakat 2019; Kubat 2020; Moaveni and Tahmasebi 2020; Nagar and Shirazi 2019; Roy 2020). Resources are central to the contemporary social world and the processes that sustain and organize it.

The second idea underpinning this volume is that *doing critical resource geography involves ongoing reflection on how, why, and for whom the academic production of knowledge about resources matters.* This Handbook is grounded in social science critical theory that acknowledges, challenges, and seeks to change the status quo that tends to promote the well-being of some at the cost of the dehumanization and devaluation of others (Agger 1998; Fay 1987; Horkheimer 1972; see Peet 2000 for a distinction between radical and critical geography). Critical scholarship is diverse in its intellectual alignments and territories—Noel Castree (2000, 956) calls it an "umbrella" term—but is oriented by a common interest in working at the "political edge" (Blomley 2006, 88) of research, teaching, and advocacy practices. Nonetheless, critical scholarship can lose this political edge when it becomes normalized and institutionalized. As Ryan Cecil Jobson (2020) argues, many established modes of doing academic work—from how scholars perform research to where and how they seek professional validation—can narrow how this knowledge production matters and to whom. One result can be a "positional superiority" (Said 1979; Smith 2012) that locates authority with researchers rather than the people and communities being researched (Tuck 2013). Recognizing these risks, this Handbook prioritizes a "restless form of thinking" (Buchanan 2010) that is attentive to who carries out and validates studies on resources and how, to the ways in which scholarly "conceptions of the world" (Gramsci 1985) matter to the study resources, and to the question of for whom this knowledge is produced. This restless approach entails resisting the urge to codify or "fix" the character of practical (and political) responses to socioecological inequities and acknowledging that universalist conceptions of liberation can risk reproducing rather than changing the status quo (Burkhart 2016; Robinson 2000).

In its totality, this Handbook offers a toolkit of critical, reflexive, and speculative approaches for studying resources and the socioecological systems with which they are co-constitutively entangled. The remainder of this introductory chapter sets the stage for what follows. The next section situates the descriptor "critical" in critical resource geography and outlines key trajectories of intellectual-political analysis within this body of scholarly work. The following section of the chapter offers an overview of the objectives and contents of the Handbook's four main sections, while the concluding section elaborates on the motivations that prompted us to produce this volume and extends an invitation to readers to engage with the world of critical resource geography.

Situating the *critical* in critical resource geography

We use the descriptor "critical" to refer to an intellectual-political method of identifying how things become resources, as well as the work that these resources do in the world. The following three broad questions animate critical resource geography:

- How do resources matter to the material organization of human societies?
- How do resources become meaningfully present in the world and what makes this possible?
- What would another world of, or without, resources look like?

Dictionary entries offer a useful entry point into how, in practice, critical analysis works in resource studies. Definitions provided in dictionaries are imbued with archival authority and reflect knowledge as a curated product of its geopolitical time (Lepore 2013). These entries are what French-Algerian philosopher Jacques Derrida ([1974] 2016) called dominant forms of "coded" reality. For Derrida, how people write about and narrate the world—for instance, in defining words and in recording these definitions in dictionaries—both represents the world and bounds what is possible to know as real.

For example, in the *Oxford English Dictionary* (*OED*), aimed for an Anglo audience, "resource," in the singular, may refer to "a means of supplying a deficiency or need; something that is a source of help, information, strength, etc."; and, in the plural form, the term may refer to "stocks or reserves of money, materials, people, or some other asset, which can be drawn on when necessary" or to "the collective means possessed by a country or region for its own support, enrichment, or defence." In these definitions, resources are assumed to have a fixed presence in the world and a universal relationship to people. Moreover, there is an implied singular audience: these definitions seem to suggest that any person in this world, regardless of their histories, beliefs, and experiences, would relate to resources in a similar way—that is, as things-in-the-world that have been (or could be) abstracted from their existing relations and used according to perceived individual or social needs.

How resources exist in the world is coded in the language used to represent them. For example, words laced within the *OED* definitions—enrichment, materials, means, possessed, reserve, source, stock—are central to common understandings of what one gets when "nature" and "resources" are fused into "natural resources." It is into this category that "societies place those components of the non-human world that are considered to be useful or valuable in some way," as Gavin Bridge (2009, 1219) notes. Natural resources, then, according to popular or default conception, are materials or phenomena "found in nature"—gold, natural gas, squid, water, wind, etc.—that people make use of (once possessed) to satisfy some want or need, including that of turning a profit by selling to others (enrichment). Sardinian intellectual Antonio Gramsci (1971, 347) referred to such default ideas as *senso comune*. Translated into English as "common sense," these are held-in-common beliefs—kinds of collective knowledge—that people encounter as already self-evident truths. And as is further implied by other terms woven into the OED definitions—reserve, source, stock—common-sense understandings of natural resources lean toward the static and ahistorical: these are things that exist as results of physical processes outside of human thought and intervention.

Just as some ideas about what resources *are* have become common sense, so too have certain notions about the social *effects* of their existence and exploitation. For instance, one common-sense idea is that patterns of wealth and poverty in the world can be explained by the uneven geographical distribution of natural resources. The idea here is that environmental factors—namely, the presence or absence of resources in any given territory—shape, if not determine, the economic-development possibilities of the territory's peoples. Such conclusions would seem to flow logically if one operates from the default position that natural resources are things "found in nature" that have value and/or utility. And the seemingly self-evident hinging of economic development to resource presence and exploitation that characterizes this line of thinking has been present in much mainstream post-WWII development theory (e.g., Rostow 1960).

Yet, evidence challenges such common-sense ideas. While one can point to examples of countries that have (apparently at least) had success in "mobilizing" their resource endowments to promote economic growth—Botswana, Canada, and Norway are oft-cited examples—there are myriad cases that tell different stories: stories of intensive resource extraction linked to environmental degradation, poverty, and marginalization, rather than broad-based social and economic development (Bebbington et al. 2008; Bunker 1985; Galeano 1971; Himley 2019; Valdivia 2018). Examples like these latter ones have led to other ideas that have themselves become common sense—at least within certain academic and policy circles—and that fall under the moniker of the "resource curse." As its name suggests, this concept is meant to convey the idea that the presence of valuable resources may be—and often is—a bane not blessing, and research in this area ties large-scale resource exploitation to a variety of economic, political, and social problems, from civil unrest to corruption to economic "underperformance" (Auty 1993; Karl 1997; for reviews, see Bebbington et al. 2008; Le Billon 2005). In this line of thinking, resource "abundance" *can* benefit a country economically, including by enhancing the flow of capital goods and investment, but only *if* governments manage and tax resource exploitation effectively (Auty 1993); alternatively, cases in which resource exploitation neither sustains economic growth nor decreases poverty signal a problem of "mismanagement." The "paradox of plenty" (Karl 1997)—the apparent inability of many developing countries to effectively use resource wealth to boost their economies—has informed policies focused on liberalizing resource sectors, enhancing the competitiveness of non-resource sectors, and reducing the influence of "weak" rent-seeking institutions (Badeeb, Lean, and Clark 2017; Karl 2005).

Critical resource scholarship questions and destabilizes these sorts of common-sense understandings of resources, including by stripping resources themselves as well as ideas about them of their taken-for-grantedness, their very "naturalness." The aim is to deconstruct what seems familiar about resources—to make resources *strange* (Li 2014)—in order to gain a fuller and thus more responsible understanding of their complex role in the world. Deconstruction can make visible how the dominant graphing of the world elevates some groups of people as authorized producers of knowledge—for example, of resources, and their curses and cures—while others are devalued or excluded entirely (Derrida [1974] 2016; Smith 2012). Critical resource scholars do so by tracing the material and discursive conditions under which resources are produced, and by asking questions like: from whose position are certain components of the world identified and partitioned as resources? Who governs the making of things as resources? Whose interests are served by these practices of resource-making?

In the case of the so-called resource curse, for example, Michael Watts and Nancy Lee Peluso (2013) acknowledge that resources can be a powerful entry point to examine the organization of societies. However, they caution against "commodity determinism," and rather than assume the inevitability of state pathologies and failures in countries "rich" in resources, they call for an approach that traces how resources and societies are *co-constituted*. This relational approach is "sensitive to the histories and geographies of the political economic settings" of resources (Watts and Peluso 2013, 192). Adopting it brings analytical attention to the mechanisms and power relations through which resources and societies are recursively constructed, as well as to the tensions and contradictions emerging in the process. In a similar vein, other critical scholars eschew functionalist analyses to detail the historically and geographically specific ways in which the making of resources is imbricated with the making of diverse sociopolitical dynamics and phenomena, from finance to expertise to infrastructures to democratic communities (Himley 2019; Koch and Perreault 2019; Marston 2019; Mitchell 2011; Valdivia 2008).[1]

This example of the resource curse and approaches to deconstructing it exemplify two important points about critical analyses of resources and resource systems. First, critical resource scholarship questions the hegemony of certain resource knowledges and the concomitant erasure of experiences and voices that typically results from the knowledge-making practices of dominant actors. Second, in studies about resources and their role in social life, critical inquiry reveals the "artificial unity" (Foucault 1978, 143) of taken-for-granted understandings of resources. In doing so, critical inquiry demonstrates that neither hegemonic ways of knowing nor the social orders these forms of knowledge function to sustain are inevitable. No idea, belief, desire, politics, or definition is conclusive, since we are unable to grasp the totality of relations and dimensions of life that make possible the world as we know it (Derrida [1974] 2016; see also Harris 2015).

Looking for the complicated roots of things and ideas that have become accepted as self-evident is more than a scholarly pursuit. It is a project of social transformation that entails an intentional listening for dissent and difference—what Catherine Walsh (2019) calls *gritos* and *grietas* (yells and cracks)—in the ontology of resource-thinking, in order to carve out space for ways of knowing and being that are currently subjected by existing relations of power (Grosfoguel 2019; Leyva Solano and Icaza Garza 2019). In other words, critical resource scholarship examines and learns from the tensions of looking at the world—and the role of resources in it—from both dominant and subaltern positions of knowing, with the ultimate aim of contributing to more just and equitable socioenvironmental futures.

Next, we offer a curated "tour" of critical resource scholarship from geographers and allied scholars that identifies key approaches in this body of work. Together, the following two subsections outline a heuristic device, which we call "resource-making/world-making," that builds from and further develops the relational forms of thinking that have long been central to resource geography. The concept of "resource-making," emerging out of anthropology and informed by science studies, focuses attention on the practices and political projects through which specific parts of complex, heterogeneous physical worlds "are rendered into knowable and exploitable resources" (Kama 2020, 335; see also Li 2014; Richardson and Weszkalnys 2014). By "world-making," we refer to the ways in which socioecological worlds are (un)made in and through the making, circulation, consumption, and disposal of resources (de la Cadena 2015; Tsing 2015). By placing these in dialectical relation—that is, by examining the interplay and coproduction of resource-making and world-making—we seek to deconstruct and make strange both the fixity of resources *and* the worlds in which resource ontologies make sense.

In the first of these subsections, we explore work that seeks to understand what we call the World of Resources, including the logics, tendencies, and dynamics of the resource systems through which things (materials, processes, phenomena, ideas) are abstracted from their original relations to be inserted—as "resources"—into new ones, principally for the purpose of wealth accumulation. This analytical project is underlain by two interrelated ideas: first, as noted earlier, that the critical analysis of historical and contemporary resource geographies is integral for making sense of the world and, second, that knowing the World of Resources requires *un*knowing common-sense notions of resources and their role in society. Then, in the following subsection, we turn to scholarship that begins from an analytical position of ontological pluralism—that is, from the idea that the World of Resources is but one of the worlds that exist (or might exist in the future). From this analytical position, this scholarship derives a political proposition that the epistemological and ontological limits of the World of Resources must be "unbounded" to allow for other ways of organizing socioecologies to flourish. Together, we consider these sections and the scholarship they contain to offer a "double perspective" (Jarvis 1998, 23) for examining resources and resource worlds.

(Un)knowing the World of Resources

Critical-geographical studies of resources and the systems responsible for their making, circulation, consumption, and disposal—what we call the World of Resources—mobilize a relational analytic. Relational analysis of socioecological dynamics is not new in geography or specific to critical-geographical scholarship on resources. Thinking relationally is a hallmark of geographic analysis that insists on the openness of the relations, ideas, and beliefs that constitute what we know as "space" (Jones 2009; Massey 2005) and is a tool for resisting and countering positivist science that splits an external nature from human society (see, e.g., Smith and O'Keefe 1980). In relational analyses, material conditions (and ideas about them) are assumed to be constantly in flux, and if there is one constant, it is that the world is not stable but always changing (Marx and Engels 1970).

In resource geography, relational approaches are often traced to the work of Erich Zimmermann, nearly a century ago. Zimmermann's (1951, 15, emphasis in original) much-cited aphorism, "Resources *are* not, they *become*," has become a shorthand for signaling that resources are neither fixed nor finite. Rather, resources emerge through material and ideational processes that in turn are linked to social understandings of utility and value (see also Bakker and Bridge 2006; Bridge 2009; Furlong and Norman 2015; Harvey 1974). From this perspective, nothing simply exists as a resource, waiting to be encountered and—à la one of the *OED* definitions cited previously—"drawn on when necessary." Rather, the "resourceness" of any given material or phenomena—the very identity of the thing as a resource—is historically and geographically contingent. It is subject to change over time and space in relation to evolving and conflict-laden debates over what is valuable, and as a result of the positioning of the thing—with its particular biophysical qualities—in relation to other things and processes: capital, infrastructure, knowledge, laws, markets, norms, technology, etc. (Hennessy 2019; Tsing 2005). Key to this perspective, then, is that resources are "irreducibly social rather than simply 'natural'" (Banoub 2017, 1).

Zimmermann's notion of resource becoming also has limits. For one, it may hide the design, intentionality, agency, and work that go into stabilizing what Tania Murray Li (2014) refers to as "resource assemblages." In geography, the term "assemblage" signals a constructionist perspective that emphasizes the emergence, multiplicity, and indeterminacy of things that appear to have a unity—things like land, crude oil, or minerals. Assemblages are collectives of things that not only come together (e.g., human, organic, inorganic, and technical things) and retain heterogeneity, but also appear to hold together into a provisional (revisable) socio-spatial formation that blurs divisions between social and material and structure and agency (Collier and Ong 2005; DeLanda 2006, cited in Anderson and McFarlane 2011).

The notion of "resource-making" used in this introductory chapter draws on this sort of assemblage thinking. Studying resources-as-assemblages makes visible the various agents, interests, and practices that intersect in the making of resources, or that are necessary to make such an effort thinkable in the first place. Resource-making scholarship places significant emphasis on the "materiality" of resources, a term that "can elide different and even incompatible ontological commitments" (Bakker and Bridge 2006, 6), but that can also function to interrogate how the World of Resources works, and for whom.

Thinking with the materiality of resources is, to put it simply, conflictive and complicated. On the one hand, acknowledging that nonhuman entities participate in social outcomes through their presence and characteristics, without acknowledging the resource systems that capture these, echoes dangerous commodity and environmental determinisms that reproduce colonial extractive logics (Harvey 1974)—recall our discussion of the "resource curse." On the other hand, acknowledging the capacity and availability of nonhuman entities in their own right and to form bonds of intelligence and intention with humans opens the possibility of asking about the locus of agency (and what and whom is recognized as having personhood) and to devise other, less damaging, world-making projects (Burkhart 2016; de la Cadena 2015; Grefa 2020). As Karen Bakker and Gavin Bridge (2006, 8) urge, quoting Elizabeth Grosz (1994: xi), resource scholars cannot escape the sticky question of materiality; critical resource geographers need to think about materiality "in ways that are simultaneously physical and cultural, that admit the significance of the physical but which also recognize that 'materiality is uncontainable in physicalist terms alone.'" Acknowledging the materiality of resource-making reveals how other-than-human entities redistribute and decenter what is recognized as agency, and reconfigures understandings of a biophysically and culturally heterogeneous world (Braun and Whatmore 2010; Fitzsimmons 1989; Gandy 2003; Lave 2012).

Starting with how resources are assembled, instead of how they simply "become," also provides an entry point for thinking through the other things—or, other worlds—that are made in and through the making of resources. For instance, through an ethnography of the creation of the Crater Mountain Wildlife Conservation Area in Papua New Guinea, Paige West (2006) examines how biodiversity was made into a resource, and how the resource-making process in turn constructed and reconfigured ideas of space, place, environment, and society. Amy Braun's (2020) study of the algae biotech sector reveals resource-making and the malleable idea of "sustainability" as coproduced through techno-scientific, institutional, and discursive practices. This coproduction has enabled algal product developers to leverage spectacular—if unrealized—sustainability stories (e.g., microalgae biofuels that replace fossil fuels, algal based aquaculture feed that can reduce wild fisheries depletion) to attract speculative investment, financial capital, and public funds. Likewise, Matthew Fry and Trey Murphy (under review) trace how coupled narratives of "potential" and visual "conjurings"—such as maps and graphs depicting locations, volumes, and values of the subsurface—not only suggested possible modes of existence for Mexico's national geologic resources, but also laid groundwork for transforming Mexico's historically state-owned, controlled, and regulated extraction regime into a liberalized sector that is open to and relies on private investment.

The nexus of resource-making and state-making has also been of keen interest to critical resource scholars (see Bridge 2014). Scholarship in this vein has examined how states enable resource mobilization and capital accumulation through scientific and political practices and has revealed how resource-making shapes socioecological relations around which "the state" coheres (Braun 2000; Hecht and Cockburn 2010; Peluso and Vandergeest 2001; Swyngedouw 1999). In one influential example, Fernando Coronil (1997) describes how oil appears to have "magical" properties that afford the Venezuelan state legitimacy via resource extraction wealth. Materiality also figures into the resource-state nexus. State power brokered via resource sovereignty, often described as "resource nationalism," is not national alone and is often intimately linked to the nature of the resource in question (and, of course, to how this resource is known via science, mapping, and other techniques of legibility), as well as to citizens' relationships with the resource. For example, in his study on oil and resource sovereignty in the Ecuadorian Amazon, Angus Lyall (2020) shows how consent to oil production in Kichwa territories is mediated by infrastructure and services that meet state-sanctioned markers of citizenship— electrification, roads, and urban-like housing—a relationship to resources that underscores the legacies of colonial capitalism deeply rooted into the social fabric of the contemporary nation-state. In the case of subterranean minerals, state sovereignty over resources can be expressed when landlord states develop property regimes for the subsurface, but sovereignty is only actualized when the property regime attracts transnational mining capital to undertake extraction (Emel, Huber, and Makene 2011). In cases such as regional oilfields (Valdivia 2015), transboundary conservation parks (Lunstrum 2013), and highly migratory fish populations (Havice 2018) in which a resource is "shared" or bridges discrete national territory, resource sovereignty is negotiated among states, entangled in regional geopolitics, and materializes with and through the object of desire and its rendering through scientific knowledge, extractive practices, and regulations.

The notion of resource "becoming" may also conceal the forms of exclusion, marginalization, oppression, and violence that enable the making and continuity of resource assemblages. For example, drawing on anti-racist intellect, Kathryn Yusoff (2018) traces how extraction has become a defining idiom and grammar of resource-making. Yusoff starts with Blackness as a subjectivity and a knowledge position from which to identify how geology, as an epistemic regime, has shaped notions of the human and the inhuman. Specifically, her analysis zeros in on how hegemonic geologic knowledge—what she calls "White Geology"—matters to the making of resources. She channels Aimé Césaire (1972) and Denise Ferreira da Silva (2007) to draw attention to the colonial violence and anti-Blackness that underpin relations of "exchange" and "property" in global resource trade, "from the cut hands that bled the rubber, the slave children sold by weight of flesh, the sharp blades of sugar, all the lingering dislocation from geography, dusting through diasporic generations" (Yusoff 2018, 32). And, focusing on a concept that has come to dominate some resource scholarship of late, the Anthropocene, Yusoff (2018, xiii) draws on Christina Sharpe (2016) to caution against universalizing narratives that make absent the racialized violence inherent to resource extraction: "If the Anthropocene proclaims a sudden concern with the exposures of environmental harm to white liberal communities, it does so in the wake of histories in which these harms have been knowingly exported to black and brown communities under the rubric of civilization, progress, modernization, and capitalism."

Resources, then, do not innocently "become," and neither do the common-sense meanings attached to the term "resources." So far, we have reviewed works that illustrate the making of resource-entities (land, sugar, biodiversity, etc.) and the imbrication of their resource-making systems with processes of world-(un)making. Another important line of analysis traces the emergence of "resources" as a social category imbued with particular meanings, including the "default" meanings we noted earlier. Here, an especially useful reference is Vandana Shiva's (2010) essay titled "Resources" in *The Development Dictionary: A Guide to Knowledge as Power*. Like the *OED* entries cited earlier, this is a dictionary entry, though this time the archive of terms is explicitly organized to level a critique of how "development-as-growth strains human relations and fundamentally threatens the biosphere" (Sachs 2010, vi).

Shiva identifies a radical shift over time in the meaning of "resource." Whereas in the early modern period, the concept "highlighted nature's power of self-regeneration and called attention to her prodigious creativity," with the advent of colonialism and industrialism, its meaning changed, coming to refer to "those parts of nature which were required as inputs for industrial production and colonial trade" (Shiva 2010, 228; see also Merchant 1980). As a result, "[n]ature has been clearly stripped of her creative power; she has turned into a container for raw materials waiting to be transformed into inputs for commodity production" (Shiva 2010, 228). This shift in the meaning of resource was propelled by dramatic changes to the systems of access to and control over the "natures"—forests, agricultural lands, aquatic environments, and the like—that people relied upon for their material well-being, a transformation Shiva (2010) refers to as the destruction of nature as commons. Shiva's relational analysis ties shifting connotations of the term "resource" to long-term political-economic transformations. This allows her to identify violent dispossession as an inherent and ongoing characteristic of the material history of the category of resource, rather than as a contingent side effect of the historical exploitation of nature by humans (see, also, Bridge 2009; Hall 2013; Simpson 2019).

Shiva's analysis dovetails with historical materialist analyses that focus on the dynamics and tendencies of capitalism and that place resources and resource-making at the center of capitalist societies' "metabolic" relationships with nonhuman natures (see, e.g., Swyngedouw 1999). For instance, critical scholarship has highlighted how the geopolitics and geoeconomics of globalized capitalism are rooted in the violent reorganization of webs of life, and that contemporary ecological crises are borne out of the intersection of race, class, and gender in capitalism's environmental history (Moore 2010a,b). This form of critique builds from Karl Marx's ([1867] 1976, 637-38) insights into capitalism's tendency to undermine, in his words, "the original sources of all wealth—the soil and the worker," and his conceptualization of "primitive accumulation," a process Marx considered to involve the expropriation and enclosure of the means of production (land and resources included) for purposes of capital accumulation (see De Angelis 2001; Federici 2004). Driving these critiques is the idea that the capitalist exploitation of resources and labor is inherently extractive, vampiric, and incompatible with human flourishing and equity.

The scholarship reviewed in this subsection revolves around two broad questions: what *are* resources and how can we understand their *role* in the world? Driving these questions is the idea that understanding the World of Resources *matters*. And yet, we also recognize that even a relational approach to resources that aims to deconstruct dominant or common-sense understandings risks reproducing the idea that resources are the undeniable material basis of contemporary life (Gibson-Graham 2008). There is a tension, then, within critical resource geography, a project that places at the core of its intellectual-political mission a concept—*resource*—that is intrinsically tied to and embodies the colonial and capitalist subjugation of people and environments. Rather than trying to resolve this tension, the scholarship grouped under the next subsection moves toward a different goal: to make space for worlds beyond resources.

Unbounding the World of Resources

While the works reviewed thus far call into question the historical-geographical fixity of resources, this section features scholarship that starts from a position of ontological pluralism to unsettle the idea of a singular, universal world. To be clear, this position is not supplemental to that described in the previous subsection. Rather it is rooted in a commitment to overcome the extractive colonialism inherent to the World of Resources by elevating other ways of knowing/being altogether. Broadly, this scholarship seeks to "unbound" the World of Resources by provincializing the history and geography of capitalist worlds. This work starts from the proposition that resource ontologies make sense within what Arturo Escobar (2018, following Law 2015) calls One-World World (OWW), or the idea that everyone lives within a single world, made up of *one* underlying reality, *one* nature, and *many* cultures. The problem with this "imperialist notion" (Escobar 2018, 86) is that it negates the knowledges and experiences of those excluded by dominant modes of world-knowing. Put differently, if we design worlds and worlds design us back (Willis 2015), then to move beyond the OWW requires decentering the dualist ontology of separation, control, and appropriation that has progressively become dominant in the colonial geologics of patriarchal capitalism.

How does this unbounding of the World of Resources work? We interpret this happening in two moves. Similar to Dipesh Chakrabarty's (1992) proposition to "provincialize" Europe, the first step is writing into the historical geographies of resources the ambivalences, contradictions, uses of force, tragedies, and ironies that attend the World of Resources (see subsection earlier). This critique aims to transcend the totality of liberal constructions of the state, race, and citizenship, as well as the naturalness of capital relations. The next move is speculative, and transformative. It starts with acknowledging that the end of the World of Resources has *already happened*—prefigured in apocalyptic visions, speculative fiction, disaster films, and biblical texts—and that the time is here to create a new world atop its ruins (e.g., Baldy 2014; Fiskio 2012; Gumbs 2018; Roy 2006; Saunders 2013; Tsing et al. 2017). Acknowledging the intellectual debts of those who have long challenged resource ontologies in social movements, as well as in decolonial, Indigenous, feminist, and post-humanist scholarship, among others, is fundamental to this orientation (Todd 2016).

Among those creating space for other ontologies after the end of the World of Resources is the Zapatista National Liberation Army (Ejército Zapatista de Liberación Nacional, or EZLN). As Reyes (2015) describes, the EZLN has conducted a profound systematic analysis of the structural crisis of capitalism—in both Mexico and the rest of the world—that has led them to declare a "war against the geopolitics of knowledge" (Leyva Solano 2017, 161). In short, the EZLN conceives of capitalist systems as forms of life that have genocidal tendencies (Reyes 2016). According to the Zapatistas, to change this world might be too hard, maybe impossible. Instead, it may be better to build a new one where many worlds fit. To counter capital's genocidal tendencies, the Zapatistas have conceptualized a world beyond capitalism, what they call "another geography" (Reyes 2015), built upon Indigenous forms of institutionality that germinate from their *semilleros*—their own seedbanks of experience—and that unbounds the colonizing practices of state and capital that rule over peoples and territory. Rather than try to fix or make demands upon the existing world and its political institutions, the Zapatistas are actively working to create different forms of governance, justice, health, economic sustainability, as well as new forms of global solidarity (Grubacic and O'Hearn 2016). For the EZLN, "theoretical concepts" and critical reflection serve as a step toward deciphering and imagining what it might take to end the dynamics and tendencies of contemporary capitalism (Ejército Zapatista de Liberación Nacional (EZLN) 2015), and with it, the modes of resources described in the prior section.

Another example of challenges to the World of Resources is the "Land Back" movement, a collective of Indigenous scholars, communities, and activists, primarily based in North America, who demand their rightful place in keeping land alive and spiritually connected. The idea behind Land Back is that a return to territorial self-affirmation and self-determination is fundamental to Indigenous livelihoods; getting back in touch with Indigenous languages, ancestral land relations, and traditional familial and governing systems is a step toward healing kin relations with other beings. In order for this to happen, the territorial logistics of the World of Resources—such as treaties, private property, the policing of private ownership, and geopolitical borders—must be interrupted. Land Back is not a new concept of resistance; it is the continuation of hundreds of years of Indigenous struggle against land theft. Like the case of the Zapatista *semilleros*, the preservation and recognition of Indigenous ways of knowing and Indigenous intelligence is central to challenging the World of Resources (Grefa 2020; Kohn 2013; Simpson 2017; Somerville 2017; Tricot 2009; Whyte 2013). In these "conceptions of the world" (Gramsci 1985), how we understand ourselves in relation to each other, to the world, and in the ordinary aspects of social life, is fundamental to socioecological transformation. "Indigenous knowledge and land co-constitute each other," and it is time to "bust open" the settler legal systems that make the logistics of Land Back difficult to imagine and put into action (Longman et al. 2020, no page).

Academic studies drawing on Black, Indigenous, and decolonial epistemologies echo the need for unbounding the World of Resources, challenging the primacy of "resources" as a way of organizing human thought and action, and imagining and experimenting with ways to think with resources otherwise, or not at all. For example, contributors to a special issue of *Environment and Society* called "Indigenous Resurgence, Decolonization, and Movements for Environmental Justice" underscore that resource systems and politics must be unbound to, first, make visible the worlds that coproduce them and, second, to counter dominant positions that claim that, in matters of global environmental change, "we are all in this together." As Jaskiran Dhillon (2018, 1) puts it, inclusionary politics in matters of environmental crises must be received with suspicion: "[I]n the wake of a planet-wide movement riddled with idioms about 'saving our home,' there has been a tidal wave of interest in Indigenous knowledge(s) about the land, water, and sky—a desire to 'capture and store' the intergenerational wisdom that speaks to the unpredictable path lying ahead." Dhillon's intentional use of the idiom of resources—capture, store, usefulness, adaptive solutions—offers an "anticolonial counterscript" that troubles the politics and effects of unreflexive inclusivity. It puts front and center the conditions of global colonization, elimination, and disenfranchisement upon which knowledge is built (Spice 2018; Yazzie 2018).

These decolonial engagements, in turn, demand answers to important questions. Are critical resource scholars ready to unbound the World of Resources? What would this unbounding look like for a body of academic scholarship that positions resources as its primary analytical concern? Has "resources," as a language, concept, and method for studying the world, reached its useful limit? These questions get at the crux of debates surrounding critical resource geography, and critical inquiry more generally. As examined throughout this introductory chapter, critical scholarship aims to effect change through transformative insight—for example, by making "strange" the World of Resources. Critical resource geography also champions progressive praxis that unbounds some parts of this World of Resources—for example, by aligning with the thinking and being of social movements and oppressed peoples to challenge the status quo of capitalist resource systems. As Noel Castree and Melissa Wright (2005, 2) remind critical scholars, "to bring the undiscussed into discussion; to stray beyond established perimeters of opinion" and "to render the familiar … strange" are a hallmark of critical scholarship, but it is not enough.

It cannot be its end. To remain engaged in restless thinking means that making familiar forms strange is but one step toward producing worlds where these familiar forms and consequences of resource-systems are unacceptable.

What is in this Handbook?

This Handbook is a polyvocal collection of scholarship that mobilizes the terms and framings of resource geography, while at the same time working to make sense of the deep-rooted tensions within this arena of research, teaching, and praxis. In this section, we outline the conversations that emerge from the contents of the 36 chapters that make up the remainder of this volume. As we considered how to introduce the contents of this rich collection, brief descriptive summaries of each chapter seemed unsatisfying. Such summaries could never do justice to the intellectual-political project of each chapter nor effectively contribute to advancing the propositions and objectives of this field of study. Thus, we have chosen to introduce the Handbook's contents by inductively drawing on chapters in each of the volume's four sections to identify overarching themes as well as points of generative tension. In each individual chapter, readers will find references to other contributions in the volume; by following these clues, the reader can engage with the productive debates and intersections that unfold across chapters. We hope that this approach to introducing the contents of the Handbook motivates readers to explore the chapters that follow, to engage them on their own terms, and to consider how they individually and collectively contribute to critical resource geography.

Chapters in Section I, "(Un)Knowing Resources," offer distinct entry points for thinking critically about the ontological status of resources (i.e., what resources *are*) and the epistemological frameworks through which we come to know them. Broadly, the chapters advance one of the long-standing intellectual-political objectives of critical resource scholarship: to make resources strange by destabilizing dominant or common-sense understandings of them. To do this, the contributing authors draw on diverse social science scholarship (e.g., materiality, feminist and queer approaches to political economy, Indigenous epistemologies, geohumanities, science and technology studies) and employ a range of methods, including literature reviews, reflections on research experiences, and empirical analysis. The chapters offer different ways for (un)knowing resources, in terms of their material and discursive make-up as well as the practices and forces that are responsible for their construction. While there are complementarities across some of the chapters' framings for (un)knowing resources, there are also tensions. One of these is between, on the one hand, calls to unbound resource geographies through analytical approaches that capture the multiple ontologies of resources and the non-inevitability of the worlds that are spun through their making and, on the other, calls to reject resource ontologies altogether—and, thus, "resources" as an operational concept—as a necessary step toward liberation. Chapters in this section are written by Helene Ahlborg and Andrea Joslyn Nightingale (Chapter 2), Kolson Schlosser (Chapter 3), Karen Bakker and Gavin Bridge (Chapter 4), Kärg Kama (Chapter 5), Skye Naslund and Will McKeithen (Chapter 6), and Andrew Curley (Chapter 7).

Chapters in Section II, "(Un)Knowing Resource Systems," examine the nature, components, and logics of resource systems. Some chapters emphasize system-wide compulsions and tendencies characteristic of the World of Resources, for example, in the relationships of resources and resource-making to dominant structures and ideologies of economic, political, and social organization (e.g., capitalism, gender, race, nationalism). Others examine resource systems through the vantage point of a particular location, whether that be an actor (e.g., corporations), a site (e.g., municipalities, borders), or an object or figure of study (e.g., dogs). Throughout, authors are attuned to how historical and contemporary resource geographies have been made through the social transformation of biophysical environments for particular purposes. Collectively, they highlight the need to conceptualize resource-making as a spatial *and* temporal process with specific logics, regularity, and ends. Through their emphasis on questions of agency and power vis-à-vis resources and resource-making, and through the use of various analytical categories (discourse, enclosure, geopolitics, plantation, profit, and value, among others), the chapters underscore the co-constitutive role that resources play in unequal power relations, exploitation, marginalization, and oppression at multiple scales. Chapters in this section are written by Sharlene Mollett (Chapter 8); Ashley Fent (Chapter 9); Danielle M. Purifoy (Chapter 10); Tom Perreault (Chapter 11); Kathryn Furlong, Martine Verdy, and Alejandra Uribe-Albornoz (Chapter 12); Heidi J. Nast (Chapter 13); Matthew T. Huber (Chapter 14); Paul S. Ciccantell (Chapter 15); and Liam Campling (Chapter 16).

Chapters in Section III, "Doing Critical Resource Geography: Methods, Advocacy, and Teaching," reflect on what constitutes scholarly engagement in critical resource geography. The contributing authors recognize their situated knowledges and reflect on their relative privilege in knowledge production to examine their experiences as academics, advocates, and teachers embedded in diverse social locations of resource-making (and un-making). Together, these chapters reveal that there is no singular way to do critical resource geography and that the geographical study of resources becomes "critical" at least in part through the form and nature of praxis. The first three chapters—written by Eloisa Berman-Arévalo (Chapter 17); Emily Billo (Chapter 18); and Christine Biermann, Stuart N. Lane, and Rebecca Lave (Chapter 19) — reflect on the ethics of knowledge production and methodological choices therein. The next six chapters trace the institutional relationships that shape, enable, and constrain resource scholarship and advocacy. Contributors consider their relationships with corporations, farmers, government agencies, social movements, and other actors as they reflect on their political commitments as critical resource scholars. Contributing authors include Elizabeth Shapiro-Garza, Vijay Kolinjivadi, Gert Van Hecken, Catherine Windey, and Jennifer J. Casolo (Chapter 20); Anthony Bebbington, Ana Estefanía Carballo, Gillian Gregory, and Tim Werner (Chapter 21); Patrick Bond and Rahul Basu (Chapter 22); Christopher Courtheyn and Ahsan Kamal (Chapter 23); Adrienne Johnson (Chapter 24); and Elvin Delgado (Chapter 25). Section III concludes with two chapters focused on pedagogical practice in field- and classroom-based learning environments (respectively) designed to engage students in the critical analysis of resource systems. The authors of these two chapters are Nicole Fabricant (Chapter 26) and Conor Harrison and Kathryn Snediker (Chapter 27).

Chapters in Section IV, "Resource-Making/World-Making," examine case studies of the co-constitution of resource-making and world-making in diverse historical-geographical contexts. These chapters analyze various types of resources—e.g., carbon, human tissues, salmon, and soy—in relation to diverse dimensions of world-making—e.g., citizenship, colonialism, knowledge production, conservation, mapping, and identity formation. Together, these chapters challenge the boundaries around which resources are defined, emphasizing resource-making as a historical and historically contingent process. This serves as an important reminder that resources—and the worlds in which these are recursively enmeshed—are not just "made" but constantly "remade," or made differently. In showing what is and what has been, these authors implicitly or explicitly raise possibilities of what future worlds might be. Chapters in Section IV are written by Gustavo de L. T. Oliveira (Chapter 28); Annah Zhu and Nancy Lee Peluso (Chapter 29); Elizabeth Lunstrum and Francis Massé (Chapter 30); Beatriz Bustos-Gallardo (Chapter 31); Elspeth Probyn (Chapter 32); Maria Fannin (Chapter 33); Wim Carton and Karin Edstedt (Chapter 34); Anna Zalik (Chapter 35); Lisa M. Campbell, Noella J. Gray, Sarah Bess Jones Zigler, Leslie Acton, and Rebecca Gruby (Chapter 36); and Joe Bryan (Chapter 37).

Our invitation

Throughout the production of this Handbook, we, as editors, have been motivated by a series of restless concerns. The first has been a desire to prompt ongoing reflection on the ethical and normative dimensions of working within the arena of critical resource geography, including on how our work can contribute to transitions toward more just worlds. Our hope is that this Handbook can nurture what Sofia Villenas (2019, 156) calls "vigilant, critical love," or "the obligation to open up rather than close off the possibilities for response by others, in order to imagine a more humanizing and just world." Critical resource scholarship is about identifying the limits of how we make sense of resources, unlearning the dominance of universalizing truths that subjugate other ways of existing, and making room for other possible ways of knowing and being in the world (Ranganathan and Bratman 2019). This disruption is not meant to "cancel" already existing geographic approaches to studying resources but to build with and against widely used concepts and frameworks so as to generate a more attuned understanding of the work resources do, and whose experiences are centered in resource geography. We understand persistent critical reflection on the "conduct" (Foucault 1982, 789) of resource geography, including in regards to how we work with (or against) assumptions, ideas, and framings, as central to the field.

Second, as editors, we spent time thinking through the boundaries of critical resource geography as a scholarly field as well as its cannon—one in which we are all steeped and that we recognize and name as having developed within a primarily heteropatriarchal, white-male space for commoditizing knowledge production. This structural-architectural formation favors particular understandings of space, place, and relationality and has created certain ideas about the knowledges and topics that have come to be seen as "foundational." While, throughout our editorial process, we have worked with many of the field's principal concepts and analytics, we are also aware—increasingly so through our engagements with Handbook authors—of the limits of some of these, of the tensions that reside within and among them, and of potential lines of flight.

Third, and related to the previous point, while this Handbook proposes that there is something we can call *critical* resource geography, our goal is not to bound critical practice. Rather, our graphing of the World of Resources and its unbounding aims to make space for unlearning and decentering dominant or common-sense ideas in order to make room for other geographies (Eaves 2020). Thus, this collection does not lead us to the declaration of a "field" or a set of norms or principles about what resource geography is, or what is should be, or to a singular notion of how to "do" resource geography. Rather, it is a book about the possibilities of resource geography, an approach that involves opening the category "resource" for examination and inquiry.

Finally, as Natalie Oswin (2020) notes, academic change is not everything. While creating the space for intellectual conversations about resources and resource-making is one part of the critical work we have identified here, it is even better if these conversations generate a response in realms of action beyond academic production. This desire for more-than-academic change is the explicit normative aim of many engaged in this field (and many authors in this Handbook) and is rooted in a desire to do work that contributes to a broader rethinking/reshaping of the academy and the world. This includes pedagogy wherein, as teachers, critical resource geographers work with students to address habits of thought, reading, writing, observing, and being in the world to understand the deep meanings, causes, social contexts, and consequences of any action or event. We hope this Handbook can contribute to this goal. Its contents demonstrate the importance of multiple forms of engagement—in research, writing, teaching, service, and activism, among others—to critical resource geography as a field of practice.

We invite you to use this Handbook to engage with critical resource geography and work with or against its assumptions, ideas, and framings.

Note

1 Another critique of work on the "resource curse" highlights that its analytical focus on the nation-state ignores the multiscalar economic networks in which resource-making activities are enmeshed, and fails to attend to how the uneven distribution of costs and benefits across these networks shapes the development implications of resource exploitation in any locale (Bridge 2008).

References

Agger, Ben. 1998. *Critical Social Theories: An Introduction*. Boulder, CO: Westview Press.

Auty, Richard M. 1993. *Sustaining Development in Mineral Economies: The Resource Curse Thesis*. London: Routledge.

Anderson, Ben, and Colin McFarlane. 2011. "Assemblage and Geography." *Area* 43 (2): 124–127.

Badeeb, Ramez Abubakr, Hooi Hooi Lean, and Jeremy Clark. 2017. "The Evolution of the Natural Resource Curse Thesis: A Critical Literature Survey." *Resources Policy* 51: 123–134.

Bakker, Karen, and Gavin Bridge. 2006. "Material Worlds? Resource Geographies and the 'Matter of Nature." *Progress in Human Geography* 30 (1): 5–27.

Baldy, Cutcha Risling. 2014. "Why I Teach *The Walking Dead* in my Native Studies Classes." Nerds of Color, April 24. https://thenerdsofcolor.org/2014/04/24/why-i-teach-the-walking-dead-in-my-native-studies-classes/

Banoub, Daniel. 2017. "Natural Resources." In *The International Encyclopedia of Geography: People, the Earth, Environment and Technology*, edited by Douglas Richardson, Noel Castree, Michael F. Goodchild, Audrey Kobayashi, Weidong Liu, and Richard A. Marston. New York, NY: John Wiley & Sons. doi: 10.1002/9781118786352.wbieg0496.

Bebbington, Anthony, Leonith Hinojosa, Denise Humphreys Bebbington, María Luisa Burneo, and Ximena Warnaars. 2008. "Contention and Ambiguity: Mining and the Possibilities of Development." *Development and Change* 39 (6): 887–914.

Blomley, Nicholas. 2006. "Uncritical Critical Geography?" *Progress in Human Geography* 30 (1): 87–94.

Braun, Amy. 2020. "Producing 'Sustainability' through Ocean Resource Making." PhD diss., University of North Carolina at Chapel Hill.

Braun, Bruce. 2000. "Producing Vertical Territory: Geology and Governmentality in Late Victorian Canada." *Cultural Geographies* 7 (1): 7–46.

Braun, Bruce, and Sarah J. Whatmore. 2010. "The Stuff of Politics: An Introduction." In *Political Matter: Technoscience, Democracy and Public Life*, edited by Bruce Braun, and Sarah Whatmore, ix–xl. Minneapolis, MN: University of Minnesota Press.

Bridge, Gavin. 2008. "Global Production Networks and the Extractive Sector: Governing Resource-Based Development." *Journal of Economic Geography* 8 (3): 389–419.

Bridge, Gavin. 2009. "Material Worlds: Natural Resources, Resource Geography and the Material Economy." *Geography Compass* 3 (3): 1217–1244.

Bridge, Gavin. 2014. "Resource Geographies II: The Resource-State Nexus." *Progress in Human Geography* 38 (1): 118–130.

Buchanan, Ian. 2010. "Negative Dialectics." In *A Dictionary of Critical Theory*. Oxford: Oxford University Press. https://www.oxfordreference.com/view/10.1093/acref/9780199532919.001.0001/acref-9780199532919-e-469

Bunker, Stephen G. 1985. *Underdeveloping the Amazon: Extraction, Unequal Exchange, and the Failure of the Modern State*. Chicago, IL: The University of Chicago Press.

Burkhart, Brian. 2016. "'Locality is a Metaphysical Fact'—Theories of Coloniality and Indigenous Liberation Through the Land: A Critical Look at Red Skin, White Masks." *Indigenous Philosophy* 15 (2): 2–7.

Castree, Noel. 2000. "Professionalisation, Activism, and the University: Whither 'Critical Geography'." *Environment and Planning A* 32 (6): 955–970.

Castree, Noel, and Melissa W. Wright. 2005. "Home Truths." *Antipode* 37 (1): 1–8.

Césaire, Aimé. 1972. *Discourse on Colonialism*. New York, NY: Monthly Review Press.

Chakrabarty, Dipesh. 1992. "Provincializing Europe: Postcoloniality and the Critique of History." *Cultural Studies* 6 (3): 337–357.

Collier, Stephen J., and Aihwa Ong. 2005. "Global Assemblages, Anthropological Problems." In *Global Assemblages: Technology, Politics, and Ethics as Anthropological Problems*, edited by Aihwa Ong, and Stephen J Collier, 3–21. Malden, MA: Blackwell Publishing.

Coronil, Fernando. 1997. *The Magical State: Nature, Money, and Modernity in Venezuela*. Chicago, IL: The University of Chicago Press.

da Silva, Denise Ferreira. 2007. *Toward a Global Idea of Race*. Minneapolis, MN: University of Minnesota Press.

De Angelis, Massimo. 2001. "Marx and Primitive Accumulation: The Continuous Character of Capital's 'Enclosures'." *The Commoner* 2: 1–22.

de la Cadena, Marisol. 2015. *Earth Beings: Ecologies of Practice Across Andean Worlds*. Durham, NC: Duke University Press.

DeLanda, Manuel. 2006. *A New Philosophy of Society: Assemblage Theory and Social Complexity*. New York, NY: Continuum.

Derrida, Jacques. (1974) 2016. *Of Grammatology*. Baltimore, MD: John Hopkins University Press.

Dhillon, Jaskiran. 2018. "Introduction: Indigenous Resurgence, Decolonization, and Movements for Environmental Justice." *Environment and Society* 9 (1): 1–5.

Du Bois, W. E. B. 2009. *The Souls of Black Folk*. edited by Brent Hayes Edwards, Oxford: Oxford University Press.

Dunbar-Ortiz, Roxanne. 2014. *An Indigenous Peoples' History of the United States*. Boston, MA: Beacon Press.

Eaves, LaToya. 2020. "Fear of an Other Geography." *Dialogues in Human Geography* 10 (1): 34–36.

Ejército Zapatista de Liberación Nacional (EZLN). 2015. *El Pensamiento Crítico Frente a la Hidra Capitalista: Participación de la Comisión Sexta del EZLN*. Chiapas: EZLN.

Emel, Jody, Matthew T. Huber, and Madoshi H. Makene. 2011. "Extracting Sovereignty: Capital, Territory, and Gold Mining in Tanzania." *Political Geography* 30 (2): 70–79.

Erakat, Noura. 2019. *Justice for Some: Law and the Question of Palestine*. Stanford, CA: Stanford University Press.

Escobar, Arturo. 2018. *Designs for the Pluriverse: Radical Interdependence, Autonomy, and the Making of Worlds*. Durham, NC: Duke University Press.

Fay, Brian. 1987. *Critical Social Science: Liberation and Its Limits*. Ithaca, NY: Cornell University Press.

Federici, Silvia. 2004. *Caliban and the Witch: Women, the Body, and Primitive Accumulation*. New York, NY: Autonomedia.

Fiskio, Janet. 2012. "Apocalypse and Ecotopia: Narratives in Global Climate Change Discourse." *Race, Gender, and Class* 19 (1): 12–36.

Fitzsimmons, Margaret. 1989. "The Matter of Nature." *Antipode* 21 (2): 106–120.

Foucault, Michel. 1978. *History of Sexuality Volume One: An Introduction*. Translated by Robert Hurley. New York, NY: Random House.

Foucault, Michel. 1982. "The Subject and Power." *Critical Inquiry* 8 (4): 777–795.

Fry, Matthew, and Trey Murphy. Under review. "The Geo-imaginaries of Potential in Mexico's Burgos Basin." *Political Geography*.

Furlong, Kathryn, and Emma S. Norman. 2015. "Resources.". In *The Wiley Blackwell Companion to Political Geography*, edited by John Agnew, Virginie Mamadouh, Anna J. Secor, and Joanne Sharp, 424–437. Malden, MA: Wiley Blackwell.

Galeano, Eduardo. 1971. *Las Venas Abiertas de América Latina*. Mexico City: Siglo Veintiuno Editores.

Gandy, Matthew. 2003. *Concrete and Clay: Reworking Nature in New York City*. Cambridge: MIT Press.

Gibson-Graham, J. K. 2008. "Diverse Economies: Performative Practices for 'Other Worlds'." *Progress in Human Geography* 32 (5): 613–632. doi: 10.1177/0309132508090821.

Grefa, Fredy. 2020. "Sacha Runa Values and Resurgence: Implications on Payments for Ecosystem Services – the Programa Socio Bosque in the Northern Ecuadorian Amazon." PhD diss., University of North Carolina at Chapel Hill.

Gramsci, Antonio. 1971. *Selections from the Prison Notebooks*. London: Lawrence and Wishart.

Gramsci, Antonio. 1985. *Selections from Cultural Writings*. London: Lawrence and Wishart.

Grosfoguel, Ramón. 2019. "Epistemic Extractivism: A Dialogue With Alberto Acosta, Leanne Betasamosake Simpson, and Silvia Rivera Cusicanqui.". In *Knowledges Born in the Struggle: Constructing the Epistemologies of the Global South*, edited by Boaventura de Sousa Santos, and Maria Paula Meneses, 203–218. New York, NY: Routledge.

Grosz, Elizabeth. 1994. *Volatile Bodies: Toward a Corporeal Feminism*. Bloomington, IN: Indiana University Press.

Grubacic, Andrej, and Denis O'Hearn. 2016. *Living at the Edges of Capitalism: Adventures in Exile and Mutual Aid*. Berkeley, CA: University of California Press.

Gumbs, Alexis P. 2018. *M Archive: After the End of the World*. Durham: Duke University Press.

Hall, Derek. 2013. "Primitive Accumulation, Accumulation by Dispossession and the Global Land Grab." *Third World Quarterly* 34 (9): 1582–1604.

Harris, Vance. 2015. "Hauntology, Archivy, and Danditry: An Engagement With Derrida and Zapiro." *Critical Arts* 29 (1): 13–27.

Harvey, David. 1974. "Population, Resources, and the Ideology of Science." *Economic Geography* 50 (3): 256–276.

Havice, Elizabeth. 2018. "Unsettled Sovereignty and the Sea: Mobilities and More-Than-Territorial Configurations of State Power." *Annals of the American Association of Geographers* 108 (5): 1280–1297.

Hecht, Susanna, and Alexander Cockburn. 2010. *The Fate of the Forest: Developers, Destroyers, and Defenders of the Amazon*. Chicago, IL: Chicago University Press.

Hennessy, Elizabeth. 2019. *On the Backs of Tortoises: Darwin, the Galápagos, and the Fate of an Evolutionary Eden*. New Haven, CT: Yale University Press.

Himley, Matthew. 2019. "Extractivist Geographies: Mining and Development in Late-Nineteenth- and Early-Twentieth-Century Peru." *Latin American Perspectives* 46 (2): 27–46.

Horkheimer, Max. 1972. "Traditional and Critical Theory." In *Critical Theory; Selected Essays*. New York, NY: Herder and Herder. 188–243

Huber, Matthew. 2018. "Resource Geographies I: Valuing Nature (Or Not)." *Progress in Human Geography* 42 (1): 148–159.

Inikori, Jospeh E. 2020. "Atlantic Slavery and the Rise of the Capitalist Global Economy." *Current Anthropology* 61 (S22): S159–S171.

Jarvis, Simon. 1998. *Adorno: A Critical Introduction*. New York, NY: Routledge.

Jobson, Ryan Cecil. 2020. "The Case for Letting Anthropology Burn: Sociocultural Anthropology in 2019." *American Anthropologist* 122 (2): 259–271.

Johnson, Walter. 2018. "To remake the World: Slavery, Racial Capitalism, and Justice." *Boston Review*, 20 February. http://bostonreview.net/forum/walter-johnson-to-remake-the-world

Jones, Martin. 2009. "Phase Space: Geography, Relational Thinking, and Beyond." *Progress in Human Geography* 33 (4): 487–506.

Kama, Kärg. 2020. "Resource-Making Controversies: Knowledge, Anticipatory Politics and Economization of Unconventional Fossil Fuels." *Progress in Human Geography* 44 (2): 333–356.

Karl, Terry Lynn. 1997. *The Paradox of Plenty: Oil Booms and Petro-States*. Berkeley, CA: University of California Press.

Karl, Terry Lynn. 2005. "Understanding the Resource Curse." In *Covering Oil: A Reporter's Guide to Energy and Development*, edited by Svetlana Tsalik, and Anya Schiffrin, 21–30. New York, NY: Open Society Institute.

Koch, Natalie, and Tom Perreault. 2019. "Resource Nationalism." *Progress in Human Geography* 43 (4): 611–631.

Kohn, Eduardo. 2013. *How Forests Think: Toward an Anthropology Beyond the Human*. Berkeley, CA: University of California Press.

Kubat, Amoke. 2020. "A Moment of Silence: 'Rest' by Amoke Kubat." *Star Tribune*, September 13. https://www.startribune.com/a-moment-of-silence-rest-by-amoke-kubat/572385552/

Lave, Rebecca. 2012. *Fields and Streams: Stream Restoration, Neoliberalism, and the Future of Environmental Science*. Athens: University of Georgia Press.

Law, John. 2015. "What's Wrong With a One-World World?" *Distinktion: Scandinavian Journal of Social Theory* 16 (1): 126–139.

Lawhon, Mary, and James T. Murphy. 2012. "Socio-Technical Regimes and Sustainability Transitions: Insights from Political Ecology." *Progress in Human Geography* 36 (3): 354–378.

Philippe, Le Billon, ed. 2005. *The Geopolitics of Resource Wars: Resource Dependence, Governance and Violence*. London: Frank Cass.

Lepore, Jill. 2013. *The Story of America: Essays on Origins*. Princeton, NJ: Princeton University Press.

Leyva Solano, Xotchtil. 2017. "Part 1: Movementscapes: Geopolitics of Knowledge and the Neo-Zapatista Social Movement Networks." In *The Movements of Movements: Part 1: What Makes Us Move?*, edited by Jai Sen, 161–184. Oakland, CA: PM Press.

Leyva Solano, Xochitl, and Rosalba Icaza Garza, eds. 2019. *En Tiempos de Muerte: Cuerpos, Rebeldias, Resistencias*. Buenos Aires: Consejo Latinoamericano de Ciencias Sociales.

Li, Tania Murray. 2014. "What Is Land? Assembling a Resource for Global Investment." *Transactions of the Institute of British Geographers* 39 (4): 589–602.

Longman, Nickita, Emily Riddle, Alex Wilson, and Saima Desai. 2020. "'Land Back' Is More than the Sum of Its Parts: Letter from the Land Back Editorial Collective." *Briarpatch*, September 10. https://briarpatchmagazine.com/articles/view/land-back-is-more-than-the-sum-of-its-parts

Lowe, Lisa. 2015. *The Intimacies of Four Continents*. Durham, NC: Duke University Press.

Lunstrum, Elizabeth. 2013. "Articulated Sovereignty: Extending Mozambican State Power Through the Great Limpopo Transfrontier Park." *Political Geography* 36: 1–11.

Lyall, Angus C. 2020. *The Millennium City: Oil Politics and Urbanization in the Northern Ecuadorian Amazon*. PhD diss., University of North Carolina at Chapel Hill.

Marston, Andrea. 2019. "Strata of the State: Resource Nationalism and Vertical Territory in Bolivia." *Political Geography* 74: 102040.

Marx, Karl. 1867. 1976. *Capital: A Critique of Political Economy Volume I*. Translated by Ben Fowkes. London: Penguin Books.

Marx, Karl, and Friedrich Engels. 1970. *The German Ideology: Part One*. New York, NY: International Publishers.

Massey, Doreen. 2005. *For Space*. London: Sage.

Merchant, Carolyn. 1980. *The Death of Nature*. London: Wildwood House.

Mintz, Sidney Wilfred.. 1986. *Sweetness and Power: The Place of Sugar in Modern History*. New York, NY: Penguin.

Mitchell, Timothy. 2011. *Carbon Democracy: Political Power in the Age of Oil*. Brooklyn, NY: Verso.

Moaveni, Azadeh, and Sussan Tahmasebi. 2020. "End US Sanctions Against Iran so that We Can Fight Coronavirus with All Our Might." *The Guardian*, March 21. https://www.theguardian.com/commentisfree/2020/mar/21/end-us-sanctions-iran-fight-coronavirus-pandemic

Moore, Jason W. 2010a. "Amsterdam is Standing on Norway' Part I: The Alchemy of Capital, Empire and Nature in the Diaspora of Silver, 1545–1648." *Journal of Agrarian Change* 10 (1): 33–68.

Moore, Jason W. 2010b. "Amsterdam is Standing on Norway' Part II: The Global North Atlantic in the Ecological Revolution of the Long Seventeenth Century." *Journal of Agrarian Change* 10 (1): 188–227.

Nagar, Richa, and Roozbeh Shirazi. 2019. "Chapter 44: Radical Vulnerability." In *Keywords in Radical Geography: Antipode at 50*, edited by Antipode Editorial Collective, 236–242. Hoboken, NJ: John Wiley & Sons.

Oswin, Natalie. 2020. "An Other Geography." *Dialogues in Human Geography* 10 (1): 9–18.

Peet, Richard. 2000. "Commentary: Celebrating Thirty Years of Radical Geography." *Environment and Planning A* 32: 951–953.

Peluso, Nancy Lee, and Peter Vandergeest. 2001. "Genealogies of the Political Forest and Customary Rights in Indonesia, Malaysia, and Thailand." *The Journal of Asian Studies* 60 (3): 761–812.

Pulido, Laura. 2018. "Racism and the Anthropocene.". In *The Remains of the Anthropocene*, edited by Gregg Mitman, Robert Emmett, and Marco Armiero, 116–128. Chicago, IL: University of Chicago Press.

Ranganathan, Malini, and Eve Bratman. 2019. "From Urban Resilience to Abolitionist Climate Justice in Washington, DC." *Antipode*. Advance online publication. doi: 10.1111/anti.12555.

Reyes, Álvaro. 2015. "Zapatismo: Other Geographies Circa 'The End of the World'." *Environment and Planning D: Society and Space* 33 (3): 408–424.

Reyes, Álvaro. 2016. "The Zapatista Challenge: Politics After Catastrophe." *Cultural Dynamics* 28 (2): 143–168.

Richardson, Tanya, and Gisa Weszkalnys. 2014. "Introduction: Resource Materialities." *Anthropological Quarterly* 87 (1): 5–30.

Robbins, Paul. 2002. "Obstacles to a First World Political Ecology? Looking Near Without Looking Up." *Environment and Planning A* 34 (8): 1509–1513.

Robinson, Cedric J. 2000. *Black Marxism: The Making of the Black Radical Tradition*. Chapel Hill, NC: University of North Carolina Press.

Rostow, Walt Whitman. 1960. *The Stages of Economic Growth: A Non-Communist Manifesto*. New York, NY: Cambridge University Press.

Roy, Arundhati. 2006. *An Ordinary Person's Guide to Empire*. Delhi: Penguin Books India.

Roy, Arundhati. 2020. "The Pandemic is a Portal." *Financial Times*, May 23. https://www.ft.com/content/442546c6-9c10-11ea-adb1-529f96d8a00b

Sachs, Wolfgang. 2010. "Preface to the New Edition." In *The Development Dictionary: A Guide to Knowledge as Power*, 2nd ed., edited by Wolfgang Sachs, xix–xxvii. London: Zed Books.

Said, Edward Wadie. 1979. *Orientalism*. New York, NY: Vintage.

Saunders, Robert A. 2013. "'Zombies in the Colonies: Imperialism and Contestation of Ethno-Political Space in Max Brooks' The Zombie Survival Guide." In *Monstrous Geographies: Places and Spaces of the Monstrous*, edited by Sarah Montin, and Evelyn Tsitas, 19–46. Leiden: Brill.

Sharpe, Christina. 2016. *In the Wake: On Blackness and Being*. Durham, NC: Duke University Press.

Shiva, Vandana. 2010. "Resources." In *The Development Dictionary: A Guide to Knowledge as Power*, 2nd ed., edited by Wolfgang Sachs, 228–242. London: Zed Books.

Simpson, Leanne Betasamosake. 2017. *As We Have Always Done: Indigenous Freedom Through Radical Resistance*. Minneapolis, MN: University of Minnesota Press.

Simpson, Michael. 2019. "Resource Desiring Machines: The Production of Settler Colonial Space, Violence, and the Making of a Resource in the Athabasca Tar Sands." *Political Geography* 74: 102044.

Smith, Linda Tuhiwai. 2012. *Decolonizing Methodologies: Research and Indigenous Peoples*. 2nd ed. London: Zed Books.

Smith, Neil, and Phil O'Keefe. 1980. "Geography, Marx and the Concept of Nature." *Antipode* 12 (2): 30–39.

Somerville, Alice Te Punga. 2017. "The Great Pacific Garbage Patch as Metaphor: The (American) Pacific You Can't See." In *Archipelagic American Studies*, edited by Brian Russell Roberts, and Michelle Ann Stephens, 324–338. Durham, NC: Duke University Press.

Spice, Anne. 2018. "Fighting Invasive Infrastructures: Indigenous Relations Against Pipelines." *Environment and Society* 9 (1): 40–56.

Swyngedouw, Erik. 1999. "Modernity and Hybridity: Nature, *Regeneracionismo*, and the Production of the Spanish Waterscape, 1890–1930." *Annals of the Association of American Geographers* 89 (3): 443–465.

Todd, Zoe. 2016. "An Indigenous Feminist's Take on the Ontological Turn: 'Ontology' Is Just Another Word for Colonialism." *Journal of Historical Sociology* 29 (1): 4–22.

Tricot, Tito. 2009. "El Nuevo Movimiento Mapuche: Hacia la (Re)Construcción del Mundo y País Mapuche." *Polis: Revista Latinoamericana* 24. http://journals.openedition.org/polis/1584

Tsing, Anna Lowenhaupt. 2005. *Friction: An Ethnography of Global Connection*. Princeton, NJ: Princeton University Press.

Tsing, Anna Lowenhaupt. 2015. *The Mushroom at the End of the World: On the Possibility of Life in Capitalist Ruins*. Princeton, NJ: Princeton University Press.

Tsing, Anna Lowenhaupt, Heather Anne Swanson, Elaine Gan, and Nils Bubandt, eds. 2017. *Arts of Living on a Damaged Planet: Ghosts and Monsters of the Anthropocene*. Minneapolis, MN: University of Minnesota Press.

Tuck, Eve. 2013. "Commentary: Decolonizing Methodologies 15 Years Later." *AlterNative: An International Journal of Indigenous Peoples* 9 (4): 365–372.

Valdivia, Gabriela. 2008. "Governing Relations Between People and Things: Citizenship, Territory, and the Political Economy of Petroleum in Ecuador." *Political Geography* 27 (4): 456–477.

Valdivia, Gabriela. 2015. "Oil Frictions and the Subterranean Geopolitics of Energy Regionalisms." *Environment and Planning A* 47 (3): 1422–1439.

Valdivia, Gabriela. 2018. "'Wagering Life' in the Petro-City: Embodied Ecologies of Oil Flow, Capitalism, and Justice in Esmeraldas, Ecuador." *Annals of the American Association of Geographers* 108 (2): 549–557.

Villenas, Sofia A. 2019. "Pedagogies of Being With: Witnessing, Testimonio, and Critical Love in Everyday Social Movement." *International Journal of Qualitative Studies in Education* 32 (2): 151–166.

Walsh, Catherine E. 2019. "(Decolonial) Notes to Paulo Freire: Walking and Asking." In *Educational Alternatives in Latin America*, edited by Robert Aman, and Timothy Ireland, 207–230. Cham, Switzerland: Palgrave Macmillan.

Watts, Michael, and Nancy Lee Peluso. 2013. "Resource Violence." In *Critical Environmental Politics*, edited by Carl Death, 184–197. New York, NY: Routledge.

West, Paige. 2006. *Conservation Is Our Government Now: The Politics of Ecology in Papua New Guinea*. Durham, NC: Duke University Press.

Whyte, Kyle Powys. 2013. "On the Role of Traditional Ecological Knowledge as a Collaborative Concept: A Philosophical Study." *Ecological Processes* 2 (1): 7. https://doi.org/10.1186/2192-1709-2-7. doi: https://doi.org/10.1186/2192-1709-2-7.

Willis, Anne-Marie. 2015. "Transition Design: The Need to Refuse Discipline and Transcend Instrumentalism." *Design Philosophy Papers* 13 (1): 69–74.

Wolf, Eric R. 1982. *Europe and the People Without History*. Berkeley, CA: University of California Press.

Wolfe, Patrick. 2006. "Settler Colonialism and the Elimination of the Native." *Journal of Genocide Research* 8 (4): 387–409.

Yazzie, Melanie K. 2018. "Decolonizing Development in Diné Bikeyah." *Environment and Society* 9 (1): 25–39.

Yusoff, Kathryn. 2018. *A Billion Black Anthropocenes or None*. Minneapolis, MN: University of Minnesota Press.

Zimmermann, Erich W. 1951. *World Resources and Industries: A Functional Appraisal of the Availability of Agricultural and Industrial Materials*. New York, NY: Harper.

Section I
(Un)knowing resources

2

Chimeras of resource geographies

Unbounding ontologies and knowing nature

Helene Ahlborg and Andrea Joslyn Nightingale

Introduction

Resource geographies have a long tradition of bringing together the social and natural sciences. Indeed, the interdisciplinary character of the field in the 1990s allowed Nightingale to ask research questions about the mutual constitution of ecologies and societies, which were otherwise difficult to do within most other disciplinary contexts. Now, interdisciplinary "socionatural" research has not only become commonplace, but it also reflects the cutting edge within a range of disciplines. In this chapter, however, we show that this increased work across the social and natural sciences has not erased the tensions and ambiguities that arise from crossing disciplinary boundaries. We use the figure of the chimera, a mythical creature of destruction and incommensurate parts, to argue for a plural—as opposed to hybrid—approach to resource geographies. This approach allows us to hold in tension multiple interpretations, to work with ontological frictions while letting go of pressures for consensus, thus promoting a kaleidoscopic view of our research problems that sparks new insights.

To begin, it is first crucial to query what "resources" are. Many resource geographers take for granted that the methodological objects are obvious: forests, water, fossil fuels, minerals, etc. In contrast, a critical approach unpacks how these materials and habitats come to be defined as useful for human societies in the first place. "Socionature" signals a theorization of nature and society as inseparable, contingent, and dynamic: in short, as co-emergent. From this stance, resources are never unproblematically separated from the social-political relations through which they are isolated, extracted, and used. Critical resource geography investigates how particular boundaries are drawn between society and nature in various sectors and practices, including in academia, boundaries which, as we will show later, have significant consequences for our scholarship.

Discussions of resources also raise difficult issues of ontology, or what we conceptualize the world to *be*. At the heart of these issues are questions of knowledge and power. Framing forests as resources, for example, conceals other dimensions of forests. It allows for particular kinds of extraction and conservation practices and authorizes some people as having the right knowledge for managing them while others are marginalized. While not necessarily always negative, the concern here is the reduction of very complex, socionatural relations into an object: forest. For critical resource geographers, reducing complexity in this manner is never innocent but rather serves to validate some ways of knowing and being over others. Attention to ontologies (being or becoming) in this manner forces one to ask difficult questions about how our knowledge practices (epistemology, or how we attempt to know the world) are complicit in making the world. It is not, therefore, a case of simply doing research on exploitation of resources but rather to critically examine how those resources come into being in the first place and what that means across scales, communities of knowing, and for exploitation.

The framing of climate change as a problem of carbon dioxide in the atmosphere, for example, has created "carbon" as a resource that can be traded, sequestered, and emitted. While we are not discounting the effects of carbon dioxide in the atmosphere, we are questioning the politics and practices that arise when we distill a problem as complex as climate change to carbon. It is only through such a simplification that imagining trading carbon becomes possible. There are multiple ways to conceptualize what the problem of climate change is (Nightingale 2016), and when we accept this, then our knowledge making must expand to embrace *multiple* ontologies. At stake are possibilities to scrutinize different sets of relations and to imagine complex problems in ways that can open up possibilities for new ways of being (de la Cadena and Blaser 2018; Viveiros De Castro 2013).

In this chapter, we show how attention to ontology is necessary in order to "unbound" resource geographies. Our own experiences of disciplinary boundary crossing have convinced us that the closure of knowledge communities (like disciplines or a conversation within a discipline), while often productive, is ultimately limiting. We propose that far from being a position of "no discipline," crossing boundaries allows scholars to embody multiple ways of knowing and to query how they relate and translate. The chimera symbolizes this embodiment of plural knowing, and in company with her unruly siblings—cyborgs, monsters, and mutants (Haraway 1991)—she inspires us to think through disciplinary disorder, communication breakdown, and chaos. The chimera suggests that working in multiple epistemic (knowledge-making) communities opens up a fruitful, multilayered terrain of meaning, where new questions and insights come into view, emerging from the process of boundary crossing itself. Rather than being afraid of "epistemic pollution" (Tsing 2017) or the mixing of knowledge-making practices, we find this a highly productive terrain yet one that meets with stubborn resistance from the scientific community. We probe the implications of such resistance for critical resource geographers and counter it with the plurality and ambiguity of the chimera.

We structure our argument around the practices of boundary making and crossing in resource geographies. Disciplinary boundary crossing necessarily causes friction from which arise three quite different outcomes. First, epistemic closure wherein a community of scholars outright rejects challenges to its understanding from boundary crossers. Second, epistemic stickiness, meaning the unwillingness to shift position, causing boundary crossers to be subjected to more stringent standards of validity and argumentation in order for their ideas to be accepted. Third, epistemological sparks, which help to trigger leaps in understanding and more imaginative insights. We will first expand on ontological frictions in critical resource geographies, and then introduce the chimera in her multiple shapes. She urges us to move beyond hybrids and the search for a shared language. This guides our discussion in the subsequent section of how we may "unbound" resource geographies (see also Furlong et al., Chapter 12 this volume). Boundary making and crossing often encounter resistance and antagonism in a variety of scientific knowledge-making contexts. We illustrate our experiences of closure, stickiness, and sparks and let the chimera inspire our responses in the concluding section. As a result, we are not seeking to dissolve boundaries, but rather, like Haraway (1991, 15), we find "pleasure in the confusion of boundaries and [try to take] responsibility in their construction."

The border war between nature and society

One of the first moments of challenging epistemic closure occurs when boundary crossers question the Cartesian separation of mind and matter. Donna Haraway described this as the "border war" that holds separate nature from society, machine from organism, and allows for the appropriation of nature as a resource along with other forms of domination (Haraway 1991). These dualisms—and the mechanistic, patriarchal, colonial logic that form their historical basis—have been thoroughly exposed by scholars in many fields, who argue that this hegemonic framing not only limits our thinking but also underpins current problems of overexploitation. Today many argue that the boundaries between society-nature and human-machines are meaningless and should be defied. Yet, most current disciplines are premised on precisely these social-natural boundaries and continue to ontologically and epistemologically pry these boundaries apart.

In resource geographies, recognition of this border war extends back to at least the 1990s with attention to questions of knowledge within resource conflicts (Peet and Watts 1996). Political ecologists emphasized how the exclusion of local ways of knowing in development interventions resulted in uneven access to and control over resources. Yet, within these conversations, a split has developed between those who have embraced the challenges of rejecting the Cartesian dualism and those who left it intact in their academic practice. This manifests most clearly in the conflict between realist and relational ontologies. (Constructivist ontologies are ambiguous in that they refute the realist position but may or may not leave the Cartesian dualism intact.) Realist ontologies start with a conceptualization that assumes objects in the world exist in a form that is intrinsic and not changed by human interactions. Scholars therefore set out to discover relationships and properties in the world (as it *is*) and document how complex social and technological processes impact upon and transform environments. This kind of ontology retains the separation of nature from society since the "environment" or "resources" can be identified as preexisting our conceptualizations of them. In contrast, relational ontologies assume that the world is constantly being made. Scholars in this tradition seek to understand how framings are complicit in creating particular relationships and properties in nature and society, such that framings are always part of making the world. Here, a separation of nature from society is rejected, and rather it is assumed that environments and resources emerge from the knowledges and practices through which they are entangled in human affairs. Most scholars fall somewhere in the middle between these positions, accepting an external reality but acknowledging that we can never know it outside of human societies. Many who work with relational ontologies recognize that how the world comes into being is shaped by nonhumans too. So, it is not simply our knowledges that make the world. Rather, in a relational framing, the entanglements of more-than-human interactions and knowledges of those entanglements make the world (Blaser 2012; de la Cadena 2010).

Yet despite these debates and advances in resource geographies, the nature-society divide continues to cause trouble. For the two of us, we often find ourselves in the middle of this border war as we shift between communities who work with realists and those who work with relational understandings of the world. This causes significant friction as the starting points for investigating resource dilemmas can be very different depending on which ontologies prevail. In response, many scholars, once they have defined their own position on the realist-relational spectrum, react to other ontological positions either with epistemic closure or stickiness. This manifests in disputes over validity, relevance, and even legitimacy of the others' problem formulations and research questions. Due to the interdisciplinary nature of our research, we are unwilling to simply reject one stance and adopt another. It is frustrating to witness communication breakdown or blocking of fruitful conversations across disciplines as a result of differences in framing; yet in our experience, it is very common and makes boundary crossing troublesome.

An example of how productive multiple ontologies can be for critical resource geographies comes from de la Cadena's (2010) work on resource extraction and resistance to mining by Indigenous peoples in the Andes. De la Cadena finds that people who oppose mining ground their opposition in multiple reasons. These reasons defy an easy categorization into eco- or anthropocentric views. Indigenous peoples see the mountain as sacred or not and may support resource extraction under certain conditions. For some people who live in Andean communities, the mountain is not an object but an *apu*, a living "earth being" who obliges them by providing and caring for its people and in return demands offerings and appropriate behavior. Knowing their landscape as living beings sits at odds with science, which assumes the mountain and surroundings are composed of inert matter and which negates the *apu's* value and political agency in matters of well-being politics. For de la Cadena, to study the conflict over mining only from a political economy lens or using a framework of class and ethnicity may be accurate but insufficient. It silences other layers of the conflict. In Andean resource extraction politics, earth beings are ontologically present (not for everyone but for many) as actors in themselves, not just as nature *represented by* science and taken into politics by people, or just as symbols in the domain of knowledge. Rather, they are political actors themselves. These conflicts therefore take place on multiple political levels: knowing, acting, and being with consequences for what resource extraction can be acceptable under what conditions and relationships (de la Cadena and Blaser 2018). Working with multiple ontologies, the mountain is no longer only a mountain but a site of equivocation where multiple worlds meet, collide, and align. There is no neat translation available to the researcher (see also Bebbington et al. Chapter 21 this volume). For de la Cadena, it is not up to the researcher to resolve the tension or collapse the worlds but rather to bring them into view and expose these multilayered politics (de la Cadena 2010; Viveiros de Castro 2004).

These efforts at decolonizing science have untied the straitjacket of a singular ontology that limits geographers as they attempt to understand resource politics around the world. Yet, the growing interest in and efforts at trans- and interdisciplinary boundary crossing have not yet unsettled academic hierarchies, funding systems, and career opportunities at a deeper level. It seems that the disciplining tendency in Western science is too sticky to get rid of, manifesting as it does in influence, careers, and recognition. Indeed, the upper hand of disciplinary science is something we struggle with on a daily basis, and as interdisciplinary feminist scholars within the society-nature and human-technology border wars, we are often confronted with attempts at disciplining us. From our stubborn attempts to communicate across fields, we have learned a few things about boundary making and disciplining practices that block the conversations that we *wish to have*. Our response to these frictions is to bring out the chimera—a creature who can move across worlds and inhabit multiple ontological realms—while refusing to please, to be disciplined, or ease the palpable tension.

We find that the chimera thus holds much promise for taking another step away from epistemic closure. The chimera embraces ambitions to decolonize scientific practices by not reducing our knowing of a problem into one consistent frame. By using a symbol of destruction and creative chaos, we invite dispute, irreconcilable differences, and ontological friction as *productive* for our scholarship and our politics.

The embodied plurality of chimeras

Chimera | kɪˈmɪrəkəˈmɪrə | (also chimaera) (noun)
    1 (Chimera) (in Greek mythology) a fire-breathing female monster with a lion's head, a goat's body, and a serpent's tail.

–any mythical animal with parts taken from various animals.

2 a thing that is hoped or wished for but in fact is illusory or impossible to achieve.

3 (Biology) an organism containing a mixture of genetically different tissues, formed by processes such as fusion of early embryos, grafting, or mutation

–a DNA molecule with sequences derived from two or more different organisms, formed by laboratory manipulation.

Source: *New Oxford American Dictionary*, 3rd ed.

As a practice of multiplicity, chimeric boundary crossing strives not for hybrids or integrative science. Haraway (1991) established the salience of cyborgs and hybrids as productive images to think through the anxieties of modern society-nature transformations, but our concern is to insist again on the disorder she advocated. Our chimeric science challenges researchers to leave tensions unresolved. Building from the previous definitions, the chimera offers to our imaginations simultaneously: the personification of chaos, a challenge to modernity by defying classification and shifting across established species and nature-society boundaries, and the harbinger of illusion or even disaster. In Greek mythology, she is an enemy of the gods, a fire-breathing female with a lion's head, a goat's body, and a serpent's tail. She threatens order with pandemonium and is an omen of storms, shipwrecks, and natural disasters, particularly volcanic eruptions. Over the centuries, she has also come to signify a phantasm, something hoped for but impossible to achieve: a *vision*. In biology, the chimera image has been taken up to describe a single organism composed of cells with separate genotypes. They occur among plants, animals, and humans, are usually fertile, and are far more common than previously thought.

Like the chimeras of mythology and fiction, as well as illusory chimeras of everyday life, academic boundary crossing that takes place under the labels of trans- and interdisciplinary science may bring chaos or be an unrealistic endeavor and, regardless, is sure to cause disruption to the disciplinary order (Pooley, Mendelsohn, and Milner-Gulland 2013). Fears of such chimeric friction—whether recognized or not—mean many scientific communities are satisfied with bringing together disciplinary knowledges in a manner that does not disrupt the existing order, or when different ontologies are acknowledged, scholars seek to resolve the tension and establish consensus as a basis for collaboration (see Shapiro-Garza et al., Chapter 21 this volume). Team members whose ideas continue to fit uncomfortably with others' framings may be quietly deleted from email lists or not invited to participate in the next grant writing round. The first kind of interdisciplinary inquiry sidesteps the problem of whether researchers are talking about the same methodological object (ontology) when they approach it from their disciplinary stances (epistemology) and therefore edits out the possibility for a plurality of meanings. The second kind fears misunderstanding and tries to find a shared language. We find both approaches unsatisfying and suggest that a chimeric science can work with the friction of boundary crossing to spark surprise and transformation in the face of complex research problems.

We therefore argue that it is important to make a strategic shift from hybrids to chimeras in how we understand boundary making and crossing. Chimeras are *not* hybrids, an insight we derived by thinking through biology. In hybrids, different genetics fuse, whereas biological chimeras reflect several genetic types in one body. Hence, a chimera *embodies* multiple creatures that coexist, without internal conflict. Rather than being monsters, these biological chimeras suggest the possibility of peaceful—or at least livable—multiplicity. Haraway's cyborg (1991), while described as a chimera, was also explicitly a hybrid. It was an attempt to hold together, even if uncomfortably, contradictions, and human-machine intermingling. In contrast, our chimera's particular incarnation of plurality captures the possibility of ambiguity and uncomfortable knowledge tensions. The chimera thus embodies uneasy contradictions that are not resolved but rather exist in parallel, allowing for multiple meanings (Domosh 2017) and leaving the equivocation intact (de la Cadena 2010). Holding such tensions dynamically is required for critical resource geography to spark disciplines into new creative terrain.

A chimeric approach to boundaries and discipline

Monsters have often defined the limits of communities in Western imagination (Haraway 1991, 180): the same and yet different enough to not be us. Yet, boundary making is an ambiguous practice, and as many have argued, the debate in one discipline is shaped by its distinction from other disciplines. For us, focusing on the parameters of a conversation in one field compared to another is one of the first moments of embracing interdisciplinarity. Boundary making can enable a constructive crossing when it is contingent and emergent rather than excluding. This suggests careful listening and humble curiosity in relation to other fields as well as nonacademic ways of knowing. In our experience, when scholars believe they know what the debate is in a given field, they ignore, or are not sufficiently receptive to, changes in nuance that drawing on a parallel debate can provide.

We have both confronted this kind of epistemic stickiness on more than one occasion. A recent peer-review process is illustrative. One of Nightingale's papers speaks across parallel debates on resource governance in political ecology, anthropology, and development studies in order to open up new insights about the socionatural character of state formation, using the conceptual device of boundary making (Nightingale 2018). By focusing on boundary making and working across different disciplines, the paper unpacks what resources, the state, and governance *are* and how they come into *being*. One reviewer reacted to the attempt to speak across these different debates by saying, "[t]hese explanations of the concept are not 'wrong' in themselves, but it is clear from them that the term 'socioenvironmental state' holds together several quite specific claims that, in the literature, are each elaborated with significantly greater precision." The reviewer went on to explain how Nightingale ought to have engaged closely with one or two theorists and narrated her case study narrowly through the resulting lens. While obviously the paper needed some revision to be more convincing, it was how the reviewer engaged with the work that concerns us. Rather than embracing the project of explicitly working with different understandings as a way to spark new life into resource geographies of the state, s/he simply chastised the author for not using in more depth the debates the reviewer already knew. This kind of critique closes down the possibility to think about the state and resource governance differently by embracing an alternative starting point for not only resource governance, but also the state itself. Such judgments on relevant objects of study are part of peer-review practice but are inevitably tied up with general patterns of power in society (Barkan and Pulido 2017) and serve to close down the possibility to think about familiar problems in new ways. Without humility, disciplinary comfort misleads individuals and collectives to believe their judgments are "correct" rather than being based on a limited, particularistic understanding.

Richard Lewontin (1991) discussed this kind of epistemic stickiness years ago in relation to the natural sciences. He showed how when new results did not fit established paradigms, they were subjected to significantly greater burdens of proof. Often, they were simply rejected (resulting in closure) rather than leading to a questioning of existing ontological commitments (opening up). Our chimera is precisely the figure through which we show the intellectual traction gained from opening up. We argue for the need to engage in boundary making with a focus on sparking exchange and translation in order to free our imagination and concepts from the disciplining practices of individual fields. Hence, the chimera is a creature who does not play by one set of rules but with multiple rules, and rather than wreaking havoc and excluding new meanings, she works hard to understand each discipline.

The chimeric approach may be likened to the polyglot's relation to languages. Disciplining, as chimeric practice, is to acquire new communication skills and abilities to perceive nuances and shifts in meaning and to recognize the importance of the underlying assumptions and processes that signify concepts. For resource geographers, this implies engaging with other fields and ontological positions on equal terms, not in order to settle for one view on "resources," but to allow the discrepancies to force us to reconsider what our "objects" could be such that friction gives way to a multilayered understanding. What comes into view through such engagement is new conceptual terrain at the multiple political levels of epistemology and ontology.

Crossing boundaries through the chimera thus reduces the hold of one discipline over epistemology therefore "unbounding" our ways of knowing. Like biological chimeras, boundary crossers need to absorb multiple codes to accept parallel relations that make familiar objects and concepts foreign and to keep them connected but separate. Despite these rewards, boundary crossing itself comes with a number of challenges. Even for established scholars, it always entails elements of personal risk and, in our experience, rejection as we recount more fully later on. As travelers who may not have an obvious disciplinary home, chimeric scholars are highly vulnerable to aggressive epistemic stickiness and closure by those who believe they already know what the debate should be, whether at conferences, in publishing, or in job-hiring practices.

Boundary crossing, vulnerability, and sparks

To illustrate the epistemic closures and stickiness that boundary crossers confront but also the productive space that opens up as we move beyond insistence upon a shared vocabulary to a chimeric engagement with languages and disciplines, we recount a few of our incidents with colleagues. As any interdisciplinary scholar learns, what counts as evidence or a valid form of argumentation differs across traditions. And while we are especially careful to validate our empirical data from the traditions within which we collect it, it is our attempts to speak across boundaries with our analysis that seem to be the most threatening. Chimeric scholars face more than usual vulnerability in the dissemination process and have a greater pedagogical burden to communicate constructively with reviewers who react to new ideas by disqualifying our scholarship.

Let us start with a spark. We have together explored the concept of "scale" in relationship to knowledge making and resource politics. In a joint paper, we compared the understanding of scale in geography, ecology, and social-ecological systems and found that conflicting assumptions, definitions, and meanings of scale strongly shaped the diverging understanding of the problems, situations, and society-environment interactions in focus. From reflecting on the tensions, a new insight emerged around the underlying epistemological scale of various types of knowledge. We realized that scale mismatches at work in knowledge production helped explain some of the divergent ontological understandings of forests, which Nightingale had previously observed in Nepal.

This interdisciplinary work met with stickiness as one reviewer levied a strong critique of its scientific merits in a manner that has become oppressively familiar to us. S/he wrote, "First and foremost, I cannot see this paper as a SCIENTIFIC PAPER. The first sections provide something of a literature review [...], followed by what is labelled as a 'case study.' Yet, this section lacks everything that could classify it as a scientific paper. No method is presented (at least not in the level that would make it reproducible), no verifiable results are presented, and neither method nor results are discussed, or their validity presented. In short, I read anecdotes, not scientifically sound material." The work was disqualified using capital letters and words like "classify" and "validity." Here, the reviewer took a particular view on what counts as valid knowledge, and since our data did not comply with that, s/he could not read any data in the paper. Our polite clarification of the empirical work our argument was based upon placated the reviewer who was able to see past his/her own epistemic stickiness and did not object on the second round.

More disappointing is our common experience of being excluded outright. Editors in particular exercise boundary policing, granting or denying access to the very audiences we want to engage. Ahlborg and colleagues (Ahlborg et al. 2019) struggled to publish a paper that bridges across social-ecological and socio-technical perspectives, conceptualizing the technological mediation of human-environment relationships. The paper explicitly highlights the disciplinary closures that need to be overcome for fruitful engagements. Yet, the editor of the chosen journal rejected it, because "we have tried with several of our subject editors to take on your interesting manuscript [...] but have failed." This editor chose to desk reject rather than finding a guest editor who had an interest in and knowledge of technological dimensions of social-ecological change. Two more journal editors also rejected the paper without sending it out for review, as it did not fit their ideas of what debates were worth having.

This kind of epistemic closure and stickiness serves to bound communities and conversations—which is sometimes useful to push forward a debate—but for an individual, it can be highly uncomfortable. Boundary crossing makes you vulnerable and exposed, not least to your own presumptions. Conferences, PhD courses, and research workshops are all occasions where the most stimulating—and upsetting—encounters happen. A dinner discussion goes out of hand when a new colleague suddenly reacts forcefully to something you said with an unexpected personal attack that reflects a clash between scientific paradigms. A visit in another field may also bring the embarrassing realization that you have uncritically (and arrogantly) dismissed entire fields of study because valued colleagues in your own discipline see that work as problematic. Thus, our concern is not only with a relatively abstract idea of epistemic friction that results in closure or stickiness but also with the embodied experiences for individual scholars who bravely engage in boundary crossing. While both of us have launched on such a path on our own, over time we have found it vital to have a small community to which we can retreat, gain support, and wherein our most productive, unbounded conversations can take place.

We want to recount a final example of epistemic stickiness based on the notion that communication cannot work if there is ambiguity or disagreement (stop being so difficult!). Interdisciplinary events often involve attempts to establish consensus on the definitions of key concepts, resulting in stalemate as participants refuse to let go of their disciplinary understandings. One of us was told by a colleague, who attempted to be supportive and provide mentoring, that she had been overly disruptive in a research team meeting, because she questioned how the core objects of the research were talked about. We have recognized that the search for consensus and a "shared base" reflects an understanding of interdisciplinarity as a hybrid, when, in fact, we need the chimera. Once we let go of the idea that a scientific exchange has to start from a joint language and shared definitions of problems and key concepts, we are able to work with epistemic frictions instead of against them. In the spirit of taking responsibility for how we make and undo boundaries, we now take more care to not disrupt communication, while continuing to engage in a campaign of deliberate plurality and continual questioning of how key concepts are used.

Debates within climate change are currently facing such dilemmas. The field has been dominated by modeling and physical scientists keen to fill up knowledge gaps and integrate their findings so as to understand the climate system and predict future changes. Yet, this work has run into major stumbling blocks over the past ten years. Global climate models do not integrate readily with regional models because of differences in scale of observation and the assumptions that underpin them. Social scientists are concerned about the impulse to integrate because of the reduction of multiple ways of knowing into models that typically privilege realist ontologies and silence other perspectives. Many of the important observations and understandings of climate change that emerge in different parts of the world cannot be put into models, which by necessity are reductionist. Critical resource geographers have focused on the political nature of adaptation and mitigation interventions by showing how the kinds of knowledge produced in models or the solutions generated from consensus decision-making are never adequate to mediate changing climatic conditions and the social politics of marginality. Several critical resource geographers have thus argued for the need to embrace multiple ontologies and to engage with modeling while holding in view quite different understandings of the climate problem in order to imagine new, creative responses across scales (Lövbrand et al. 2015).

By accepting multiple ontologies and seeking to embody them through the chimera, we have found new life and inspiration breathed into old and tired concepts. Ambiguity and friction—not the least in heated debates between the two of us when we disagree over an idea or interpretation—have resulted in insights neither of us could have reached without the other. However, the shift from one language to multiple does not justify the sloppy use of concepts. In contrast, the chimeric approach requires more careful and precise choice of words and conceptualizations; some measure of stickiness must remain. But it also requires the development of new understanding, which brings us closer both to our theoretical commitments and empirical data while focusing our attention on communication as relational practice.

Conclusion

Our aim in this chapter has been to unbound critical resource geographies by exposing the frictions that come from crossing disciplinary boundaries. The chimera has been our guide in thinking through epistemic closures and stickiness, which serve to maintain boundaries and create exclusions, while also helping us to find creative sparks within the disorder, ambiguities, and possibilities that working with multiple ontologies brings. While we do not reject entirely the impulse to integrate and bring consensus across different fields, we find more creative potential in defying purity of form. Our argument is thus not about getting rid of disciplines or dissolving all boundaries but rather to work with contingent closure, explore the creative frictions that result when we engage caringly with ontological difference, and treat different knowledges as equal but always different.

In our experience, it is the process of shifting and translating between multiple disciplines that generates something new: concepts are re-scrutinized, assumptions are queried from positionalities that are impossible to imagine from within one's own discipline, and taken-for-granted relationships can appear implausible. From these encounters, oscillating back and forth, translating and contrasting, and the experience of being a guest, a visitor, and yet unfamiliar with vernacular meanings of concepts, we make ourselves vulnerable and gain new insights. Out of the corner of our eyes, we glimpse mythical creatures we are *not sure of*: they bring destructive-visionary uncertainty that opens up our methodological objects to creative scrutiny by posing the questions of what/who it could be.

The chimera of interdisciplinarity thus suggests a productive space in the making. For us, a chimeric approach to resource geographies, thus, is a zone of tension wherein scholars challenge the impulse to seek consensus or comfort and focus instead on livable coexistence. On an intellectual level, this means taking a polyglot approach to disciplines, seeking to understand them on their own terms and allowing them to challenge our ontological starting points. On a personal level, it requires a steadfast commitment to friendliness and generosity: to remain open when faced with conflict or disagreement so that rather than producing chaos, we are able to create new beings, like the biological chimeras. So, while the chimeric scholar should expect to be misunderstood on a regular basis, there are also sparks of deep inspiration to be had.

Resource geographers are accustomed to engaging with resource conflicts, but recent debates between relational and critical realist perspectives within the field signal an entrenchment of epistemic communities. The chimera insists upon a different reaction to the contested terrain of "resources": one where we are willing to be unsure of the most fruitful research approach, reframe our questions in unfamiliar ways, and query how we define methodological objects. Our approach offers inspiration to read widely and experiment with what comes into view when our core concerns are framed in new ways. Like our attempts at reframing the socioenvironmental state or social-technical systems, sparks of insight ignite not only from asking what resources or landscapes *are* in different ontologies but also by holding them in tension. This chimeric approach suggests an explicit effort to create pluralist research processes, e.g., to see the forest as temporal change in vegetation cover through the lens of satellite imagery; a terrain of ecosystem services and differing degrees of biodiversity; and then allowing these views to sit uneasily with how people experience the forest as home, source of life, place of worship, commodity for the market, and a signifier of status and labor relations. Rather than resolving the tension between these layered realities and ways of knowing and being, one can use the ruptures and glitches between them as places to dwell and ponder while listening for the footsteps of the chimera. Her presence sparks new insights into how resource and knowledge politics are entangled with the plural landscapes we know, sense, and anticipate.

References

Ahlborg, Helene, Ilse Ruiz-Mercado, Sverker Molander, and Omar Masera. 2019. "Bringing Technology into Social-Ecological Systems Research—Motivations for a Socio-Technical-Ecological Systems Approach." *Sustainability* 11 (7): 2009. doi: 10.3390/su11072009.

Barkan, Joshua, and Laura Pulido. 2017. "Justice: An Epistolary Essay." *Annals of the American Association of Geographers* 107 (1): 33–40. doi: 10.1080/24694452.2016.1230422.

Blaser, Mario. 2012. "Ontology and Indigeneity: On the Political Ontology of Heterogeneous Assemblages." *Cultural Geographies* 21 (1): 49–58. doi: 10.1177/1474474012462534.

de la Cadena, Marisol. 2010. "Indigenous Cosmopolitics in the Andes: Conceptual Reflections Beyond 'Politics.'" *Cultural Anthropology* 25 (2): 334–370.

de la Cadena, Marisol, and Mario Blaser. 2018. *A World of Many Worlds*. Durham, NC: Duke University Press.

Domosh, Mona. 2017. "Radical Intradisciplinarity: An Introduction." *Annals of the American Association of Geographers* 107 (1): 1–3. doi: 10.1080/24694452.2016.1229596.

Haraway, Donna. 1991. *Simians, Cyborgs, and Women: The Reinvention of Nature*. New York, NY: Routledge.

Lewontin, Richard C. 1991. "Facts and the Factitious in the Natural Sciences." *Critical Inquiry* 18 (Autumn): 140–153.

Lövbrand, Eva, Silke Beck, Jason Chilvers, Tom Forsyth, Johan Hedrén, Mike Hulme, Rolf Lidskog, and Eleftheria Vasileiadou. 2015. "Who Speaks for the Future of Earth? How Critical Social Science Can Extend the Conversation on the Anthropocene." *Global Environmental Change* 32: 211–218. doi: 10.1016/j.gloenvcha.2015.03.012.

New Oxford American Dictionary, 3rd ed., s.v. "chimera," accessed May 3, 2021, https://www.oxfordreference.com/view/10.1093/acref/9780195392883.001.0001/m_en_us1232673.

Nightingale, Andrea Joslyn. 2016. "Adaptive Scholarship and Situated Knowledges? Hybrid Methodologies and Plural Epistemologies in Climate Change Adaptation Research." *Area* 48 (1): 41–47. doi: 10.1111/area.12195.

Nightingale, Andrea Joslyn. 2018. "The Socioenvironmental State: Political Authority, Subjects, and Transformative Socionatural Change in an Uncertain World." *Environment and Planning E: Nature and Space* 1 (4): 688–711. doi: 10.1177/2514848618816467.

Peet, Richard, and Michael Watts, eds. 1996. *Liberation Ecologies. Environment, Development, Social Movements.* London: Routledge.

Pooley, Simon P., J. Andrew Mendelsohn, and E. J. Milner-Gulland. 2013. "Hunting Down the Chimera of Multiple Disciplinarity in Conservation Science." *Conservation Biology* 28 (1): 22–32. doi: 10.1111/cobi.12183.

Tsing, Anna Lowenhaupt. 2017. *The Mushroom at the End of the World: On the Possibility of Life in Capitalist Ruins.* Princeton, NJ: Princeton University Press.

Viveiros de Castro, Eduardo. 2004. "Perspectival Anthropology and the Method of Controlled Equivocation." *Tipití: Journal of the Society for the Anthropology of Lowland South America* 2 (1): 3–22.

Viveiros De Castro, Eduardo. 2013. "The Relative Native." *HAU: Journal of Ethnographic Theory* 3 (3): 473–502. doi: 10.14318/hau3.3.032.

3

Knowing the storyteller

Geohumanities and critical resource geography

Kolson Schlosser

Introduction

Geohumanities refers to geography's renewed engagement with the humanities beginning roughly with the publication of the volumes *Envisioning Landscapes, Making Worlds* (Daniels et al. 2011) and *Geohumanities: Art, History, Text at the Edge of Place* (Dear et al. 2011), as well as the launch of the journal *Geohumanities* by the American Association of Geographers in 2015. I say *renewed* because, in fact, geography has always had a humanistic strain. From Aristotle to von Humboldt, geography predominantly focused on describing regional differentiation of both landscapes and peoples. To do geography was to engage in "earth writing" (its literal translation) from the position of a purportedly unbiased observer. As has been well chronicled (Blunt 2009), the move to quantitative modeling and (later) some forms of orthodox Marxism in the mid-twentieth century upended the humanistic flavor of geographic practice; what was once largely *descriptive* became *predictive* and *analytical* in various ways.

But as early as the 1970s, humanistic geography was reinvigorated through the work of Doreen Massey, Pierce Lewis, and Yi-fu Tuan, among others. This strain of humanistic geography (not yet called geohumanities per se) no longer assumed an unmediated relationship between observer and things observed. It distanced itself from quantitative modeling and orthodox Marxism, however, in terms of *how* that mediation is understood. Where orthodox (though far from all) Marxism saw that mediation largely in economic terms, humanistic geography tended to see it in more fluid and dynamic ways—be it culturally, affectively, symbolically, or what have you. This concept of mediation turns out to be crucial to how the "earth-writing" tradition in humanistic geography was redefined; for instance, by the 1990s, Barnes and Duncan (1992) demonstrated that to "write" the earth was to potentially change it, as all forms of representation inevitably involve social power. This understanding of the power involved in representation was indicative of geography's post-structural turn, bringing attention to the productive effects of discourse throughout the 1990s (Blunt 2009).

So why the 2011 invocation of the term geohumanities? Humanistic geography advanced and diversified from the 1990s onward by, for instance, further engaging with literary theory (Brousseau 1994) and advancing visual methodologies (Rose 2001). As the field developed, critical human geographers further explored how we understand the social role of stories and narratives—advancing, for instance, nonrepresentational, performative, or normative understandings of stories (Cameron 2012). As Hones (2015) suggests, the field of literary geographies (a subset of geohumanities) has been so infused by cross-disciplinary collaboration that defining its boundaries is increasingly difficult. We might think of geohumanities, then, as a codification and recognition of this increasing diversification in humanistic geography—not to impose order on it but to recognize what geography has always been, at least in certain times and places. Doing so has arguably broadened its analytical potential as well. For instance, much of the work done under the geohumanities moniker has engaged with the normative potential or limitations of stories as humans craft new visions of life in the Anthropocene (Daniels and Endfield 2009; Braun 2015; Schlosser 2018).

The purpose of this chapter is to explore the current and potential utility of geohumanities to a more specific aspect of anthropocentric life: the geography of natural resources. Understanding how *nature* becomes *resource*, not to mention all the consequences that follow, is predicated on how we conceptualize such categories as nature, resources, and nature-society metabolism. Thus, knowing what questions to ask—that is, which questions yield the most liberatory or just answers—requires a theoretical engagement with the mediated relationship between our questions and the world to which they refer. This mediation involves *ontological* questions about what the world is understood to be, *epistemological* questions of how one comes to know about the world (and who is doing the knowing), and the *methodological* implications that flow from this.[1]

To be sure, the epistemological, ontological, and methodological are overlapping, necessarily interlocked categories, which inevitably mediate the relationship between the world, the story, and the storyteller (more on that below). This mediation creates a fair amount of abstraction, but the broadened analytical potential of geohumanities that I mentioned above allows it to grapple with this abstraction in productive ways. This chapter shows how geohumanities can and has done this with regards to the study of natural resources. Borrowing from Cameron's (2012, 574) description of research on stories, it also examines how a "concern with the ways in which personal experience and expression interweave with the social, structural, or ideological" can further inform critical resource geographies.

The chapter is structured around three broad, interrelated themes. First, it explores the epistemological light geohumanistic inquiry sheds on grain production in North Africa and natural gas extraction in British Columbia and expert knowledges; that is, it foregrounds the question of how and by whom expert knowledge is created. Second, the chapter explores some of the ontological questions of how stories and narrative connect to, or "interweave with" (Cameron 2012, 574), natural resource regimes, in this case mining in Bolivia, Poland, and perhaps even outer space. Third, the chapter discusses the methodological implications that might flow from geohumanistic inquiry, including both the use of storytelling as a method itself and the use of stories as objects of analysis. To the extent that geohumanistic methodologies inform those very epistemological and ontological questions previously discussed, we can characterize them as helping to build what Cresswell (2011, 75) terms a critical geosophy, or "an account of geographical ideas and the roles they play in production, reproduction, and transformation of power." The chapter concludes with some reflections on the normative implications of geohumanistic inquiry.

The construction of "expert" knowledges

Robbins (2015) refers to political ecology as the "trickster science," because while it borrows methods and concepts from the physical sciences to inform its analysis of ecological change, it also works to undermine the hegemony of positivist science. He thus employs the folkloric figure of the trickster to describe the potential subversiveness of political ecology: "By turns, Trickster performs as a boastful clown or jester, occasionally acts as a thief, and constantly undermines haughty heroes. But Trickster also protects the weak, transports the seeds of culture, and provides gifts for humanity" (92). Robbins acknowledges that this double-edged quality of political ecology is exactly what its most ardent critics find flawed about it but counters that political ecology's most productive insights—indeed, the rewards to the disadvantaged that the Trickster ultimately brings—produce an important creative tension. In other words, the "trick" that political ecology plays on positivist science is an epistemological one; political ecology *can* yield objective knowledge at the same time that it constantly holds in question how it is that we arrive at such truths.

Much the same can be said of the use of geohumanities in critical resource studies. For example, Davis's *Resurrecting the Granary of Rome* (2007) critiques the French colonial declensionist narrative of North African ecological change often used to justify colonial control. In this narrative, North Africa served as the primary grain supplier of ancient Rome and only fell into disrepair and desertification as native Africans and swarms of Arab nomads later abused its natural resource bounty. French colonial conservation efforts were thus proposed as a solution. Davis does in fact empirically falsify the narrative, showing that the region was never quite "the granary of Rome" and that colonial policies themselves brought the most ruinous practices. In that sense, her study takes as its object of analysis this historical narrative, frequently replicated in literary texts and paintings. But more importantly, Davis's analysis offers epistemological lessons regarding historical reconstruction; that is, her study of ecological change in North Africa shows how historical narrative is a social product, as the objects of investigation left available in archives are relics of power and social conflict. The way the archive is constructed itself often reflects histories of domination. Davis's work both constructs an alternate narrative and shows how our ability to put order to the events of the past is inevitably mediated through the histories of unequal power that precede us. Her focus on narrative thus sheds light on epistemology. The "trick" that Davis's work provides is that it unsettles putatively apolitical analysis by providing historical depth (Davis 2015); while her work is not typically categorized as geohumanities, its trickster qualities exemplify the geohumanistic threads running through it.

Epistemology speaks to the question of how knowledge is derived, which is one component of a broader theoretical framework. Theoretical frameworks guide judgments about what research questions we should ask and what methods we should use to answer them. Geohumanities often entail literary-geographic methods (more on this later), but it is important to note that methods often feed back into and inform theoretical frameworks. For instance, Milligan and McCreary's (2018) fascinating comparison of an environmental assessment of a liquefied natural gas terminal in the Haisla territory of British Columbia and the novel *Monkey Beach*, authored by Haisla novelist Eden Robinson (2000), bears both empirical and theoretical insights. Their innovative method addresses a persistent, vexing problem facing studies of Indigeneity and resource extraction: that is, reliance on state recognition of Indigenous rights potentially reproduces the colonial relationship in the first place, but rejecting state recognition runs the risk of justifying yet more abuses of Indigenous peoples (Coulthard 2014). The question becomes, then, how might we characterize Indigeneity in such a way that does not reproduce the state's disciplinary need for concrete, static intelligibility? Milligan and McCreary (2018, 46) seek to answer this question by bringing the two texts into "productive interference." As they put it:

> Whereas the Kitimat LNG report on Haisla land use conveys great certainty about the integrity of its articulations of Indigenous difference, Robinson unpacks how knowledge morphs as it moves across cultural registers, challenging the presumption that the qualities

of Haisla being can be encapsulated within the frames of regulatory discourse on tradition. Rather than rendering Indigeneity legible, Robinson ruptures the colonial interpellation of Indigenous being as traditional, instead highlighting continuing dynamism of Indigenous territorial relations unfolding within and against the colonial present.

(Milligan and McCreary 2018, 46–47)

Their approach does not settle the question of the legibility of Haisla resource futures, but rather finds the flaw in the question in the first place. Indeed, following Hawkins (2015), Milligan and McCreary (2018, 50) refer to their approach in terms of an "ethos" to constantly disrupt the taken-for-granted. Rather than using literature strictly as an object of study, they use literary techniques "to analyze the geography of extractive resource economies and the workings of difference within resource governance texts" (2018, 50). By highlighting their research results as one of "productive interference" (46) rather than a definitive answer, Milligan and McCreary's geohumanistic approach yields valuable epistemological insight, simply by helping us ask better questions.

The ontological status of stories in resource-making

Is the flow of "nature" into the various matrices of human economic, cultural, and political systems best understood as a material process, with stories (and other types of discourse) understood as mere reflections of it, or do stories themselves bear some sort of power or influence in this process? For example, Jack London frequently wrote about the Yukon Gold Rush in which he participated; are his stories, then, *reflections* of a historical time period, useful for gleaning information about it, or are they in fact *actors* in it, shaping the very history they chronicle? This is why ontology and epistemology are so integrally linked: if we answer the question in the latter, our relationship to those stories, as objects of study, is by definition changed. Given that history matters materially for the present, we become participants in its shaping, no matter what critical distance we assume. This is what I meant in the Introduction by suggesting that epistemology, ontology, and methodology "mediate" the relationship between ourselves and the resources we study: geohumanities helps us *know* that mediation. This is the case whether we are telling stories about how nature becomes resources, or stories about stories that have been told about how nature becomes resources.

As is the case with most contemporary geographic scholarship, the existing literature on geohumanities and natural resource geography demonstrates how the conversion of nature into resources is both a material and discursive phenomenon. Kuchler and Bridge (2018), for instance, demonstrate how socio-technical imaginaries of coal production are co-constructive with resource materialities in Poland. Illustrating how natural resources are simultaneously discursive and material is something Cameron (2011, 2015) does brilliantly throughout her work on Arctic Canada. In her review of the geographic literature on stories and storytelling, Cameron (2012) elucidates at least three broad strands in which geographers have conceptualized the ontological status of stories. These range from a focus on "small stories"—those not conceived as reflective of a broader discursive stratum but worthy of study nonetheless; to a focus on stories as performative ontological politics (following Gibson-Graham [2008])—those that attempt to prefigure said discursive stratum; to a focus on storytelling not as object of study but rather as geographic method. She also points out that not all work on stories necessarily sees them as discursive formations reflecting an ever-shifting discourse, as they might have typically been understood in the post-structural wave of the 1990s. Research on what she refers to as "small stories," or "stories to attend to the small, the personal, the mundane and the local" (575) rather than metanarratives, would typically fall within the vein of "non-representational theory" (most typically following Lorimer [2003]). Nonrepresentational theory typically holds the connection between particular stories and a broader discursive stratum as outside of what is ontologically knowable. It is "non-representational" because it neither asserts nor denies such a connection but rather suggests that such an ontology cannot be usefully represented in our study of it.

But again, I would argue that stories do, in one way or another, "interweave with the social, structural, or ideological" (Cameron 2012, 574). Kearnes and van Dooren (2017, 179), for instance, describe how current discussions of what an ethical mining regime in outer space might look like are "fully laden with cosmic dreaming, theological wonderings, and science fiction fabulations." As a more current and earthly example, Perreault (2018) analyzes historical narratives of mining in Bolivia with respect to their meaning for contemporary politics. He argues that such historic symbols as murals, monuments, etc., "memorialize miners and place them squarely in the national story" (Perreault 2018, 233). None of this, however, is a simple matter of the past, as this narration invokes a populist vision of resource nationalism, reproduces gender norms, and neglects the toxic legacy of mining. In other words, as Perreault argues, memory is always to some degree selective as it intersects with, and finds its purchase in, the political context of the present. Thus, the stories of Bolivian mining history may or may not reflect a more enduring discursive context, but insomuch as they have material impacts in the present, specifically because their multiple meanings are taken up and made knowable by the demands of the present, they are still imbricated in the operation of social power (see also Perreault, Chapter 11 this volume). Stories are valuable sites of study because they quite literally *know something* about the world (Saunders 2010), regardless of how one conceptualizes the manner of their production (discursive, material, etc.). Thus, the manner in which stories interweave with the knowable world is an enduring question of ontology, and one that geohumanities can help bring into clarity and focus.

Stories as technique and stories as object of analysis

The third broad strand of literature on stories outlined by Cameron (2012) involves not the study of stories per se, but the use of narrative as a method of transmitting information or motivating action. For example, Kohl, Farthing, and Muruchi's (2011) *From the Mines to the Streets: A Bolivian Activist's Life* autobiographically details, in narrative form, the life of Bolivian miner and union leader Félix Muruchi. In tracing Muruchi's experience in the tin mines, to his experience being imprisoned and tortured, to the eventual inauguration of Evo Morales as president, the first-person narrative paints a vivid picture of the intersection of resource extraction and Bolivian politics. Likewise, Hern, Johal, and Sacco (2018) use a narrative structure, in conjunction with critical commentary, to bring the nuanced politics of tar sands development in Alberta into clearer view. Such approaches recognize that geography's earth-writing tradition is not simply descriptive or mimetic (following Barnes and Duncan [1992]) and likewise seek to use the normative power of writing to work toward social justice. Marston and de Leeuw (2013, iv) in fact advocate the use of creative expression not only methodologically but also epistemologically as creative work can "challenge the normative spaces and practices of disciplinary knowledge-making."

Besides creative expression as a technique, it can be used productively as an object of analysis, as was the case with my work on Jack London's literary description of nature-society metabolism (Schlosser 2015). My reading of London's work addresses what is often seen in literary studies as his central contradiction—that he espouses a Marxist class politics along with a social Darwinist understanding of human conflict. This is seen as a contradiction in light of twentieth-century efforts (by some Marxists) to distinguish naturalist epistemology (particularly those of Herbert Spencer) from Marx's understanding of historical materialism, but in fact in the late nineteenth century, it was commonplace to conflate naturalism and materialism. My method was to read London's descriptions of personal strength (which riddle his work) as a proxy for value, and upon doing so, I found that his understanding of nature-society dialectics was more naturalist than materialist (which is possibly due to how Engels tried to popularize Marx's writing after his death). Far from espousing Marxian class politics, in other words, London reproduced what Smith (2008) referred to as the "ideology of nature."

Without reproducing the entire analysis here, it is worth noting that my study of London's work is also deeply informed by, and I hope to some degree informs, the "production of nature" literature in geography (Castree 2000; Smith 2008; Ekers and Loftus 2013). The production-of-nature literature typically understands use value as rooted in labor, while exchange value operates as a hegemonic abstraction specific to capitalism. The dialectic of note is that labor's metabolism of nature fundamentally changes as the hegemony of exchange value strips workers of the autonomy of their labor and accelerates nature-society metabolism (Castree 2000). The putative contradiction in London's literature is resolved when we see his understanding of dialectics as always between material entities (which comes into view when we read strength as a proxy for value), rather than between the material and the abstract. And perhaps more importantly, my reading of London's work recognizes how literature operates as a "spatial event" emerging "at the intersection of social practices and geographical contexts" (Hones 2011, 247). For example, deep reading as a method "helps historicize the particular forms of produced nature in capitalism—particularly the frontier. This literary intervention foregrounds the frontier as a process of domination... London's work naturalized an imperial project that in fact had historically and culturally specific motivations" (Schlosser 2015, 159).

Why might that matter for a critical understanding of natural resource geography? First, it matters because it sheds light on the cultural politics of resource extraction in the Progressive Era, as London participated in and often wrote about the Yukon Gold Rush. Second, and more importantly for my purposes in this chapter, it matters because the methodology used here (deep reading of fictional texts) bears implications for the epistemological and ontological issues described above. That is, the tendency to adhere to theoretical frameworks that emphasize either the material or the discursive, or perhaps to understand stories either as particularities unmoored from structure or as representative of some broader discursive flow, begins to appear as a problematic dualism. This is the case at least in part because the very questions we ask about the world are inherently mediated through our notions of ontology and epistemology. The point of the London work was not simply to resolve an esoteric debate about what London did or did not believe; the *object of study* might be books, but the problem the study addresses is quite literally that space of mediation between knowing subject (literary authors and readers) and knowable object (the frontier, in this case). In other words, ontologies of the frontier cannot be understood separately from epistemology. That space of mediation is as much "the field" as a physical space of extraction.

In other words, what we call theory, empirics, ontology, methods, and so on can never be surgically removed from the other concepts. Rather, I think of them as layers of abstraction that we need to work through if we are to get to any sort of truth in the world. We cannot escape the abstract and theoretical, but we can try to make sense of it and avoid misleading, dualistic understandings of universality and particularity. I illustrate the importance of avoiding such dualistic framings in my examination of George Romero's zombie movies and Anne Rice's vampire novels (Schlosser 2017), but in short, doing so helps build what Cresswell (2011, 75)—as noted above—refers to as a critical geosophy, or "an account of geographical ideas and the roles that they play in the production, reproduction, and transformation of power." Geohumanities are, in other words, reflexive: they help us think not just about how nature becomes resources but to think about how we think about this process. The examples described above can be applied to geographic inquiry, at least insomuch as they demonstrate the necessity of keeping our epistemological, ontological, and methodological presuppositions constantly in tension, never fully settled.

Conclusion

A number of geographers have argued that a critical understanding and deployment of literary and other fictive geographies has important political ramifications (Sharp 2000; Marston and de Leeuw 2013; Andrew 2018). For instance, the second critical approach to stories in geographic scholarship outlined by Cameron (2012) involved the use of story and storytelling to enact a performative ontological politics—that is, an engagement with the humanities that prefigures new forms of politics (rather than uncritically reacting to old ones). Strauss (2015), for instance, argues in favor of the political utility of narratives of climate apocalypse, suggesting that rather than closing down debate, they push the boundaries of what types of politics and policies people might consider feasible. As this chapter has shown, the point is not just contributing political answers but opening up new questions in the first place.

As discussed throughout this chapter, geohumanities can help keep our epistemological and ontological frameworks in productive states of uncertainty, thus holding open space for new lines of inquiry and critical self-reflection. Much of this, of course, can be said about feminist geography, Marxist geography, postcolonial studies, and more. But humanistic inquiry is a thread of analysis that can run through any of these analytic frameworks, and in focusing our attention on the mediated relationship between research subjects and objects—on the very humanness of our research praxis—it can prevent the closing down of potentially productive avenues of thought. This, I suggest, is a crucial part of ensuring that we are asking good questions about natural resource geography, a precursor to ensuring we are providing just or liberatory answers.

Note

1   Ontology is an especially slippery concept, sometimes implying "what there is" in a philosophical sense, and other times what alternative worlds do or might exist in a more tangible, less representational sense. I mean ontology mostly in the former sense in this chapter, but in discussing it in conjunction with epistemology and methodology, the chapter also posits that knowledge construction itself can change what the world "is." Thus, even in a philosophical sense, ontology should not be understood in the singular (rather, "ontologies") (see also Ahlborg and Nightingale, Chapter 2 this volume).

References

Andrew, Laurie McRae. 2018. "Towards a Political Literary Geography." *Literary Geographies* 4 (1): 34–37.
Barnes, Trevor, and Jim Duncan, eds. 1992. *Writing Worlds: Discourse, Text and Metaphor in the Representation of Landscape*. New York, NY: Routledge.
Blunt, Alison. 2009. "Geography and the Humanities Tradition." In *Key Concepts in Geography*, edited by Nicholas Clifford, Sarah Holloway, Stephen Rice, and Gill Valentine, 66–82. London: Sage.
Braun, Bruce. 2015. "Futures: Imagining Socioecological Transformation—An Introduction." *Annals of the Association of American Geographers* 105 (2): 239–243.
Brousseau, Marc. 1994. "Geography's Literature." *Progress in Human Geography* 18 (3): 333–335.

Cameron, Emilie. 2011. "Copper Stories: Imaginative Geographies and Material Orderings of the Central Canadian Arctic." In *Rethinking the Great White North: Race, Nature and the Historical Geographies of Whiteness in Canada*, edited by Andrew Baldwin, Laura Cameron, and Audrey Kobayashi, 169–190. Vancouver: University of British Colombia Press.

Cameron, Emilie. 2012. "New Geographies of Story and Storytelling." *Progress in Human Geography* 36 (5): 573–592.

Cameron, Emilie. 2015. *Far Off Metal River: Inuit Lands, Settler Stories, and the Making of the Contemporary Arctic.* Vancouver: University of British Columbia Press.

Castree, Noel. 2000. "Marxism and the Production of Nature." *Capital and Class* 24 (3): 5–35.

Coulthard, Glen. 2014. *Red Skin, White Masks: Rejecting the Colonial Politics of Recognition.* Minneapolis, MN: University of Minnesota Press.

Cresswell, Tim. 2011. "Race, Mobility and the Humanities: A Geosophical Approach." In *Envisioning Landscapes, Making Worlds: Geography and the Humanities*, edited by Stephen Daniels, Dydia DeLyser, J. Nicholas Entrikin, and Douglas Richardson, 74–83. Abingdon: Routledge.

Daniels, Stephen, and Georgina Endfield. 2009. "Narratives of Climate Change: Introduction." *Journal of Historical Geography* 35 (2): 215–222.

Daniels, Stephen, Dydia Delyser, J. Nicholas Entrikin, and Douglas Richardson, eds. 2011. *Envisioning Landscapes, Making Worlds: Geography and the Humanities.* Abingdon: Routledge.

Davis, Diana. 2007. *Resurrecting the Granary of Rome: Environmental History and French Colonial Expansion in North Africa.* Athens, OH: Ohio University Press.

Davis, Diana. 2015. "Historical Approaches to Political Ecology." In *The Routledge Handbook of Political Ecology*, edited by Tom Perreault, Gavin Bridge, and James McCarthy, 263–275. London: Routledge.

Dear, Michael, Jim Ketchum, Sarah Luria, and Douglas Richardson, eds. 2011. *Geohumanities: Art, History, Text at the Edge of Place.* Abingdon: Routledge.

Ekers, Michael, and Alex Loftus. 2013. "Revitalizing the Production of Nature Thesis: A Gramscian Turn?" *Progress in Human Geography* 37 (2): 234–252.

Gibson-Graham, J.K. 2008. "Diverse Economies: Performative Practices for 'Other Worlds'." *Progress in Human Geography* 32 (5): 613–632.

Hawkins, Harriet. 2015. "Introduction: What Might Geohumanities Do?" *GeoHumanities* 1 (2): 211–232.

Hern, Matt, Am Johal, and Joe Sacco. 2018. *Global Warming and the Sweetness of Life: A Tar Sands Tale.* Boston, MA: MIT Press.

Hones, Sheila. 2011. "Literary Geography: The Novel as Spatial Event." In *Envisioning Landscapes, Making Worlds: Geography and the Humanities*, edited by Stephen Daniels, Dydia Delyser, J. Nicholas Entrikin, and Douglas Richardson, 247–255. Abingdon: Routledge.

Hones, Sheila. 2015. "Literary Geographies, Past and Future." *Literary Geographies* 1 (2): 1–5.

Kearnes, Matthew, and Thom van Dooren. 2017. "Rethinking the Final Frontier: Cosmo-Logics and an Ethic of Interstellar Flourishing." *GeoHumanities* 3 (1): 178–197.

Kohl, Benjamin, Linda Farthing, and Félix Muruchi. 2011. *From the Mines to the Streets: A Bolivian Activist's Life.* Austin, TX: University of Texas Press.

Kuchler, Magdalena, and Gavin Bridge. 2018. "Down the Black Hole: Sustaining National Socio-Technical Imaginaries of Coal in Poland." *Energy Research and Social Science* 41: 136–147.

Lorimer, Hayden. 2003. "Telling Small Stories: Spaces of Knowledge and the Practice of Geography." *Transactions of the Institute of British Geographers* 28: 197–217.

Marston, Sallie, and Sarah de Leeuw. 2013. "Creativity and Geography: Towards a Politicized Intervention." *The Geographical Review* 103 (2): iii–xxvi.

Milligan, Richard, and Tyler McCreary. 2018. "Between Kitimat LNG Terminal and *Monkey Beach*: Literary-Geographic Methods and the Politics of Recognition in Resource Governance on Haisla Territory." *Geohumanities* 4 (1): 45–65.

Perreault, Tom. 2018. "Mining, Meaning and Memory in the Andes." *The Geographical Journal* 184: 229–241.

Robbins, Paul. 2015. "The Trickster Science." In *The Routledge Handbook of Political Ecology*, edited by Tom Perreault, Gavin Bridge, and James McCarthy, 89–101. London: Routledge.

Robinson, Eden. 2000. *Monkey Beach.* Toronto: Vintage Canada.

Rose, Gillian. 2001. *Visual Methodologies*. London: Sage.

Saunders, Angharad. 2010. "Literary Geography: Reforging the Connections." *Progress in Human Geography* 34 (4): 436–452.

Schlosser, Kolson. 2015. "Nature-Society Dialectics and Class Struggle in Selected Works of Jack London." *Historical Geography* 43: 158–174.

Schlosser, Kolson. 2017. "Critical Geosophies: A Psychotopological Reading of Rice's Vampires and Romero's Zombies." *Environment and Planning D: Society and Space* 35 (3): 533–549.

Schlosser, Kolson. 2018. "Geohumanities and Climate Change Skepticism." *Geography Compass* 12 (10): 1–11.

Sharp, Joanne. 2000. "Towards a Critical Analysis of Fictive Geographies." *Area* 32 (3): 327–334.

Smith, Neil. 2008. *Uneven Development*. 3rd ed. Athens, GA: University of Georgia Press.

Strauss, Kendra. 2015. "These Overheating Worlds." *Annals of the Association of American Geographers* 105 (2): 342–350.

4

Material worlds redux

Mobilizing materiality within critical resource geography

Karen Bakker and Gavin Bridge

Introduction

Materiality has become a core concept for critical resource geographers and is invoked in multiple contexts by scholars working from distinctly different perspectives. How does such a concept—simultaneously unifying yet multivalent—become mobilized and evolve? What work might materiality perform to cohere critical resource geography as a subfield? This chapter reviews how the concept of materiality has been taken up in critical resource geography and cognate disciplines, focusing on the last 15 years. Our starting point is our co-authored article published in 2006 in *Progress in Human Geography* ("Material Worlds: Resources geographies and the 'matter of nature'") that has now become one of the seminal references on this topic (despite being initially rejected for publication).

The chapter is divided into three parts. We begin by revisiting the original article and characterizing the goal and circumstances of our 2006 intervention. We consider the degree to which contemporary resource geographies center (or not) on three arguments we made for "materiality": its capacity to (a) radically redistribute and decenter agency, (b) revitalize the concept of "construction," and (c) reconfigure political-economic analyses of a biophysically heterogeneous world. We also explore some new lines of inquiry, including work on posthumanism and the Anthropocene, Indigenous, and decolonizing geographies, and the potentiality of nonliving (abiotic) matter.

In the second part of the chapter, we analyze how our 2006 article has been taken up and explore how the concept of materiality has morphed and evolved (considerably). We identify four motivations behind the contemporary use of the concept of materiality in critical resource geography, while acknowledging that materiality is not the only way to approach resource geography. In doing so, we underscore a point we made in 2006: the concept of materiality sits across several distinctive conceptual agendas and has been taken up within sharply different theoretical traditions. In the final part of the chapter, we reflect on whether the concept of materiality still has the capacity to "revive" resource geographies and consider future avenues for research.

Looking back: championing materiality to "revive" resource geography

"Material Worlds" was an effort to introduce to the subdiscipline of resource geography a body of conceptual work—under the umbrella of materiality—that could help move the field beyond some of its inherited default positions. These included both the instrumental categories and vocabulary of resource management (Rees 1991) and repeated assertions of the irreducibly social character of "natural resources" (Harvey 1974; Hudson 2001). We began the paper with an overview of Anglo-American research on the geographies of resources that (in a deliberate oversimplification) categorized a schism between two types of research: first, an established literature (and an allied teaching mission) dedicated to the practices of resource and/or environmental management and, second, a critical literature that does not necessarily self-identify as "resource geography" but which is nonetheless centrally concerned with the identification, appropriation, and management of biophysical materials and processes. In identifying this schism, we argued that research on resource geographies was divided (unnecessarily, in our opinion) between two distinct and, to some degree, incommensurate goals: improving or optimizing resource allocation on the one hand and, on the other, critiquing modes of production and resource allocation, particularly in the context of provisioning capitalist economies. In claiming that the former (arguably dominant) perspective could learn much from the latter, we welcomed the resurgence of research on resource geographies, which was anticipated by Margaret Fitzsimmons' (1989) call to place the "matter of nature" squarely within the sights of a politically engaged human geography. Like Fitzsimmons, we decried the "peculiar silence on the … geographical and historical dialectic between societies and their material environments" (Fitzsimmons 1989, 106). In short, we championed the concept of materiality as a means to engage in robust critique.

Equally, however, we acknowledged that invoking "the material" is not unproblematic. As urban geographers had already pointed out, concepts of materiality can be enlisted to shore up, as well as radically challenge, conventional modes of explanation (Lees 2002). This ambivalence hinges on an underlying debate over the definition of materiality. Should materiality be conceptualized in predominantly biophysical (and dualistic) terms: as a raw substrate, an immutable reality counterposed to social, cultural, and textual factors? Or should materiality be defined in a non-dualist manner that, as Elizabeth Grosz puts it, is "uncontainable in physicalist terms alone" (Grosz 1994, xi)? If one accepts that nonhuman entities have active capacities "in making history and geography" (Castree 1995, 13), one needs a way to express these capacities without straying into object fetishism or without attributing intrinsic qualities to entities/categories, the boundaries of which are "extrinsic" (i.e., are defined, at least in part, socioculturally). So, although we championed the concept of materiality, we were aware of the pitfalls: the resurgence of the material after a decade of social constructionism risked raising worn out dualisms, resurgent physicalism, object fetishism, and perhaps even environmental determinism.

In an effort to avoid this pitfall, our 2006 paper drew on insights from scholars engaged with the "social construction of nature" and the "production of nature," both of which wrestle with the relationship between the natural and the social, and with related questions of agency. For example, we explored how Smith's (1984) "production of nature" thesis repositions "nature" as an outcome of social relations rather than an asocial input to the economy. Its lasting achievement, we argued, had been to demonstrate how the geographically uneven character of capitalist development rests upon a more fundamental process through which exchangeable values are produced by transforming or "metabolizing" nature. We also argued, however, that the historical materialism informing the production of nature thesis struggled with how to represent the active capacities of biophysical processes dialectically without invoking an external nature. Furthermore, this literature floated uneasily between understanding materiality as the socio-ecological conditions of production (i.e., as the motor of history) and as empirical stuff coursing through commodity chains and provisioning systems.

Given these concerns, we turned to an examination of debates over the social construction of nature to find possible conceptual synergies and paths forward. Our analysis parsed the (already large) literature by examining three distinct approaches to materiality: commodity stories, textuality and corporeality, and hybridity. Here, our analysis drew on work in social and cultural geography, anthropology, and feminist geographies. Building on this discussion, we attempted to answer the thorny question of "how matter matters" or, in other words, to offer novel answers to a persistent and deceptively simple question: how and why do things other than humans make a difference in the way socio-natural relations unfold?

The core analytical challenge for resource geography, we suggested, was to mobilize an emergent and decentered conceptualization of agency. We advanced the claim that the production of nature and construction of nature frameworks offered (in our view) an impoverished view of agency. We explored work on the then-nascent research agenda on embodiment, which we argued offered new approaches to analyzing the shaping of social relations by material conditions without falling prey to environmental or biological determinism. And we urged resource geographers to grapple with the decentering of agency upon which actor-network theory (ANT) rests, where new approaches to the subject/object binary had recently been published, shedding light on ways to transcend the nature/society divide that had characterized much research on the geographies of resources.

Our assessment that resource geography needed revitalizing was a twofold critique of the subdiscipline. First, critical social theory had breathed life into other parts of human geography over the previous couple of decades but had bypassed a substantial portion of resources geography, much of which remained steadfastly focused on the instrumental management of resources and environment. Second, within the part of resources geography that had engaged with critical political economy (here Fitzsimmons [1989] was a touchstone for us), the progressive development of novel conceptualization had reached an impasse. In retrospect, our article did not "revive" the field but, rather, gave voice to a set of trends that complemented (but did not displace) conventional resource management analyses. Within and beyond geography, some of the key points we argued for have indeed been taken up: a more nuanced and expansive approach to agency; a revitalization of the concept of construction (drawing on STS); deeper exploration of the political-economic implications of biophysical heterogeneity (e.g., waste economies); and acknowledgment of the significance of materiality in studies of commodification, financialization, assetization, and the dynamics of rent. Equally, however, several new conceptual arguments have been mobilized within critical resource geographies that we did not discuss in our 2006 intervention: work on posthumanism and the Anthropocene as well as postcolonial insights focused on encounter, including work on animal geographies. Critical resource geography as a field is undoubtedly revitalized, although not in ways we had anticipated. Resource geographers (and other fellow travelers) have harnessed materiality to take forward several distinct conceptual agendas, so a single term now reflects the impact on the field of different theoretical traditions. We made this point in 2006, but it has now become much clearer. This necessarily raises the question, explored later, of whether materiality is still a useful and generative concept today.

Taking stock: four critical motivations for mobilizing materiality in contemporary resource geography

The concept of materiality has traveled and proliferated over the last decade, evolving into something of a polymorphous signifier. Materiality is frequently invoked as part of a critical strategy of differentiating and positioning new research. Implicit, and sometimes explicit, is the claim that attending to materiality affords greater analytical insight or addresses weaknesses in existing research. In invoking the concept of materiality, however, researchers are pursuing disparate agendas. Indeed, part of materiality's appeal is that its gesture to the constitutive role of the nonhuman in social and economic life easily aligns with different strands of social theory, from well-established "materialisms" (e.g., feminism, historical materialism, phenomenology) to STS, ANT, assemblage theory, and geophilosophy.

The extent to which our own piece has been put to work by others since publication as an intervention in resource geography offers a window on this broader proliferation and expansion of materiality. Initially cited in geography and environmental studies and primarily in relation to work on commodification, our piece was subsequently taken up by researchers publishing in economics, sociology, political science, planning, environmental science, and the interdisciplinary fields of water resources as well as sustainability and technology (Figure 4.1). The patterns in Figure 4.1 reflect an interdisciplinary migration and, simultaneously, a multiplication of how the concept of materiality (and our piece) is being used. Specifically, (a) our initial concern with the "matter of nature" in geographical political economy is now layered with interests in political theory and geophilosophy (Elden 2017; Bruun 2018; Peters et al. 2018); (b) an earlier focus on the "recalcitrance" and "resistance" of

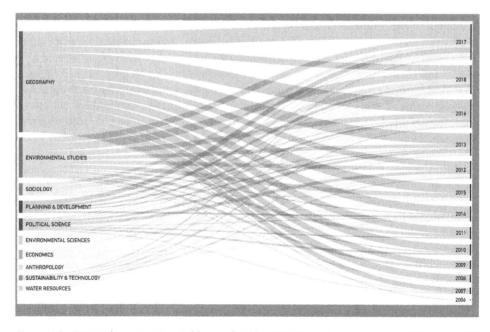

Figure 4.1 Research areas citing Bakker and Bridge (2006) each year between 2006 and 2018 (n = 684): shows the growing interdisciplinary reach of the paper over time. Citation data generated via Web of Knowledge and classified by research area (based on primary disciplinary affiliation of the journal).

biophysical materials to commodification has evolved into a fuller appreciation of how the liveliness and mutability of materials challenge existing modes of thought across economy, politics, and law and ethics (Braun and Whatmore 2010; Braverman 2015); (c) a growing interest in assemblage and territorialization/deterritorialization derived from Deleuze and Guattari has enriched the relational repertoire, supplementing actor-network approaches based on the work of Latour, Callon, and Law (Bear 2013; Müller 2015; Forman 2020); and (d) the vitality, potentiality, and affective capacity of *nonliving* entities are increasingly central to a field drawn initially—and perhaps too readily—to the "fleshy exuberance of biological life" (Clark and Yusoff 2017, 13).

This proliferation of discernibly different stands of work on materiality arises from human geography's famously polyvocal character, and the emergence of parallel debates in cognate disciplines. As time has gone on, researchers citing our piece are not always following through on our specific argument for a revived resource geography. Citations in journals with an arguably instrumentalist, policy-oriented focus, such as *Energy Policy* for example, speak to the wider, polymorphic possibilities of materiality as an alternative perspective, one capable of illuminating the limitations of current conceptual frameworks while also overcoming them.

We acknowledge materiality's ongoing salience owes little to our own writing on the subject and has more to do with the unfolding historical conjuncture. Our original intervention came at a moment when human geography was turning to relational ontologies in an effort to rid the field of inherited habits of thought (Castree 2003; Braun 2008; Lorimer 2012; Greenhough 2016). Appeals to materiality—including our own—need to be situated, therefore, in the context of this search by human geographers (and other social scientists) for ways to move beyond three foundational dualisms: nature-society, mind-matter, and life/nonlife. More broadly, the turn to materiality reflects a growing critique of anthropocentrism and the legacy of liberal humanism in the critical social sciences that is arising, in part, because of the ongoing conjunction of three elements. First, a growing desire in the wake of the cultural turn to admit the "massive materiality" of social existence and the role of biological, technological, and other physical "stuff" in social relations (Coole and Frost 2010, 2). Second, new developments in life sciences that suggest a profound "ecological turn" within capitalism centered on the capitalization of life itself that, at the same time, undercuts liberal notions of the autonomous human subject and its singular capacity for agency (Hayles 1999; Collard 2014; Goldstein and Johnson 2015). And third, a growing planetary catastrophe in which the elevation of human subjects as a geological force is collapsing philosophical distinctions between human-nonhuman and between life/nonlife: as Yusoff (2016, 4) explains, for example, "If humans now author the rocks, atmosphere and oceans with anthropogenic signatures then the inhuman (as nature, earth, geology) becomes decidedly changed as a category of differentiation." For others, however, the Anthropocene disturbs received categories but does not collapse them in a final or settled sense: Malm (2017), for example, argues that in a warming world it becomes more important than ever to distinguish between the natural and the social.

Together these currents have propelled a critical appraisal of the anthropocentric (and ethnocentric) knowledge systems' derivative of the Enlightenment for the way they "feed human hubris and…earth-destroying fantasies of conquest and consumption" (Bennett 2010a, ix) and encouraged a flourishing of posthumanist and "new materialist" perspectives which, in different ways (as we show later), "allow for a place for the force of things" (Braun and Whatmore 2010, x). In the remainder of this section, we distinguish four motivations behind the turn to materiality. These are present unevenly within work on critical resource geography: the first and second are well-established; the third formed a core element of our original article but has rapidly evolved in interesting directions; and the fourth is more tentative and emergent.

Materiality as the constitutive copresence of the nonhuman

Those who invoke materiality often express a desire to acknowledge the copresence and heterogeneity of nonhuman entities (species and things) within human-centered worlds. This amounts to an ontological claim about the dissimilar and variegated character of the nonhuman and its significance for social theory. While materiality here is ostensibly a bland recognition of "difference" within nonhuman "others," its insistence on the constitutive character of the nonhuman acquires analytical force because conventional social science has largely stripped out this heterogeneity. (Theories of value are a case in point, as is the way critical social theory uncovers "always something human" [Bennett 2010a, xv].) In relation to animal geographies, for example, making the "corporeal, creative and consequential…presence [of animals] felt in our accounts of the social presents serious epistemological and practical problems that science (social and natural) has barely begun to admit" (Whatmore 2002, 32).

"Rematerializing" accounts of society, economy, politics, or infrastructure, then, can be a profoundly unsettling process. In many cases, admitting the nonhuman means incorporating the "constituent outside" by which social science has conventionally defined its categories, terrain, and expertise, as work on the role of nature in economic geography has pointed out (Bridge 2011a; Bakker 2012). The impulse to move beyond a "lumpen nature" (Whatmore 2002, 14) by recognizing the heterogeneous and constitutive character of the nonhuman is, then, more than a mere consequence of undertaking grounded analysis and more substantive than a primordial acknowledgment that "context matters." It is, instead, the necessary outworking of a relational perspective, adopted as an alternative to dualist ontologies, that highlights the "co-constituted" character of phenomena conventionally labeled either social or nonhuman. A relational conceptualization of materiality underpins research in critical resource geography on, for example, the constitutive role of nonhuman capacities within commodity production and circulation (Collard 2014; Kaup 2014), and the ethics of care conducive to flourishing and conviviality in a more-than-human world (Haraway 2008; Krzywoszynska 2019).

Materiality as ontological politics

A second motivation for invoking materiality within critical resource geography is to draw attention to the socio-technical practices through which the presence, form, character, and meaning of resources are brought into being. Here materiality opens a window to plural ontological possibilities—the world can be assembled in a range of possible ways—and directs attention to the role of knowledge in establishing presence and meaning of materials and, ultimately, their inherent mutability. It aligns well, for example, with notions of governmentality as it draws attention to the devices and calculative tools through which environments, bodies, and materials are known and "made actionable" (Braun 2000; Bruun 2018). This perspective evolves a long-standing tendency within critical resource geography to "de-naturalize" the status of natural resources by highlighting the economic, political, and cultural conditions that give rise to their material form. Recent work on "resource-making" doubles down on this lineage, harnessing the conceptual tools of STS to "examine how heterogeneous…substances are rendered into knowable and exploitable resources" and become targeted for investment (Bridge 2011b; Richardson and Weszkalnys 2014; Kama 2019, 2; Kama, Chapter 5 this volume).

This work draws heavily on STS in accounting for how practices of scientific investigation enact and stabilize particular worlds (that enable control, intervention, productivity), but it is also crosscut by several different strands of theorization. Some researchers highlight the performative character of models and instruments, drawing attention to the materiality, situatedness, and partiality of practices of knowing. Others harness Foucauldian understandings of power-knowledge and governmentality to highlight models' political and economic effects. And some draw on Stenger's notion of "informed materials" to stress the social processes of "informational enrichment" through which materials become "encas(ed)…in an array of figures, traces and samples" and are made the subject of knowledge controversies (Barry 2013, 141, 16).

Across this work, materiality is framed as a state of becoming/unbecoming dependent on the enactment and stabilization of knowledge claims. Materiality serves as "a form of 'ontological politics' … (directing attention to) how certain enactments of the resource are reinforced at the expense of other possible 'resource ontologies', which have become either obscured in the past…or can still be detected amidst contemporary disputes" (Kama 2019). Li (2013), for example, shows how social groups contesting gold mining in Peru construct the materiality of a mountain in radically different ways: as valuable mineral deposit, water reserve, or sacred and sentient being, the mountain's materialities give rise to associations of people and things that condition political life.

An appeal to materiality as ontological politics may also be mobilized as an anti-colonial strategy aimed at the structures and practices through which the materiality of "others" is constituted. Indigenous scholars have been at the forefront of research identifying ontological differences in understanding both the nature of matter itself, and in the socio-economies and political ecologies via which human exchanges with the natural world are constituted (see, for example, Kimmerer 2013; Coulthard 2014). Indigenous conceptions of resources, while heterogeneous, often position both biotic and abiotic entities as part of a living planet, which is often framed as sacred. Indeed, many Indigenous languages are predicated on the interconnections of these entities with ecological processes (Schreyer et al. 2014). Coulthard and Simpson (2016) offer the concept of "grounded normativity" to frame the materiality of Indigenous place-based practices and associated forms of knowledge; this, they argue, provides a substantively distinct ontological foundation for Indigenous critiques, grounded in place-based solidarity.

Ontological debates over materiality thus at times intersect with debates over posthumanism and decolonizing the discipline of geography. On the one hand, as Sundberg (2014) notes, posthumanism offers powerful tools to critique nature/culture dualisms that underpin Eurocentric knowledge and colonialism. Simultaneously, however, she critiques the way posthumanist scholarship tends to reproduce colonial modes of thought by subordinating other ontologies, sparking reflection on the ambivalences and challenges of decolonizing geographies (Radcliffe 2017; Naylor et al. 2018). Scholars have responded to this call to "unsettle" geographical knowledges (Daigle and Sundberg 2017; De Leeuw and Hunt 2018; Daigle 2019). For example, Collard et al.'s (2015, 322) *Manifesto for Abundant Futures* embraces decolonization as a "political sensibility" in order to confront the legacy of nature/society dualisms in the context of biodiversity conversation in the Anthropocene. However, as Tuck and Yang (2012) argue in the inaugural issue of *Decolonization: Indigeneity, Education and Society*, decolonization is more than a mere metaphor. They warn against the overly easy or ubiquitous adoption of decolonization in advocacy and scholarship, insofar this makes possible a set of evasions, or "settler moves to innocence," that reconcile settler guilt and complicity while entrenching settler colonialism. Acknowledging this, Collard frames materiality as a means for challenging received ontologies and forms of control, a resource for "political strategies to reckon with colonial-capitalist ruins" (Collard et al. 2015, 322); invoking materiality within a decolonizing framework has, as its goal, the repatriation of Indigenous land and life and is not a mere conceptual device for analyzing conceptual inconsistencies or incommensurabilities (see also Neville and Coulthard 2019).

Materiality as generative friction

A third motivation behind invocations of materiality is a desire to acknowledge the generative effects of the Earth's biophysical and geophysical qualities on socio-natural relations, an effort to capture the "physical, chemical, biological, and cultural acting…(and) sociotechnical mediation" of materials (Swyngedouw 2015, 28). Here the motivation is to create more symmetrical forms of analysis in which the capacity for "agency" is not solely human. Materiality here is less directly about advancing a relational ontology and accounts of co-constitution, and more about introducing points of friction—i.e., a recognition of difference as generative potential—into existing analyses to show how matter matters economically and politically (Boyd, Prudham, and Schurman 2001; Prudham 2005; Hudson 2008).

Critical resource geography has shown a long-standing interest in how "the productivity and resilience of matter" (Coole and Frost 2010, 7) complicates commodification, creates opportunities for rent capture, or requires certain forms of labor (see also, Fannin, Chapter 33 this volume). Grounded initially in "nature-facing" sectors such as forestry, fisheries, mining, and agriculture, this work has been supplemented with a rapidly growing focus on the metabolic and affective dimensions of "animal labor" and more-than-human modes of innovation. Research by critical resource geographers on bio-capital, such as the growing interest of corporate and military interests in the capabilities of biomimicry (Johnson and Goldstein 2015), significantly extends the field's primary claim that "production can no longer be thought of in strictly anthropocentric terms," while also suggesting how such critical insights and their progressive possibilities become "strategically contained" (Braun 2015).

These engagements with the generative qualities of matter pivot on the knotty problem of thinking about agency outside of a liberal frame, which constructs agency as willful and conscious action and, as such, the exclusive property of (some) humans. As a way to "decenter" agency and level up matter's role in unfolding social relations, some critical resource geographers have mobilized the flatter ontologies of ANT and assemblage theory (Greenhough 2011; Müller 2015). Others have turned to the repertoires of feminist technoscience, queer ecologies, and posthumanism for critical analytical tools (Ingram 2010); or appropriated concepts from diverse anthropological traditions, such as the notion of "affordances" borrowed from and work on human-material interactions in environmental psychology. Affordance draws attention to what "environments, organisms, and things furnish for the purpose of a subject" (Knappett 2004; Barua 2016, 730). The term has traveled into critical resource geography via the work of anthropologists like Ingold (2002), where it has been used to think about the scope of different bodies, ecologies, and geologies present for capital circulation (Collard 2014; Castree 2015; Kama 2019). Barua's (2016, 2017, 284) work on elephants and lions, for example, highlights the "lively affordances" these charismatic species present for spectacular forms of tourist encounter and, consequently, how the physiology and behaviors of these species are an active constituent of their commodification. Attention to materiality in this work, then, exposes the humanism of labor theories of value and the possibility of alternative "relational grammars" for thinking about agency and capacity outside of a liberal frame (Barua 2019).

The search for ways of thinking about the dynamic character of materials has led some to reject the language of materiality altogether. While matter and materiality are often used interchangeably, Ingold (2007) advocates focusing on the properties of materials rather than the materiality of objects. The problem, he argues, is that materials largely disappear from view when attention is on materiality, missing the way materials "continue to mingle and react as they always have done, forever threatening the things they comprise with dissolution" (2007, 9). Accordingly, Ingold champions a view in which humans "swim in an ocean of materials. Once we acknowledge our immersion, what this ocean reveals to us is... a flux in which materials of the most diverse kinds... undergo continual generation and transformation" (2007, 7). Whitt's (2018) work on Bolivian mudflats, for example, draws on Ingold to highlight the movements and multiplicity of materials that comprise the materiality of mud. Thinking through materials, as a critical antidote to the materiality of objects, has been deployed to good effect in work on economies of waste disposal: Gregson, Watkins, and Calestani (2010) and Gregson et al. (2010), for example, show how material objects (like large, oceangoing ships) are "at best partially stable...endlessly being assembled, always becoming something else somewhere else" (Gregson et al. 2010, 846).

Materiality as the potentiality of nonlife

A fourth motivation extends an interest in the generative effects of materiality (motivation 3) but elevates it from a specific social ordering effect to the level of an "active principle" intrinsic to matter (Bennett 2010b, 47). This "vital materialism" explores materiality as immanent potential, a "quivering effervescence" comprising "forces with trajectories, propensities, (and) tendencies" (Bennett 2010a, 55) that endow matter with a vitality and productive power beyond human intention. Here materiality is not so much an actant comprising a certain social order but involves potentiality, becomingness and the possibilities of shifting composition (see Greenhough 2011, 2016). It marks a shift away from STS's strict empiricism of network associations to a fuller embrace of a Deleuzean perspective, where the vitality of materials is not a network effect but "intrinsic to materiality as such" (Bennett 2010a, xiv). What Bennett is getting at here is the potential of nonhuman forms to be affective; and, critically, she extends this potentiality not only to nonhuman forms of life (e.g., to animals and plants) but to nonlife as well. Her work signals a move toward a much richer understanding of the potentiality of materials and "a reckoning with the forces of mute matter in lively bodies: a corporeality that is driven by inhuman forces" (Yusoff 2013, 790).

Critical resource geographers' explorations of liveliness have expanded from the *bio* to the *geo* in the last few years, engaging with the capacity of nonliving materialities to "change in unforeseeable ways" so that "potentiality can no longer be defined, even ambivalently, as the provenance of life, but rather as something that characterizes nonlife entities such as Earth systems" (Lehman 2016, 120). For example, research on the subjectivities and cultural forms created in and around fossil fuels highlights how "the biopolitics of life has a more expansive mineralogical geography that needs attention" (Yusoff 2013, 780). Huber's (2013) work on oil's constitutive role in American culture, for example, foregrounds this biopolitical dimension of fossil fuels while also drawing attention to the resource geographies through which consumer culture and the "American Way of Life" are sustained (a gesture toward what Yusoff [2013, 780] terms the "inhuman dimensions of subjectivity").

More broadly, attention to materiality in this work serves its familiar role of troubling a received dualism by drawing new lines of difference, in this case repartitioning potentiality in a way that undercuts the distinction between life and nonlife (Povinelli 2016). Here invocations of materiality gesture to the potential of the Earth itself, shifting focus from either the flat space of territory or the vertical fixity of geological strata to the perpetual immanence and indeterminacy of geophysics (Steinberg and Peters 2015). Used in this way, materiality becomes a way to explore relations between the geophysical and the geopolitical, such as how dynamic physical environments (characterized by ice, silt, or sand, for example) complicate political and legal geographies of territory (Peters et al. 2018). Elden (2017, 223) has recently reworked the hoary concept of terrain to account explicitly for these dynamic and mobile properties of the Earth, defining it as "where the geopolitical and the geophysical meet," or the "materiality of territory." Attention to the material foundations of political power has also animated calls for a more "earthly geopolitics" characterized by the intrinsic potentiality of matter (Clark 2011). In response, researchers have turned to Grosz's concept of "geopower" as its focus on the "primordial interface" between "forms of life and forms of the earth" offers a more active "geo" within geopolitics (Yusoff et al. 2012; Yusoff 2017); and to the geontopolitics of Povinelli (2016, 5) who shows how biopolitics "has long depended on a subtending geonto-power" derived from adjudicating the distinction between the lively (life) and the inert (nonlife). Povinelli's analysis arises from her long-term engagement with resource politics in the settler-colonial context of Australia's Northern Territory where, as in other settler-colonial societies, the distinction between *bios* (aboriginal existence) and *geos* (mineral resources) characterizes the governance of difference (Povinelli 2016; Povinelli, Coleman, and Yusoff 2017). Such analyses raise discomforting questions for resource geography, a field constituted through such distinctions: from facile categorizations (e.g., biotic vs. abiotic resources) to the way the language of "resources" (conveying inherent potential) dramatizes the possibility of life via the transformation of inert materials into human flourishing.

Looking forward: surprises, opportunities, and new directions

Materiality refers to the manifestations of (colonial) knowledge systems as the Earth's spaces, depths, and materials are mapped, measured, and claimed. As this framing of materiality helps us to uncover the more-than-human character of the relations, processes, and actions shaping environments, and experience in the epoch of the Anthropocene, it allows us to engage with the indeterminate properties of both living and nonliving things (and indeed the Earth itself), and their capacity to surprise and subvert economic and political strategy. How should critical resource geography respond to materiality's different analytical trajectories? There are efforts outside of geography to bring some of these distinctive elements together within a more-or-less unifying framework. The "resource materialities framework" offered by Richardson and Wesz-kalnys (2014), for example, seeks to hold together the temporal and spatial "distributedness" of resources and their ontological multiplicity. Our approach here has been different and prioritizes a forensic rather than synthetic approach: we have sought not to corral these different motivations or bring them into alignment but to unpack their differences by considering what invocations of materiality aim to achieve within each body of work.

We conclude with four brief observations. First, our survey of recent work illustrates how materiality has proliferated widely within critical resource geography as an important conceptual device, but its application is far from universal. Much work within the subdiscipline/field does not mobilize this term or orientate itself by reference to one of the four motivations we have outlined here. We suggest, then, that materiality has not yet attained the status of a "master concept" in any meaningful sense and, indeed, is unlikely to do so, given the polyvocal character of human geography. Second, critical resource geography is alive and well and its diverse conceptual and empirical orientations have certainly enlivened the subdiscipline of resource geography. Our hope in 2006 was that a richer engagement with materiality would bring this about. In practice, the influences on the field have been more numerous and more diverse than we anticipated, so that much of the new thinking in critical resource geography has come from pursuing the intersections of materiality with posthumanism, postcolonialism, geophilosophy, and critical engagements with the Anthropocene. Third, materiality opens up possibilities for creative forms of interdisciplinary research and not simply because of the apparent ease with which it is able to align with different epistemological traditions. Rather, materiality conveys an openness to the multiplicity and alterity of the world and the need, therefore, to reckon with different ways of knowing and acting within it. Working with and through the concept of materiality can offer a renewed impetus to some familiar forms of interdisciplinary research (such as collaborations among social scientists, resource managers, and physical scientists) and, at the same time, points to new opportunities. Finally, some of the most promising scholarship (for example, debates at the intersection of postcolonial, posthumanist, and Indigenous geographies) is challenging the ontological bases of the concept of materiality itself. Many of these authors seek to work through or directly against the dualisms and naïve vitalism which materiality has often implied. Although materiality served its purpose for a time, we welcome the thought that the discipline of geography may be growing beyond it as a concept.

References

Bakker, Karen. 2012. "The 'Matter of Nature' in Economic Geography." In *The Wiley-Blackwell Companion to Economic Geography*, edited by Trevor J. Barnes, Jamie Peck, and Eric Sheppard, 104–117. Malden, MA: Wiley-Blackwell.

Bakker, Karen, and Gavin Bridge. 2006. "Material Worlds? Resource Geographies and the 'Matter of Nature'." *Progress in Human Geography* 30 (1): 5–27.

Barry, Andrew. 2013. *Material Politics: Disputes Along the Pipeline*. Malden, MA: John Wiley & Sons.

Barua, Maan. 2016. "Lively Commodities and Encounter Value." *Environment and Planning D: Society and Space* 34 (4): 725–744.

Barua, Maan. 2017. "Nonhuman Labour, Encounter Value, Spectacular Accumulation: The Geographies of a Lively Commodity." *Transactions of the Institute of British Geographers* 42 (2): 274–288.

Barua, Maan. 2019. "Animating Capital: Work, Commodities, Circulation." *Progress in Human Geography* 43 (4): 650–669.

Bear, Christopher. 2013. "Assembling the Sea: Materiality, Movement and Regulatory Practices in the Cardigan Bay Scallop Fishery." *Cultural Geographies* 20 (1): 21–41.

Bennett, Jane. 2010a. *Vibrant Matter: A Political Ecology of Things*. Durham, NC: Duke University Press.

Bennett, Jane. 2010b. "A Vitalist Stopover on the Way to a New Materialism." In *New Materialisms: Ontology, Agency, and Politics*, edited by Diana Coole, and Samantha Frost, 47–69. Durham, NC: Duke University Press.

Boyd, William, W. Scott Prudham, and Rachel A. Schurman. 2001. "Industrial Dynamics and the Problem of Nature." *Society & Natural Resources* 14 (7): 555–570.

Braun, Bruce. 2000. "Producing Vertical Territory: Geology and Governmentality in Late Victorian Canada." *Ecumene* 7 (1): 7–46.

Braun, Bruce. 2008. "Environmental Issues: Inventive Life." *Progress in Human Geography* 32 (5): 667–679.

Braun, Bruce. 2015. "New Materialisms and Neoliberal Natures." *Antipode* 47 (1): 1–14.

Braun, Bruce, and Sarah J. Whatmore. 2010. "The Stuff of Politics: An Introduction." In *Political Matter: Technoscience, Democracy and Public Life*, edited by Bruce Braun, and Sarah J. Whatmore, ix–xl. Minneapolis, MN: University of Minnesota Press.

Braverman, Irus, ed. 2015. *Animals, Biopolitics, Law: Lively Legalities*. Abingdon, UK: Routledge.

Bridge, Gavin. 2011a. "The Economy of Nature: From Political Ecology to the Social Construction of Nature." In *The SAGE Handbook of Economic Geography*, edited by Andrew Leyshon, Roger Lee, Linda McDowell, and Peter Sunley, 217–230. Los Angeles: SAGE.

Bridge, Gavin. 2011b. "Resource Geographies I: Making Carbon Economies, Old and New." *Progress in Human Geography* 35 (6): 820–834.

Bruun, Johanne M. 2018. "Grounding Territory: Geoscience and the Territorial Ordering of Greenland During the early Cold War." PhD diss., Durham University.

Castree, Noel. 1995. "The Nature of Produced Nature: Materiality and Knowledge Construction in Marxism." *Antipode* 27 (1): 12–48.

Castree, Noel. 2003. "Environmental Issues: Relational Ontologies and Hybrid Politics." *Progress in Human Geography* 27 (2): 203–211.

Castree, Noel. 2015. "Capitalism and the Marxist Critique of Political Ecology." In *Routledge Handbook of Political Ecology*, edited by Tom Perreault, Gavin Bridge, and James McCarthy, 279–292. Abingdon, UK: Routledge.

Clark, Nigel. 2011. *Inhuman Nature: Sociable Life on a Dynamic Planet*. London: SAGE Publications.

Clark, Nigel, and Kathryn Yusoff. 2017. "Geosocial Formations and the Anthropocene." *Theory, Culture & Society* 34 (2–3): 3–23.

Collard, Rosemary-Claire.. 2014. "Putting Animals Back Together, Taking Commodities Apart." *Annals of the Association of American Geographers* 104 (1): 151–165.

Collard, Rosemary-Claire, Jessica Dempsey, and Juanita Sundberg. 2015. "A Manifesto for Abundant Futures." *Annals of the Association of American Geographers* 105 (2): 322–330.

Coole, Diana, and Samantha Frost, eds. 2010. *New Materialisms: Ontology, Agency, and Politics*. Durham, NC: Duke University Press.

Coulthard, Glen Sean. 2014. *Red Skin, White Masks: Rejecting the Colonial Politics of Recognition*. Minneapolis, MN: University of Minnesota Press.

Coulthard, Glen Sean, and Leanne Beasamosake Simpson. 2016. "Grounded Normativity/Place-Based Solidarity." *American Quarterly* 68 (2): 249–255.

Daigle, Michelle. 2019. "The Spectacle of Reconciliation: On (the) Unsettling Responsibilities to Indigenous Peoples in the Academy." *Environment and Planning D: Society and Space* 37 (4): 703–721.

Daigle, Michelle, and Juanita Sundberg. 2017. "From Where We Stand: Unsettling Geographical Knowledges in the Classroom." *Transactions of the Institute of British Geographers* 42 (3): 338–341.

De Leeuw, Sarah, and Sarah Hunt. 2018. "Unsettling Decolonizing Geographies." *Geography Compass* 12 (7): e12376.

Elden, Stuart.. 2017. "Legal Terrain—the Political Materiality of Territory." *London Review of International Law* 5 (2): 199–224.

Fitzsimmons, Margaret. 1989. "The Matter of Nature." *Antipode* 21 (2): 106–120.

Forman, Peter J. 2020. "Security and the Subsurface: Natural Gas and the Visualisation of Possibility Spaces." *Geopolitics* 25 (1): 1–24.

Goldstein, Jesse, and Elizabeth Johnson. 2015. "Biomimicry: New Natures, New Enclosures." *Theory, Culture & Society* 32 (1): 61–81.

Greenhough, Beth. 2011. "Assembling an Island Laboratory." *Area* 43 (2): 134–138.

Greenhough, Beth. 2016. "Vitalist Geographies: Life and the More-Than-Human." In *Taking-Place: Non-Representational Theories and Geography*, edited by Ben Anderson, and Paul Harrison, 51–68. Abingdon, UK: Routledge.

Gregson, Nicky, Mike Crang, Farid Uddin Ahamed, N. Akhter, and Raihana Ferdous. 2010. "Following Things of Rubbish Value: End-of-Life Ships, 'Chock-Chocky' Furniture and the Bangladeshi Middle Class Consumer." *Geoforum* 41 (6): 846–854.

Gregson, Nicky, Helen Watkins, and Melania Calestani. 2010. "Inextinguishable Fibres: Demolition and the Vital Materialisms of Asbestos." *Environment and Planning A* 42 (5): 1065–1083.

Grosz, Elizabeth. 1994. *Volatile Bodies: Toward a Corporeal Feminism*. Bloomington, IN: Indiana University Press.

Haraway, Donna. 2008. *When Species Meet*. Minneapolis, MN: University of Minnesota Press.

Harvey, David. 1974. "Population, Resources, and the Ideology of Science." *Economic Geography* 50 (3): 256–277.

Hayles, N. Katherine. 1999. *How We Became Posthuman: Virtual Bodies in Cybernetics, Literature, and Informatics*. Chicago, IL: University of Chicago Press.

Huber, Matthew T. 2013. *Lifeblood: Oil, Freedom, and the Forces of Capital*. Minneapolis, MN: University of Minnesota Press.

Hudson, Ray. 2001. *Producing Places*. New York, NY: Guilford Press.

Hudson, Ray. 2008. "Cultural Political Economy Meets Global Production Networks: A Productive Meeting?" *Journal of Economic Geography* 8 (3): 421–440.

Ingold, Tim. 2002. *The Perception of the Environment: Essays on Livelihood, Dwelling and Skill*. London: Routledge.

Ingold, Tim. 2007. "Materials Against Materiality." *Archaeological Dialogues* 14 (1): 1–16.

Ingram, Mrill. 2010. "Keeping Up with the *E. coli*: Considering Human-Nonhuman Relationships in Natural Resources Policy." *Natural Resources Journal* 50 (2): 371–392.

Johnson, Elizabeth R., and Jesse Goldstein. 2015. "Biomimetic Futures: Life, Death, and the Enclosure of a More-Than-Human Intellect." *Annals of the Association of American Geographers* 105 (2): 387–396.

Kama, Kärg. 2019. "Resource-Making Controversies: Knowledge, Anticipatory Politics and Economization of Unconventional Fossil Fuels." *Progress in Human Geography* 44 (2): 333–356.

Kaup, Brent Z. 2014. "Divergent Paths of Counter-Neoliberalization: Materiality and the Labor Process in Bolivia's Natural Resource Sectors." *Environment and Planning A: Economy and Space* 46 (8): 1836–1851.

Kimmerer, Robin Wall. 2013. *Braiding Sweetgrass: Indigenous Wisdom, Scientific Knowledge and the Teachings of Plants*. Minneapolis, MN: Milkweed Editions.

Knappett, Carl. 2004. "The Affordances of Things: A Post-Gibsonian Perspective on the Relationality of Mind and Matter." In *Rethinking Materiality: The Engagement of Mind With the Material World*, edited by Elizabeth DeMarrais, Chris Gosden, and Colin Renfrew, 43–51. Cambridge: University of Cambridge.

Krzywoszynska, Anna. 2019. "Caring for Soil Life in the Anthropocene: The Role of Attentiveness in More Than Human Ethics." *Transactions of the Institute of British Geographers* 44 (4): 661–675.

Lees, Loretta. 2002. "Rematerializing Geography: The 'new' Urban Geography." *Progress in Human Geography* 26 (1): 101–112.

Lehman, Jessica. 2016. "A Sea of Potential: The Politics of Global Ocean Observations." *Political Geography* 55: 113–123.

Li, Fabiana. 2013. "Relating Divergent Worlds: Mines, Aquifers and Sacred Mountains in Peru." *Anthropologica* 55 (2): 399–411.

Lorimer, Jamie. 2012. "Multinatural Geographies for the Anthropocene." *Progress in Human Geography* 36 (5): 593–612.

Malm, Andreas. 2017. *The Progress of This Storm: Nature and Society in a Warming World*. London: Verso Books.

Müller, Martin. 2015. "Assemblages and Actor-Networks: Rethinking Socio-Material Power, Politics and Space." *Geography Compass* 9 (1): 27–41.

Naylor, Lindsay, Michelle Daigle, Sofia Zaragocin, Margaret Marietta Ramírez, and Mary Gilmartin. 2018. "Interventions: Bringing the Decolonial to Political Geography." *Political Geography* 66: 199–209.

Neville, Kate J., and Glen Coulthard. 2019. "Transformative Water Relations: Indigenous Interventions in Global Political Economies." *Global Environmental Politics* 19 (3): 1–15.

Peters, Kimberley, Peter Steinberg, and Elaine Stratford, eds. 2018. *Territory Beyond Terra*. London: Rowman and Littlefield.

Povinelli, Elizabeth A. 2016. *Geontologies: A Requiem to Late Liberalism*. Durham, NC: Duke University Press.

Povinelli, Elizabeth A., Matthew Coleman, and Kathryn Yusoff. 2017. "An Interview With Elizabeth Povinelli: Geontopower, Biopolitics and the Anthropocene." *Theory, Culture & Society* 34 (2–3): 169–185.

Prudham, W. Scott. 2005. *Knock on Wood: Nature as Commodity in Douglas-Fir Country*. New York, NY: Routledge.

Radcliffe, Sarah A. 2017. "Decolonising Geographical Knowledges." *Transactions of the Institute of British Geographers* 42 (3): 329–333.

Rees, Judith. 1991. *Natural Resources: Allocation, Economics and Policy*. London: Routledge.

Richardson, Tanya, and Gina Weszkalnys. 2014. "Introduction: Resource Materialities." *Anthropological Quarterly* 87 (1): 5–30.

Schreyer, Christine, Jon Corbett, Nicole Gordon, and Colleen Larson. 2014. "Learning to Talk to the Land: Online Stewardship in Taku River Tlingit Territory." *Decolonization: Indigeneity, Education & Society* 3 (3): 106–133.

Smith, Neil. 1984. *Uneven Development: Nature, Capital, and the Production of Space*. Hoboken, NJ: Blackwell.

Steinberg, Philip, and Kimberley Peters. 2015. "Wet Ontologies, Fluid Spaces: Giving Depth to Volume Through Oceanic Thinking." *Environment and Planning D: Society and Space* 33 (2): 247–264.

Sundberg, Juanita. 2014. "Decolonizing Posthumanist Geographies." *Cultural Geographies* 21 (1): 33–47.

Swyngedouw, Eric. 2015. *Liquid Power: Contested Hydro-Modernities in Twentieth-Century Spain*. Cambridge, MA: MIT Press.

Tuck, Eve, and K. Wayne Yang. 2012. "Decolonization Is Not a Metaphor." *Decolonization: Indigeneity, Education & Society* 1 (1): 1–40.

Whatmore, Sarah. 2002. *Hybrid Geographies: Natures Cultures Spaces*. London: SAGE.

Whitt, Clayton. 2018. "Mudflats: Fluid Terrain: Climate Contestations in the Mudflats of the Bolivian Highland." In *Territory Beyond Terra*, edited by Kimberley Peters, Philip Steinberg, and Elaine Stratford, 91–106. London: Rowman and Littlefield.

Yusoff, Kathryn. 2013. "Geologic Life: Prehistory, Climate, Futures in the Anthropocene." *Environment and Planning D: Society and Space* 31 (5): 779–795.

Yusoff, Kathryn. 2016. "Anthropogenesis: Origins and Endings in the Anthropocene." *Theory, Culture & Society* 33 (2): 3–28.

Yusoff, Kathryn. 2017. "Geosocial Strata." *Theory, Culture & Society* 34 (2–3): 105–127.

Yusoff, Kathryn, Elizabeth Grosz, Nigel Clark, Arun Saldanha, Kathryn Yusoff, Catherine Nash, and Elizabeth Grosz. 2012. "GeoPower: A Panel on Elizabeth Grosz's Chaos, Territory, Art: Deleuze and the Framing of the Earth." *Environment and Planning D: Society and Space* 30 (6): 971–988.

5

Temporalities of (un)making a resource

Oil shales between presence and absence

Kärg Kama

Liminal resources: between presence and absence

Two pivotal moments provoked me to ask what makes something a resource, if the qualities and significance of this resource are not evident and present, but strikingly absent.[1] The first comes from a warm September day in 2010, when, by lucky coincidence, I joined a small group of people on their field trip to an exclusive resource-processing complex in northeast Estonia, just a few miles from the Russian border (Figure 5.1). The delegation included elite state officials from Jordan, led by the Minister of Energy, and representatives of the Estonian state energy company responsible for operating the complex. The Jordanians had come here to be convinced of Estonia's energy independence, achieved with the help of a little-known fossil fuel resource called oil shale. They understood that oil shale exploitation is a protracted business: Jordan's vast shale deposits have been explored for many decades, but so far failed to become amenable to commercial production. By witnessing the Estonian industry firsthand, the delegation hoped to discover the secret to unlocking the "resource of the future," which their government so eagerly awaited. After a long day of visiting the mines, sorting facilities, and the world's largest shale-fired power plant, our exhausted group arrived at its destination—a small bespoke factory that produced shale oil. Suddenly, the delegation filled the plant, excited, upbeat, and hopeful. One of the retorts was slowly rotating with a loud sound, milling a content that no one could actually see, but could otherwise sense. It smelled like oil.

A year later, in 2011, I witnessed a different moment of the resource in the making, as part of another fortuitous field trip, this time in Utah, United States. Here, a group of American industry experts was cautiously standing on the verge of a bottomless shaft that opened deep into the oil shale deposits of the Rocky Mountains (Figure 5.2). The shaft was drilled during a previous exploration boom in the wake of the 1973 oil crisis, but it had become inaccessible following a fatal methane explosion and flooding of the underground mine. After decades of twists and turns under different license-holders, the concession had recently been acquired by the same Estonian state company mentioned earlier, which promised to finally make the deposit's oil potential a reality for Americans, as they had promised for the Jordanians. Without a trace of oil to be smelled, however, the new owner rendered its intentions credible otherwise. In addition to maintaining the mine shaft and drilling more boreholes to study the resource, it had

Figure 5.1 The delegation of Jordanian officials visiting an opencast oil shale mine in Narva, Estonia, September 2010. Photo by author.

even erected a small weather station to measure baseline environmental data for the impending times when production would take place. These vertical infrastructures appeared to compress the industry's doomed past and its deferred futures (cf. Bridge 2009), invoking different kinds of affect around the prospect of extraction. Unlike the optimism shared in Estonia, this delegation looked skeptical. There had been too many booms and busts over the last century, and too many failed promises (for an overview, see Hanson and Limerick 2009), manifested in a popular saying that "oil shale is the fuel of the future, and always will be." Even as the production of other nontraditional fossil fuels has grown in recent years, expert circles in the United States are increasingly doubtful whether the vast deposits of the Rockies will ever become exploited.

As these two moments suggest, the "resourceness" of oil shales is neither intrinsic nor socially fixed (Li 2014; Richardson and Weszkalnys 2014). Rather, in the absence of large scale, commercially proven production, the essence and significance of the resource lies in the extended promise of *future* supply. Oil shales are "not as yet" resources. They are nonetheless performative: oil shales lure both national politics and speculative capital as an oil economy to come, once conventional hydrocarbons become increasingly scarce and expensive to render lower grade alternatives more viable. So, what makes oil shales a resource?

In this chapter, I start out with a premise that the distinction between resources and non-resources is contingent. As geographers have long acknowledged, not all earth substances are resources by default, but supposedly "natural" things become resources through specific cultural and economic appraisals which may vary between different geographical locations and time periods (e.g., Bakker and Bridge 2006). More recently, critical resource scholars have adopted the concept of "resource-making," piloted in environmental anthropology, to emphasize that

Figure 5.2 The American group looking into an abandoned mine shaft at the White River oil shale concession, Utah, October 2011. Photo by author.

the becoming of resources is an active, albeit indeterminate process that requires both affective work and strategic interventions (Ferry and Limbert 2008; Bridge 2011; Kama 2013; Li 2014; Richardson and Weszkalnys 2014). Despite acknowledging that resources are *processual* becomings, the scholarship is still preoccupied with key primary commodities that are well entrenched in global networks of production and consumption, and for which evident physical presence is taken for granted. Little notice is taken of "liminal resources," where the qualities and value of the resource have been established only provisionally (Kama 2020). Indeed, there is a notable omission of forms of *speculative* resource-making that are controversial and short-lived, but nonetheless persistent, such as the protracted histories of oil shales exploration suggest.

In the following sections, I examine how the contingent resourceness of oil shales is perpetuated through contending techniques of time (Bear 2016), or what Weszkalnys (2015) calls "gestures of resource potentiality." Orientated toward both the future and the past, such temporal techniques are used to envisage and speculate on absent possibilities and render them actionable in the present. Here, they enable oil shale to be constituted and retained as a discrete global resource in the face of its persistent absence in the world energy mix. In practice, the resource has taken on multiple meanings and material states in association with the disparate geo-economic conditions of producing countries. By sketching oil shales' multilayered modes of existence, I propose that resources can simultaneously be made as much as *unmade*, in a transient and nonlinear fashion (see also Kneas 2018). My aim is to make the case for critical resource geography to engage with the question of "resource temporalities" in tandem with resource materialities, following recent conceptual efforts that draw on Science and Technology Studies (STS) and wider social science theories of time and extractive capitalism

(Ferry and Limbert 2008; Richardson and Weszkalnys 2014; Weszkalnys 2015; Kama and Weszkalnys 2017; D'Angelo and Pijpers 2018; Kama 2020). The next section details the case of oil shales as a liminal resource, before outlining the key analytical questions raised by their processual and transient nature.

Assembling resourceness: from materiality to temporality

Oil shales raise the question of temporality for critical resource geography, because their "resourceness" stems from their anticipated use value as a future energy source, rather than from their petrochemical or lithological qualities (see also Carton and Edstedt, Chapter 34 this volume). In the geoscientific literature, the term "oil shale" is used to denote a wide range of fine-grained sedimentary rocks that contain kerogen: a solid, insoluble organic matter that can be transformed into liquid hydrocarbons when heated to high temperatures. The presence of kerogen is the common denominator that enables oil shales to be categorized as a *singular*, global resource. However, kerogen is not a uniform substance and it is hosted by various rocks, with different geological origins and mineral compositions, most of which are not even true shales. Moreover, while kerogen-rich rocks have been found all over the world and tested for various purposes since the early-nineteenth century, their "geological potential" (Weszkalnys 2015) remains largely unrealized. Commercial production has recurrently failed to gain traction and become part of the mainstream hydrocarbon industry. Any experiments with converting the minerals into heavy crude oil and other products remain confined to particular locations and historical periods, such as the shale oil and chemical industries of the West Lothian county in Scotland (1851–1962), New South Wales in Australia (1865–1952), and the naval oil factories of the Närke district in Sweden (1923–1966). Others have periodically been closed and resurrected again, including several laboratories of leading international oil firms in the American West (for the best historical overview, see Russell 1990). There are only a few small-scale projects that remain in continuous operation; these are located in Estonia, Brazil, and China. Compared to other unconventional fossil fuels, oil shales are thus especially peculiar because they have lain dormant for such a long time, the entire duration of the oil era.

As kerogen-rich sedimentary rocks are found across the world, they continue to captivate the geopolitical imagination of many countries, especially those that lack traditional fossil fuel deposits. The desire to find domestic alternatives to imported hydrocarbons has been particularly strong during times of imposed scarcity, such as the fuel shortages of the two world wars and the 1970s energy crisis. More recently, the promise of oil shales has once again been revived in response to the popularity of "peak oil" concerns at the turn of the millennium. Driven by scarcity fears and volatile market prices, oil shale development has returned on the agenda of many resource-holding states, especially in the United States, the Levant region, and the former Soviet bloc. The most recent wave of interest is augmented by the efforts of international oil corporations to secure their resource holdings at a time of resurgent geopolitical constraints, which in combination with cheap finance capital has enabled the "success story" of other unconventional fuels, notably the exponential growth of shale gas and tar sands exploitation in North America. Oil shale proponents aspire to this eventual breakthrough, in the hope that the prolonged experiments of state-owned companies and junior mining firms will likewise pave the way to full-scale commercialization by major corporations. Yet, the situation is far more complicated than this vision of linear progress suggests. In a remarkable twist, shale gas has quickly turned from a blessing to a curse for the other shale. Despite reviving interest in the extractive sector in general, especially following the 2008 financial crash, it has turned investors away from experiments with oil shale retorting, while oil majors have also shifted their efforts to fracking and deep-sea drilling. Worse still, the shale gas boom has appropriated the very name of the older resource, *shale oil*, which is now widely used to denote the tight crude accessed via fracking, not the synthetic crude distilled from kerogen.

The struggles of the oil shale industry to attract investment capital are exacerbated by public concerns over climate emergency. Indeed, being one of the most polluting and inefficient forms of energy production, the promissory future resource is increasingly called into doubt as a marginal resource of the past. Given the need to rapidly cut cumulative emissions in the atmosphere, there is a growing pressure to keep most fossil fuels in the ground, particularly those with such a high carbon content. In line with wider efforts to decarbonize the energy sector, unconventional fuels have thereby become subject to attempts to proactively *unmake* these resources, by de-economizing extraction through both regulation and divestment (Kama 2020). The anticipation of post-carbon futures thus adds another layer to long-standing controversies around both existing and prospective oil shale industries. Yet, as explained later, unmaking is not limited to decarbonization but is rather a flip side of the temporal techniques through which resources are constituted in general.

To account for the processual (un)making of oil shales across these diverse registers of time, I now turn to nascent conceptual trends on the fringes of critical resource studies, which are informed by both STS-led assemblage thinking and resurgent social science interests in temporality. The key merit of the STS perspective is that it enables us to move away from our fixation with resources as a material given. As the opening vignettes imply, the geological potential of oil shales does not provide a firm ground of external materiality (cf. Ingold 2011), upon which technoscientific innovation and political consensus over future energy systems can be rested. Rather, the industry's protracted operations beg the question of how the resource might eventually *materialize*—particularly as a significant global resource—not least because its potential qualities continue to be revealed differently across geographically disparate locations and actors. Oil shale cannot thus be defined once and for all. Instead, its resourceness is gradually "assembled" through competing scientific, corporate, and geopolitical practices, so that the resource becomes "available for some purposes to the exclusion of others" (Kama 2013; Li 2014, 592; Richardson and Weszkalnys 2014). STS scholarship has traditionally emphasized that the material world "out there" is not available to us in the form of universal objectivity, but rather it is actively created with the help of various research routines, apparatuses, and devices. The assembling of resources can thus be observed not only in institutionalized "centres of calculation" (Latour 1999), such as national geological surveys, but also in more vernacular forms of knowledge production, which exceed the practices of geological prospecting, technological experiments, and investment decisions. Moreover, it is acknowledged that such practices are locally, culturally, and materially situated, and hence highly contingent (Haraway 1998; Shapin 1998; Law and Mol 2001).

Developing this perspective further, resources can equally have a conceptual and affective presence, not just a material one, as their making mobilizes various imaginations, desires, and speculations (Kama and Weszkalnys 2017). Following Latour (2005a), the scientific re-presentation of earth substances as discrete physical volumes awaiting exploitation cannot be separated from their "re-presentation" in the political and economic sites of assembly, where divided views about their anticipated use value and significance are weighed and operationalized, often by force. Resources should thus neither be essentialized as part of external reality nor reduced to mere social conventions that are no longer subject to change. Rather, we can speak of precarious resource assemblages that might still fail, depending on the specific arrangements through which geological potential is abstracted, rendered valuable, and stabilized (compare with Latour 2005b, 91). Similarly, for Stengers, the potentiality of substances cannot be subjugated under a single objective definition, such as "oil shale," but rather it is always distributed through multiple modes of engagement with the matter in question. However, since material phenomena serve as the "cause for thinking," their re-presentation also generates active witnesses to the objectivity of knowledge claims (Stengers 2010). What one can derive from this theory of "collective objectivity" (Braun and Whatmore 2010) is that resourceness is always in flux and may take multiple forms, rather than being a static state of affairs. This is not just because our engagements with the geological reality are plural, but because material phenomena themselves multiply in the process of enacting reality (Mol 2003; Law 2004).

Therefore, the starting point for a more processual account of resourceness is to recognize that the essence and qualities of resources could have been and may still be constituted differently. This requires us to examine not just how resource ontologies reflect choices made at certain historical junctures (Richardson and Weszkalnys 2014), but also how contemporary extractive struggles reveal competing attempts to define what counts as resources, what their role was in the past, and what they ought to become in the future (Kama 2016). As the case of oil shales shows, resource assemblages have temporal conjectures. Their making operates through the affective conjuring of impending possibilities as much as it relies upon tangible evidence of the inevitability of production, beyond the irrefutable presence of geological accumulations. The quest of critical research is to analyze these contending registers and calculations of time and to render visible the multiple resource realities that emerge as a result. In brief, we should approach resources as distributed in both material and temporal senses. Indeed, materiality incorporates also the immaterial and the virtual (see also Bakker and Bridge, Chapter 4 this volume), which in turn makes it possible to render distant times and places actionable in the here and now, as increasingly acknowledged in geographical theory (Anderson and Wylie 2009).

To illustrate this approach, the next two sections provide a brief sketch of how the resourceness of oil shales is assembled, transformed, and multiplied through an interplay of the industry's future projections and past legacies. These include estimates of the global resource base and examples of purportedly successful, yet limited instances of previous production.

Future projections: calculating abundance

Oil shales become a resource in relation to hydrocarbon scarcity, whether predicted or already occurring. The Earth's crust contains multiple sedimentary layers that are to some degree rich in kerogen. For oil shales to be constituted as a resource, however, the sheer scale of these geological deposits needs to be portrayed by industry proponents as a wasted opportunity: an "untapped fuel" awaiting exploitation, once rising market prices for oil have rendered lower grade alternatives more viable. Being promoted as part of future energy solutions, the exploitation of organic-rich rocks purports to match the widening gap between the world's hydrocarbon demand and supply, and thus mitigate the looming energy crunch in many oil-importing countries. Aside from peak oil concerns, oil shale development is fundamentally tied to states' aspirations of resource-led economic modernization and energy sovereignty, in line with long-standing efforts to explore "vertical territory" for the sake of nation-building (Braun 2000). Since the 1973 oil crisis, such aspirations have converged with industry's needs to secure their reserves and production flows, for the sake of capital accumulation. Despite difficulties in utilizing the rocks in a manner that would not consume more energy than gained, the geological endowment of oil shales continues to be perpetuated in both national energy policies and companies' investment portfolios as petroleum's timely successor.

Imaginaries of vast geological abundance are therefore key to the making of oil shales' resourceness. One only needs to search the Internet for "oil shale resources" to find numerous global estimates that exceed conventional crude reserves by many orders of magnitude. Such figures and maps may occur in various shapes, but they all convey a unanimous message: while remaining stocks of petroleum are limited, the impending shortage can be overcome by another source, "shale oil," which is impossible to exhaust in the foreseeable future. These gestures of resource potential enable oil shales to be assembled not just as any resource, but as a *singular* global resource on a par with hydrocarbons, notwithstanding the heterogeneity of world deposits, the incompatible geological assessments of resource-holding countries, and their substantially different pathways to exploitation.

In practice, projections of world oil shales as an infinite supplement to hydrocarbons conceal long-term problems in defining and calculating such mineral deposits *as oil*. As Mitchell (2011, 252) contends, this requires "opening up anew the politics of nature" through which petroleum has been assembled thus far. At a closer look, it is evident that current estimates of world oil shales are simply based on summing up various calculations of state geological surveys and other local experts, without any agreed-upon standard in place. The caveat is that the methods and quality of geological prospecting vary hugely, even across the same basin, whereas only a few estimates are informed by experience with actual production. Any projections of the global resource therefore suffer from a lack of reliable and comparable local data. Moreover, what renders these figures especially speculative is that they are juxtaposed with a fundamentally distinct category: the global *reserves* of crude oil. Rather than accounting for the total scale of the resource base, reserves are accumulations that can be recovered through *present* technological means and economic parameters, expressed with different degrees of probability. By contrast, the making of oil shales is primarily orientated toward *future* extraction. Due to a dearth of production data, it is virtually impossible to predict the conditions under which such "in-place" geological occurrences may one day become exploitable. Hence, any attempts to incorporate oil shales into analyses of future hydrocarbon supply have been futile.

It is common for any extractive resources to be progressively upgraded from total in-place resources to a more pragmatic figure of reserves, when progress in geoscientific research and verified production flows make it possible to more accurately evaluate the recoverability of the subsurface deposit. A hydrocarbon reserve is always a *virtuality*, not a representation of total physical stocks, since it incorporates projections of future discoveries by inference to those accumulations already in production, "without necessarily ever knowing precisely what exists in the present" (Barry 2013, 14). What is striking about oil shale projections, however, is that this temporal shift is not even desired by industry proponents. Oil shale resources are also virtualities, but in a distinct way. They account for the absolute geological abundance; "the total amount of shale oil in a deposit, including those proportions that are not economically or technologically recoverable at present" (Dyni 2004, 739). This imaginary work is indispensable to rendering the haphazard industry a step closer to "reality." After all, kerogen-rich deposits clearly have geological potential, even if there is little agreement on the future worlds in which they may (or may not) become more widely exploited (see also Zalik, Chapter 35 this volume).

As virtualities, oil shale estimates are not just incomplete representations of the subsurface (Valdivia 2015) but also, more fundamentally, re-presentations that invoke distinct *states of materiality* (cf. Anderson and Wylie 2009). The tangible qualities of oil shales are evident in the shape of solid rocks, such as the ones I witnessed in the impressive outcrops of the Rocky Mountains or in the heaps of feedstock at Estonian mines. Yet, for oil shales to resemble a hydrocarbon resource, their mineral form of existence needs to be erased. Oil shales are strategically assembled as a liquid fuel, shale oil, which is calculated by analogy to conventional crude: in barrels of "oil equivalent." The transubstantiation of solid rocks into hydrocarbons is thus another hurdle in (un)making the resource. These divergent states of materiality further exacerbate difficulties in estimating the global resource, as they arise from specific technological inventions and market arrangements, which again vary geographically. For example, whereas the US Geological Survey measures the oil shales of the Rockies (allegedly the world's largest deposit) as barrels of in-place oil, other countries tend to account for total geological stocks with respect to their mineral form of existence, reflecting other purposes of use. Whether assembled as accumulations of mineral ores or hydrocarbon reservoirs, oil shale denotes "a coexistence of multiple entities that go by the same name" (Mol 2003, 151).

Past legacies: transposing technological experiments

Oil shales do not exist as merely future possibilities. The (un)making of their resourceness is also haunted by the industry's drawn-out past. These contending registers of time are negotiated, in particular, through continued technological experiments in harnessing the geological potential of kerogen-rich rocks in the face of repeated delays. In order to maintain the "economy of appearances" necessary for government support and capital investments to keep flowing into such speculative ventures (Tsing 2000), oil shale proponents need to continuously feed the promise with evidence of actual production. Such evidence is available in the form of distant industry histories or faraway geographical locations, leading to the exchange of technological expertise between countries such as Estonia, Jordan, and the United States. Alongside resource estimates, the alignment of past and prospective industries therefore incrementally adds to the existence of oil shales as a global, singular resource. At the same time, it raises the specter of failure, because existing industries have endured in vastly different circumstances. So, the onto-logical politics of what counts—or used to count—as a resource in one place may not easily transpose to another geo-economic setting (see also Bakker and Bridge 2006; Kneas 2018; Kama 2020).

The exercise of technology transfer between resource-holding countries straddles not just distinct materialities but also temporalities, which result from divergent ways of assembling the resource. When oil shales are doubly accounted for in both solid and liquid forms, this discrepancy reflects locally specific pathways to production. Historical shale oil factories, including the ones in operation in Estonia, have been based on a robust technique of heating the mineral substance on the ground, in bespoke retorts, after the rocks have been excavated through traditional methods. In brief, the transubstantiation of oil shales into hydrocarbons takes place post-mining, leaving behind huge amounts of waste material. Better known, how-ever, are the capital-intensive experiments of Shell, Exxon, and other major oil companies in the United States since the late 1970s, which are aimed at converting the rocks in situ, thou-sands of feet underground, without disposing the spent shale on the surface. Such experiments essentially compress millennia of geological time of oil formation into only three or four years. Unlike surface retorting, in situ conversion has never been implemented beyond labora-tory experiments and it is equally problematic for environmental reasons. Yet, by keeping the grubby realities of the retorting process out of sight, it appears better suited to the affective work of rendering oil shales conceivable and sensible as liquid fuel, thus overstepping its solid presence. Indeed, the resource-making practices of the US Geological Survey correspond to the in situ pathway as they account for the absolute potential "oiliness" of the subsurface, as if shales would have always existed in their desired, liquid state. As a virtuality, shale oil conflates the temporal scales of the resource.

At the same time, the conjectures of oil shales as a global oil resource obscure the fact that most world production is still being utilized as solid fuel. Instead of being converted into syn-thetic crude, oil shales are commonly used as feedstock in electric power generation, similarly to the direct combustion of brown coal. Estonia is the leading world producer precisely for this reason, with only a fraction of its output being kerogen oil that is, however, not upgraded but used as heating and bunker fuel. In addition to Estonia, there are shale-fueled power plants in China and Israel. Most strikingly, Germany uses the high-carbon substance as a raw material for cement production, not as an energy resource at all. By contrast, the American West and other regions that have recently sought to utilize their shale deposits, in line with the peak oil rhetoric of mitigating import dependency and volatile prices, have little to do with the resource ontologies of solid fuel. For the industry proponents of these countries, the resource needs to be untangled from its previous assemblages, and reassembled all over again, in order to give way to its liquid materiality as something akin to hydrocarbons. Nevertheless, concrete evidence of shale's economic viability in the form of solid fuel is still indispensable for the making of the hydrocarbon resource. After all, it is a historical fact that millions of tons of kerogen rock have already been burned in power plants. Likewise, a fleet of ships has been fueled by the heavy crude product of retorting, even if this stuff has never been upgraded into high-octane transport fuel in commercial quantities. These legacies are not to be forsaken. Quite the contrary, they become strategically and selectively mobilized by industry for the sake of perpetuating the economy of appearances and rendering the promise of imminent production more plausible for investors and governments alike.

Conclusions

In 2019, nine years after my field trip with the government delegation, Jordan finally mined its first one million tons of oil shale, with the help of Estonian know-how and Chinese capital. Yet, there is still no scent of oil. Instead of using the mined shale to launch domestic crude production, the state government looks forward to opening its first shale-powered electricity plant, based on direct combustion, and expected to cover 15 percent of the country's demand. Liquid fuel production has once again been deferred, allegedly due to both technological failures and rapid changes in the global market. The Estonian project for surface retorting in Utah has not made much progress either, other than gradually adding value to the concession through ongoing geological prospecting and environmental monitoring. Likewise, the in situ experiments afoot in the United States have all been suspended in the last few years, as the oil majors moved into fracking. While kerogen-rich rocks continue to be utilized in smaller quantities across the world, shale oil remains largely absent: an anticipatory economy that may one day turn into a reality, or it may not.

Oil shales have remained a short-lived, not fully made, and localized resource, unlike established commodities. It is precisely for that reason, I argue, that this case has important implications for critical geography research. First, insofar as resources are made, they have potential to also be unmade. Strictly speaking, there is no such thing as oil shale. Instead of adhering to the ideal of a discrete, self-explanatory geological category, oil shales are assemblages of both solid and liquid resources, which are being shaped by specific technological histories and socioeconomic collectives. The dispersed nature of oil shales therefore affirms that resources are not only relational, but their assembling exhibits multiple, relational materialities that cannot be defined solely in biophysical terms (Bakker and Bridge 2006). Instead of taking resourceness for granted, critical scholars need to inquire "how the properties and/or capacities of materialities thereafter become effects of that assembling" (Anderson and Wylie 2009, 320). Although oil shales are marginal to mainstream fossil fuel industries, their protracted development suggests that resourceness should be understood in a more processual sense, as being neither intrinsic nor socially fixed. Rather, such resource assemblages unfold in multiple and dynamic forms and may easily fail to become realized.

Second, despite their liminal and transient nature, the conjectures of oil shales as a future oil resource also show remarkable endurance. After all, hopes for shale oil's eventual global significance have never been forsaken, despite its evanescence. While the process of reassembling diverse earth substances as new fossil fuel supplies may fall apart, it may also persevere against the odds. Through strategic projections of infinite geological potential, and the affective showcasing and witnessing of existing projects, diverse resource materialities come to "hang together" (Mol 2003) and become conflated in strikingly interdependent ways. To render kerogen-rich rocks equal to hydrocarbons, their potential needs to be both calculated and sensed as petroleum. We must therefore also pay attention to "the work it takes to pull a resource assemblage together and make it cohere" (Li 2014, 600), so that heterogeneous materialities come to perform as one resource, rather than many.

Finally, the endurance of shale oil development indicates the wider significance of temporality, or the contending registers and techniques of time inherent in resource speculations. Oil shale developers combine select constructs of the industry's past with projections of future supply, in order to herald imminent production and accumulate value, while continuously deferring their operations. However, at a time of peak oil and climate emergency concerns, it is increasingly contested whether the industry's legacies serve as an index of an alternative energy economy to come—or, conversely, something to be avoided at any cost. Specifically, it remains to be seen whether synthetic crude will eventually turn out to be the future fuel, or become obsolete, a marginal resource of the past. This suggests that resources may be not only materially, but also temporally dispersed. Their making is defined by absence and virtuality as much as by tangible geological substances and infrastructures. Yet, even as immaterial entities, such not-yet resources are performative: they have real-world effects through their conceptual and affective presence, without necessarily ever being materialized as such. Even if the much-deferred event never happens, other realities are shaped in the process, involving significant infrastructural investments and energy policy decisions. Whether oil shales eventually become a resource for distant times and places, or do not, their making is already transforming the carbon economy and politics of the present.

Note

1 While this paper has grown out of my doctorate research (Kama 2013), it is also informed by close collaboration with Gisa Weszkalnys on "resource temporalities," to the extent that some of our ideas have become inseparable.

References

Anderson, Ben, and John Wylie. 2009. "On Geography and Materiality." *Environment and Planning A* 41 (2): 318–335. doi: 10.1068/a3940.

Bakker, Karen, and Gavin Bridge. 2006. "Material Worlds? Resource Geographies and the 'Matter of Nature'." *Progress in Human Geography* 30 (1): 5–27. doi: 10.1191/0309132506ph588oa.

Barry, Andrew. 2013. *Material Politics: Disputes Along the Pipeline*. Oxford: Wiley Blackwell.

Bear, Laura. 2016. "Time as Technique." *Annual Review of Anthropology* 45: 487–502. doi: 10.1146/annurev-anthro-102313-030159.

Braun, Bruce. 2000. "Producing Vertical Territory: Geology and Governmentality in Late Victorian Canada." *Cultural Geographies* 7 (1): 7–46. doi: 10.1177/096746080000700102.

Braun, Bruce, and Sarah J. Whatmore. 2010. *Political Matter: Technoscience, Democracy, and Public Life*. Minneapolis, MN: University of Minnesota Press.

Bridge, Gavin. 2009. "The Hole World: Scales and Spaces of Extraction." *New Geographies* 2: 43–48.

Bridge, Gavin. 2011. "Resource Geographies I: Making Carbon Economies, Old and New." *Progress in Human Geography* 35 (6): 820–834. doi: 10.1177/0309132510385524.

D'Angelo, Lorenzo, and Robert J. Pijpers. 2018. "Mining Temporalities: An Overview." *Extractive Industries and Society* 5: 215–222. doi: 10.1016/j.exis.2018.02.005.

Dyni, John R. 2004. "Oil Shale." In *Encyclopedia of Energy*, edited by Cutler J. Cleveland, 739–752. London: Elsevier.

Ferry, Elizabeth Emma, and Mandana E. Limbert. 2008. *Timely Assets: The Politics of Resources and Their Temporalities*. Santa Fe, NM: School for Advanced Research Press.

Hanson, Jason L., and Patty Limerick. 2009. *What Every Westerner Should Know About Oil Shale: A Guide to Shale Country*. Boulder, CO: Center of the American West, University of Colorado at Boulder. https://www.centerwest.org/projects/energy/oil-shale/why-oil-shale-3.

Haraway, Donna. 1998. "Situated Knowledges: The Science Question in Feminism and the Privilege of Partial Perspective." *Feminist Studies* 14 (3): 575–599.

Ingold, Tim. 2011. *Being Alive: Essays on Movement, Knowledge and Description*. London: Routledge.

Kama, Kärg. 2013. "Unconventional Futures: Anticipation, Materiality, and the Market in Oil Shale Development." DPhil thesis, University of Oxford.

Kama, Kärg. 2016. "Contending Geo-Logics: Energy Security, Resource Ontologies, and the Politics of Expert Knowledge in Estonia." *Geopolitics* 21 (4): 831–856. doi: 10.1080/14650045.2016.1210129.

Kama, Kärg. 2020. "Resource-Making Controversies: Knowledge, Anticipatory Politics and Economization of Unconventional Fossil Fuels." *Progress in Human Geography* 44 (2): 333–356. doi: 10.1177/0309132519829223.

Kama, Kärg, and Gisa Weszkalnys. 2017. "Resource Temporalities: Anticipations, Retentions and Afterlives." Working paper presented at the "Anticipating Abundance" workshop, University of Durham, May 12.

Kneas, David. 2018. "Emergence and Aftermath: The (Un)Becoming of Resources and Identities in Northwestern Ecuador." *American Anthropologist* 120 (4): 752–764. doi: 10.1111/aman.13150.

Latour, Bruno. 1999. *Pandora's Hope: Essays on the Reality of Science Studies*. Cambridge, MA: Harvard University Press.

Latour, Bruno. 2005a. "From Realpolitik to Dingpolitik or How to Make Things Public." In *Making Things Public: Atmospheres of Democracy*, edited by Bruno Latour, and Peter Weibel, 14–43. Cambridge, MA: MIT Press.

Latour, Bruno. 2005b. *Reassembling the Social: An Introduction to Actor-Network-Theory*. Oxford: Oxford University Press.

Law, John. 2004. *After Method: Mess in Social Science Research*. London: Routledge.

Law, John, and Annemarie Mol. 2001. "Situating Technoscience: An Inquiry into Spatialities." *Environment and Planning D: Society and Space* 19 (5): 609–621. doi: 10.1068/d243t.

Li, Tania Murray. 2014. "What Is Land? Assembling a Resource for Global Investment." *Transactions of the Institute of British Geographers* 39 (4): 589–602. doi: 10.1111/tran.12065.

Mitchell, Timothy. 2011. *Carbon Democracy: Political Power in The Age of Oil*. London: Verso.

Mol, Annemarie. 2003. *The Body Multiple: Ontology in Medical Practice*. Durham, NC: Duke University Press.

Richardson, Tanya, and Gisa Weszkalnys. 2014. "Resource Materialities." *Anthropological Quarterly* 87 (1): 5–30.

Russell, Paul L. 1990. *Oil Shales of the World: Their Origin, Occurrence and Exploitation*. Oxford: Pergamon.

Shapin, Steven. 1998. "Placing the View from Nowhere: Historical and Sociological Problems in the Location of Science." *Transactions of the Institute of British Geographers* 23 (1): 5–12.

Stengers, Isabelle. 2010. "Including Nonhumans in Political Theory: Opening Pandora's Box?." In *Political Matter: Technoscience, Democracy, and Public Life*, edited by Bruce Braun, and Sarah J. Whatmore, 3–33. Minneapolis, MN: University of Minnesota Press.

Tsing, Anna. 2000. "Inside the Economy of Appearances." *Public Culture* 12 (1): 115–144.

Valdivia, Gabriela. 2015. "Oil Frictions and the Subterranean Geopolitics of Energy Regionalisms." *Environment and Planning A* 47 (7): 1422–1439. doi: 10.1177/0308518X15595764.

Weszkalnys, Gisa. 2015. "Geology, Potentiality, Speculation: On the Indeterminacy of 'First Oil'." *Cultural Anthropology* 30 (4): 611–639. doi: 10.14506/ca30.4.08.

6

Brave new worms

Orienting (non)value in the parasite bioeconomy

Skye Naslund and Will McKeithen

Introduction

Parasites have long been disdained as a drain to be eliminated and eradicated. From the moment modern parasitology crystallized as a scientific field in the early-nineteenth century, parasites have been seen as a blight on human life and capitalist productivity. The profession of parasitology was itself a colonial project aimed at maintaining a productive colonized labor force and protecting European profits (Foster 1965). Within medical discourse, the number of parasitic worms (known as helminths) living inside a person is called a *worm burden*, highlighting the toll they take on their human hosts. By definition, the parasite benefits at the expense of its host. Now, a nascent but growing industry is questioning this commonsense and imagining alternative human-helminth relationships.

Helminthic therapy refers to the practice of ingesting parasitic worms into the human body to treat a range of autoimmune diseases and disorders like severe allergies or Crohn's disease. Helminthic therapy reframes the parasite and, importantly, the parasitic relationship between helminths and human as a valuable resource. In some circles, parasites have become a hot commodity, worthy of being bought, sold, grown, traded, and gifted (see also Fannin, Chapter 33 this volume). Indeed, a loose network of biohackers, patient advocates, and medical researchers has emerged around this practice and proto-industry. The visions and tactics deployed by each differ, however, when it comes to helminths, helminthic therapy, and the future of this newly valued resource. As these contests unfold, shifting biosocial relations push and pull helminthic therapy in multiple, often contradictory directions. Regulate. Outlaw. Share freely. Test. Synthesize. Provision as a public good. Enshrine as a human right. Eradicate. Conserve. Patent. Commodify. What is to be the future of helminthic therapy, the people who seek it out, the worms on which it depends, or the burgeoning bioeconomy it has created?

In this chapter, we ask *what worlds are being oriented by humans and helminths?* As an answer, we present a case study of helminthic therapy,[1] applying a relational framework of biosocial relations that extends beyond capitalist value and beyond the resource-as-commodity and makes visible the *multiple* power-laden worlds in which humans and helminths dwell together. We then present multiple practices of helminthic therapy currently taking root and explore their shifting relations of (non)value as well as their implications for human and more-than-human ways of life. In our conclusion, we discuss what insights critical resource geographers might take from this research for the study of world-making.

Value and its others

Biocapitalism refers to the accumulation of capital through the commodification and sale of what Donna Haraway calls "lively capital" (2008): vital materials like blood plasma, lively critters like exotic pets, or ecosystem functions like CO_2 sequestration (Collard 2014; Collard and Dempsey 2013; Cooper 2008; Rajan 2006; see also Probyn, Chapter 32 this volume). Surveying the late twentieth–century rise of the biotechnology industry and genomic science, it would seem as though biocapitalism has only accelerated. New breeds of cattle. New strains of corn. New means of organ transplantation. Indeed, helminthic therapy seems to tell a similar tale—a strange and sensational story wherein capital commodifies another untapped nature, a natural resource newly minted.

Yet, this narrative proves insufficient in both analytical and political terms. It carves life into a binary: valued commodities and everything else. If we focus only on commodification, we risk painting capitalism as discrete and totalizing, as though all life were either inside or outside its walls (Gibson-Graham 1996). The battles over helminthic therapy, however, are more than a two-sided struggle over whether *to commodify or not to commodify*. They unfold through a complex web in which relations of capitalist value are both made possible and exceeded by relations of capitalist *non-value*. Non-value refers not to value's mirror opposite but to its diverse alternatives, its queer relatives.

As J. K. Gibson-Graham (1996) argue, capitalism has historically been conceived in cisheterosexist terms. Champions and detractors alike talk about capitalism as "penetrating" new markets or "capturing" nonmarket life, as though capitalism were itself a sovereign, masculine body. Yet, if capitalism *were* a body, then capitalism, like all bodies, would be both penetrating and penetrated. To think queerly about capitalism means understanding this relationality, mutual vulnerability, porosity, and, most importantly for our purposes, multiplicity. Some life-forms are valued as a resource. Some resources are commodified. Yet, life also takes other forms under capitalism. Rather than being commodities, some life-forms may be seen as not-yet-but-soon-to-be-commodities, as useful but not formally valued, as inconsequential, or as dangerous. These relations constitute non-value under capitalism. They are what Collard and Dempsey call "value's necessary others" (2017, 78).

Rosemary-Claire Collard and Jessica Dempsey (2017) craft a nonbinary relational framework for conceptualizing life under capitalism. They outline five *orientations* that humans and nonhumans can take in relation to capitalist value. Borrowing from Sara Ahmed, they define an *orientation* as the way in which one thing is "positioned in relation to something else" (Collard and Dempsey 2017, 81). Orientations are performative, depending on "the repetition of norms and conventions, of routes and paths taken" (Ahmed 2006, 16). Nothing just *is* a resource. No resource just *is* a commodity (Bakker and Bridge 2006; Richardson and Weszkalnys 2014; Smith 2010; see also Kama, Chapter 5 this volume). Commodity-ness is a consolidated effect. Thus, for something to be *oriented as a commodity*, it must be *repeatedly related to as a commodity*. Bought and sold. Day after day. As Collard and Dempsey outline, however, *commodity*, or what they term *officially valued*, is just one of five possible orientations humans and nonhumans can take in relation to capital accumulation.[2] These include:

1. *Officially valued*: lives, labors, and bodies directly valued as inputs for capital accumulation; e.g., wage laborers, slaves, commodified animals, and titled land;

2. *Reserve army*: not yet directly valued but promising future exchange value and profit; e.g., temporarily unemployed laborers, seeds for planting, "biodiversity," and gene banks;
3. *Underground*: useful to, but officially unvalued by, capitalism—commonly referred to as social (or biosocial) reproduction; e.g., unpaid domestic labor and ecosystem functions;
4. *Outcast surplus*: inconsequential to capitalist accumulation; e.g., paupers and feral dogs;
5. *Threat*: hindering or endangering capital accumulation, or perceived to be; e.g., union organizers and mad cow disease.

Importantly, Collard and Dempsey include seemingly inanimate entities such as land within their framework, thereby suggesting their refusal of the life/not life binary so characteristic of settler colonial US society and other post-Enlightenment contexts (TallBear 2017). Indeed, whether it be made of flesh or stone, under the logic of capital "there is not one way that nature is appropriated or exploited or dominated to produce capitalist value" (Collard and Dempsey 2017, 94).

This approach, we argue, offers several important tools and insights for doing critical resource geographies. First, by turning our attention to the broader field of biosocial relations that extend beyond capitalist value and beyond the resource-as-commodity, we are able also to expand our analytical frame beyond capitalism's most obvious and violent sites of transformation. We are able to excavate the "hidden abodes" of capitalism (Fraser 2014) to see how processes of capital accumulation have world-making effects both within and beyond spaces of formal value. If we focus only on the forest felled, hewn, bought, and sold, then we miss the pine beetle, the soil, the undergrowth laid waste, and the ties binding commodified lumber to each.

Second, centering the multiple orientations of (non)value illustrates the constitutive relation between resource-making and world-making. Commodification, for example, always also involves making and remaking whole ways of living, which are themselves entangled within, but irreducible to, capitalism. The relational framework of (non)value thus makes visible the *multiple* power-laden worlds in which humans and helminths dwell together. Big Pharma. Self-help. Gut communes. Worm banks. Parasite conservation. Black markets. Rather than just inaugurating and commodifying a novel resource, helminthic therapy creates, however tentative or liminal, new forms of socio-environmental organization, new ways of getting on together, new ways of (de)valuing one another. This approach thus makes clear how the creation and commodification of novel resources always requires the making, unmaking, and remaking of whole worlds.

Third, the concept of orientation decenters anthropocentric notions of agency, highlighting how nonhumans play a vital role in (re)orienting life and (non)value, co-shaping with other actors and forces their existence as resource, waste, underground, or otherwise. Worms are not a passive resource used and molded by human whim (compare to Bakker and Bridge 2006). As we will demonstrate below, worms actively shape those paths taken and not taken within helminthic therapy.

Finally, by tracing the way orientations direct us toward "some ways of living over others" (Ahmed 2006, 44), this approach lays bare the contingency and thus instability and mutability of resource creation and capitalist enclosure. As Gibson-Graham argue, capitalist relations are partial, mutually vulnerable, and interpenetrated with noncapitalist relations (1996). Thus, while we offer no programmatic prescription for altering life's orientations under capitalism, we illustrate how even the most commodified relations can contain the germs of their own undoing.

The many pasts, presents, and futures of helminthic therapy

Helminths, at various points in time and in different geographic and social contexts, have taken on all five of the orientations presented by Collard and Dempsey (2017). Starting as a threat to be eliminated by eradication programs, they have since been oriented as outcast surplus (irrelevant to capital accumulation) by public health authorities in the United States, as underground (necessary, but not formally valued) by those growing their own helminths, as officially valued by helminth vendors, and as reserve army (having some future value) by biomedical and pharmaceutical researchers studying them.

Indoor plumbing, socialized waste management, refrigeration, antimicrobial products, and personal hygiene, all projects of nineteenth- and twentieth-century state-sponsored public health, radically altered relations between humans and a host of microbes, especially in wealthy countries of the Global North. Helminths, like bacteria and other "germs," were seen as threats to modernity, to the centralized state, to the social order, and to capitalist progress. Thus humans endeavored to sanitize them from the ecology. The American hookworm—*Necator americanus*—was first identified and described in 1902 (Rockefeller Foundation 1922). At the turn of the twentieth century in the southern United States where both the climate and social practices supported hookworm life, as much as 40 percent of people suffered from hookworm infection. Infections were concentrated among the poor, the working class, and children (Elman, McGuire, and Wittman 2014; Ettling 1981; Rockefeller Foundation 1922). By the 1920s, however, after a massive campaign to build latrines, educate school children on the importance of wearing shoes, and treat affected populations, hookworm infection became largely nonexistent. The eradication campaigns of the 1910s and 1920s explicitly targeted hookworm infection as a scourge on worker productivity and the cause of perceived economic and cultural backwardness in the South. These campaigns, however, focused specifically on recapacitating the White working class, while overtly deprioritizing Black communities (Wray 2006). Hookworm infection was termed "lazy cracker disease" and was blamed for the stereotypes associated with poor rural Whites (Ettling 1981; Wray 2006). This population, and thus the hookworms infecting them, was seen as an internal threat to the White race and the cultural logics of White supremacy (Wray 2006). Hookworm eradication, like other public health campaigns of the time period, was thus aimed at mass social change while maintaining the dominant racial order.

Given the largely successful eradication campaigns leveraged against hookworms and other parasitic infections in the United States and other Global North countries, helminths were insignificant to the workings of capital in that part of the world throughout much of the twentieth century. That changed in 1989 when British epidemiologist David Strachan (1989) first proposed what became known as the "hygiene hypothesis." The hypothesis links this history of environmental over-sanitation to the rise in autoimmune and allergic disorders in the Global North. Without the microbes and parasites that populated humans for millennia and with whom humans have coevolved, our immune systems no longer develop properly. According to the hygiene hypothesis, this dispossession of parasites explains the precipitous rise in autoimmune disorders among Global North populations in the latter half of the twentieth century. While scientific debate about the hygiene hypothesis and its practical implications continues, some researchers have proposed deliberately reintroducing parasites as one potential remedy to historical over-sanitation (Velasquez-Manoff 2012). This intentional reintroduction of helminths is called helminthic therapy. Research on helminthic therapy continues to develop and multiply and the practice of helminthic therapy itself is highly variegated and contested. The deliberate reintroduction of parasites to the human body has taken many different forms, each with its own orientation of (non)value.

First tested in humans in the early 2000s, early clinical trials showed tremendous promise for helminthic therapy. Jasper Lawrence, an Anglo-British man living in California, suffered from severe asthma but had failed to find relief through conventional medicine. Lawrence read about the nascent helminthic therapy research and wanted in. No clinical trials were accepting patients, so Lawrence decided to conduct his own informal experiment. Lawrence traveled to Belize, where he walked barefoot around open latrines hoping to contract the same hookworms that had been eradicated from the United States a century earlier. As Lawrence tramped through the feces-contaminated soil, dozens of hookworm larvae sensed the mammalian warmth of human skin. The larvae burrowed through the soles of Lawrence's feet and into his bloodstream. From there, they migrated to his lungs and, after irritating his lungs enough to be coughed up and swallowed, they reached Lawrence's small intestine, where they attached themselves to his intestinal lining, bloodletting their sustenance. Once attached, hookworms can survive two to five years, continuously excreting eggs that then pass into the stool of their hosts and back into the soil where they hatch and seek out new hosts. After considerable trial and error, Lawrence found himself asthma-free. He was hooked. Lawrence began his own business *Autoimmune Therapies* selling hookworms, and later whipworms, to fellow allergy, asthma, and autoimmune disease sufferers. Using his own body to gestate worm larvae, Lawrence extracts eggs from his feces, processing, packaging, and selling them online.

Today, clinical trials are still underway, showing mixed results. While these trials have established the safety of helminthic therapy, its efficacy remains inconclusive (Feary et al. 2009; Fleming et al. 2017; Summers et al. 2003). Given the long timelines for biomedical research and government approval, not to mention the desperation of those individuals suffering from Crohn's disease or multiple sclerosis for whom conventional biomedicine has failed, others are following in Lawrence's footsteps. A community of patients and biohackers is experimenting with helminthic therapy. Helminthic therapy users can cultivate worms themselves from their own or donated feces or they can purchase worms from one of several online vendors. This burgeoning bioeconomy is largely illicit, however. Most users live in the United States, but in 2009, the US Food and Drug Administration banned the commercial sale of parasitic worms. Online retailers like Lawrence have either moved their operations to countries where their business remains unregulated (e.g., Mexico) or gone underground to undisclosed locations.

Four species of helminths are currently used in helminthic therapy. Two are human parasites; the helminths complete their life cycle by colonizing human hosts and viable eggs are excreted in human feces (hookworm and whipworm). The other two are nonhuman parasites—one's life cycle passes through pig hosts (pig whipworm) and the other through a combination of rats and grain beetles (rat tapeworm). While still illegal to sell in the United States, the nonhuman parasites have been legally regulated in Thailand and the United Kingdom where vendors of those helminths are located. No country has yet explicitly regulated human helminths, and thus the helminths used by many helminthic therapy users in the United States are unregulated by any governmental organization.

As of 2015, it was estimated that approximately 7000 people worldwide were using helminthic therapy (Cheng et al. 2015), though that number is quickly rising. The four species of helminth that are in use today have been carefully selected by researchers and users based on their life cycles and interactions with the human body. These worms do not reproduce within the human body, they can be effectively cultivated, and, when used in relatively small numbers, they do not produce severe side effects or illness in their human hosts. The worms' biology and life cycle pose little threat to humans. For some users, the worms have no apparent effect. For others, however, these worms are lifesaving, justifying a steep price. Helminthic therapy can run anywhere from $30 to $1000 depending on the species, the number of worms per dosage, the number of doses, and the degree of vendor support provided. "It was a lot of money to potentially throw away," as one user put it, "but it was worth it for a decent chance of it helping, which it happens it has entirely been worth it" (Personal interview 2017). For such users, parasites can, with care-full human collaboration, become symbionts, both nourished and nourishing.

Within the helminthic therapy user community, however, debate rages over the ethics and economics of this growing industry. Some are happy to pay vendors for worms, assured by the security and accountability retailers claim to offer. Others insist that helminths should be free. Within this latter camp, some individuals self-cultivate their own worms, secreting, sterilizing, and ingesting their own larvae. Other users have developed small gifting economies, cultivating and donating worms peer-to-peer. As one such individual put it, "I give it away for free at no cost. I just give it to people" (Personal interview 2017). These networks are often small and proximate, usually only sharing worms amongst friends and family. Sharing via the Internet does occur, but it has proven limited. These gut communes often develop their own standards for membership, requiring sharers to publish their means of self-cultivation and disclose their HIV status. (Vendors also commonly do both.) Some helminthic therapy users, however, worry about the safety of parasites cultivated within these collectives and oppose such sharing. The safety of helminthic therapy depends on dosage—the number of worms ingested. For this reason, some individuals refuse to obtain worms from anyone other than a well-established vendor. As the crowdsourced Helminthic Therapy Wiki warns, "[T]here are risks attached to accepting larvae from a private grower" (2019). For vendors, proving that you can accurately count eggs under a microscope (and thus secure against the perceived and actual threat of overdose) has become essential to ensuring the value of your worm stock. Even self-cultivators will freeze excess eggs or fecal matter, to kill their larvae and ensure that worms thrown in the garbage cannot spread in the wild.

Some scientific researchers, however, insist that unregulated do-it-yourself (DIY) helminthic therapy is inherently dangerous. Live worms, they claim, will never be safe. As one researcher explained, using live worms "is a bad idea for several reasons…[Parasites] are not benign organisms. [Parasites] cause serious pathology and have to be monitored under very carefully regulated circumstances" (Personal interview 2018). In response, research is underway to remove the worm from helminthic therapy altogether. These researchers believe the future of autoimmune disease treatment lies with synthetically manufactured drugs that mimic the compounds worms secrete during infection (Helmby 2015). The eventual goal is to develop a pharmaceutical product or drug modeled on the biological mechanisms of helminthic therapy. Naslund's interviews with DIY users, however, revealed that most are skeptical of this research direction. The biomedical establishment failed them before, they feel. Now they worry Big Pharma will capture, enclose, and deprive them of helminthic therapy. Such a drug, they argue, would make something that is currently relatively cheap or free much more expensive. After all, pharmaceutical products, unlike worms, can be patented.

As researchers have worked to isolate the compounds responsible for calming host immune system attacks against parasites, they have encountered other potential avenues for capital accumulation. Research is currently underway to see whether hookworms might provide the base for a new class of antibiotics. Hookworms burrow into the intestines of their hosts for years, exposing abrasions to fecal matter and other microbes without ever producing infections. As one researcher explained, helminths "must be controlling the bacterial environment in their vicinity, because theoretically, you should be making the person more prone to bacterial infection, but it's the opposite" (Personal interview 2018). While research remains preliminary, a recently published study identified an antibiotic compound in soil-dwelling nematode worms (Pantel et al. 2018), distant relatives of those worms used in helminthic therapy.

Worlds in the making

Each of these world-making projects responds to the world diagnosed by David Strachan and his acolytes. The hygiene hypothesis flips previous thinking on its head, reconstructing the *lack* of parasites as the real threat to capitalist ways of life. It reconceptualizes parasites (and, more specifically, a moderate number of particular species of worms living inside each human) as an ecological underground—not explicitly valued or commodified but a condition of possibility for a range of sources of capital accumulation. In doing so, it argues for a world in which parasites are not removed from the body but rather reintroduced to it. Individuals suffering from severe allergies and autoimmune disorders do not make healthy humans or productive workers. The problem of life and (non)value under capitalism must be solved anew.

Though born from the same hygiene hypothesis, each of these potential world-making projects adopts a different orientation among worms, humans, and (non)value. Online retailers and DIY helminthic users officially value helminths as commodities to be bought and sold. The worms present a steep price to many, but the use value they present in terms of (self-reported) health merits the high cost. Though the gray market of worm commerce presents the most public and likely largest segment of the helminthic therapy bioeconomy, worms are oriented in many ways beyond commodity. Self-cultivators and gift economies both aim to decommodify worms, gifting and sharing them at no cost beyond the social price of knowledge and intimacy (e.g., HIV status disclosure). Within these DIY spaces, we can understand parasites from both capitalist and noncapitalist vantages. On the one hand, gift economies orient worms as a capitalist underground. Parasites become a biological commons that individuals or communities share. Parasites, and the immune system modulation they prompt, recapacitate people who had previously been unable to work. People once unable to get out of bed can now commodify their labor. Health is wealth. Like bee pollination and industrial agriculture, parasites provide an ecosystem service and thereby make capitalist life possible. On the other hand, as with all social reproduction under capitalism, commoning sustains not just capitalist relations but all human life. In this way, helminthic therapy both supports *and* exceeds capitalist relations. Helminthic therapy users do more than sell their rehabilitated labor power. While helminthic therapy as a practice and industry remains relatively small in size, parallel tensions exist between capitalist orientations (e.g., worker productivity and profit) and noncapitalist orientations (e.g., health, well-being, and a good life) in, for instance, the Paleo diet and other Silicon Valley–inspired biohacking (Leiper 2017), Wal-Mart employee "self-improvement" campaigns (Tveten 2017), and socialized medicine (Fine 2018).

Medical researchers conducting clinical trials also officially value parasites. They purchase living worms to use as experimental inputs. Officially valued, commodified, and traded, these worms become, for researchers, a clinical tool. Researchers write parasites into their budgets, buying them from vendors or cultivating them in the lab from donated feces.

Helminths in general, however, have become a reserve army of future exchange value and profit (on the exchange and use value of lively commodities, see Nast, Chapter 13 this volume). Indeed, within the field of environmental conservation, a growing movement has called for the conservation of parasites, jump-started by Donald Windsor's evocative call for "Equal Rights for Parasites" (1995). Within this discourse, parasites represent a wealth of understudied and untapped species biodiversity. Almost every plant and animal on earth has its own parasites. Little is known about what potential benefits parasites may have for their hosts, ecological neighbors, or biotech companies. This future wealth might come in the form of intact organisms, synthesized proteins, or antibiotics developed through biomimicry.

Researchers seeking a synthetic alternative, however, see parasitic worms as inherently patho-genic, as ontologically oriented to threaten human (capitalist) life. They aim to make a com-modity devoid of its liveliness, a commodity less likely to reorient itself away from the circuit of patent, profit, and capital accumulation. As Goldstein and Johnson argue, biomimicry itself "pro-duces 'nature' through the well-worn logics of resource enclosure and privatization" in its reori-entation of "nature as intellectual property" (2015, 61). While self-cultivators also acknowledge the threat that live worms pose if reintroduced into the wild, they reorient these worms from threats to waste by freezing and killing excess larvae before disposal. DIY users see researchers as too fixed on the construction of parasites as a threat. They see drug development not as some beneficent reorientation of the parasite away from threat but as an insidious effort to eliminate their commons in three steps: substitute the unruly non-commodifiable agency of worms with a controllable and scalable synthetic, enclose the parasite as intellectual property, and outlaw its unpatented, unregulated existence.

Conclusion

Helminthic therapy is a field still wildly at play, messy, and undetermined, multiple and even contradictorily oriented. The long history of helminth eradication, the equally deep associa-tions between helminths and harm, helminthic therapy's liminal legal status, the desperation and discord among autoimmune sufferers and worm users, and the vital agency of hookworms themselves all entangle to produce multiple orientations within this web of humans and hel-minths, life and (non)value. At times, these orientations coexist. At others, they compete. We offer the framework of orienting as one approach to understand resource-making as world-making. As multiple actors (re)orient themselves to worms and (re)orient worms to value, they spin into being whole worlds and whole ways of organizing social and biological life. Sara Ahmed describes orientations as "paths well trodden" (2006, 16). In the case of helminthic ther-apy, some paths are more trodden than others. Some paths carry the power of law or corporate might. Some paths resist. Other paths cut entirely new lines of flight. The future for worms and humans remains uncertain. By offering our multiple diagnosis of this messy present, however, we hope to grow our shared analytical capacity to read, hear, and tell stories of resource-worlds not yet written.

This complex case, we argue, carries several key insights for critical resource geographers and allied scholars of capitalist nature. First, as Collard and Dempsey (2017) argue, orientations are dynamic. They can change slowly or precipitously. An entirely new orientation can take hold or, as we illustrate here, multiple, conflicting orientations may compete or coexist. When a resource is commodified, it is never simple. Logging might assign capitalist value to some trees, but such valuation does not preclude other relational orientations toward trees. Trees continue to be valued as habitat, as CO_2 sinks, or for their esthetic beauty. By examining the birth of the helminthic therapy bioeconomy through the lens of multiple orientations, we can see the insta-bility of resource creation and capitalist enclosure. Following Gibson-Graham (1996), capitalist relations are partial and mutually vulnerable, interpenetrated by noncapitalist relations. "[B]odies move or are moved through [orientations] and even the orientations themselves may change" (Collard and Dempsey 2017, 94). Rather than seeing the world as divided between capitalism and everything else, this approach makes visible the multiple and dynamic worlds that are made possible through different orientations.

Second, we demonstrate how nonhumans play a vital role in (re)orienting life and (non) value, co-shaping with other actors and forces their existence as resource, waste, underground, or otherwise. Worms are not a passive resource, used and molded by human whim (Bakker and

Bridge 2006). Worms actively shape those paths taken and not taken within helminthic therapy. Worms' liveliness frustrates the efforts of biomedical researchers while simultaneously offering important insights. Worms' life cycles make peer-to-peer gifting risky. Worms commingle with human immune responses to produce benign or null effects, shifting both use value and exchange value. As life-forms that can themselves move, infect, and evolve, worms can be simultaneously cure and poison, valued and threat.

Finally, parasitic worms push us to extend Collard and Dempsey's framework of orientation, life, and (non)value. We argue that, at least in the case of helminths, but likely in the case of other "companion species" that entangle human and nonhuman ways of life (Haraway 2003), human and nonhuman orientations are relational and dialectically interdependent. Helminthic therapy is inherently relational. There is no parasite without a host. Value depends on non-value, not-yet-value, anti-value, and vice versa. Different worms become (non)valued because different humans become (non)valued and vice versa. Early twentieth–century eradication campaigns in the US South oriented parasites as threats precisely because US racial capitalism demanded productive White workers. Present-day campaigns in the Global South adopt a parallel logic, enrolling eradication efforts within projects of economic development. Today, helminthic therapy promises to recapacitate the sick. By contrast, gifting economies (re)orient worms as a biological commons, emphasizing human life over labor. As one user professed, sharing worms is simply "completing your biome" (Personal interview 2017). Parasites are valued because of the value assigned to their hosts, which is in turn dependent on the lively workings of parasites themselves.

Helminthic therapy, in all its orientations, reminds us of the relational nature of life, as well as the far-reaching networks of valuation within which so many ways of life and life-forms are tangled. Commodification extracts and abstracts resources from their broader context in order to assign them value. Yet, as we have shown, this is not the whole story. By considering (non) value itself as also relational, we argue that critical resource geographers can more consciously attend to the ways in which resource-making is always also a process of making and unmaking many worlds.

Notes

1   The data that underlies this chapter consists of interviews with people in the helminthic therapy community and a digital ethnography of the online spaces in which helminthic therapy is discussed and practiced. This data was collected by Skye Naslund between 2016 and 2018.
2   Collard and Dempsey (2017) clarify that their typology is not meant to draw a simple analogy or moral equivalency between human and nonhuman exploitation, appropriation, or domination. Rather, each orientation highlights the similarities in how human and nonhuman lives become *structurally* valued or devalued. We echo this caution.

References

Ahmed, Sara. 2006. *Queer Phenomenology: Orientations, Objects, and Others*. Durham, NC: Duke University Press.

Bakker, Karen, and Gavin Bridge. 2006. "Material Worlds? Resource Geographies and the 'Matter of Nature'." *Progress in Human Geography* 30 (1): 5–27. doi: 10.1191/0309132506ph588oa.

Cheng, Anna M., Darshana Jaint, Steven Thomas, Janet K. Wilson, and William Parker. 2015. "Overcoming Evolutionary Mismatch by Self-Treatment with Helminths: Current Practices and Experience." *Journal of Evolutionary Medicine* 3: 1–22. doi: 10.4303/jem/235910.

Collard, Rosemary-Claire. 2014. "Putting Animals Back Together, Taking Commodities Apart." *Annals of the Association of American Geographers* 104: 151–165. doi: 10.1080/00045608.2013.847750.

Collard, Rosemary-Claire, and Jessica Dempsey. 2013. "Life for Sale? The Politics of Lively Commodities." *Environment and Planning A* 45: 2682–2699. doi: 10.1068/a45692.

Collard, Rosemary-Claire, and Jessica Dempsey. 2017. "Capitalist Natures in Five Orientations." *Capitalism Nature Socialism* 28 (1): 78–97. doi: 10.1080/10455752.2016.1202294.

Cooper, Melinda. 2008. *Life as Surplus: Biotechnology and Capitalism in the Neoliberal Era.* Seattle, WA: University of Washington Press.

Elman, Cheryl, Robert A. McGuire, and Barbara Wittman. 2014. "Extending Public Health: The Rockefeller Sanitary Commission and Hookworm in the American South." *American Journal of Public Health* 104 (1): 47–58. doi: 10.2105/AJPH.2013.301472.

Ettling, John. 1981. *The Germ of Laziness: Rockefeller Philanthropy and Public Health in the New South.* Cambridge, MA: Harvard University Press.

Feary, Johanna R., Andrea J. Venn, Alan P. Brown, Doreen Hooi, Franco H. Falcone, Kevin Mortimer, David Pritchard, and John Britton. 2009. "Safety of Hookworm Infection in Individuals with Measurable Airway Responsiveness: A Randomized Placebo-Controlled Feasibility Study." *Clinical & Experimental Allergy* 39 (7): 1060–1068. doi: 10.1111/j.1365-2222.2009.03187.x.

Fine, Michael P. 2018. *Health Care Revolt: How to Organize, Build a Health Care System, and Resuscitate Democracy—All at the Same Time.* Oakland, CA: PM Press.

Fleming, John. O., Gianna Hernandez, Leslie Hartman, Jane Maksimovic, Sara Nace, Benjamin Lawler, Todd Risa, et al. 2017. "Safety and Efficacy of Helminth Treatment in Relapsing-Remitting Multiple Sclerosis: Results of the HINT 2 Clinical Trial." *Multiple Sclerosis Journal* 25 (1): 81–91. doi: 10.1177/1352458517736377.

Foster, William Derek. 1965. *A History of Parasitology.* Edinburgh: E. & S. Livingstone Ltd.

Fraser, Nancy. 2014. "Behind Marx's Hidden Abode: For an Expanded Conception of Capitalism." *New Left Review* 86: 55–72.

Gibson-Graham, J. K. 1996. *"The" End of Capitalism (as We Knew It): A Feminist Critique of Political Economy; with a New Introduction.* Minneapolis, MN: University of Minnesota Press.

Goldstein, Jesse, and Elizabeth Johnson. 2015. "Biomimicry: New Natures, New Enclosures." *Theory, Culture & Society* 32 (1): 61–81. doi: 10.1177/0263276414551032.

Haraway, Donna J. 2003. *The Companion Species Manifesto.* Chicago, IL: Prickly Paradigm Press.

Haraway, Donna J. 2008. *When Species Meet.* Minneapolis, MN: University of Minnesota Press.

Helmby, Helena. 2015. "Human Helminth Therapy to Treat Inflammatory Disorders – Where Do We Stand?" *BMC Immunology* 16: 12. doi: 10.1186/s12865-015-0074-3.

Helminthic Therapy Wiki. 2019. *Helminthic Therapy Wiki.* Accessed January 1, 2019. http://helminthictherapywiki.org/wiki/index.php/Helminthic_Therapy_Wiki.

Leiper, Chelsea. 2017. "Re-wilding the Body in the Anthropocene and Our Ecological Lives' Work." *Society and Space* (Online), November 14, 2017. http://societyandspace.org/2017/11/14/re-wilding-the-body-in-the-anthropocene-and-our-ecological-lives-work/.

Pantel, Lucile, Tanja Florin, Malgorzata Dobosz-Bartoszek, Emilie Racine, Matthieu Sarciaux, Marine Serri, Jessica Houard, et al. 2018. "Odilorhabdins, Antibacterial Agents that Cause Miscoding by Binding at a New Ribosomal Site." *Molecular Cell* 70 (1): 83–94. doi: 10.1016/j.molcel.2018.03.001.

Rajan, Kaushik Sunder. 2006. *Biocapital: The Constitution of Postgenomic Life.* Durham, NC: Duke University Press.

Richardson, Tanya, and Gina Weszkalnys. 2014. "Resource Materialities." *Anthropological Quarterly* 87 (1): 5–30. doi: 10.1353/anq.2014.0007.

Rockefeller Foundation. 1922. *Bibliography of Hookworm Disease.* Baltimore, MD: Waverly Press.

Smith, Neil. 2010. *Uneven Development: Nature, Capital, and the Production of Space.* Athens, GA: University of Georgia Press.

Strachan, David P. 1989. "Hay Fever, Hygiene, and Household Size." *British Medical Journal* 299: 1259–1260. doi: 10.1136/bmj.299.6710.1259.

Summers, Robert W., David E. Elliott, Khurram Qadir, Joseph F. Urban, Robin A. Thompson, and Joel V. Weinstock. 2003. "Trichuris Suis Seems to be Safe and Possibly Effective in the Treatment of Inflammatory Bowel Disease." *The American Journal of Gastroenterology* 98 (9): 2034–2041. doi: 10.1111/j.1572-0241.2003.07660.x.

TallBear, Kim. 2017. "Beyond the Life/Not Life Binary: A Feminist-Indigenous Reading of Cryopreservation, Interspecies Thinking and the New Materialisms." In *Cryopolitics: Frozen Life in a Melting World*, edited by Joanna Radin, and Emma Kowal, 179–200. Cambridge, MA: MIT Press.

Tveten, Julianne. 2017. "Our Bosses, Ourselves." *Mask Magazine*, October. http://www.maskmagazine.com/the-body-issue/struggle/our-bosses-ourselves.

Velasquez-Manoff, Moises. 2012. *An Epidemic of Absence: A New Way of Understanding Allergies and Autoimmune Diseases*. New York, NY: Scribner.

Windsor, Donald A. 1995. "Equal Rights for Parasites." *Conservation Biology* 9 (1): 1–2. doi: 10.1046/j.1523-1739.1995.09010001.x.

Wray, Matt. 2006. *Not Quite White: White Trash and the Boundaries of Whiteness*. Durham, NC: Duke University Press.

7

Resources is just another word for colonialism

Andrew Curley

Introduction

Resources shape the political (Huber 2019, 554), and the political is colonial. To turn nature into resources is to violently abstract from complex and interconnected ecological processes for purposes of extraction and exploitation. To borrow and repurpose from Métis scholar Zoe Todd (2016), *resources* is just another word for colonialism. This is not to say colonialism is only about exploiting resources, but the idea of resources is colonial constructions consistent with genocide, displacement, exploitation, and capitalism. Colonialism creates *colonialscapes* and displaces Indigenous ontologies (Hunt 2014a, 2014b).

Colonialism is the dispossession of Indigenous lands. It is a project intent on turning the stuff of nature into the raw commodities that are needed for global capitalism (Coulthard 2014). One cannot talk about resources as constitutive of today's political practices without accounting for the dispossession of Indigenous lands (see Bryan, Chapter 37 this volume). For the last 500 years, resources served as ideological basis of anti-Indian violence in the "New world," from Argentina to Canada. In these continents, settler states erased Indigenous peoples from the land and transformed Indigenous understandings of place, water, and nonhuman relatives. Colonial regimes worked to erase kinship networks with notions of property and to view the world as a repository of "resources."

In this chapter, I summarize the role of resources in shaping our understanding of tribes in Native North America, both in terms of development practices and tribal governance. I consider notions of energy transition in the Navajo Nation as a way to understand the subtle and contradictory ways ontologies of resources are foundational for projects described as sustainable and working toward "transition." I conclude with a consideration of "decolonization" and its implications for the future understandings of resources.

Resources and colonization

Colonial appropriation of Indigenous lands in the Americas started in 1492. For European colonialists, both land and precious metals were coveted. At first, colonialists used religion to contrive legal and political justifications to undermine the inherent authorities of tribes. For

European settlers, the power of the church provided the moral framework necessary to enslave and displace Indigenous peoples. In 1493, Pope Alexander VI issued the famous "Papal Bulls" that divided the Americas between Spain and Portugal under the ostensible purpose of spreading Christianity (Miller 2010). In practice, missionary work meant the murder, rape, slavery, and genocide for Indigenous peoples of North America and Africa (Dunbar-Ortiz 2014). Colonial planners conscripted the best navigators, oceanographers, geographers, and ethnographers to accomplish this task. European kingdoms competed with each other to claim the largest and most profitable parts of the world that they could conquer through violence.

The idea of resources brought extractive industries to Indigenous homelands. The environmental costs of these industries are an important area of research for scholarship on Indigenous peoples (Churchill and LaDuke 1986; Geisler et al. 1982; Jorgensen 1978). Initially understood as paths toward modernization and development, subterranean coal (Allison 2015; Curley 2019; Needham 2014; Powell 2017), uranium (Brugge, Benally, and Yazzie-Lewis 2006; Voyles 2015), oil, and natural gas have come to exacerbate colonial inequalities between setter communities and Indigenous ones. Increasingly oil and gas infrastructures threaten Indigenous lands and environments. Indigenous scholars demonstrate that these projects are another example of colonial dispossession and environmental racism (Spice 2018; Whyte 2017; Wood and Rossiter 2017).

For the history of "settler-colonial" states like Canada and the United States, Indigenous peoples have stood between "resources" and profits (Coulthard 2007; Estes 2016). The lands upon which centuries of lifeways were constructed were confiscated, plundered, and directed toward farming, mining, and White settlement. Settler states have used the idea of resources to not only dispossess Indigenous nations of land, but also water and water rights (Daigle 2018; Wilson 2019; Yazzie and Baldy 2018). Indigenous nations were displaced onto reserves, reservations, or went into hiding.

Tribal governing institutions and mineral energy development

The idea of resources works through colonial institutions. Ontologically turning nature into resources placed Indigenous peoples on top of the best lands, waters, and sources of mineral wealth. For many of the world's Indigenous peoples, the mental work of converting nature into resources is fundamental to colonial expansion and integration of Indigenous lands into settler states (Nadasdy 2017; see also Mollett, Chapter 8 this volume). The more Indigenous nations stipulated to these frameworks, the more entangled they became (Dennison 2012, 2017).

Within the United States, colonial institutions organized Indigenous communities into "tribal" governments. Today many tribal governments are shaped by legacies of mineral and energy development in and around reservation lands (Smith and Frehner 2010). Despite neoliberal devolution of federal authorities (Corntassel and Witmer 2008), US colonial laws provide tribes with limited authorities to develop mineral resources within their territories. Research on Indigenous participation in extractive industries has focused on the structural limitations of colonialism and capitalism. This scholarship focuses on Indigenous notions of "sovereignty" and self-determination as they are practiced and understood through tribal institutions (Barker 2005; Carroll 2014; De la Cadena and Starn 2007; Deer 2009; Duarte 2017; Goeman and Denetdale 2009; Lewis 2019).

Federal law–incentivized mining on the reservation and tribal governments were organized to operate consistently with federal mining laws (Allison 2015; Powell and Curley 2008). Since the 1930s, tribes have been organized as semiautonomous governments on reservations that the United States considers "federal lands." Many of these governments were established with the explicit interest in accessing Indigenous mineral rights (Smith and Frehner 2010; Voggesser 2010). In the Navajo Nation, "the Navajo Tribe" was first organized as a government in 1922 to facilitate the approval of oil leases (Powell and Curley 2008). This government was soon replaced with an Indian Reorganization Act government (IRA) in 1934 that centralized power in tribal governments across reservation communities (Wilkins 2013). With IRA, the scale and purpose of government changed from inherent Indigenous governing practices to those that resembled patriarchal nation-states. It became a representation of all tribal members under the limitations of federal-Indian law. Following the New Deal, Congress passed the Indian Mineral Leasing Act of 1938 to provide clearer law for mining within Indian reservations (Allison 2015; Voggesser 2010).

The Bureau of Indian Affairs (BIA) encouraged this form of environmental management and governance by providing money, legal authorities, and expertise to develop oil, uranium, natural gas, and eventually coal in reservations. Although oil leases existed in the Navajo Nation since the 1920s (Chamberlain 2000), it was not until uranium was found in the 1940s when companies with government contracts introduced industrial mining onto the reservation (Brugge, Benally, and Yazzie-Lewis 2006; Eichstaedt 1994; Voyles 2015). This intrusion on Diné understandings about the land, water, and animals ushered in "natural resources" as a permanent category of statecraft and development. Monies tied to extractive industries provided for the expansion of tribal services to an underserved population.

Uranium mining generated revenues and expanded social services. Coal would do the same in the coming decades. Historian Peter Iverson, who spent his career documenting Diné tribal institutions, wrote that it was during the 1950s at the height of uranium mining when the Navajo Tribe became "the Navajo Nation"—although the official name change would not happen until 1968 (Iverson and Roessel 2002). Alongside Johnson's Great Society Program, the Navajo Nation focused on poverty relief, "modern" housing, urban relocation, and the development of mining sector in the reservation. In the 1960s and 1970s, four coal mines opened across the reservation. These mines employed thousands of workers at relatively good salaries. The tribe initially agreed to coal mining for fear of losing out to nuclear power (Ambler 1990).

Starting in the 1970s, Indigenous organizers and activists challenged extractive industries in their lands as sources of environmental destruction and "dependency" (Dunbar-Ortiz 1979; Weiss 1984; White 1983). Recent scholarship has focused on the role of environmental groups in contesting extractive industries (Powell 2017; Sherry 2002). Labor history in Native North America and Indigenous anthropology has shown the importance of wage-labor work for tribes, even in the realm of extractive industries (Allison 2015; Dennison 2012; Hosmer, O'Neill, and Fixico 2004; Needham 2014; O'Neill 2005; Smith and Frehner 2010). A focus on the intersection of tribal governance and the idea of natural resources complicates how we think about Indigenous people's view of the environment and development and about Indigenous participation in capital-intensive industries and economies.

This national movement has led to a number of high-profile events, including the Occupation of Alcatraz in 1969 and the Standoff at Wounded Knee in 1973 (Smith and Warrior 1996). The Indigenous movement toward self-determination led to several changes in federal-Indian law. The Indian Self-Determination and Education Assistance Act of 1975 gave tribes new authorities over BIA institutions operating in reservations. A few years later, the Indian Mineral Development Act of 1982 gave tribes more decision-making power over resources. These new laws enhanced the capacity of tribes to enter into independent contracts with energy interests and extractive industries (Allison 2015; Wilkinson 2005). In combination with existing federal laws, these acts became the legal-political basis by which tribes exerted a new language of "sovereignty" over their lands and resources.

In the 1980s, independent researcher Philip Reno, working out of the Navajo Community College, showed that revenues from extractive industries dominated the tribal budget (Reno 1981). Between 1975 and 2005, four coal mines operated like clockwork, supplying the essential fuel for the region's power plants that were built around the reservation. The power generated from these plants supplied the energy needs for desert cities including Albuquerque, Phoenix, Los Angeles, Las Vegas, and Tucson. Diné coal supplied 90 percent of the power required to pump water from the Colorado River to Phoenix and Tucson. These relational dynamics of development were the consequence of centuries of colonialism (see also Purifoy, Chapter 10 this volume). Power, figuratively and literally, rested squarely in the region's utilities, both public and private, and fueled by tribal coal (Needham 2014; Ross 2011). By 2013, I verified that coal, oil, and land made up more than half of the tribe's non-federal revenues.

Tribal sovereignty was an ideological response to the limitations placed on Indigenous governing authorities and practices inherent in colonial regimes. In effect, tribal sovereignty became a negotiation between the colonizer and the colonized over the power to decide what happens on tribal lands. Consequently, Indigenous ideas of labor, development, and modernization are made consistent with colonial understandings and practices. Indigenous critics of tribal sovereignty claim it is an "inappropriate concept" (Barker 2005), and that there is a colonial "cunning" (Coulthard 2007; Povinelli 2002) in state recognition of tribal groups. Mohawk anthropologist Audra Simpson critically interrogates definitions of borders and boundaries that were imposed on Mohawk communities in Canada and the United States as a form of recognition politics (Simpson 2011, 2014a). Osage scholar Jean Dennison shows how Osage identities were shaped by oil and revenues depending on year they were enrolled into the nation (Dennison 2012). For Osage, some members are entitled to oil revenues and others are not dependent on when their ancestors were recorded as tribal members. For "resource rich" tribes, particularly in the US west, extraction, labor, and identity are closely linked (Allison 2015; Fixico 2012; Needham 2014; Smith and Frehner 2010). Choctaw anthropologist Valerie Lambert shows how ideas of nation-building, institutional evolution and ideas of "natural resource management"—particularly around water rights—contribute to new expressions of tribal identity and nationalism (Lambert 2007). Today Indigenous environmental organizers and activists are challenging notions of sovereignty that exploit the land and convert the earth into resources. But often they are articulating notions of sovereignty and sustainability that both build upon and defy the colonial ontology of resources (on mapping and Indigenous ontologies, see Bryan, Chapter 37 this volume).

Indigenous environmentalism

By the late 1970s and early 1980s, scholars challenged the kind of mineral leasing contracts the Department of Interior facilitated between tribes and energy interests (Dunbar-Ortiz 1979; Hall and Snipp 1988; Snipp 1988; Weiss 1984; White 1983; Wilkins 1993). Using the language of environmentalism, Indigenous activists and community members highlighted the disproportionate risk of industries to vulnerable populations (Jorgensen 1978; Robbins 1978). Indigenous activists critiqued the presence of extractive industries on tribal lands not only because of their unfavorable terms in leases, but also for the damage they caused to the local environment (Ambler 1990; Churchill and LaDuke 1986; Gedicks 1993; LaDuke 1992, 1999).

Indigenous environmentalism merged discourses of environmentalism with the cultural practices and knowledges unique to Indigenous communities. Activists and organizers challenged the Eurocentric foundations of tribal sovereignty, as nominal "democratic" regimes representing a "tribe" while negotiating lucrative energy contracts. Today Indigenous activists and organizers are challenging how tribal communities relate with the world around us.

Through an examination of the Navajo Green Jobs Act and emerging ideas of "transition," Diné organizers and activists contest core practices of the tribal government and its continued uses of "natural resources" for economic development. Diné historian Jennifer Denetdale showed that the abuse of "tradition," strongly associated with ideas of Navajo culture is reifying exploitative governance in the Navajo Nation (Denetdale 2006, 2009). Diné scholar Lloyd Lee has asked about the future of Navajo nationalism in today's restrictive practices of tribal governance (Lee 2007). Anthropologist Dana Powell and I interviewed Diné environmental activists in 2008 and identified distrust in tribal institutions as parts of a larger colonial-state apparatus. Such concerns echo Diné scholar Melanie Yazzie who describes extraction and resources on Diné lands as a form of "biopolitics" (Yazzie 2018). Instead they put forward the Diné notion of K'é, which speaks to the inherent relatedness Diné people have with each other and Diné lands (Powell and Curley 2008).

A troubled "transition" in the Navajo Nation

Diné environmental groups introduced the language of "transition" in the mid-2000s to challenge the hegemony of coal and other fossil fuels on the reservation. The work of Diné environmentalists changed the framework and approach of Indigenous environmental politics. It initiated thinking and planning around alternative development proposals. These proposals combined with political desires to reverse the impact of the Indian Reorganization Act of 1934 on Diné governance. As mentioned in the previous section, the IRA centralized governing power into the tribal government. Instead, alternative development proposals worked to decenter development discourse on the reservation.

Although transition appears as a break from previous practices around extraction, what remains consistent in these proposals is an understanding of land, water, air, and nonhuman relatives as "resources." This new discourse conforms with prevailing ideas of sustainable energy technology and development. Transition referred to the particular energy technologies used, *not* the colonial logics at work. It is an undergirding colonial logic that complicates how green is translated into the Diné political experience.

Consider the range of ideas and activities understood as "green" in Table 7.1. The emphasis on sheep and wool are meant to assist Diné people into returning to a preindustrial subsistence economy when natural resources were not the basis of tribal governance. It is a way of life that worked in relationship with the land, water, and animals based on notions of kinship, not exploitation. In the same list, exist ideas associated with non-Indigenous understandings of "green" that are fundamentally organized around the exploitation of natural resources, presumably in the expansion of modern states with laws, policies, and bureaucracies that are consistent with capitalist economies. Conceptually these ideas are very different from each other, but for environmental activists and organizers working though the real political terrain of tribal governance, both are necessary.

The 2009 Navajo Green Jobs Act—passed as a first version of a Green New Deal a decade before the Green New Deal became a national conversation—created policy and opportunity to develop alternative energy resources, focusing on the role of local communities to build these industries (Curley 2018). When the Act passed in 2009, organizers created an accompanying website that explained in simple terms its intent. The website is no longer active—another indication of the program's mortality. But it once highlighted the program's central ideals and projects.

Table 7.1 Navajo Green Jobs ideals and projects

Navajo Green Jobs ideals	Navajo Green Jobs projects
Fair wages, Navajo economic self-sufficiency, and a transition to a sustainable economy	Community renewable energy
Opportunity for families, chapters, and individuals to establish local green businesses	Green manufacturing, such as wool mills
More jobs for youths, students, veterans, fathers, and mothers to work close to home on the Navajo Nation	Energy efficiency projects, such as weatherizing homes and sustainable water projects
	Local business ventures, such as weavers' co-ops and green construction firms
	Traditional agriculture, such as farmers, markets, and community gardens
	Green Job training programs, such as workforce development, green contractors, and public service projects

Source: The Navajo Green Jobs website was at http://navajogreenjobs.com/, last accessed 2014. Alternatively, these same points were found in the accompanying Navajo Nation Council Resolution (0179-09).

Table 7.1, taken from the Navajo Green Jobs website, summarizes alternative energy initiatives in the Navajo Nation. For clarity, I organized the program's bullet points, once on the website, into two categories: one group that defined Navajo Green Jobs ideals and the second column dedicated to the kinds of projects it envisioned. The points and projects were an attempt to translate the Diné language into *green*. During the lead-up to passing the Navajo Green Jobs Acts in 2009, Diné environmental organizers talked about how environmental notions of "green" might reflect the unique social and political contexts of reservations. In July 2008, one organizer said to me, "I like jobs because that's what gets people interested [in the campaign] on the reservation. Talking about jobs and jobs creation" (personal notes, July 7, 2008). Another organizer said, "The need for job creation is really important. The [Navajo] Nation is going to say we are doing this and doing that, but we need to be more in line with traditional way of life in order to keep people on the land" (ibid). For the first time as a policy sanctioned by the tribal government, the end of the Navajo coal industry entered the discourse of Navajo politics.

Transition became a part of the Navajo Nation's energy policy in 2012. The following year, Navajo Nation President Ben Shelly made "transition" a major point in his presidential report to the Navajo Nation Council. The Navajo Nation Council mandated energy transition in the creation of the Navajo Transitional Energy Company, LLC (NTEC), a company designed to administer a 60-year-old coal mine on the eastern end of the reservation. According to NTEC's enabling legislation, the company was required to invest 10 percent of its profits into renewable and alternative energy research. The mandate was meant to move the Navajo Nation away from fossil fuels and toward renewable energy technologies.

Members of Diné environmental groups have promoted transition in the form of sustainable energy technologies and green jobs as a pragmatic alternative to the social and political challenges extractive industries have created. Diné activists are aware of the need for jobs in Diné communities and revenues for the tribal government. When Diné environmental organizers and activists first talked about transition in 2005, coal workers and tribal officials interpreted their proposals as a threat to the Navajo coal economy. With the rapid decline of coal over the last 15 years, these proposals are now the only alternative to coal in political discussion. Elected officials in the Navajo Nation grafted ideas of transition into Navajo energy policy. But in the background of all these developments is the larger notion of the environment as resource that has been thoroughly integrated via historical, colonial institutions of federal and tribal-federal relations and that sets the terms for development in the Navajo Nation and across reservation communities.

Conclusion

Indigenous peoples and communities have struggled for decades with exploitative understandings and practices around "natural resources." When we look critically at the ontological work of "resources" in shaping worlds, we see patterns of colonialism and intervention. When did "coal" become something to exploit and turn into money? What were the conditions of this transformation in perspective? For Indigenous peoples in the Americas, we see this transformative process play out again and again in similar ways. From the Spanish slave labor in Inca silver mines 500 years ago to oil pipelines snaking through Canada's homelands today, colonialism and a particular genre of "natural resource" use have gone hand-in-glove.

The logic of resource use is tied to this project. The land is valued in what it contains in untapped energies for export and profit over life and sustainability. The language of "green jobs" and "transitions" attempts to rectify exploitative relationships between tribal communities and capital-intensive extractive industries. It addresses two fundamental social changes wrought through decades of Diné participation in these industries: a need for working-class jobs and regular revenues for the tribal government. But these needs are only solved through a particular understanding of the environment that subjugates portions of it into categories of "natural resources" meant solely for human benefit above concern for long-term ecological sustainability and the needs of nonhuman relatives. Fundamentally, decolonization requires the rejection of "resources" as an operational concept of development and governance. We have to recenter Indigenous philosophies in ways that do not replicate crude essentializations and stereotypes. It is to take seriously what Diné scholar Melanie Yazzie called "radical relationality" (Yazzie and Baldy 2018).

Many Indigenous thinkers are trying to recover inherent philosophies, ways of understanding our relationship with the land and its other inhabitants. Using the term "resurgence," Leanne Simpson writes that the land is the key to recovering our relational systems of world understandings (Simpson 2014b). But rearticulating Indigenous traditions and philosophies to a non-Native audience is a challenge. For decades, non-Native media has stereotyped Indigenous peoples as primitive, premodern, and inherent environmentalists. Non-Native journalists often reduce complex Indigenous thinking into silly characterizations that portray Indigenous peoples talking to animals or crying over discarded trash. Such stereotyping is both racist and a reification of dominant understandings about the environment that see modern life as only possible through a utilitarian division of society from its wealth in "natural resources."

The premise of "resources" brings challenges to Indigenous peoples and ways of life. Calling something a resource contributes to how we think about and relate to it. Indigenous environmental researchers challenge the colonial practices of state resource managers in departments of forestry, fish and wildlife, and land. Whyte and Reo show how Ojibwe hunters in colonial Wisconsin continue to rely on "traditional ecological knowledge" (TEK) and moral codes when hunting and killing animals (Goldman, Nadasdy, and Turner 2011; Reo and Whyte 2012). TEK has been one way to account for different understandings of the natural world at work between Indigenous peoples and institutions built for the management of the state's resources (Nadasdy 1999; Reo 2011; Whyte 2013).

What this brief essay suggests is that "resources" are a violent project of world making for Indigenous peoples and always have been. But beyond that blunt critique, the idea of resources generates its own complicated understandings of work, livelihood, and identity on the ground and within communities, as was quickly shown in the example of Navajo Green Jobs and transition. Put differently, the idiom of "natural resources" simplifies the world into ways that are destructive and counterproductive for the people and the planet but necessary for profit. And impacted communities intuit and act upon the complicated and contradictory landscapes in tireless efforts to remain in place and create the conditions of survival for future generations. Decolonization requires the rejection of "resources" as an operational concept for both future development and governance practices. To be critical cannot assume the colonial ontology of "resources."

References

Allison, James Robert III. 2015. *Sovereignty for Survival: American Energy Development and Indian Self-Determination*. New Haven, CT: Yale University Press.

Ambler, Marjane. 1990. *Breaking the Iron Bonds: Indian Control of Energy Development*. Lawrence: University Press of Kansas.

Barker, Joanne. 2005. *Sovereignty Matters: Locations of Contestation and Possibility in Indigenous Struggles for Self-Determination*. Lincoln: University of Nebraska Press.

Brugge, Doug, Timothy Benally, and Esther Yazzie-Lewis, eds. 2006. *The Navajo People and Uranium Mining*. Albuquerque, NM: University of New Mexico Press.

Carroll, Clint. 2014. "Native Enclosures: Tribal National Parks and the Progressive Politics of Environmental Stewardship in Indian Country." *Geoforum* 53: 31–40.

Chamberlain, Kathleen. 2000. *Under Sacred Ground: A History of Navajo Oil, 1922–1982*. Albuquerque, NM: University of New Mexico Press.

Churchill, Ward, and Winona LaDuke. 1986. "Native America: The Political Economy of Radioactive Colonialism." *Critical Sociology* 13 (3): 51–78.

Corntassel, Jeff, and Richard C. Witmer II. 2008. *Forced Federalism: Contemporary Challenges to Indigenous Nationhood*. Norman: University of Oklahoma Press.

Coulthard, Glen. 2007. "Subjects of Empire: Indigenous Peoples and the 'Olitics of Recognition' in Canada." *Contemporary Political Theory* 6: 437–460.

Coulthard, Glen. 2014. *Red Skin, White Masks: Rejecting the Colonial Politics of Recognition*. Minneapolis, MN: University of Minnesota Press.

Curley, Andrew. 2018. "A Failed Green Future: Navajo Green Jobs and Energy 'Transition' in the Navajo Nation." *Geoforum* 88: 57–65.

Curley, Andrew. 2019. "T'áá hwó ají t'éego and the Moral Economy of Navajo Coal Workers." *Annals of the American Association of Geographers* 109 (1): 71–86.

Daigle, Michelle. 2018. "Resurging Through Kishiichiwan: The Spatial Politics of Indigenous Water Relations." *Decolonization: Indigeneity, Education, & Society* 7 (1): 159–172.

Deer, Sarah. 2009. "Decolonizing Rape Law: A Native Feminist Synthesis of Safety and Sovereignty." *Wicazo Sa Review* 24 (2): 149–167.

De la Cadena, Marisol, and Orin Starn, eds. 2007. *Indigenous Experience Today*. Oxford: Berg.

Denetdale, Jennifer Nez. 2006. "Chairmen, Presidents, and Princesses: The Navajo Nation, Gender, and the Politics of Tradition." *Wicazo Sa Review* 21 (1): 9–28.

Denetdale, Jennifer Nez. 2009. "Securing Navajo National Boundaries: War, Patriotism, Tradition, and the Diné Marriage Act of 2005." *Wicazo Sa Review* 24 (2): 131–148.

Dennison, Jean. 2012. *Colonial Entanglement: Constituting a Twenty-First Century Osage Nation*. Chapel Hill, NC: University of North Carolina Press.

Dennison, Jean. 2017. "Entangled Sovereignties: The Osage Nation's Interconnections with Governmental and Corporate Authorities." *American Ethnologist* 44 (4): 684–696.

Duarte, Marisa Elena. 2017. *Network Sovereignty: Building the Internet Across Indian Country*. Seattle, WA: University of Washington Press.

Roxanne, Dunbar-Ortiz, ed. 1979. *Economic Development in American Indian Reservations*. Albuquerque, NM: Native American Studies, University of New Mexico Press.

Dunbar-Ortiz, Roxanne. 2014. *An Indigenous Peoples' History of the United States*. Boston, MA: Beacon Press.

Eichstaedt, Peter H. 1994. *If You Poison Us: Uranium and Native Americans*. Santa Fe, NM: Red Crane Books.

Estes, Nick. 2016. "Fighting for Our Lives: #NODAPL in Historical Context." The Red Nation, September 18.

Fixico, Donald Lee. 2012. *The Invasion of Indian Country in the Twentieth Century: American Capitalism and Tribal Natural Resources*. 2nd ed. Boulder, CO: University Press of Colorado.

Gedicks, Al. 1993. *The New Resource Wars: Native and Environmental Struggles Against Multinational Corporations*. Boston, MA: South End Press.

Geisler, Charles, R. Green, D. Usner, and P. West, eds. 1982. *Indian SIA: The Social Impact Assessment of Rapid Resource Development on Native Peoples*. Monograph No. 4. Ann Arbor, MI: University Michigan National Resources Sociology Resource Lab.

Goeman, Mishuana R., and Jennifer Nez Denetdale. 2009. "Native Feminisms: Legacies, Interventions, and Indigenous Sovereignties." *Wicazo Sa Review* 24 (2): 9–13.

Goldman, Mara J., Paul Nadasdy, and Mathew D Turner, eds. 2011. *Knowing Nature: Conversations at the Intersection of Political Ecology and Science Studies*. Chicago, IL: University of Chicago Press.

Hall, Thomas D. and C.M. Snipp. 1988. "Patterns of Native American Incorporation into State Societies." In *Public Policy Impacts on American Indian Economic Development*, edited by C. Matthew Snipp, 23–38. Albuquerque, NM: University of New Mexico Press.

Hosmer, Brian C., Colleen O'Neill, and Donald Lee Fixico. 2004. *Native Pathways: American Indian Culture and Economic Development in the Twentieth Century*. Boulder, CO: University Press of Colorado.

Huber, Matt. 2019. "Resource Geography II: What Makes Resources Political?" *Progress in Human Geography* 43 (3): 553–564.

Hunt, Sarah Elizabeth. 2014a. "Ontologies of Indigeneity: The Politics of Embodying a Concept." *Cultural Geographies* 21 (1): 27–32.

Hunt, Sarah Elizabeth. 2014b. "*Witnessing the Colonialscape: Lighting the Intimate Fires of Indigenous Legal Pluralism*." PhD Diss., Simon Fraser University.

Iverson, Peter, and Monty Roessel. 2002. *Diné: A History of the Navajos*. Albuquerque, NM: University of New Mexico Press.

Jorgensen, Joseph G., ed. 1978. *Native Americans and Energy Development*. Boston, MA: Anthropology Resource Center.

LaDuke, Winona. 1992. "Indigenous Environmental Perspectives: A North American Primer." *Akwe:kon Journal* 9 (2): 52–71.

LaDuke, Winona. 1999. *All Our Relations: Native Struggles for Land and Life*. Boston, MA: South End Press.

Lambert, Valerie. 2007. *Choctaw Nation: A Story of American Indian Resurgence*. Lincoln, NE: University of Nebraska Press.

Lee, Lloyd L. 2007. "The Future of Navajo Nationalism." *Wicazo Sa Review* 22 (1): 53–68.

Lewis, Courtney. 2019. *Sovereign Entrepreneurs: Cherokee Small-Business Owners and the Making of Economic Sovereignty*. Chapel Hill, NC: University of North Carolina Press.

Miller, Robert J. 2010. *Discovering Indigenous Lands: The Doctrine of Discovery in the English Colonies*. Oxford: Oxford University Press.

Nadasdy, Paul. 1999. "The Politics of TEK: Power and the 'Integration' of Knowledge." *Arctic Anthropology* 36 (1/2): 1–18.

Nadasdy, Paul. 2017. *Sovereignty's Entailments: First Nation State Formation in the Yukon*. Toronto: University of Toronto Press.

Needham, Andrew. 2014. *Power Lines: Phoenix and the Making of the Modern Southwest*. Princeton, NJ: Princeton University Press.

O'Neill, Colleen. 2005. *Working the Navajo Way: Labor and Culture in the Twentieth Century*. Lawrence, KS: University Press of Kansas.

Povinelli, Elizabeth A. 2002. *The Cunning of Recognition: Indigenous Alterities and the Making of Australian Multiculturalism*. Durham, NC: Duke University Press.

Powell, Dana E. 2017. *Landscapes of Power: Politics of Energy in the Navajo Nation*. Durham, NC: Duke University Press.

Powell, Dana E., and Andrew Curley. 2008. "K'e, Hozhó, and Non-Governmental Politics on the Navajo Nation: Ontologies of Difference Manifest in Environmental Activism." *Anthropological Quarterly* 81: 17–58.

Reno, Philip. 1981. *Mother Earth, Father Sky, and Economic Development: Navajo Resources and Their Use*. Albuquerque, NM: University of New Mexico Press.

Reo, Nicholas James. 2011. "The Importance of Belief Systems in Traditional Ecological Knowledge Initiatives." *International Indigenous Policy Journal* 2 (4): 8.

Reo, Nicholas James, and Kyle Powys Whyte. 2012. "Hunting and Morality as Elements of Traditional Ecological Knowledge." *Human Ecology* 40: 15–27.

Robbins, Lyn A. 1978. "Energy Developments and the Navajo Nation." In *Native Americans and Energy Development*, edited by Joseph G. Jorgensen, 35–48. Boston, MA: Anthropology Resource Center.

Ross, Andrew. 2011. *Bird on Fire: Lessons From the World's Least Sustainable City*. Oxford: Oxford University Press.

Sherry, John Williams. 2002. *Land, Wind, and Hard Words: A Story of Navajo Activism*. Albuquerque, NM: University of New Mexico Press.

Simpson, Audra. 2011. "Settlement's Secret." *Cultural Anthropology* 26 (2): 205–217.

Simpson, Audra. 2014a. *Mohawk Interruptus: Political Life Across the Borders of Settler States*. Durham, NC: Duke University Press.

Simpson, Leanne Betasamosake. 2014b. "Land as Pedagogy: Nishnaabeg Intelligence and Rebellious Transformation." *Decolonization: Indigeneity, Education & Society* 3 (3): 1–25.

Smith, Paul Chaat, and Robert A. Warrior. 1996. *Like a Hurricane: The Indian Movement from Alcatraz to Wounded Knee*. New York, NY: The New Press.

Smith, Sherry L., and Brian Frehner, eds. 2010. *Indians & Energy: Exploitation and Opportunity in the American Southwest*. Santa Fe, NM: School for Advanced Research Press.

Snipp, C. Matthew. 1988. *Public Policy Impacts on American Indian Economic Development*. Albuquerque, NM: Native American Studies, Institute for Native American Development, University of New Mexico.

Spice, Anne. 2018. "Fighting Invasive Infrastructures: Indigenous Relations Against Pipelines." *Environment and Society* 9 (1): 40–56.

Todd, Zoe. 2016. "An Indigenous Feminist's Take on the Ontological Turn: 'Ontology' Is Just Another Word for Colonialism." *Journal of Historical Sociology* 29 (1): 4–22.

Voggesser, Garrit. 2010. "The Evolution of Federal Energy Policy for Tribal Lands and the Renewable Energy Future." In *Indians & Energy: Exploitation and Opportunity in the American Southwest*, edited by Sherry L. Smith, and Brian Frehner, 55–88. Santa Fe, NM: SAR Press.

Voyles, Traci Brynne. 2015. *Wastelanding: Legacies of Uranium Mining in Navajo Country*. Minneapolis, MN: University of Minnesota Press.

Weiss, Lawrence D. 1984. *The Development of Capitalism in the Navajo Nation: A Political-Economic History*. Minneapolis, MN: MEP Publications.

White, Richard. 1983. *The Roots of Dependency: Subsistence, Environment, and Social Change Among the Choctaws, Pawnees, and Navajos*. Lincoln, NE: University of Nebraska Press.

Whyte, Kyle Powys. 2017. "The Dakota Access Pipeline, Environmental Injustice, and US Colonialism." *Red Ink: An International Journal of Indigenous Literature, Arts & Humanities* 19 (1): 154–169.

Whyte, Kyle Powys. 2013. "On the Role of Traditional Ecological Knowledge as a Collaborative Concept: A Philosophical Study." *Ecological Processes* 2: 7.

Wilkins, David E. 1993. "Modernization, Colonialism, Dependency: How Appropriate Are These Models for Providing an Explanation of North American Indian 'Underdevelopment?'" *Ethnic and Racial Studies* 16 (3): 390–419.

Wilkins, David E. 2013. *The Navajo Political Experience*. Lanham, MD: Rowman & Littlefield.

Wilkinson, Charles F. 2005. *Blood Struggle: The Rise of Modern Indian Nations*. New York, NY: W. W. Norton & Company.

Wilson, Nicole J. 2019. "'Seeing Water Like a State?': Indigenous Water Governance Through Yukon First Nation Self-Government Agreements." *Geoforum* 104: 101–113.

Wood, Patricia Burke, and David A. Rossiter. 2017. "The Politics of Refusal: Aboriginal Sovereignty and the Northern Gateway Pipeline." *The Canadian Geographer/Le Géographe canadien* 61 (2): 165–177.

Yazzie, Melanie K. 2018. "Decolonizing Development in Diné Bikeyah: Resource Extraction, Anti-Capitalism, and Relational Futures." *Environment and Society* 9 (1): 25–39.

Yazzie, Melanie K., and Cutcha Risling Baldy. 2018. "Introduction: Indigenous Peoples and the Politics of Water." *Decolonization: Indigeneity, Education & Society* 7 (1): 1–18.

Section II
(Un)knowing resource systems

8

Resistance against the land grab

Defensoras and embodied precarity in Latin America

Sharlene Mollett

Introduction

On March 14, 2018, Marielle Franco was assassinated. Franco was a Rio de Janeiro city councilor, born and raised in the low income, informal neighborhood of Mare, among the largest of Rio's *favelas*. Mare is an important place of urban restructuring. Like many *favelas* and informal settlements across the globe, Mare is targeted by the government's thirst for foreign direct investment as part of an urban imaginary. Such an imaginary (dis)places Mare on a transformative elite urban pathway from informality and gang violence to a space of formal urban renewal and elite real estate development, seemingly without the poor (Phillips 2018). Franco, an Afro-Brazilian feminist, lesbian woman, unsettled a history of a predominately white, mestizo, and heteronormative city council. As a councilor, and years before, her advocacy included public denouncements of police brutality, anti-black racism, and genocide, and more recently, Franco helped to organize land-registration programs for land- and housing-insecure *favela* residents. Franco's multipronged advocacy and heighted visibility aligned with NGO efforts to make *favelas*, like Mare, legible on official maps of Rio (The Economist 2019). As a councilor, Franco participated in a commission responsible for investigating militias (paramilitary gangs) and their violent attempts to control urban land as part of extralegal real estate schemes worth millions. Franco's defense of disenfranchised urban residents challenged the pseudo-authority of the militias and ultimately cost her, her life. Indeed, in 2019, eight people were arrested in connection to Franco's murder. Among them were former and active members of the military, the police, and a city councilor, who together were involved in "over 80 cases of land grabbing fraud in regions dominated by paramilitary groups ... *where Marielle and her advisers were helping in the development of land regularization programs*" (Nogueira 2019, emphasis mine; Mollett 2020).

Latin America is a "dangerous" place to be a land defender. In fact, more than 50 percent of all the world's murders perpetrated against land defenders take place in Latin America (Global Witness 2018). The work of land defenders grows increasingly perilous as states incorporate and expand extractive capitalist regimes as part of national development plans, even in the context of "sustainable" development (Global Witness 2018; Mollett 2018). In addition, many Latin American states criminalize social movements and public dissent through legislative measures limiting the rights and the freedoms of those who seek to defend land, natural resources, and ultimately the lives of their communities. For many Latin American states, both defenders and their communities are often imagined as "in the way" of development, an imaginary that materializes in the perilous environments in which they are compelled to resist land grabbing (Gies 2017; Global Witness 2018). In this chapter, I reflect on a major theme in critical resource geography, land grabbing, and question the way this literature centers public and large-scale resource grabs as part of "rational" state development policies. Indeed, it is true that recent interventions in critical resource geographies illustrate the "unintended" and "extra-economic" consequences burdened onto local communities as a direct result of extractive development ventures within Indigenous, *campesino*, and Black communities (Hall et al. 2015; Berman-Arévalo 2019). However, I challenge prevailing tendencies within the literature that elide small-scale grabs, colonial forms of land grabbing, and the plural forms of power imbued in land-grabbing processes. I seek to make clear that grabbing-induced-land displacement is not a universal experience among "the poor," nor is it *simply* about land.

A "matrix of power" imbued in land-grabbing processes shape both access to and control over land and bodies (Collins 2000; Mollett 2016; see also Fent, Chapter 9 this volume). Patriarchal power -shaping gender relations are co-constituted with racial and carnal logics that together shape the lived experiences of *Defensoras* and their communities throughout the region. To begin, I explain how the prevailing working definition of "large-scale land grabbing," as a novel phenomenon, is questionable and obscures micro land grabs. Second, I place in conversation a fusion of decolonial, postcolonial, and black feminist thinking with feminist political ecology (FPE) to extend recent analyses of gender, land, and resource grabbing, in a way that moves beyond simply how women and men experience dispossession *differently* (Mollett and Faria 2013; see also Chung 2020). To do so, I briefly reflect upon and entwine two important conceptual insights—"postcolonial intersectionality" (Mollett and Faria 2013) and "corporeal-spatial precarity" (Cordis 2019). These critical feminist conceptualizations together are useful in laying bare how "gender does not act alone" in struggles over land and natural resources (Mollett and Faria 2013, 123). Moreover, "postcolonial intersectionality" and "corporeal-spatial precarity" attend to the ways patriarchy, indigeneity, and blackness and their entanglements, take on symbolic and embodied meanings as part of the *longue durée* of coloniality embedded in contemporary extractive development processes veiled under "sustainable development" schemes and the like (Mollett 2017; Cordis 2019; Zaragocin 2019). Third, to illustrate these conceptual discussions, I highlight the struggles of Garifuna *Defensoras* in Honduras who, like the late Mareille Franco, lead resistance against the violence of resource grabs and the accompanying disruptions to lives and livelihoods in the name of tourism development. Finally, I argue that intersectional forms of power inform land grabbing in symbolic and material ways and reveal how land grabs, for Indigenous and Afro-descendant communities, consist of more than land and territorial loss, but rather actualize grave disruptions to land *and* body entanglements.

In search of a more expansive understanding of land grabbing: gender and the land grab

Land grabbing, according to Borras and Franco (2013, 1725) is defined as

> the capturing of control of relatively vast tracts of land and other national resources through a variety of mechanisms and forms, carried out through extra-economic coercion that involves large-scale capital, which often shifts resource use orientation into *extraction*, whether for international and domestic purposes

Often touted by international development organizations and state governments as an effective poverty reduction strategy, there is overwhelming evidence that land grabbing efforts, on their own and in combination with other kinds of resource grabs, are "extractive." Thus, rather than ameliorate poverty, impoverishment is entrenched through dispossession (Hall et al. 2015). Land grabbing disrupts people's relations with land and resources and the means by which subsistence production and access to other resources, like water and trees, are possible (Cardenas 2012; Mollett 2016). Land grabbing, however, is about more than rural resources and access to them.

The scholarship on land grabbing, more recently, increasingly attends to "a multiplicity of smaller land deals and the involvement of a large variety of actors" (Verma 2014; Steel et al. 2017, 133). Such a shift opens space to link rural grabbing with urban land grabs such as militia land fraud in Rio (mentioned earlier). Accordingly, a focus on multiple and smaller land transactions (legal or extralegal) highlights "more fragmented, gradual and therefore less visibly-outstanding" examples of land grabbing (Steel et al. 2017, 133). In similar ways, while patriarchal power operates at all levels, a more grounded definition and micro-level foci are well suited for understanding the gender dynamics of land and land loss within households and across communities, urban and rural alike. Feminist insights on land grabbing offer robust understandings of the complexity of power in land-grabbing processes. Indeed, recent scholarship shows how patriarchal power is imbued in land grabbing and concomitant dispossession (Verma 2014; Lamb et al. 2017; Chung 2020). As Behrman, Meinzen-Dick, Quisumbing note (2012, 73), "the available evidence thus far indicates that large-scale land deals have tended to overlook the rights, needs and interests of women [vis-à-vis men] and as a result tended to aggravate gender inequalities in affected communities." Attention to gender is a welcome contribution to understandings of land grabbing. Still, I join a small group of critical feminists who bolster critiques of extractive development, including land grabbing, by attending to how gender is co-constituted with other kinds of power and status imbued in extractive regimes that grab land and other resources customarily held by Indigenous, Black, and informal communities, such as *favelas* (Perry 2013; Hernandez 2019; Zaragocin 2019).

Furthermore, attention to the micropolitics of land grabbing discloses its embodied materialities. For instance, with a focus on food security in Ghana, Nyantakyi-Frimpong and Bezner Kerr (2017) disclose how gender and customary factors differentiate households and show that household members do not absorb the shock of land loss in uniform ways. In Ghana, women's rights to land are increasingly less secure with the expansion of mining operations. Such rights, fastened through marriage and their roles as mothers and daughters, are now insecure. Similarly, in their study of Maasai women in Tanzania, Goldman, Davis, and Little (2016, 782) question the neoliberal logic of land registration programs that assume with private property rights for women comes protection against land grabbing. Rather, they argue that access is curated through the intersectional, subjective positioning of pastoral Maasai women in Tanzania. Indeed, what matters more is the "structural and relational access to mechanisms" which informs how women secure access and control over land *whether or not they are in possession of a private ownership title* (Goldman et al. 2016, 782). As Goldman et al. show, this knowledge comes with more micro-level inquiry. Indeed, Perry's salient ethnography of black women community leaders, as defenders, lays bare how gender, race, and carnal ideologies craft particular and pejorative representations of black women in Brazil and are entangled with their fight to defend against land dispossession and home demolitions in the historical urban neighborhood of Gamboa de Baixo, Salvador, Brazil (Perry 2013). Together, such insights reveal that the entanglement of gender, race, sexuality, marriage, custom, and colonial history illuminates how land grabbing is a deeply *colonial* and not a "novel" phenomenon. Thus, a focus on place histories and how they endure in the present help punctuate critical geographic discussions on land grabbing and concomitant dispossession (Verma 2014; see also Mollett 2016; Safransky 2019, Purifoy, Chapter 10 this volume). Moreover, as feminist inquiry notes, an overstated focus on large-scale land grabs, particularly as new, occurs "at the expense of acknowledging gendered micro-political grabs unabated over long periods of time" and elides the loss of Indigenous and Afro-descendant customarily held lands that are not "legible" because of state refusals and delays to formalization (Verma 2014, 53; Mollett 2016). When aggregated together, such intersectionally gendered and embodied differences shape dispossession in significant ways.

Defensoras en peligro: "postcolonial intersectionality" and "corporeal-spatial precarity"

In Latin America, land defenders live with everyday reminders that their advocacy on behalf of their communities comes with wide-scale material and symbolic violence to their bodies, including murder (see also Berman-Arevalo, Chapter 17 this volume). This violence is part of an economic development model that foments extreme inequality and an erosion of human rights across communities. *Defensoras,* in particular, live in peril as their public work violates gender and cultural norms that often relegate Indigenous and Afro-descendant women to the margins. The decision to denounce the pillage and appropriation of customary resources intensifies state and elite campaigns that target *Defensoras* for stigmatization, repression, and a climate of hostility (IM-Defensoras 2015). Such violence and threats are not simply because *Defensoras* are working against extractive development practices in their communities. Moreover, they become targets because they are Indigenous and Black women, imagined through dehumanizing colonial narratives long employed by states and elites to justify Indigenous and Black dispossession, a racialized, gendered, and carnal violence perpetrated on land and body.

I conceptualize the predicament of *Defensoras* in Latin America through a fusion of decolonial, postcolonial, and black feminist thinking in conversation with insights from FPE as one of many critical geographic approaches to resources (on FPE, see also Fent, Chapter 9 this volume). I draw from the fused insights to strengthen feminist political-ecological engagements with bodies and embodiments through the concept of "postcolonial intersectionality" (Mollett and Faria 2013, Mollett 2017). Namely, "postcolonial intersectionality acknowledges the way patriarchy and racialized processes are consistently bound in a postcolonial genealogy that embeds race and gender ideologies within nation-building and international development processes" (Mollett and Faria 2013, 120). Furthermore, postcolonial intersectional thinking acknowledges how land and territorial struggles, imbued with racial, patriarchal, and carnal ideologies, travel through *the long durée* of colonial thinking, shaping widespread extractivist violence to lands and bodies since the conquest, and conditioning the contemporary lives of *Defensoras* (Hernandez Reyes 2019; Mollett 2017).

Postcolonial intersectionality also draws insight from the framework of Maria Lugones's "coloniality of gender" (Lugones 2007). This framework demands that we think from an embodied experience paying particular attention to how gender also has a history that serves to dehumanize Indigenous and Black women while feminizing and elevating white women as belonging to humanity (see McClintock 1995; Lugones 2007). Thus, woven into colonial expansion and modern forms of land accumulation on behalf of white European elites are a set of presuppositions that disavow Indigenous and Black humanity and continue as justification for the territorial dispossession and embodied disposability of Indigenous and Black communities, and particularly women, in the Americas (McKittrick 2013; Saldaña-Portillo 2016; Simpson 2016; King 2019). Such thinking is useful to counter the scarce acknowledgment in land-grabbing scholarship of *ongoing* forms of colonial power and embodied violence (see

Verma 2014). As a result, *Defensoras* live in what Shanya Cordis refers to as a "corporeal-spatial precarity" (2019, 22). Such a spatiality is marked by the way the state's adoption of global sustainable development agendas aims to advance economic and environmental development policies as benign, even while they bolster environmental destruction and deepen poverty and social dislocation. Such degradation disproportionately and violently imposes upon impoverished Indigenous and Black peoples forced to sacrifice in the face of extractive appropriation of their lands, resources, territories, and too often their bodies (Mollett 2017; Cordis 2019). Together, postcolonial intersectionality and corporeal-spatial precarity help articulate how the work of *Defensoras* is not simply about protecting resources, but about defending Indigenous and Black bodies against violence and securing their *human* rights to customary and legally held resources as sanctioned under international law (ILO 1989; United Nations 2007). In such a landscape, their precarity is constitutive of extractive regimes couched in the myth of "sustainable" development. I briefly illustrate with the case of Honduras.

Garifuna *Defensoras* in Honduras

> In Honduras, like in the rest of Latin America and the Caribbean, women are in the frontline when it comes to fighting for our rights, against racial discrimination and to defend our environment and survival. We don't just fight with our own bodies; we also provide strength, our ideas and our proposals. We don't just give birth to children, but also ideas and actions
>
> (Miriam Miranda, *Defensora*, Honduras, Oxfam 2020)

One of the most perilous countries for defenders is Honduras. Since the 2009 military coup and the exile of democratically elected President Mel Zelaya, the erosion of human rights, a lack of human security and widespread violence, including femicide, ravaging both rural and urban environments, punctuates everyday life for ordinary Hondurans. In many cases, this violence is linked to the notion that Honduras is "Open for Business," a mantra initiated by postcoup President Porfirio Lobo and continued by current present Juan Orlando Hernandez. This pro-business initiative advanced a series of business-friendly legislation couched in the language of sustainable development that sought to enhance the country's industrial production and competitiveness (Loperena 2017). In May 2011, the Honduran Ministry of Foreign affairs organized a conference dubbed, "Honduras is Open for Business" (HOB). Invited were key regional and international executives and pro-business leaders offering global investors unprecedented opportunities for capitalist growth, a move that strengthened financial support from the United States to Honduras and helped to improve the country's crumbling economy and fragile legitimacy after the coup (Shipley 2016; Loperena 2017). After 2011, foreign direct investment in mining and tourism expanded deep within the territorial homelands of the country's Indigenous and Afro-descendant peoples.

Later that same year, OFRANEH (Black Fraternal Organization of Honduras) organized the Forum on Land Grabbing in Africa and Latin America (*Foro sobre el acaparamiento de tierras y territorios en África y América Latina*). This meeting brought together leaders from Garifuna communities across the north coast who opposed the extractivist directions of national development plans (PROAH 2011). From the meeting, a final declaration challenged the state's HOB mantra and development policy. An excerpt from this declaration reads:

> We reject megaprojects, such as hydroelectric dams, the REDD+, oil exploration concessions, enclave tourism, monoculture, mining projects and all that leads to the displacement

of the communities and their inhabitants. We demand once and for all that Honduras creates the conditions necessary to respond to the basic needs of the majority of the population, which lives under a regimen of exploitation and exclusion

(quoted in Loperena 2017)

Within this declaration, delegates not only aligned tourism development with more traditional extractive industries such as mining and African Palm production, but also disclosed the ways violence runs through such extractive practices. In particular, the Forum made clear how the Garifuna were targeted for erasure, as tourism development, in the words of a Garifuna woman from Triunfo de la Cruz, "is violating our rights as human beings, as [Indigenous] peoples, and as women" (quoted in Trucchi 2011). Despite ongoing resistance and a multiethnic coalition protesting state complicity in the violent development practices that grab resources from Indigenous and Afro-descendant territories, extractive kinds of development remain at the heart of national foreign direct investment strategies (often framed as a poverty reduction strategy) in Honduras. In fact, in 2018, the Honduran state renewed a ministerial agreement that seals the content of environmental permits awarded to companies and extraction corporations for five years (United Nations 2019). Moreover, "information on the type and location of approved concessions or projects… [are] not available to the public" (United Nations 2019). Such an agreement contradicts Honduras's international obligations to seek consent from communities regarding extractive development on their lands and territories and lays bare the disregard for local people in the search for foreign profits (ILO 1989). Such pro-extraction development priorities shape embodied precarity for *Defensoras* and their communities.

The Honduran north coast

In the fall of 2019, the Inter-American Commission on Human Rights outlined the precarious conditions facing the lives of Garifuna *Defensoras*. The Special Rapporteur writes:

> *Defensoras* "are in a situation of multiple risks due to their gender, ethno-racial origins and work…we are concerned that the attacks against Garifuna and Indigenous leaders seek to prevent their activities to defend human rights, particularly in the contexts of megaproject development in the country."
> (Inter-American Commission on Human Rights, Visit to Honduras, 2019)

Since the military coup in 2009, incidences of violence against women have soared, marking attention to the growing rates of femicide in Honduras. According to Jokela-Pansini (2020), femicide rates grew by 250 percent between 2010 and 2014. The rates of femicide and accompanying impunity for these crimes also reflect prevailing state patriarchal violence that informs a "corporeal-spatial precarity" facing *Defensoras*. Indeed, in September 2019, four *Defensoras* were murdered because of their work in leading community resistance against evictions from coastal villages for tourism and land appropriation for African Palm monocropping on the north coast. *Defensoras* occupy a central role in their communities' resistance against extraction. However, this is not just about mobilizing against land grabbing. In fact, *Defensoras* seek to bolster democracy on behalf of social movements (Ardon and Flores 2017). As such, the space of *Defensora* resistance moves beyond questions of land displacement and dispossession. Rather, their work to protect the land includes protecting bodies on the land.

On the Atlantic coast of Honduras, the region of Tela Bay—the customary homelands of the Garifuna peoples—is a site of tourism-induced land grabbing. Since 2011, Garifuna land evictions intensified with the clearing of the coast for large-scale resort tourism enclaves and residential tourism (Mollett 2014; Loperena 2017). In this context, many Garifuna report maltreatment and exploitation. On the beaches of Tela Bay, Garifuna women are frequently harassed and insulted by private security guards and male tourists while they walk on the beach from their villages to sell goods, braid hair for tourists, or simply to walk to and from the commercial markets located in the center of Tela. When women and young girls (en)counter racist insults or refuse sexual advances, guards reportedly threaten to report them for trespassing (even while beaches are officially public land). Some women do reluctantly engage in sexual intercourse with security guards and tourists for very little sums of money, but it is also reported that women and girls who decline advances risk rape and physical abuse (Mollett 2020). Both the occurrence and threats of sexual violence on the beaches of Tela Bay exemplify how critical feminist approaches in conversation with FPE punctuate a need to center land and body in critical geographic approaches to land grabbing. The work of *Defensoras* not only includes making public demands for formal land rights and protections from foreign investors and the state. Included also are the collective refusals of the everyday carnal practices of hotel operators and their male proxies (either guards or tourists) who not only occupy Garifuna customary lands, but also mobilize colonial scripts that imagine Garifuna women and girls as sexually available and presumably "disposable" (Razack 2016, 286).

Publicly contesting the everyday harassment of Garifuna women and girls has risks for *Defensoras* as well. Filing public complaints and police reports about rape and sexual harassment is often met with impunity. Moreover, too frequently their complaints serve as fuel for a retaliation campaign by the municipality (at the behest of tourism operators and hotel administration) threatening *Defensoras* with legal sanctions. Even without making complaints, *Defensoras* endure public condemnation as "bad mothers" and "promiscuous wives" (IM-*Defensoras* 2015, 40) and imagined as "out of place" in public battles over land grabbing and extraction (Mollett 2018; 2020).

In Honduras, the widespread criminalization of dissent confirms that Indigenous and Afro-descendant *Defensoras* and their communities are expected to sacrifice on behalf of the nation's extractive development plans. In addition, costs to defend against libel means deepened poverty and financial insecurity for *Defensoras* and their families. Furthermore, criminal records and detention, even when charged with minor infractions, permeate community relations and undermine the reputations of leaders even when the claims against them are fictitious (IM-*Defensoras* 2015). Lastly, this carnal violence is never simply just discursive, even when dismissed by municipal police as the "harmless flirting" of "bored" security guards and "adventurous" male tourists (Mollett fieldnotes 2013). As the murder of Garifuna *Defensora* Belkis Garcia illustrates, carnal violence is imbued in the violent tapestry of land grabbing, as her murdered body demonstrated "signs of sexual violence" (IACHR 2019).

The risks for *Defensoras* continue to escalate in Honduras. Since 2018, the new criminal code (Article 590) criminalizes social movement organizations and social dissent as the actions of terrorism. Whether an organization is determined as a terrorist organization is at the discretion of a judge who under the law can order a jail sentence for up to 20 years (IACHR 2019). For social movement organizations like OFRANEH, with the mission to fight for Garifuna protection and control over territorial homelands, their freedom is precarious (Mollett 2018).

Conclusion

In this chapter, I place in conversation the concepts of "postcolonial intersectionality" and "corporeal-spatial precarity" to illustrate how land grabbing, whether urban or rural, is an embodied process mutually grounded in the concrete and imagined sacrifices of Black and Indigenous women's bodies in Latin America. In drawing insights from critical feminist thinking, I argue that centering land and body in feminist political ecological analyses of land grabbing make clear an enduring practice of conquest that simultaneously shapes land and bodies (Simpson 2016; King 2019). Land grabbing by foreign investors and national elites in Honduras is justified through prevailing stereotypes that engender Black and Indigenous women (and girls) as sexually available. Such imaginaries are not new but form part of a colonial legacy of racial-patriarchal power that dehumanizes Black and Indigenous women and communities on the Atlantic coast (Morgan 2004). Similar to the carnal desires of European male explorers and travel writers, the extractive development policies of the state, on behalf of international foreign elites, condition and spatialize precarity. For *Defensoras* and their communities, "violence against their bodies and extraction-led degradation and dispossession of lands and territories share the same process" (Mollett 2018, 182). Indeed, key for critical geographic approaches to resources, I demonstrate how land is not the only resource being grabbed in land-grabbing processes. Indeed bodies, too, are grabbed—a practice justified through racial, gendered, and carnal ideologies, which have long served as symbolic resources driving colonial incursions on Indigenous and Afro-descendant territories in the Americas. As critical feminist scholars assert, conquest fuses on the bodies of Indigenous and Black women (Simpson 2016; King 2019). Thus, the intersectional logics shaping the ways in which *Defensoras* and those whom they protect, live with "corporeal-spatial precarity" (Cordis 2019), a condition that demands centering land and body relations in critical resource geographies.

References

Ardón, P. and Flores, D., 2017. Berta Lives: COPINH Continues. *SUR-Int'l J. on Hum Rts.*, *25*, p.109.

Borras Jr, S.M. and Franco, J.C., 2012. Global land grabbing and trajectories of agrarian change: A preliminary analysis. *Journal of agrarian change*, *12*(1), pp.34–59.

Behrman, Julia, Ruth Meinzen-Dick, and Agnes Quisumbing. 2012. "The Gender Implications of Large-Scale Land Deals." *Journal of Peasant Studies* 39 (1): 49–79.

Berman-Arévalo, Eloisa. 2019. "El 'fracaso ruinoso' de la reforma agraria en clave de negridad: comunidades afrocampesinas y reconocimiento liberal en Montes de María, Colombia." *Memorias: Revista Digital de Historia y Arqueología desde el Caribe Colombiano* (37, enero–abril): 117–149.

Cárdenas, Roosebelinda. 2012. Green multiculturalism: articulations of ethnic and environmental politics in a Colombian 'black community'. *Journal of Peasant Studies*, *39*(2), pp.309–333.

Chung, Youjin B.. 2020. "Governing a Liminal Land Deal: The Biopolitics and Necropolitics of Gender." *Antipode* 52 (3): 722–741.

Collins, Patricia H. 2000. *Black Feminist Thought*. New York: Routledge

Collins, P.H., 2015. Intersectionality's definitional dilemmas. *Annual review of sociology*, *41*, pp.1–20.

Cordis, Shanya. 2019. "Forging Relational Difference: Racial Gendered Violence and Dispossession in Guyana." *Small Axe: A Caribbean Journal of Criticism* 23 (3 (60)): 18–33.

Crenshaw, K., 1989. Demarginalizing the intersection of race and sex: A black feminist critique of antidiscrimination doctrine, feminist theory and antiracist politics. *u. Chi. Legal f.*, p.139.

Doshi, S., 2017. Embodied urban political ecology: Five propositions. *Area*, *49*(1), pp.125–128.

Gies, Heather. 2017. "Honduras, the Deadliest Country in the World for Environmental Defenders, Is about to Get Deadlier." *Upside Down World*, September 29. http://upsidedownworld.org/archives/honduras/honduras-deadliest-country-world-environmental-defenders-get-deadlier/

Global Witness 2018. At what cost? Irresponsible business and the murder of land and environmental defenders in 2017. *London: Global Witness.* https://www.globalwitness.org/en/campaigns/environmental-activists/at-what-cost/

Goldman, Mara J., Alicia Davis, and Jani Little. 2016. "Controlling Land They Call Their Own: Access and Women's Empowerment in Northern Tanzania." *The Journal of Peasant Studies* 43 (4): 777–797.

Hall, Ruth, Marc Edelman, Saturnino M. Borras Jr, Ian Scoones, Ben White, and Wendy Wolford. 2015. "Resistance, Acquiescence or Incorporation? An Introduction to Land Grabbing and Political Reactions 'From Below'." *Journal of Peasant Studies* 42 (3–4): 467–488.

Hernández Reyes, C.E., 2019. Black women's struggles against extractivism, land dispossession, and marginalization in Colombia. *Latin American Perspectives*, 46(2), pp.217–234.

IACHR. 2019. "IACHR Condemns the Prevalence of Murders and Other Forms of Violence against Garifuna Women in Honduras," press release no. 238/19, September 24. https://www.oas.org/en/iachr/media_center/PReleases/2019/238.asp

ILO, International Labour Organization. 1989. Convention 169, Indigenous and Tribal Peoples Convention. https://www.ilo.org/global/topics/indigenous-tribal/lang--en/index.htm

IM-Defensoras. 2015. *Violence Against Women Human Rights Defenders in MesoAmerica 2012–2014 Report.* https://im-defensoras.org/2015/12/violence-against-women-human-rights-defenders-in-mesoamerica-2012-2014-report/

Jokela-Pansini, Maaret. 2020. "Complicating Notions of Violence: An Embodied View of Violence Against Women in Honduras." *Environment and Planning C: Politics and Space* 38 (5): 848–865. doi: 10.1177/2399654420906833.

King, Tiffany Lethabo. 2019. *The Black Shoals: Offshore Formations of Black and Native Studies.* Durham, NC: Duke University Press.

Lamb, Vanessa, Laura Schoenberger, Carl Middleton, and Borin Un. 2017. "Gendered Eviction, Protest and Recovery: A Feminist Political Ecology Engagement With Land Grabbing in Rural Cambodia." *The Journal of Peasant Studies* 44 (6): 1215–1234.

Loperena, C.A., 2017. Honduras is open for business: extractivist tourism as sustainable development in the wake of disaster?. *Journal of Sustainable Tourism*, 25(5), pp.618–633.

Lugones, María. 2007. "Heterosexualism and the Colonial/Modern Gender System." *Hypatia* 22 (1): 186–219.

McClintock, A., 1995. *Imperial leather: race, gender and sexuality in the colonial contest* Routledge. New York.

McKittrick, Katherine. 2013. "Plantation Futures." *Small Axe: A Caribbean Journal of Criticism* 17 (3 (42)): 1–15.

Mollett, Sharlene. 2014. "A Modern Paradise: Garifuna Land, Labor, and Displacement-in-Place." *Latin American Perspectives* 41 (6): 27–45.

Mollett, Sharlene. 2016. "The Power to Plunder: Rethinking Land Grabbing in Latin America." *Antipode* 48 (2): 412–432.

Mollett, Sharlene. 2017. "Irreconcilable Differences? A Postcolonial Intersectional Reading of Gender, Development and Human Rights in Latin America." *Gender, Place & Culture* 24 (1): 1–17.

Mollett, Sharlene. 2018. "Embodied Histories of Land Struggle in Central America." In "Historical Geographies of, and for, the Present," Co-Authored by Levi Van Sant, Elizabeth Hennessy, Mona Domosh, Mohammed Rafi Arefin, Nathan McClintock, and Sharlene Mollett. 2020." *Progress in Human Geography* 44 (1): 168–188.

Mollett, S., 2020. Hemispheric, relational, and intersectional political ecologies of race: Centring land-body entanglements in the Americas. *Antipode*.1-21, https://doi.org/10.1111/anti.12696

Mollett, Sharlene, and Caroline Faria. 2013. "Messing With Gender in Feminist Political Ecology." *Geoforum* 45: 116–125.

Morgan, Jennifer. 2004. *Laboring Women: Reproduction and Gender in New World Slavery.* Philadelphia: University of Pennsylvania Press.

Nogueira, Gabriela. 2019. "Brazilian Crime Story: The Assassination of Marielle Franco." *Medium*, March 14. https://medium.com/@resistbrasil.scot/brazilian-crime-story-the-assassination-of-marielle-franco-d30fa56b2b88.

Nyantakyi-Frimpong, Hanson, and Rachel Bezner Kerr. 2017. "Land Grabbing, Social Differentiation, Intensified Migration and Food Security in Northern Ghana." *The Journal of Peasant Studies* 44 (2): 421–444.

Oxfam. 2020. "Women Defenders of the Land and Environment: Silenced Voices." *Oxfam America.* https://www.oxfam.org/en/women-defenders-land-and-environment-silenced-voices.

Perry, Keisha-Khan Y. 2013. *Black Women Against the Land Grab: The Fight for Racial Justice in Brazil.* Minneapolis, MN: University of Minnesota Press.

Phillips, Dom. 2018. "Marielle Franco: Brazil's Favelas Mourn the Death of a Champion." *The Guardian*, March 17. https://www.theguardian.com/world/2018/mar/18/marielle-franco-brazil-favelas-mourn-death-champion#maincontent.

PROAH. 2011. "De cumbres y contra-cumbres: el Foro sobre Acaparamiento de Territorio en América Latina y África." *Proah* (blog). August 22. https://proah.wordpress.com/2011/08/22/de-cumbres-y-contra-cumbres-el-foro-sobre-acaparamiento-de-territorio-en-america-latina-y-africa/.

Razack, Sherene H. 2016. "Gendering Disposability." *Canadian Journal of Women and the Law* 28 (2): 285–307.

Safransky, Sara. 2018. Land justice as a historical diagnostic: Thinking with Detroit. *Annals of the American Association of Geographers*, *108*(2), pp.499–512.

Saldaña-Portillo, Maria Josefia. 2016. *Indian Given: Racial Geographies Across Mexico and the United States.* Durham, NC: Duke University Press.

Shipley, Tyler. 2016. "Enclosing the Commons in Honduras." *American Journal of Economics and Sociology* 75 (2): 456–487.

Simpson, Audra. 2016. "The State is a Man: Theresa Spence, Loretta Saunders and the Gender of Settler Sovereignty." *Theory & Event* 19 (4).

Steel, G., van Noorloos, F. and Klaufus, C., 2017. The urban land debate in the global South: New avenues for research. *Geoforum*, *83*, pp.133–141.

Sultana, Farhana. 2011. "Suffering for Water, Suffering from Water: Emotional Geographies of Resource Access, Control and Conflict." *Geoforum* 42 (2): 163–172.

The Economist. "Mapping the Maré: A New Census Shows How a Brazilian Favela Really Works." *The Economist.* May 30, 2019. https://www.economist.com/the-americas/2019/05/30/a-new-census-shows-how-a-brazilian-favela-really-works.

Trucchi, Giorgio. 2011. "Perder nuestra tierra es perder nuestra madre y vida." http://www.albasud.org/noticia/es/218/perder-nuestra-tierra-es-perder-nuestra-madre-y-nuestra-vida.

United Nations. 2007. *United Nations Declaration on the Rights of Indigenous Peoples (UNDRIP).* https://www.un.org/development/desa/indigenouspeoples/declaration-on-the-rights-of-indigenous-peoples.html.

United Nations. 2019. *Report of the Special Rapporteur on the Situation of Human Rights Offenders.* https://reliefweb.int/report/honduras/visit-honduras-report-special-rapporteur-situation-human-rights-defenders.

Verma, Ritu. 2014. "Land Grabs, Power, and Gender in East and Southern Africa: So, What's New?" *Feminist Economics* 20 (1): 52–75.

Zaragocin, Sofia. 2019. "Gendered Geographies of Elimination: Decolonial Feminist Geographies in Latin American Settler Contexts." *Antipode* 51 (1): 373–392.

9

Gender in extractive industry

Toward a feminist critical resource geography of mining and hydrocarbons

Ashley Fent

Introduction

How do resource-making and resource extraction shape gendered relations of access, use, and control? This chapter traces this question through the traditional resource geography literature, where gender has been largely absent, and the more recent turn toward critical resource geography. Although gender relations and gendered effects have been analyzed in studies of land and biotic resources (Carney 1993; Rocheleau and Edmunds 1997; Schroeder 1999), as well as water (Birkenholtz 2013; Harris 2009; Sultana 2011), gender remains tangential to much of the critical resource geography literature that focuses on mining and hydrocarbons. Is this gap an epistemological flaw in how critical resource geographers have theorized and studied resource extraction? Or does it reflect how these types of resource extraction operate?

Resource extraction has been gendered over time, through various policies, ideologies, and actions that have played out differently across geographic space and social positions. As the result of these processes, certain resources, forms of extraction, and sites of value creation have been dominated almost exclusively by men and have been associated with cultural notions of masculinity. This has been naturalized and normalized as an "unmarked" category, such that gender does not *appear* to be central to an analysis of these resources or forms of resource extraction, even as it is in many ways constitutive of this empirical reality. The gendering of resource worlds both reflects and reproduces heteronormative binaries among valued and unvalued natures, bodies, and forms of labor power. This in turn influences epistemological approaches to understanding mining and hydrocarbon extraction; analyzing the role of gender requires greater attention to gaps, silences, and absences, as well as interrogation of how these have been historically produced and how extractive spaces are connected to other spaces, such as the home.

In seeking to elucidate how gender operates in and through mining and hydrocarbon extraction, I focus on four themes: (1) property, enclosure, and dispossession; (2) labor relations; (3) environmental effects and social reproduction; and (4) gender roles in activism. In exploring each, this chapter traces avenues for critical resource geography research into the role of gender in the resource worlds of minerals and hydrocarbons.

Feminist theory and gender in traditional and critical resource geographies

Scholarship has long suggested that resources are brought into being through interactions between physical matter, human societies, and technological innovation (Ackerman 1958; De Gregori 1987; Zimmermann 1933). What has become known as "traditional" resource geography focused on cataloging countries' or regions' energy, land, and water resources while offering prescriptions for how these could be managed, conserved, and used for economic growth. This work did not address gender as a relevant category of analysis. Instead, it theorized a generalized humanity's relationship to resources. When social differentiation was addressed, it took the form of national-level distinctions, civilizational or technological hierarchies, or essentialist ideas about linkages between race, climate, and culture, as they pertained to resource use (e.g., in Zimmermann 1933, 138–140).

After the early 1990s, resource geography largely disappeared as a formal subfield. The study of resources continued in two separate directions: resource management and political ecology (Bakker and Bridge 2006). The former focused on economic, political, and legal institutions and conflicts around the rational use of resources (Wescoat 1991, 1992). While engaging more with social theory than earlier work, the resource management literature did not substantially incorporate insights from feminist theory or gender analysis.

By contrast, political ecology includes a rich tradition of feminist analyses of gendered patterns of resource access and use. Access refers to bundles of means, relations, and processes through which different people gain benefits from a natural resource (Ribot and Peluso 2003). Feminist political ecology (FPE) has shown that capitalist development, combined with patriarchal institutions, often limits women's access to critical resources (Carney 1993; Rocheleau, Thomas-Slayter, and Wangari 1996; Schroeder 1999). More recent work in FPE has embraced embodiment, performativity, nonbinary understandings of gender, and post-structural and postcolonial feminisms (Elmhirst 2011; see also Mollett, Chapter 8 this volume); this work advances the understanding of gender as a process through which subjectivities and differences are produced, altered, and made to matter in certain contexts (Nightingale 2006, 166). While the insights of this scholarship appear in critical work on biotic resources and water, it is less explicitly engaged in critical resource geography literatures on minerals and hydrocarbons, with some notable exceptions (e.g., Jenkins 2014; Lahiri-Dutt 2011).

Critical resource geography is nonetheless well-positioned to address questions of gender vis-à-vis extractive industries because of its attention to materiality, its analysis of regulation and institutions, and its rootedness in political ecology and critical political economy. It aims to denaturalize "resources" and trace how their material consequences emerge through institutions and networks of actors (Bakker and Bridge 2006), and it compels attention to geographies of power, including sociopolitical struggles (Bridge and Jonas 2002). These intellectual preoccupations resonate with feminist theory's interest in demonstrating the social and political constructedness of gender and the operation of gendered power relations in particular contexts. Some of the critical resource geography literature on mining and hydrocarbons briefly addresses issues of social reproduction in the home (Huber 2013), gendered artisanal labor forces (Marston and Perreault 2017), disproportionate impacts of environmental degradation on women (Perreault 2013), and the involvement and marginalization of women in mining negotiations and activism (Jenkins 2017; Perreault 2013). Centering gender in resource extraction enriches critical resource geography's elucidation of the sociopolitical processes through which resources are made, the institutionalization of certain socio-natural relations (and the delegitimization or erasure of others), and the constitutive outsides of extractive capitalism.

In the sections that follow, I suggest four themes that are ripe for greater engagement between critical geography research on mining and hydrocarbons and broader work on gender in relation to extractive industry, land, and biotic resources; these themes are also highly applicable to critical resource geography research on other kinds of resources. These include examination of:

- How land appropriation for extraction alters gendered patterns of resource access and use.
- The production of gendered labor regimes, including historicizing the masculinization of underground work, its dependence on the institution of the heteronormative family, and its role in the invisibilization of women's direct and indirect roles in the mining sector.
- Gendered and embodied effects of environmental consequences from extractive industries on social reproduction.
- Gendered distinctions in approaches to activism around extractive industry.

Gender and extractive industry

Property, enclosure, and dispossession

Critical resource geographers' engagements with institutions, materiality, and conflicts over resource access compel stronger examination of the gendered effects of dispossession from mining and hydrocarbon extraction. Even before ground has been broken, processes of identifying and isolating valuable resources for industrial extraction may compete with or negate existing forms of access and use, as well as local knowledges. At the frontiers of capitalist production, access regimes are often gendered, such as through women's ability to collect tree products through overlapping and seasonal forms of tenure (Rocheleau and Edmunds 1997, 1351). Legal, cultural, and political economic institutions transform these complex socio-natural worlds into land and resources for capitalist investment (Braun 2000; Li 2014; Tsing 2005). In critical resource geographers' attention to these processes, a focus on gendered access illuminates the socially differentiated effects of exclusionary resource rights.

Enclosure of land and other resources—understood by Marx as a form of "primitive accumulation" and by subsequent Marxian scholars as an ongoing process of "accumulation by dispossession"—simultaneously establishes exclusive property rights over land and produces a supply of "free" laborers. Property is the form of access that is privileged within Western European and American thought and law, based on Lockean views of labor acting upon nature to produce value in land; reflecting Eurocentric biases in understandings of whose labor and what kinds of labor count, private property rights over land and resources have been historically denied to certain people based on class, race, gender, and nationality (Scott 2010).

The commodification of nature as property often negates or reshapes gendered patterns of access, control, and use of fuel, food, water, and non-timber forest resources (Holden, Nadeau, and Jacobson 2011; Rocheleau and Edmunds 1997). For example, colonial regimes in Africa imposed Victorian understandings of both gender and race that eroded culturally specific forms of women's access to resources under precolonial and customary forms of tenure (Amadiume 1987; Tsikata 2003). Within gendered labor dichotomies that uphold (certain) men's rights to land ownership and labor, the social reproduction of the labor force often falls to women, through what has been termed the feminization of subsistence (Polier 1996, 10). In these circumstances, women in particular experience the effects of dispossession on their subsistence activities and the work of social reproduction (Cohen 2014; Fernandez 2018).

Critical resource geographers' interests in dispossession and enclosure also align with concerns about how contractual terms of exclusionary resource control in extractive sectors affect broader gender relations. In cases where extractive projects involve land appropriation, men may receive monetary compensation or jobs in exchange for land concessions, while women often are not compensated for this lost access (Bose 2004; Scheyvens and Lagisa 1998). The establishment of exclusive property rights over land and resources may ultimately erode women's social and economic status and position relative to men (Bose 2004; Lahiri-Dutt 2011; compare with Polier 1996), even where land is family-owned and inherited matrilineally (Scheyvens and Lagisa 1998).

Labor relations

Critical resource geography is also well-positioned to address the gendering of labor forces in extractive industry, given existing work on cultural and political institutions and engagements with labor theory and labor organizing. For example, critical resource geographers have examined how oil has become central to cultural ideals of domesticity, home, and freedom, and how these associations allow for the penetration of capital into daily life outside the workplace (Huber 2013). This work could engage more with the relationships between resources, gender, and labor by addressing how historical, political economic, and cultural shifts in resource extraction have produced the underground (and, increasingly, the "offshore" [Appel 2012]) as a space of hypermasculinized male labor, while at the same time producing feminized spaces (such as the household) that subsidize and are increasingly subsumed by capitalism.

Historical and political processes produced the subsurface as a gendered, racialized, and classed space. Although women play important (if often overlooked) roles in artisanal and small-scale mining (ASM) labor forces, their exclusion and marginalization within industrial mining has been bound up with understandings of the normative working body and the association of certain commodities with capitalist masculinities. These processes are institutionally enforced and enacted through laws and policies; however, they are also cultural, as exemplified by the development of taboos against women in underground mines and the performances of masculinities through everyday workplace interactions in the mining, oil, and natural gas industries (Burke 2006; Miller 2004; Scott 2010).

In many precapitalist mines, women and men worked together through familial labor arrangements (Burke 2006). As mining industrialized, women were an attractive labor force, because they could be paid less, were more reliable, and were considered more docile. But with the development of shaft-sinking technology and accompanying decreases in labor demand, women were increasingly excluded from underground labor in order to be "protected" from dangerous working conditions (Lahiri-Dutt 2008; Lahiri-Dutt and Macintyre 2006). These labor distinctions produced and reproduced gender as an axis of inclusion or exclusion and hinged upon the heteronormative family unit.

A number of policies and ideologies have institutionalized this gendering of the underground. Historically, Victorian ideas of domesticity and "women's work" influenced mining policies both within Western Europe and colonized countries. In Britain, women were excluded from underground work in 1842, as part of campaigns against female and child labor (Burke 2006). Additional efforts to eradicate women's exploitation in underground labor—particularly in the global South—were initiated by the International Labour Organization (ILO) in the 1920s with the intention of "protecting" impoverished and Indigenous women from dangerous mining labor in "backward" and colonized areas (Zimmermann 2018). In 1935, the ILO passed the Underground Work (Women) Convention, which barred women from working in "any undertaking ... for the extraction of any substance from under the surface of the earth" (International Labour Organization [ILO] 1935, Article 2). Reflecting class distinctions, exceptions were made for women in management positions, temporary research studies, and "nonmanual" labor.

As a result, the underground became a masculine space, and forms of manual and mechanized labor became increasingly associated with men. This was not inevitable; in opposition to legislation that targeted women, others advocated the general improvement of labor conditions in mines for *both* women and men (Zimmermann 2018, 239–240). However, the approach of the ILO and of labor regulations passed by various nation-states around the world focused on the inappropriateness of women's bodies for underground work—although bans on underground labor were frequently suspended during conditions of labor scarcity or wartime needs (Sone 2006). At the same time, hazardous working conditions became naturalized as inherent properties of mining, rather than as the result of capitalist imperatives.

The gendering of the underground also relied on broader cultural ideologies about work and the household. Under industrial capitalist forms of resource extraction, legitimate laboring bodies are gendered as male, with women's roles being positioned as in the domestic sphere (Tallichet 2006). This dichotomy depends on and reproduces the heteronormative institution of the household. Although life-sustaining activities in the home are excluded from wage considerations, they are nevertheless instrumental to capitalist production in mining operations.

Because of this gendered division of labor, which positions men as primary income earners and women as supplementary, women's labor is cheapened and rendered dispensable. For example, although affirmative action and nondiscrimination policies in the United States in the 1970s allowed women to work underground (Lahiri-Dutt and Macintyre 2006; Tallichet 2006), women engaging in manual mining labor are often paid lower wages than men, are employed on more casual and temporary contracts, and receive few benefits. These patterns are further mediated by race, class, and education levels, with women geoscientists in the lucrative oil and gas industry having very different experiences of gender inequality than women with blue-collar mining jobs (Williams, Kilanski, and Muller 2014).

ASM also demonstrates gender differentiation. As a whole, the sector is largely unregulated and informal; within this, women's labor is often mobilized through family labor arrangements, which may carry over from agrarian contexts (Marston 2020). Women's jobs are concentrated in lower-value industrial minerals and in the least lucrative opportunities in higher value minerals, such as sifting through tailings for small gems (Hinton, Veiga, and Beinhoff 2003a). In some cases, however, women may participate in cooperatives and gain more representation, including at the national scale (Marston 2020). Many women in ASM are from Indigenous or low-caste communities, and they work in dangerous jobs, such as processing minerals with heavy chemicals (Lahiri-Dutt 2008). Because there is consistent downward pressure on labor costs, many operators have moved toward flexibilization and casualization—such as women doing hazardous processing work within the home (Lahiri-Dutt 2003, 2008). As operations expand, industrialize, and mechanize, women often lose their jobs; amid downsizing and reduced labor demands, the remaining jobs tend to go to men, due to cultural associations between men and machines and common understandings of men as legitimate wage-earners (Burke 2006; Graulau 2006; Hinton, Veiga, and Beinhoff 2003a; Lahiri-Dutt 2003). Women who have been racially, economically, or socially marginalized also engage in care work, cooking, or sex work in mining camps, engaging in the labor of social reproduction (Graulau 2006).

Engagement with feminist theories of capitalism—and specifically with often-unseen and devalued forms of care work and familial labor relations—holds the potential to build on critical resource geographers' attention to relationships between nature and value (Huber 2018; see also Naslund and McKeithen, Chapter 6 this volume) and to labor arrangements, such as mining cooperatives and unions (Marston and Perreault 2017). Similarly, critical resource geography's insights into how labor regimes are impacted by and impact wider political economies adds to the broader field of gender and extraction a recognition of how wider political economic processes at different scales impact the enactment and spatialization of gendered labor in and around extractive industry.

Environmental effects and social reproduction

Gendered experiences of embodiment in and around extractive industry—and the ability to seek compensation or visibility for these different forms of exposure—resonate with a materialist approach to resource geography, which examines "how bodies are enrolled in the construction of social places and spaces and political economic processes" (Bakker and Bridge 2006, 15). Because of the structural and symbolic gendering of spaces and activities, the environmental impacts of resource extraction are often borne by women—particularly rural, low-income, and Indigenous women. These effects are embodied, through increased time and physical toil to gain access to resources, and through gendered differences in exposure to toxicity.

One of the most important and contested aspects of resource extraction is its impacts on water resources and quality (Bebbington and Williams 2008). Within the gendered division of labor that has arisen around the mining industry, the contamination of water often adversely impacts women's abilities to provide food for their families, yet they may have little control over decision-making or water users' associations (Li 2009, 2015). Further, finding clean sources of water often involves greater travel time and labor burdens for women (Perreault 2013, 1061). Pollution, deforestation, and other ecological effects of resource extraction may also erode women's access to important tree, plant, and fish species used for household needs, economic activity, and medicinal purposes (Scheyvens and Lagisa 1998). This highlights the ways that corporate- and state-led extraction of mineral and hydrocarbon resources produces gender-specific burdens within socionatural worlds.

Critical resource geographers are well positioned to further address how the materialities of resource extraction are differentially embodied, based on these socio-historical configurations of households, work, and subsistence. The naturalization of mining work as "dangerous," discussed in the previous section, creates gendered occupational hazards. For example, the dangers of industrial coal mining have become part of cultural narratives that associate certain masculinities with hard, manual labor and with acute and chronic physical injuries (Scott 2010). Additionally, small-scale miners face embodied occupational health hazards linked to environmental toxicity. Women in ASM who work in processing are disproportionately exposed to chemicals like cyanide and mercury and to higher risks of respiratory diseases (Hinton, Veiga, and Beinhoff 2003b). These experiences of toxicity are further mediated by women's intersectional positions within the society and household, through which gender discrimination combines with racial, class-based, ethnic, and other forms of discrimination to produce different degrees of risk exposure and different levels of concern or apathy.

Gender roles in activism

Within the critical resource geography literature, scholars highlight contestation and conflict around resources and suggest that "potential conflicts are often negotiated through historically and geographically specific sociopolitical struggles that become codified as the institutions and social practices within which resource extraction activities are embedded" (Bridge and Jonas 2002, 759–760). This emphasis resonates with understandings of how gendered forms of exclusion and inclusion are institutionalized in decision-making around resource extraction as well as the material and social effects of these practices.

Gender norms influence how various identities and tactics are mobilized through activism and the extent to which different modes of engagement are recognized as politically significant. Although the gendered division of labor within industrial mining formally excluded women from movements around class solidarity and labor organizing (Lahiri-Dutt 2006), women have nevertheless played important roles in various types of activism around extractive industry.

Traditional gender roles are both upheld and disrupted through activism. For example, while women have long been political agents in sustaining strikes and opposing pit closure, they have often done so by embracing their heteronormative roles as wives and mothers (Aulette and Mills 1988; Beckwith 1996; Kingsolver 1989; Smith 2015). In addition to these movements aimed at perpetuating environmentally destructive but economically important forms of extraction, women around the world have also been at the forefront of opposition to extractive projects and land expropriation, likewise drawing on their identities as mothers and often also on their marginalization as Indigenous or peasant women (Jenkins 2017; Mollett 2015). In Appalachia in particular, women have been central in organizing against mountaintop removal, in response to the collapse of unions, the hollowing out of communities, and the degradation of local environments by earlier periods of coal mining (Bell and Braun 2010; Smith 2015).

In addition to activism, women have frequently played important roles in mining negotiations and consultations, although their contributions are often overlooked in the course of the discussions and in scholarly analysis (O'Faircheallaigh 2013; Rondón 2009). The micropolitics of public consultations may marginalize women, as when other speakers interrupt or otherwise silence them; these practices reflect particular biases against rural or Indigenous women with low levels of education (Perreault 2015, 443). As such, women's experiences of inclusion and their degrees of influence in the consultation and negotiation process are variegated geographically and socially.

Furthermore, exclusive attention to large-scale public movements and formal negotiations overlooks many strategies and tactics through which women have resisted resource extraction within the broader cultural, political, and social contexts that produce and sustain gendered forms of public engagement. In some cases, women opposing mining and other forms of resource appropriation have formed their own groups and meeting spaces (Jenkins 2017). Others have engaged in everyday forms of resilience and confrontation, by continuing to work the land, asserting their rights to stay in place, refusing gifts and bribes, embracing ambivalent subjectivities, and engaging in culturally significant practices that may not be recognized as "activism" in liberal humanist conceptions (Jenkins 2017; Jenkins and Rondón 2015; Méndez 2018; Mollett 2015, 2017; Nightingale 2006). In taking up calls for greater engagement with women's roles in activism around resource extraction (Jenkins 2014, 2017), critical resource geography scholarship has a central role to play in moving beyond the taken-for-granted invisibility of women in these movements and investigating diverse spaces and forms of political action around resources.

Conclusion

In this chapter, I have suggested that gendered practices, ideologies, and policies are central to resource extraction, even as gender remains largely under-theorized within the critical resource geography literature on mining and hydrocarbons. In large part, these gaps reflect the way that extractive industry has been structured in practice, rendering certain groups of women marginal and invisible while also producing masculinities that are precariously tied to hazardous forms of manual labor—jobs that are in steady decline in the increasingly capital-intensive mining and hydrocarbons industries. This simultaneously effaces the importance of gender analysis and reproduces normalized gender categories and binaries. Manifestations of gender in resource extraction are not homogeneous but are rather intersectional and culturally and geographically specific.

Greater theorization and analysis of gender is poised to add to critical resource geography in a number of ways. The field's attention to materiality would be enriched by attention to socially differentiated forms of exposure to toxic materials, and how displacement or enclosure from extractive projects—as well as from newer investments motivated by carbon offsets and other financialized approaches to climate change mitigation—impacts women's lives, labor, and access to other resources. In this respect, critical resource geography is also well positioned to draw upon FPE insights into how household and familial labor relations are altered by increased incorporation into capitalist systems (Carney 1993) by showing how these relations may be deployed or altered within extractive regimes (e.g., Marston 2020). Further, these considerations could be taken up to reconstruct value theory, as applied to the study of resources (Huber 2018). Additionally, examinations of diverse and gendered political actions—and how women's voices may be included or excluded through institutional and cultural norms—would broaden understandings of why conflicts and contestations arise around extraction and how resistance is articulated and practiced.

Gendered analysis is essential to critical resource geography, complementing its commitments to materiality, embodiment, institutions, and the deconstruction of taken-for-granted categories, objects, and resources. These commitments demonstrate fruitful synergies with past and present work in intersectional feminist theory and could be productively engaged to push the broader field of gender and extraction to consider theories of value, resource materialities, and the role of political economic institutions.

References

Ackerman, Edward Augustus. 1958. *Geography as a Fundamental Research Discipline*. Department of Geography Research Paper No. 53. Chicago, IL: University of Chicago Department of Geography.

Amadiume, Ifi. 1987. *Male Daughters, Female Husbands: Gender and Sex in an African Society*. London: Zed Books.

Appel, Hannah. 2012. "Offshore Work: Oil, Modularity, and the How of Capitalism in Equatorial Guinea." *American Ethnologist* 39 (4): 692–709.

Aulette, Judy, and Trudy Mills. 1988. "Something Old, Something New: Auxiliary Work in the 1983-1986 Copper Strike." *Feminist Studies* 14 (2): 251–268.

Bakker, Karen, and Gavin Bridge. 2006. "Material Worlds? Resource Geographies and the 'Matter of Nature." *Progress in Human Geography* 30 (1): 5–27.

Bebbington, Anthony, and Mark Williams. 2008. "Water and Mining Conflicts in Peru." *Mountain Research and Development* 28 (3/4): 190–195.

Beckwith, Linda. 1996. "Lancashire Women against Pit Closures: Women's Standing in a Men's Movement." *Signs* 21 (4): 1034–1068.

Bell, Shannon Elizabeth, and Yvonne A. Braun. 2010. "Coal, Identity, and the Gendering of Environmental Justice Activism in Central Appalachia." *Gender and Society* 24 (6): 794–813.

Birkenholtz, Trevor. 2013. "'On the Network, off the Map': Developing Intervillage and Intragender Differentiation in Rural Water Supply." *Environment and Planning D: Society and Space* 31 (2): 354–371.

Bose, Sharmistha. 2004. "Positioning Women within the Environmental Justice Framework: A Case from the Mining Sector." *Gender, Technology and Development* 8 (3): 407–412.

Braun, Bruce. 2000. "Producing Vertical Territory: Geology and Governmentality in Late Victorian Canada." *Ecumene* 7 (1): 7–46.

Bridge, Gavin, and Andrew E. G. Jonas. 2002. "Governing Nature: The Reregulation of Resource Access, Production, and Consumption." *Environment and Planning A* 34: 759–766.

Burke, Gill. 2006. "Women Miners: Here and There, Now and Then." In *Women Miners in Developing Countries: Pit Women and Others*, edited by Kuntala Lahiri-Dutt, and Martha Macintyre, 25–50. Hants, UK: Ashgate.

Carney, Judith. 1993. "Converting the Wetlands, Engendering the Environment: The Intersection of Gender with Agrarian Change in the Gambia." *Economic Geography* 69 (4): 329–348.

Cohen, Roseann. 2014. "Extractive Desires: The Moral Control of Female Sexuality at Colombia's Gold Mining Frontier." *The Journal of Latin American and Caribbean Anthropology* 19 (2): 260–279.

De Gregori, Thomas R. 1987. "Resources Are Not; They Become: An Institutional Theory." *Journal of Economic Issues* 21 (3): 1241–1263.

Elmhirst, Rebecca. 2011. "Introducing New Feminist Political Ecologies." *Geoforum* 42 (2): 129–132.

Fernandez, Bina. 2018. "Dispossession and the Depletion of Social Reproduction." *Antipode* 50 (1): 142–163.

Graulau, Jeannette. 2006. "Gendered Labour in Peripheral Tropical Frontiers: Women, Mining and Capital Accumulation in Post-Development Amazonia." In *Women Miners in Developing Countries: Pit Women and Others*, edited by Kuntala Lahiri-Dutt, and Martha Macintyre, 289–306. Hants, UK: Ashgate.

Harris, Leila M. 2009. "Gender and Emergent Water Governance: Comparative Overview of Neoliberalized Natures and Gender Dimensions of Privatization, Devolution and Marketization." *Gender, Place & Culture* 16 (4): 387–408.

Hinton, Jennifer J., Marcello M. Veiga, and Christian Beinhoff. 2003a. "Women and Artisanal Mining: Gender Roles and the Road Ahead." In *The Socio-Economic Impacts of Artisanal and Small-Scale Mining in Developing Countries*, edited by Gavin M. Hilson, 149–188. Lisse, The Netherlands: Swets & Zeitlinger.

Hinton, Jennifer J., Marcello M. Veiga, and Christian Beinhoff. 2003b. "Women, Mercury and Artisanal Gold Mining : Risk Communication and Mitigation." *Journal de Physique IV* 107 (May): 617–620.

Holden, William, Kathleen Nadeau, and R. Daniel Jacobson. 2011. "Exemplifying Accumulation by Dispossession: Mining and Indigenous Peoples in the Philippines." *Geografiska Annaler: Series B, Human Geography* 93 (2): 141–161.

Huber, Matthew T. 2013. *Lifeblood: Oil, Freedom, and the Forces of Capital*. Minneapolis, MN: University of Minnesota Press.

Huber, Matt. 2018. "Resource Geographies I: Valuing Nature (or Not)." *Progress in Human Geography* 42 (1): 148–159.

International Labour Organization (ILO). 1935. Underground Work (Women) Convention. www.ilo.org/dyn/normlex/en/f?p=NORMLEXPUB:12100:0::NO::P12100_ILO_CODE:C045.

Jenkins, Katy. 2014. "Women, Mining and Development: An Emerging Research Agenda." *The Extractive Industries and Society* 1 (2): 329–339.

Jenkins, Katy. 2017. "Women Anti-Mining Activists' Narratives of Everyday Resistance in the Andes: Staying Put and Carrying on in Peru and Ecuador." *Gender, Place & Culture* 24 (10): 1441–1459.

Jenkins, Katy, and Glevys Rondón. 2015. "'Eventually the Mine Will Come': Women Anti-Mining Activists' Everyday Resilience in Opposing Resource Extraction in the Andes." *Gender & Development* 23 (3): 415–431.

Kingsolver, Barbara. 1989. *Holding the Line: Women in the Great Arizona Mine Strike of 1983*. Ithaca, NY: ILR Press.

Lahiri-Dutt, Kuntala. 2003. "Not a Small Job: Stone Quarrying and Women Workers in the Rajmahal Traps in Eastern India." In *The Socio-Economic Impacts of Artisanal and Small-Scale Mining in Developing Countries*, edited by Gavin M. Hilson, 403–424. Lisse, NL: Swets & Zeitlinger.

Lahiri-Dutt, Kuntala. 2006. "Mining Gender at Work in the Indian Collieries: Identity Construction by Kamins." In *Women Miners in Developing Countries: Pit Women and Others*, edited by Kuntala Lahiri-Dutt, and Martha Macintyre, 163–181. Hants, UK: Ashgate.

Lahiri-Dutt, Kuntala. 2008. "Digging to Survive: Women's Livelihoods in South Asia's Small Mines and Quarries." *South Asian Survey* 15 (2): 217–244.

Lahiri-Dutt, Kuntala. 2011. "The Megaproject of Mining: A Feminist Critique." In *Engineering Earth: The Impacts of Megaengineering Projects*, edited by Stanley D. Brunn, 329–351. Dordrecht: Springer.

Lahiri-Dutt, Kuntala, and Martha Macintyre. 2006. "Introduction: Where Life is in the Pits (and Elsewhere) and Gendered." In *Women Miners in Developing Countries: Pit Women and Others*, edited by Kuntala Lahiri-Dutt, and Martha Macintyre, 1–22. Hants, UK: Ashgate.

Li, Fabiana. 2009. "Negotiating Livelihoods: Women, Mining and Water Resources in Peru." *Canadian Woman Studies* 27 (1): 97–102.

Li, Fabiana. 2015. *Unearthing Conflict: Corporate Mining, Activism, and Expertise in Peru*. Durham, NC: Duke University Press.

Li, Tania Murray. 2014. "What Is Land? Assembling a Resource for Global Investment." *Transactions of the Institute of British Geographers* 39 (4): 589–602.

Marston, Andrea. 2020. "Vertical Farming: Tin Mining and Agro-Mineros in Bolivia." *The Journal of Peasant Studies* 47 (4): 820–840.

Marston, Andrea, and Tom Perreault. 2017. "Consent, Coercion and *Cooperativismo*: Mining Cooperatives and Resource Regimes in Bolivia." *Environment and Planning A: Economy and Space* 49 (2): 252–272.

Méndez, Maria José. 2018. "'The River Told Me': Rethinking Intersectionality from the World of Berta Cáceres." *Capitalism Nature Socialism* 29 (1): 7–24.

Miller, Gloria E. 2004. "Frontier Masculinity in the Oil Industry: The Experience of Women Engineers." *Gender, Work and Organization* 11 (1): 47–73.

Mollett, Sharlene. 2015. "'Displaced Futures': Indigeneity, Land Struggle, and Mothering in Honduras." *Politics, Groups, and Identities* 3 (4): 678–683.

Mollett, Sharlene. 2017. "Irreconcilable Differences? A Postcolonial Intersectional Reading of Gender, Development and *Human* Rights in Latin America." *Gender, Place & Culture* 24 (1): 1–17.

Nightingale, Andrea. 2006. "The Nature of Gender: Work, Gender, and Environment." *Environment and Planning D: Society and Space* 24: 165–185.

O'Faircheallaigh, Ciaran. 2013. "Women's Absence, Women's Power: Indigenous Women and Negotiations with Mining Companies in Australia and Canada." *Ethnic and Racial Studies* 36 (11): 1789–1807.

Perreault, Tom. 2013. "Dispossession by Accumulation? Mining, Water and the Nature of Enclosure on the Bolivian Altiplano." *Antipode* 45 (5): 1050–1069.

Perreault, Tom. 2015. "Performing Participation: Mining, Power, and the Limits of Public Consultation in Bolivia." *The Journal of Latin American and Caribbean Anthropology* 20 (3): 433–451.

Polier, Nicole. 1996. "Of Mines and Min: Modernity and Its Malcontents in Papua New Guinea." *Ethnology* 35 (1): 1–16.

Ribot, Jesse C., and Nancy Lee Peluso. 2003. "A Theory of Access." *Rural Sociology* 68 (2): 153–181.

Rocheleau, Dianne, and David Edmunds. 1997. "Women, Men and Trees: Gender, Power and Property in Forest and Agrarian Landscapes." *World Development* 25 (8): 1351–1371.

Rocheleau, Dianne E., Barbara P. Thomas-Slayter, and Esther Wangari, eds. 1996. *Feminist Political Ecology: Global Issues and Local Experiences*. London: Routledge.

Rondón, Glevys. 2009. "Corporate Social Responsibility and Women's Testimonies." *Canadian Woman Studies* 27 (1): 89–96.

Scheyvens, Regina, and Leonard Lagisa. 1998. "Women, Disempowerment and Resistance: An Analysis of Logging and Mining Activities in the Pacific." *Singapore Journal of Tropical Geography* 19 (1): 51–70.

Schroeder, Richard A. 1999. *Shady Practices: Agroforestry and Gender Politics in The Gambia*. Berkeley, CA: University of California Press.

Scott, Rebecca R. 2010. *Removing Mountains: Extracting Nature and Identity in the Appalachian Coalfields*. Minneapolis, MN: University of Minnesota Press.

Smith, Barbara Ellen. 2015. "Another Place Is Possible? Labor Geography, Spatial Dispossession, and Gendered Resistance in Central Appalachia." *Annals of the Association of American Geographers* 105 (3): 567–582.

Sone, Sachiko. 2006. "Japanese Coal Mining: Women Discovered." In *Women Miners in Developing Countries: Pit Women and Others*, edited by Kuntala Lahiri-Dutt, and Martha Macintyre, 51–72. Hants, UK: Ashgate.

Sultana, Farhana. 2011. "Suffering for Water, Suffering from Water: Emotional Geographies of Resource Access, Control and Conflict." *Geoforum* 42 (2): 163–172.

Tallichet, Suzanne E. 2006. *Daughters of the Mountain: Women Coal Miners in Central Appalachia*. University Park, PA: Pennsylvania State University Press.

Tsikata, Dzodzi. 2003. "Securing Women's Interests within Land Tenure Reforms: Recent Debates in Tanzania." *Journal of Agrarian Change* 3 (1/2): 149–183.

Tsing, Anna Lowenhaupt. 2005. *Friction: An Ethnography of Global Connection.* Princeton, NJ: Princeton University Press.

Wescoat, James L. 1991. "Resource Management: The Long-Term Global Trend." *Progress in Human Geography* 15 (1): 81–93.

Wescoat, James L. 1992. "Resource Management: Oil Resources and the Gulf Conflict." *Progress in Human Geography* 16 (2): 243–256.

Williams, Christine L., Kristine Kilanski, and Chandra Muller. 2014. "Corporate Diversity Programs and Gender Inequality in the Oil and Gas Industry." *Work and Occupations* 41 (4): 440–476.

Zimmermann, Erich W. 1933. *World Resources and Industries.* New York, NY: Harper.

Zimmermann, Susan. 2018. "Globalizing Gendered Labour Policy: International Labour Standards and the Global South, 1919–1947." In *Women's ILO: Transnational Networks, Global Labour Standards, and Gender Equity, 1919 to Present,* edited by Eileen Boris, Dorothea Hoehtker, and Susan Zimmerman, 227–255. Geneva: International Labour Organization.

10

The plantation town

Race, resources, and the making of place

Danielle M. Purifoy

Introduction

Pinehurst is a luxury golf resort in the Sandhills region of North Carolina. Hosting nine major golf championships since 1936, the village boasts over 2000 acres of golf courses and pine woods, dozens of restaurants, hotels, equestrian centers, and shopping plazas (Pinehurst Resort 2019). Pinehurst is a white municipality in southern Moore County, consistently ranked as one of the wealthiest counties in North Carolina; the village is aspirational for many tourist towns across the state. A 2005 feature article in *The New York Times*, however, revealed a social arrangement obscured by Pinehurst's representations of independent white resourcefulness. Surrounding the village is a cluster of historic Black communities dating back to the postbellum era, which, despite their immediate proximity, bear none of Pinehurst's markers of wealthy development. As Pinehurst hosted the 2005 US Open, a multimillion-dollar event with an international following, the Black communities launched a campaign opposing the tournament through a series of lawsuits and media stories drawing attention to the fact that the resort town and its neighboring golf towns were exploiting Black labor and blocking their access to potable water, wastewater sanitation, and land autonomy (Dewan 2005).

Pinehurst officials engaged in a process geographers call municipal underbounding, the systematic exclusion of communities of color from residence, political participation, and town-financed services in largely white municipalities through the routine practices of redrawing and expanding their physical boundaries for growth and development (Aiken 1987; Johnson et al. 2004; Wilson et al. 2008). Many municipalities in North Carolina, and in many states in the United States, are assigned an extraterritorial jurisdiction (ETJ) immediately outside of their municipal boundaries for the purpose of future town expansion via annexation. Underbounding occurs when municipalities selectively annex parts of their ETJ during town growth in a manner that prevents the increase of populations of color into largely white local polities. Black communities like Jackson Hamlet and Monroe Town, located in Pinehurst's ETJ, were denied municipal annexation as Pinehurst grew all around them in area, population, and wealth (Dewan 2005; Joyner and Christman 2005; Strom 2005). Inaccessible municipal resources— drinking water and sanitation—are a primary consequence of this spatial exclusion (Joyner and Christman 2005). The denial of these resources in Black communities is reinforced by Pinehurst's land use control within its ETJ, which prevents underbounded communities from seeking independent development, while prohibiting their political participation in municipal elections and most planning decisions (Joyner and Christman 2005). Because Pinehurst is not obligated to annex any specific territory within its ETJ, the land use restrictions in underbounded communities often become a permanent feature of arrested development and political repression.

Aided by a team of civil rights attorneys and researchers, Black communities outside of Pinehurst, along with others located in the ETJ of the neighboring golf towns of Southern Pines and Aberdeen, organized locally to fight for basic resources and infrastructure, especially as these Black communities comprise much of the golf resorts' service economy—the caddies, the housekeepers, the groundskeepers, and the waitstaff (UNC Center for Civil Rights 2006).

The coexistence of such wealth and luxury with resource-dispossessed communities contending with raw sewage and contaminated water; the simultaneous reliance of Pinehurst on Black physical and affective labor; and the coproduction of these entangled worlds through the extraction, technological transformation, and hoarding of land and water, paint a picture of classic environmental racism manifested through plantation colonialism—the localized extraction of labor and natural resources for white wealth development at the political, economic, and environmental expense of Black people and Black places (McKittrick 2013; Woods 2017). Black places embedded within this white plantation structure, however, extend beyond surviving and resisting extraction—their sustained "livingness" (McKittrick 2016) insists on a different mode of creating place than is legible by the plantation itself.

The municipality—a configuration of persons, infrastructures, and various forms of nonhuman life mediated through politicized spatial boundaries—remains a critical site to observe how towns are accomplished through the plantation tradition of using racial social structure and natural resources—particularly land and water—as instruments of development. Though the municipality is not a precise replica of the plantation, it inherits socioeconomic systems that perpetuate white planter societies. I focus specifically on how the plantation mode of development influences the use and distribution of land and water in Pinehurst, how it manipulates environmental risk and political power in the service of building white wealth and status, and how Black communities interrupt extraction and protect their own ways of living.

Race, resources, and plantation town development

The case of municipal underbounding in Pinehurst, NC, demonstrates the endurance of plantation colonialism in contemporary battles by Black residents for environmental justice (EJ), which in this case is mostly about access to land and water resources, the foundation of any town. White-founded and majority white municipalities, which represent most municipalities in the United States, develop through processes that reproduce environmental racism, the targeted disposability of communities of color, and clarify its purpose, which is to create new forms of white wealth with extracted resources from those communities (Purifoy 2019; Seamster and Purifoy 2020). Though the term "environmental racism" was first coined during the EJ struggles of the 1980s, its place-based resource extraction finds its roots in the plantation economics of the seventeenth to nineteenth centuries, which are also the socioeconomic foundations of modern towns (McKittrick 2013). Thus, as I argue, contemporary processes of transforming natures into development resources are inextricably linked to the transformation of the United States (and much of the global West) from the agricultural plantation to modern local industrial and service economies (see also Ciccantell, Chapter 15 this volume). Racial social structures are reproduced to accommodate the changing formations and usages of those resources.

The municipality is a formalized process of place development in the United States, granting state-sanctioned authority to manifest the socioeconomic desires of a local body politic (Lawrence and Millonzi 1982; Rice, Waldner, and Smith 2014). Those desires are asserted through physical structures, social and economic regulations, and the iterative establishment of spatial boundaries, choosing where the municipality may physically expand, and what and who to absorb within the municipal body. The municipality reconstitutes ecosystems, extracting and transforming nature into development resources, the building blocks of capitalist wealth creation (Bakker 2010; Gandy 2004; Swyngedouw, Kaïka, and Castro 2002). Land and water resources in particular are adjudicated, rearranged, and technologized to create new landscapes according to spatial control concerns and evolving place aesthetics; they are at times eliminated to make way for more convenient, more profitable orderings of matter. The spatial orderings of matter are predicated on the social arrangements of the municipality, particularly racial social structures (Seamster 2015, 2016).

Though EJ scholarship does not center the municipality as a primary unit of analysis, its substantive empirical findings—which trace the spatial disparities of environmental quality based on race and other social demographics—demonstrate the role of the municipality as a form of environmental management (Bullard 2008; Wilson et al. 2008). In the abstract, EJ literature is partly predicated upon the reconstituted ecologies of the municipality, with consequences that are differently experienced within the body politic and extend far beyond its boundaries.

The practice of environmental racism challenges traditional resource conservationism by demonstrating the select protection of resources and landscapes for white consumption and place development at the expense of the resources, cultural landscapes, and living spaces of communities of color, particularly Black, Latinx, and Indigenous places (Taylor 2002, 2009). Creating such "sacrifice zones"—from the expulsion and massacre of Indigenous peoples to place vast natural resources under white "stewardship," to the bulldozing of Black communities and landscapes to build interstate highways for seamless transit from one white place to another, to the siting of waste incinerators in Latinx neighborhoods to dispose of the by-products of white town development—not only undermines universalist notions of protecting the Earth, but also uncovers intimate connections between resource extraction and race (Jacoby 2001; Park and Pellow 2011; Vasudevan 2019).

By demonstrating vast racial disparities in the control of natural resources and environmental quality, EJ scholars laid the foundation for a more comprehensive framework of connecting natural resources, place development, and social power structures. The central premise of the framework is that race structures societal relations, which are also spatial relations, which are both tied to the exploitation and distribution of natural resources most acutely at the local level.

The prototype for the intersections of race, place development, and natural resources within this framework in the United States—and much of the Western Hemisphere—is the plantation, "a method of colonization that imposes upon social landscapes a distinct regime of political, economic, and ethnic regulation" (Woods 2017, 41). Plantation colonialism uses the domination of natural resources and Black persons to produce white wealth and to build white-dominated economies. This form of racial capitalism tied specific persons to roles and to spaces—white masters and mistresses to their estates, Black enslaved nursemaids to their second-tier quarters within the estate, the Black enslaved overseers, and field workers on the land and in their cabins on the perimeter of the estate (Ellis and Ginsburg 2010). These roles and spaces were entrenched in legal, social, and political institutions, reinforcing racial hierarchy both socially and materially.

The distribution of resources were most stark within this system, the planters extracting wealth and sustenance from every facet of the plantation—the land, the livestock, the harvest, the bodies of the enslaved exploited for labor and sold at market (Beckford 1999). The enslaved were burdened with substandard housing, overwork, family separation, physical abuse, sexual exploitation, and malnourishment: Their food was frequently the refuse of the white planters. At the height of the plantation economy prior to the US Civil War, the fields tended by sometimes thousands of enslaved Black people were degraded by intensive monoculture farming. By the War's end in 1865, much of the soil's nutrients had been exhausted, leaving thousands of acres barren (Nelson 2010).

Beckford's (1999) "plantation economy thesis" argues the plantation endures through the replication of these socioeconomic relations and the institutions that support them onto different landscapes. I argue that the municipality is one iteration of plantation colonialism, possessing different microfunctions, but a similar tethering of racialized persons to roles and landscapes in the service of building white-centered places and economies.

"[T]he plantation is often defined as a 'town,' with a profitable economic system and local political and legal regulations" (McKittrick 2013, 8). The municipal body politic operates on the premise of what Mills (2014) calls the "racial contract," a commitment to white place development that builds and maintains landscapes, infrastructures, and economies that serve the interests of white domination at the expense of non-whites, and particularly Black people. Mills articulates that the racial contract racializes spaces, constructing physical limits between white and non-white spaces that determine resource distribution. He contends that Blackness is defined out of the normative boundaries of white political space, such that even the legal boundaries of a municipality are insufficient to understand who belongs and who does not (Mills 2001, 2014).

The development of white towns can therefore be understood as a process of possessing and controlling land—often through Black land dispossession—and of capturing and systematizing water resources through infrastructural technologies in the service of controlling the terms of economic development and the social mobility of communities of color, who are kept segregated and "underdeveloped" (Woods 2017). Municipal underbounding is a tangible development practice of the body politic through the legal redrawing of its own boundaries to demarcate which spaces to invest resources and which spaces to extract resources, as observed in the Village of Pinehurst. Just as central to the replication of the plantation in the modern town is the starkly different, but related, places that form from its processes.

Pinehurst, a plantation town

Pinehurst is a site of fastidiously curated pleasure and consumption in North Carolina's Sandhills region. Its landscape is miles of manicured green rolling hills adjacent to smooth-paved roads, multimillion-dollar estates, and quaint shops selling monogrammed attire and handcrafted confections. The sentiment of the village of 16,000 is relaxed, almost effortless, an affect accomplished by the rigorous construction and maintenance of a clean-swept playground infused with abundant natural and built environments and far removed from wear or decay.

A trip to the village archives is a reminder of Pinehurst's central product—the golf champion. Flags line every wall of the large archival room, commemorating professional championship games across the 124-year history of the golf resort. Gleaming gold trophies sit enclosed in cramped glass cases; still others spill across the space, on floors and shelves. In the far-right corner of the room, standing adjacent to the employee's kitchen, is a different kind of trophy—a five-foot-tall wooden statue of a Black mammy, her paint-weathered face with fixed vacant eyes and palms folded upward (Figure 10.1).

The Black mammy is an archetype of plantation society, specifically, but not limited to, the US South. It represents an imagined mutual intimacy between whites and Blacks, the patronage of the former in exchange for the latter's domestic labor (Sewell 2013). The figure is a distortion

Figure 10.1   Black mammy figure at the Tufts Archives, Pinehurst, NC. March 8, 2018. Photo by author.

of white racial power relationships to reinforce white comfort and morality. It is a repository for white racial sentimentality, an inexhaustible resource to reinforce (and justify) white power (Collins 2002; Sharpe 2009). The visible "keeping" of the mammy figure within the public archive—and the placement of the figure adjacent to the archive's kitchen, a primary space of domestic production—signals to visitors that the space of the archive, the village, is created and maintained for white power and comfort. The mammy figure also signals in this village comprising 94 percent white residents that Blackness is firmly contained elsewhere, emerging only where necessary as a resource to relieve them of unnecessary labor, serving as golf caddies, as housekeepers, as course keepers, and as waitstaff; that "elsewhere" can be avoided through specific thoroughfares, though it is merely walking distance from Pinehurst's fertilized landscape (UNC Center for Civil Rights 2006).

Pinehurst persists not only via the labor of Black people in the surrounding communities of Monroe Town, Jackson Hamlet, and Waynor Road, but also through the strategic usage of its municipal boundaries to control the land and water resources of the Black communities, denying them access to viable water and wastewater sanitation services. Municipal underbounding is thus not only a mechanism to maintain white political power by prohibiting local Black political participation, but it also asserts economic domination over Black communities to sustain white economies (Aiken 1987; Joyner and Christman 2005; Woods 2017). In this sense, the Black mammy figure in the Pinehurst archive represents Black labor as an extractable resource, which, like land and water, has been integral to producing the physical landscape of the resort, the pleasure of the village lifestyle, and the prestige of the golf game.

Mills (2014) articulates racialization as a dichotomizing process of mutually constituted Blackness and whiteness. This dichotomy creates parallel and opposing social, political, and economic realities for each racial group, though only the reality manifested by whiteness becomes hegemonic (Mills 2014). Discursively, Mills argues, because Blackness and whiteness are co-constitutive, so too are the places where white and Black persons inhabit. If whiteness translates into "civilized" space, as persistently represented in Western philosophy, then Black space is construed as "wilderness," space inhabited by "humanoid" beings who "are continuous with the flora and fauna…the objects rather than the subjects of the distinctively human process of molding nature to human ends, which [social] contractarianism presumes" (Mills 2001, 78). The psychic configuring of Black space around Pinehurst as a wilderness, a political vacuum upon which civilized white space can act at will, translates into material opportunism for the municipality, which extracts from Black communities what it wants and disposes in those communities what it does not need (UNC Center for Civil Rights 2006). Embedded within these contemporary spatial relations are imprints of previous iterations of conquest and colonialism, along with Black configurations of space that are illegible to those structures, because they lie firmly outside the confines of the white-imagined civilization-wilderness binary. Instrumental to the maintenance of these power structures, past and present, is the exploitation and hoarding of land and water as development or "civilizing" resources.

Land

The social-spatial configuration of contemporary Pinehurst arises from economic relations to the longleaf pine—a vast forest ecosystem that anchored the antebellum economy of North Carolina's Sandhills. Mass extraction of turpentine, pitch, and lumber from the forests' pre-mechanization was made possible through enslaved Black labor. So critical was the system of enslaved labor and extraction of longleaf pine to the colonial economy that it produced approximately 70 percent of the tar and 50 percent of the turpentine exported to support British naval industries in the late-eighteenth century (Bleeding Pines of Turpentine 2012; Outland 1996). This system also reproduced and entrenched at the local level the racial social structure that slavery disseminated across the globe, guaranteeing its endurance in future iterations, regardless of the particular configuration of resources and economies.

Pinehurst itself was formed in the post–Civil War from 5800 acres of decimated longleaf pine forest, then described as a "barren sandy wasteland" (Pinehurst 2019). Even as the ecological devastation forever altered the landscape of the Sandhills region (the longleaf pine remains a threatened tree species), the subsequent devaluation of the land presented a lucrative opportunity for James Walker Tufts, a soda fountain magnate from Massachusetts. Tufts purchased the land for approximately $1/acre to establish a health resort for middle-class workers in the North, believing the waves of tuberculosis and other ailments stemming from Northern urbanization could be cured "in a mild and healthy climate" (Pinehurst 2019; UNC Center for Civil Rights 2006). This purchase signaled a large-scale shift in the regional economy, refocusing labor and resources from agriculture to hospitality and recreation—a "civilizing" project aimed at creating and serving a wealthier, more elite community.

The production of leisure and sport is an assemblage of high-valued experiences—a calming ambience, unfettered access to amenities and indulgences, persistent satiation. Perhaps above all, the leisure space created must be imbued with a sense of social belonging, a reproduction of the social relations of the home. Such reproduction of domestic and social life for the white middle classes of the late-nineteenth to early-twentieth centuries typically entailed both physical isolation from and some physical or affective labor by Black people (Glymph 2008; Jones 2009). The physical landscape of Pinehurst is thus a white space designed with Black satellite spaces as intimate segregated neighbors. This near-replica of the plantation requires control of land both within and outside of white space, accomplished historically through legal prohibitions of

Black landownership or residence within Pinehurst, and in the present day through municipal underbounding and exploitation of the village's Black ETJ (Joyner and Christman 2005; UNC Center for Civil Rights 2006). Tufts induced Black settlement on the outskirts of Pinehurst by selling land—taken as an opportunity for Black social and political autonomy—to newly free Blacks (UNC Center for Civil Rights 2006). Waynor Road, Monroe Town, and Jackson Hamlet can be characterized as Black freedom colonies, which have endured in Moore County for at least a century in attempts to preserve Black landownership and cultural heritage (Roberts 2017; Sitton and Conrad 2005). However, as per Mills' racial contract, Black landownership in these communities does not translate into the same monetary wealth or assets enjoyed by white landowners. As an example, Pinehurst's ETJ land use controls (NCGS 160A-360) make literal colonies of the Black communities, prohibiting them from pursuing forms of development that interfere with Pinehurst's own social and economic interests, such as building water and sewer infrastructure to advance local economies not dependent upon the golf industry (UNC Center for Civil Rights 2006).

As a consequence, Black communities have been forced to use outhouses or straight pipe their sewage from their homes onto their land by legal and financial prohibitions from installing functional septic systems, such municipal codes prohibiting septic tanks for new construction, and less than optimal acreage space to replace existing systems. Their household garbage, which they must pay expensive private services to collect, sits in piles behind their homes or is incinerated, releasing uncontrolled pollutants into the atmosphere (UNC Center for Civil Rights 2006). Though these communities form their own political and economic institutions in the absence of local governance, Pinehurst treats them nonetheless as disposable landscapes.

Even in situations where they were able to wrest control of their own communities from Pinehurst, as happened in the case of Taylortown, which formed as an independent Black municipality in 1987, Black places must still contend with white power interests at the county and state levels, which often still prioritize the development of white towns at the expense of Black towns, primarily through uneven distribution of public finance (Joyner and Christman 2005). Meanwhile, Black communities add daily value to the land and assets of Pinehurst by maintaining its extensive evergreen landscape, including over 43 public and private golf courses, by providing the physical and affective labor for residents and for the thousands of visitors to the village every year (UNC Center for Civil Rights 2006). They also paradoxically add monetary value to Pinehurst by their *de facto* exclusion from living within the village, as Black residents would undermine Pinehurst's status as a segregated white leisure space (Taylor 2014). Thus, as with the plantation, Pinehurst reproduces a schema in which whites and Blacks occupy dichotomous roles attached to co-constituted landscapes of differential value, thus perpetuating the exploitation of land and water resources through the continuous social structure of white supremacy.

Within this schema, however, Black communities develop vastly different relationships to land, which often result in the long-term maintenance of communities *despite* iterative extraction from neighboring white towns. Land in particular is treated as non-fungible, rather taking on a value of preserving family histories and extending legacies for multiple generations (Lipsitz 2011). Efforts to preserve the "home place" often results in the creation of legal title to land held in common by several generations of the family, instead of by individuals, ostensibly to prevent any one family member from selling the entire legacy. Paradoxically, these so-called heir properties are legally more vulnerable to seizure by outside interests, and subject to legal prohibitions on building essential water and sewer infrastructures, as property laws are structured to favor individual ownership (Dyer and Bailey 2008). Importantly, Black communities like

Jackson Hamlet and Monroe Town may resist annexation, even with the promise of infrastructure, because of a desire to preserve the Black-centered identities of their communities, and the institutions that have sustained them, against further intrusions of white rules and cultural production.

Water

The Sandhills region is contained within the Coastal Plain of the state, an ecosystem known as much for its vulnerability as its beauty. Though the region boasts an abundance of water resources, a critical component of successful community development, the Coastal Plain also contends with a high water table. This makes the region not only more likely to experience flooding, but also more likely to contend with infrastructure challenges, including flooded septic tanks and groundwater contamination (Mallin 2013). These specific vulnerabilities have potential for profound impacts on ecological health, creating hazards for humans and nonhumans alike (on the spatial configuration of vulnerabilities, see also Fabricant, Chapter 26 this volume). Pinehurst largely evades the risks of these impacts by utilization of Moore County's water and sewer infrastructure, which is supplied almost exclusively to incorporated municipalities, rather than to the unincorporated Black communities in the ETJ of the white municipalities or to decentralized household water wells and septic systems, which are reliant upon individual financial resources and private monitoring (Joyner and Christman 2005; UNC Center for Civil Rights 2006).

Municipal water and sewer infrastructures are highly technical systems that transform natural resources into distributed ecosystem services—potable water and human sewage treatment—to reduce the probability of health hazards in areas of higher population, such as towns and cities, and to facilitate the scaled development of larger communities (Hamid and Narendran 2004; Kaika and Swyngedouw 2000). Indeed, such services are often a central rationale for forming municipalities, as a local polity enters into a social contract to finance the necessary infrastructure to benefit the entire polity, rather than forcing individuals to support themselves. In Pinehurst, however, the social contract at the local level is still a racial contract, which produces artificial forms of resource abundance and scarcity by structuring who has access to even the most basic infrastructure, and by extension, who can pursue development (Kaika 2003; Mills 2014). As water flows freely to the village's homes and businesses and nourishes dozens of golf courses, the Black communities just outside of Pinehurst's boundaries grapple with raw sewage from failing or nonexistent septic tanks, with drinking water drawn from groundwater wells that have no external monitoring (UNC Center for Civil Rights 2006). Indeed, Pinehurst's persistent practice of underbounding precludes Black residents from accessing critical municipal infrastructure that often lies just yards from their homes.

Though there are no published investigations of public health impacts of the raw human sewage and untreated drinking water specifically impacting Black communities outside of Pinehurst, there is substantial documentation in similar contexts in the US South that ecological alterations manifested by these unmanaged human impacts are associated with a rise in various parasites, like hookworm. Nutrients disposed with human waste contaminate surface water, generating algae blooms which deprive water habitats from necessary oxygen sources, with further environmental health impacts for nonhuman animals and humans reliant on local water bodies as sources of food, habitat, and recreation (Flowers 2018; McKenna et al. 2017).

Pinehurst exploits the vulnerability of the regional ecosystem and the relative economic vulnerability of Black communities as its own form of economic insurance, hoarding viable access to water and sanitation infrastructure in a manner that prohibits the economic competition that might arise from other forms of local development (UNC Center for Civil Rights 2006). By controlling both access to its own infrastructure *and* authorization for viable independent infrastructure in these communities, Pinehurst reproduces the Black satellite communities as degraded territories, lowering the value of land and structures located there, and artificially increasing the barriers to both internal and external investment. As the golf industry evolves, and the service jobs once executed by Black residents of these communities are eradicated or transformed by technological advances (e.g., golf caddies replaced by golf carts), these Black communities are vulnerable to transform once again into land banks, preserved for Pinehurst's future physical development, one that will likely also attempt to exclude these communities. This particular stage of Pinehurst's history, like the earlier stage in the late-nineteenth century, is also predicated upon enforced devaluation of resources and people.

In 2014, nearly a decade after the 2005 US Open, Jackson Hamlet was successful in its political and legal fight for access to water and sewer infrastructure (Sinclair 2014). Monroe Town won their fight for infrastructure in 2000, but was still unable to gain access to garbage collection and other services (Dewan 2005). These successes signal some progress for the physical health of the communities; however, the gains are limited by vast attrition from the communities after years of dispossession, and by the fact that both Monroe Town and Jackson Hamlet are still contained within the colonial structures of Pinehurst's ETJ—perpetually preserved for the village's own future. This means the Black communities are still not free to pursue independent development, and that they are legally limited in their options to resist unwanted and potentially detrimental types of development imposed by Pinehurst.

Nevertheless, these communities continue by sustaining social ties and local community institutions—churches, social clubs, political action groups—which develop alternative ways of life and self-governance even as they continue to pressure government at all levels for greater independence; even as they experience the technological and normative shifts in Pinehurst's economy, and the gradual diminishing of population, as younger generations look for viable opportunities elsewhere. The persistence of life in these Black communities, even within the plantation structure, signals as per McKittrick (2013, 12), "a mode of being human that…is not victimized and dispossessed and wholly alien to the land," but seeks its own "living" outside the structure of "the fittest," or the most dominant.

Conclusion

Tracing the continuities of the plantation through the development of contemporary municipalities like Pinehurst provides a historically grounded framework to understanding the lineage of environmental racism. Far from an isolated phenomenon of racialized hazard distribution discovered in the 1980s, environmental racism has always been central to white-centered modes of development in the United States and across the West, from the creation of legal regimes of land and water management through underbounding and the colonial ETJ to the imposition of waste and pollution and denial of infrastructures to manage technologically transformed land and water resources. Though environmental racism is not as obviously linked to the plantation in all geographic contexts, the plantation's extractive mode of development is exported across the globe, with dire consequences for both natural resources and subaltern groups within any society. Given these lineages, and the lives and resources at stake, EJ necessarily requires alternative modes and measures of development, which requires alternative concepts of human life and its relationship to nonhuman life, beyond "resources."

Though Black communities and other subjugated groups are not entirely resistant to extractive modes of production in their own conceptualization of development, I argue that those communities, most disadvantaged by enduring plantation power, are best positioned to see and advance these alternatives to save their own lives and to preserve their own ways of living—as exemplified by the "home place"—which may ultimately protect the future of life on Earth. This observation does not absolve white power structures of their responsibility for abolishing the plantation; rather, it requires that any structural change in human modes of being on the planet be centered around the *living* of marginalized humans and nonhuman natures.

References

Aiken, Charles S. 1987. "Race as a Factor in Municipal Underbounding." *Annals of the Association of American Geographers* 77 (4): 564–579.

Bakker, Karen. 2010. *Privatizing Water: Governance Failure and the World's Urban Water Crisis.* Ithaca, NY: Cornell University Press.

Beckford, George L. 1999. *Persistent Poverty: Underdevelopment in Plantation Economies of the Third World.* Mona, Jamaica: University of the West Indies Press.

Bleeding Pines of Turpentine. 2012. "The Long Leaf Pine Trees of the Round Timber Tract." 2012. http://www.bleedingpines.com/pines.html.

Bullard, Robert D. 2008. *Dumping in Dixie: Race, Class, and Environmental Quality.* Boulder, CO: Westview Press.

Collins, Patricia Hill. 2002. *Black Feminist Thought : Knowledge, Consciousness, and the Politics of Empowerment.* New York, NY: Routledge.

Dewan, Shaila. 2005. "In County Made Rich by Golf, Some Enclaves Are Left Behind." *The New York Times*, June 7. https://www.nytimes.com/2005/06/07/us/in-county-made-rich-by-golf-some-enclaves-are-left-behind.html.

Dyer, Janice F., and Conner Bailey. 2008. "A Place to Call Home: Cultural Understandings of Heir Property Among Rural African Americans." *Rural Sociology* 73 (3): 317–338.

Ellis, Clifton, and Rebecca Ginsburg. 2010. *Cabin, Quarter, Plantation: Architecture and Landscapes of North American Slavery.* New Haven, CT: Yale University Press.

Flowers, Catherine. 2018. "Opinion: A County Where the Sewer Is Your Lawn." *The New York Times*, May 22. https://www.nytimes.com/2018/05/22/opinion/alabama-poverty-sewers.html.

Gandy, Matthew. 2004. "Rethinking Urban Metabolism: Water, Space and the Modern City." *City* 8 (3): 363–379. doi: 10.1080/1360481042000313509.

Glymph, Thavolia. 2008. *Out of the House of Bondage: The Transformation of the Plantation Household.* Cambridge, UK: Cambridge University Press.

Hamid, Ir Haniffa, and M Narendran. 2004. "The Role of Sewage Treatment in Public Health." *The Ingenieur* 23 (September-November), 50–53.

Jacoby, Karl. 2001. *Crimes against Nature: Squatters, Poachers, Thieves, and the Hidden History of American Conservation.* Oakland, CA: University of California Press.

Johnson, James H., Allan Parnell, Ann Moss Joyner, Carolyn J. Christman, and Ben Marsh. 2004. "Racial Apartheid in a Small North Carolina Town." *The Review of Black Political Economy* 31 (4): 89–107.

Jones, Jacqueline. 2009. *Labor of Love, Labor of Sorrow: Black Women, Work, and the Family, from Slavery to the Present.* New York, NY: Basic Books.

Joyner, Ann Moss, and Carolyn J. Christman. 2005. *Segregation in the Modern South: A Case Study of Southern Moore County.* Mebane, NC: Cedar Grove Institute for Sustainable Communities, Inc. www.mcmoss.org/CedarGrove/Docs/SMoore_Cnty_Case_Study_comprsd.pdf.

Kaika, Maria. 2003. "Constructing Scarcity and Sensationalising Water Politics: 170 Days That Shook Athens." *Antipode* 35 (5): 919–954. doi: 10.1111/j.1467-8330.2003.00365.x.

Kaika, Maria, and Erik Swyngedouw. 2000. "Fetishizing the Modern City: The Phantasmagoria of Urban Technological Networks." *International Journal of Urban and Regional Research* 24 (1): 120–138. doi: 10.1111/1468-2427.00239.

Lawrence, David M., and Kara A. Millonzi. 1982. *Incorporation of a North Carolina Town.* Chapel Hill, NC: Institute of Government, University of North Carolina.

Lipsitz, G. 2011. *How Racism Takes Place*. Philadelphia, PA: Temple University Press.

Mallin, Michael A. 2013. "Septic Systems in the Coastal Environment." *Monitoring Water Quality*, 81–102. doi: 10.1016/B978-0-444-59395-5.00004-2.

McKenna, Megan L., Shannon McAtee, Patricia E. Bryan, Rebecca Jeun, Tabitha Ward, Jacob Kraus, Maria E. Bottazzi, Peter J. Hotez, Catherine C. Flowers, and Rojelio Mejia. 2017. "Human Intestinal Parasite Burden and Poor Sanitation in Rural Alabama." *The American Journal of Tropical Medicine and Hygiene* 97 (5): 1623–1628. doi: 10.4269/ajtmh.17-0396.

McKittrick, Katherine. 2013. "Plantation Futures." *Small Axe* 17 (3): 1–15.

McKittrick, Katherine. 2016. "Diachronic Loops/Deadweight Tonnage/Bad Made Measure." *Cultural Geographies* 23 (1): 3–18. doi: 10.1177/1474474015612716.

Mills, Charles W. 2001. "Black Trash." In *Faces of Environmental Racism*, edited by Laura Westra, and Bill E. Lawson, 73–91. Oxford: Rowman & Littlefield Publishers.

Mills, Charles W. 2014. *The Racial Contract*. Ithaca, NY: Cornell University Press.

Nelson, Lynn A. 2010. *Pharsalia: An Environmental Biography of a Southern Plantation, 1780-1880*. Athens: University of Georgia Press.

Outland, Robert B. 1996. "Slavery, Work, and the Geography of the North Carolina Naval Stores Industry, 1835-1860." *The Journal of Southern History* 62 (1): 27–56. doi: 10.2307/2211205.

Park, Lisa Sun-Hee, and David N. Pellow. 2011. *The Slums of Aspen: Immigrants vs. the Environment in America's Eden*. New York, NY: New York University Press.

Pinehurst. 2019. "From Pasture to Pinehurst." *Pinehurst Resort* (blog). Accessed February 13, 2019. https://www.pinehurst.com/about/pasture-pinehurst/.

Pinehurst Resort. 2019. Great Golf Resorts of the World. Accessed February 13, 2019. http://greatgolfresorts.com/resort/pinehurst-resort/.

Purifoy, Danielle M. 2019. North Carolina [Un]incorporated: Place, Race, and Local Environmental Inequity. Advance online publication. *American Behavioral Scientist*. doi: 10.1177/0002764219859645.

Rice, Kathryn T., Leora S. Waldner, and Russell M. Smith. 2014. "Why New Cities Form: An Examination into Municipal Incorporation in the United States 1950–2010." *Journal of Planning Literature* 29 (2): 140–154. doi: 10.1177/0885412213512331.

Roberts, Andrea R. 2017. "The Farmers' Improvement Society and the Women's Barnyard Auxiliary of Texas: African American Community Building in the Progressive Era." *Journal of Planning History* 16 (3): 222–245. doi: 10.1177/1538513216657564.

Seamster, Louise. 2015. "The White City: Race and Urban Politics." *Sociology Compass* 9 (12): 1049–1065.

Seamster, Louise. 2016. *"Race, Power and Economic Extraction in Benton Harbor, MI."* PhD Diss., Duke University. doi: 10.31235/osf.io/6qx9j.

Seamster, Louise and Purifoy, Danielle M. 2020. What is Environmental Racism For? Place Based Harm and Relational Development. Advance online publication. *Environmental Sociology*. doi: 10.1080/23251042.2020.1790331.

Sewell, Christopher J. P. 2013. "Mammies and Matriarchs: Tracing Images of the Black Female in Popular Culture 1950s to Present." *Journal of African American Studies* 17 (3): 308–326. doi: 10.1007/s12111-012-9238-x.

Sharpe, Christina. 2009. *Monstrous Intimacies: Making Post-Slavery Subjects*. Durham, NC: Duke University Press.

Sinclair, David. 2014. "Village Approves Contract to Complete Jackson Hamlet Sewer Project." *The Pilot Newspaper*, January 28. https://www.thepilot.com/news/village-approves-contract-to-complete-jackson-hamlet-sewer-project/article_0ca0b344-8884-11e3-851e-001a4bcf6878.html.

Sitton, Thad, and James H. Conrad. 2005. *Freedom Colonies: Independent Black Texans in the Time of Jim Crow*. Austin, TX: University of Texas Press.

Strom, Jennifer. 2005. "In the Shadow of the U.S. Open." *Indy Week*, May 4. https://www.indyweek.com/indyweek/in-the-shadow-of-the-us-open/Content?oid=1194714.

Swyngedouw, Erik, Maria Kaïka, and Esteban Castro. 2002. "Urban Water: A Political-Ecology Perspective." *Built Environment* 28 (2): 124–137.

Taylor, Dorceta E. 2014. *Toxic Communities: Environmental Racism, Industrial Pollution, and Residential Mobility*. New York, NY: New York University Press.

Taylor, Dorceta E. 2002. "Race, Class, Gender, and American Environmentalism." *General Technical Report. PNW-GTR-534. Portland, OR: U.S. Department of Agriculture, Forest Service, Pacific Northwest Research Station*. doi: 10.2737/PNW-GTR-534.

Taylor, Dorceta E. 2009. *The Environment and the People in American Cities, 1600s1900s: Disorder, Inequality, and Social Change*. Durham, NC: Duke University Press.

UNC Center for Civil Rights. 2006. *Invisible Fences: Municipal Underbounding In Southern Moore County*. Chapel Hill, NC: The University of North Carolina Center for Civil Rights.

Vasudevan, Pavithra. 2019. "An Intimate Inventory of Race and Waste." *Antipode*. Advance online publication. doi: 10.1111/anti.12501.

Wilson, Sacoby M., Christopher D. Heaney, John Cooper, and Omega Wilson. 2008. "Built Environment Issues in Unserved and Underserved African-American Neighborhoods in North Carolina." *Environmental Justice* 1 (2): 63–72.

Woods, Clyde. 2017. *Development Arrested: The Blues and Plantation Power in the Mississippi Delta*. 2nd ed. London: Verso.

11

Materializing space, constructing belonging

Toward a critical-geographical understanding of resource nationalism

Tom Perreault

Introduction

Nationalism has returned with a vengeance.[1] From Donald Trump's self-description as an "American nationalist" to the politics of Hindu nationalism in India, to the ominous specter of resurgent nationalisms in Europe, appeals to nationalism and the nation are on the rise. If the nature and even existence of the nation itself is much debated, there is little doubt as to the power of national*ism* as a force for mobilizing people around a common (if not universally agreed upon) sense of belonging (Breuilly 2008). Nationalism takes many forms, but one of the most potent and enduring is *resource nationalism*: the political, economic, and discursive articulation of natural resources with national interests and identity. It should come as no surprise that natural resources are commonly viewed as crucial for economic development and are often a source of national pride. Resource economies have long been productive of strong regional identities—think of coal districts in Wales or southern Appalachia—and frequently motivate protectionist policies, social mobilization, and even violent conflict. But if resource nationalism is widespread, it is far from universal. Not all resources inspire nationalist sentiment, let alone legislation or collective action in their defense. Resources that are seen as tightly woven into the national fabric in one country may attract little notice in another. Similarly, resource nationalisms in a given place ebb and flow with changing economic and political contexts. In short, resource nationalism has both a history and a geography.

This raises questions of when, where, why, in what form and to what effect resource nationalisms arise. Resource nationalism takes many forms—from formal state measures of protectionist policies or nationalization to public spectacles, monuments, and murals, to banal, everyday practices and discourses such as news reports and advertising images. In this chapter, I advance two claims. First, while its manifestations are diverse, I argue that all resource nationalisms are fundamentally expressions of anxiety over control of economically, politically, or culturally important resources by a threatening other, either domestic or foreign. At times, these anxieties are felt by national elites, who find themselves thrust into competition with international capitalists. Just as often, however, it is the anxieties of the popular masses that give rise to demands for protectionist policies. In this case, it is typically not the threat of competition but rather the specter of (neo)colonial domination, the theft of national wealth, or the economic and cultural decline of once thriving resource-based communities that animate populist anxieties. Second, I argue that expressions of resource nationalism involve the intersection of political economy and cultural politics. Whereas most mainstream examinations of resource nationalism in the popular press, business-oriented trade journals, and the political science and policy literatures focus exclusively on economic factors, a critical-geographical approach to resource nationalism attends to the cultural as well as the economic. That is, resource nationalism is expressed not only through trade policy but also through art, literature, and popular mobilizations. At times, it takes the form of public spectacle; at other times, it is banal and commonplace. In other words, a critical-geographical approach to resource nationalism takes seriously the various ways that the nation and national belonging are constructed and reproduced.

While resource nationalism is manifest most commonly in relation to the strategically important extractive sectors of mining, oil, and natural gas, nationalist sentiment and other forms of collective identity can articulate with a range of natural resources. For instance, Swyngedouw (1999, 2015) has demonstrated how water management was enrolled in Spain's nationalist project of regeneracionismo ("regeneration" after decades of national economic decline). In this case, large-scale water projects and the redistribution of water from northern to central and southern Spain for purposes of agriculture, industry, and urbanization were central features of Franco's fascist regime as well as subsequent democratic governments. In a similar vein, forests are often imbued with nationalist meaning, with well-documented examples as diverse as Thailand (Vandergeest and Peluso 1995), Russia (Davidov 2015), and the United States (Biermann 2016; Kosek 2006). Even the production of Chilean salmon (Bustos 2010) and the consumption of Argentine beef (Romero 2013) are freighted with nationalist symbolism. But resource nationalism is most commonly expressed in relation to minerals, oil, and gas, and for that reason (as well as limitations of space), I limit my analysis to these subsurface resources.

The chapter proceeds as follows: the next section briefly considers the concepts of resources and nationalism and then examines the existing literature on resource nationalism, distinguishing between conservative, realist approaches (that include most popular press accounts and industry trade journals, as well as many academic analyses) and critical approaches. This is followed by an effort to outline a framework for critical-geographical analyses of resource nationalism. Here, I illustrate my argument with empirical examples from Chile, Bolivia, Venezuela, and the United States.

Resources, nationalism, and resource nationalism

I take as my starting point that natural resources are socially produced and historically and geographically contingent. Zimmermann's (1933, 15) aphorism that "[r]esources are not; they become" gestures in this direction. Zimmermann presaged by half a century Harvey's historical materialist understanding of natural resources, which argued that resources "can only be defined in relationship to the mode of production which seeks to make use of them and which simultaneously 'produces' them through both the physical and mental activity of the users. According to this view, then, there is no such thing as a resource in the abstract or a resource which exists as a 'thing in itself'" (Harvey 1974, 265). In this instance, Harvey is critiquing the neo-Malthusian "overpopulation" scares of the late 1960s and early 1970s, but he makes a broader point: in contrast to an Aristotelian view of resources as discreet things to be discovered in nature, Harvey posits a dialectical perspective in which resources exist only in relation to historically specific modes of production that produce them, imbue them with meaning, and assign them value (see also Huber, Chapter 14 this volume). In this view, resources are "cultural appraisals of nature" through which we organize our social and socio-natural relations (Bridge 2009).

Such a position works against the resource fetishism that dominates mainstream, realist accounts of resource scarcity (e.g., the "resource wars" perspectives of Kaplan [2013] and Klare [2002, 2004]) and resource abundance (e.g., the "resource curse" arguments of Ross [2012, 2015]). *Contra* the views adopted by these authors, Le Billon (2013) notes that people consume resources such as oil not simply because they need energy, but because their energy needs are rooted in historically and geographically specific relations of production and consumption. As he argues, the theoretical and political move to de-fetishize natural resources "seeks to counter naturalistic and deterministic interpretations attributing unmediated powers of influence to resources" (Le Billon 2013, 283).

This view of natural resources as socially produced and historically and geographically contingent directs our attention to the relations between resources and the state. In his analysis of the "resource state nexus," Bridge (2014, 119) points out that resources and states, as core features of capitalist modernity, are both drivers and effects of "socio-natural ordering." In this sense, then, state-making and resource-making are mutually constitutive processes (Bridge 2010, 2014; Koch and Perreault 2018). As the production of material things, rooted in national soil, resource-making is necessarily a territorializing project, what Bridge (2010, 825) refers to "the expression of social power in geographical form." Even the growing tendency to "offshore" resource acquisition (sites of extraction, processing, and distribution) only serves to reinscribe the territorial state by re-spatializing sovereignty to include overseas and offshore sources of oil, gas, and minerals (Childs 2016) and may be understood as a loosening and expansion of the spatial imaginaries of national resources.

Nationalism and the nation

A comprehensive review of the literature on nationalism and the nation is beyond the scope of this short chapter. For our purposes, however, it is worth noting how little concern the resource nationalism literature gives to the *nationalism* side of the equation. Whereas most authors writing about resource nationalism ground their analyses in the political economic and socio-natural dynamics of resource geographies, they tend to take the existence of the nation and the meaning of nationalism for granted. Such accounts fail to acknowledge that, like resources and the state, the nation is socially produced, fundamentally relational, historically specific, and never stable. Theories of the nation abound. Characterized alternately as an "imagined community" (Anderson 1991), an "invented tradition" (Hobsbawm and Ranger 1983), a "mode of constructing and interpreting social space" (Williams and Smith 1983, 502), and a collection of "myths and memories" (Smith 1986, 2), the nation is nothing if not a contested concept. It is commonly invoked in both political speech and everyday vernacular. Nevertheless, in the resource nationalism literature, the concept of "nation" is often and confusingly used interchangeably with "state." Sutherland (2012) makes a useful distinction between the *nation*, which refers to the legitimating basis for authority, and the *state*, which, she argues, embodies the territorial and institutional dimensions of that authority. For Sutherland, the nation—comprising those who share an imagined history and common geography—underpins and legitimizes the modern state and serves to justify the emplacement and enforcement of state borders. As such, "a nation need not have a state, but states need some kind of national construct to legitimate their control" (Sutherland 2012, 10). State legitimation is thus achieved through the project of nation-building, which Sutherland understands to be one of state-led nationalism. Nationalism, in this sense, both predates and helps constitute the nation.

But whereas the legitimacy and authority of a given nation, or even of the concept of the nation itself, is a matter of contention, the political and cultural power of national*ism* is something of a settled question (Breuilly 2008). As a principle that binds the political to the national, nationalism inevitably involves the articulation of history, territory, and political and/ or ethnic community (Hobsbawm 1990; Smith 1986). The production of nationalist imaginaries entails a process of highly selective remembering and forgetting and the production of spatial imaginaries, often expressed visually in the form of murals, maps, and monuments (Anderson 1991; Radcliffe and Westwood 1996). Indeed, as Billig (1995) has demonstrated, such "banal" forms of nationalism may be contrasted with the "hot" nationalism of parades, political rallies, and military campaigns. These everyday, unremarkable reminders of one's place in the world are particularly salient in discussions of resource nationalism. Notwithstanding occasional fevered claims about "resource wars" (Klare 2002), countries rarely wage actual battle over natural resources. But the constant discursive and visual imbrication of resources and the nation provides a rationale for resource policies—from subsidies to protectionist tariffs to the wholesale nationalization of resource sectors—and serves to justify the disregard, displacement, and oppression of those who would stand in the way of resource use practices or aspirations (Bebbington and Humphreys Bebbington 2011).

Contrary to most mainstream and conservative analyses, resource nationalism is not the exclusive provenance of the central state. As has been demonstrated through the politics of Ecuadorian oil (Valdivia 2008), substate and nonstate actors can be leading protagonists in the production of resource nationalism, as economic policy and political and cultural discourse. These actors are often motivated by a desire to capture a greater share of resource rents or establish a measure of political autonomy. Such substate nationalisms can take territorial form as well, as demonstrated by efforts to create a new province out of Bolivia's natural gas–producing region (Anthias 2018). Resource nationalism, in this view, is constructed across a diversity of spatial scales and administrative levels, and among an array of actors, including central governments, regional elites, social movements, and labor unions.

Resource and realism: mainstream accounts of resource nationalism

A central claim of this chapter is that a critical-geographical approach to resource nationalism must account for cultural as well as economic considerations and must attend to scales beyond the nation-state. Nevertheless, much of the mainstream literature on resource nationalism, as presented in the pages of industry-oriented trade journals, business-oriented popular press, and realist analyses by scholars in international relations, policy, and political science, is exclusively concerned with economic and national security policies emanating from the central state. Such analyses focus on various forms of state control, invariably seen as a threat to the free flow of resources from the global South to lucrative markets in the global North. Analysts in this genre warn northern capitalists and policy makers to be vigilant to the threats of taxation (Grimley 2015), expropriation (Joffé et al. 2009), rent capture (Wilson 2015), and xenophobia (Ganbold and Ali 2017). These writers frequently root their analyses in specific historical events that they retell as cautionary tales. In the view of Andreasson (2015, 310), for instance, economic nationalism has "haunted international resource companies" ever since President Lázaro Cárdenas nationalized the Mexican petroleum industry in 1938. Such analyses invariably identify resource nationalism as a pathology of the global South, with little attention paid to the policies and politics—let alone the imperial histories—of the industrialized North. Moreover, these analyses display an evident bias in favor of foreign involvement and liberal trade and investment policies, representing nationalization and protectionist policies in apocalyptic terms, as a "specter," a "weapon," and a "peril," that could "overwhelm all other considerations" for resource industries and the importing countries they serve (Hirsch 2008, 881; see also Andreasson 2015; Ganbold and Ali 2017). Joffé et al. (2009, 3), for instance, warn of the "growth of radicalism among [global South] oil producers" who pose "threats to the sanctity of contracts in Latin America." These authors go on to warn that this radicalism has spread contagion-like to North Africa, the Middle East, Russia, and Central Asia. Similarly, Vivoda (2009, 523) laments that "major [international oil companies] have full access to countries with only [six percent] of the globe's known [oil] reserves, mainly in North America and Europe." For their part, Bremmer and Johnston (2009, 150) warn that what they term "revolutionary resource nationalism" can have a "dangerous effect on international resource companies" when "[o]wnership of prized assets may be wrenched away through forced renegotiation of existing contracts, using perceived historical injustice or alleged environmental or contractual misdeeds by the companies as justification." These authors go on to warn of the economic dysfunctions brought on by "resource-drunk politicians" and the threat they pose to global economic stability (Bremmer and Johnston 2009, 152). Such statements leave little doubt as to the class sympathies and political commitments of these authors.

Beyond their unabashed affection for extractive capitalism, such mainstream, realist, and business-oriented press accounts of resource nationalism share four characteristics. The first of these is the tendency among realist authors to focus almost exclusively on economic policies in their analyses of resource nationalism, emphasizing the threat of expropriation and nationalization, tariffs, and forced renegotiation of contracts. Such analyses tend to ignore noneconomic factors such as cultural beliefs, political struggles, and shared historical memory. Many authors reduce resource nationalism to a simple function of the boom and bust cycles common to resource sectors. For instance, Joffé et al. (2009), Vivoda (2009), and Grimley (2015) all portray nationalist sentiment as cyclical, rising and falling mechanistically in response to commodity prices. In one of the more nuanced accounts in this genre, Wilson (2015) presents a typology of resource nationalism focused entirely on economic factors: rent capture, developmentalism, and market-based strategies. When political considerations are taken into account, as in Andreasson (2015), who draws primarily on Bremmer and Johnston's (2009) anemic account, they are reduced to the state's role in setting economic policy. Similarly, Ganbold and Ali (2017) make clear their sympathies: "negative" forms of resource nationalism are those policies that hinder international investment and trade in resources, while "positive resource nationalism" is that which provides some benefit for local populations while still fostering international involvement in national resource sectors. Cultural factors, such as historical memory of the colonial past and its role in shaping contemporary struggles over resource extraction (e.g., Himley 2014), are simply not considered.

A second characteristic shared by realist and conservative accounts is their focus on the central state as the sole site and scale of resource nationalism. The possibility of resource nationalisms emerging from sub or supra-state actors, social movements, or local elites—such as oil workers' movements (Valdivia 2008), Nigerian militias (Watts 2004), regional socio-territorial movements (Riofrancos 2020), or mining cooperatives (Marston and Perreault 2017)—is not accounted for. Rather, policies and policy makers of the central state are the focus of analysis and the lone source of nationalism.

This recognition leads us to a third characteristic shared by realist analyses: the analytical collapse of the nation and the state. Frequent references are made to "national goals" (Wilson 2015, 399) and "resource-rich nations" (Bremmer and Johnston 2009, 149). No attempt is made, however, to critically examine either the state or the nation or their relationship, in spite of the voluminous literatures on state theory and nationalism. Rather, the concepts are used interchangeably. The failure of these authors to distinguish between state and nation signals an inability to analytically separate the institutional forms that governance arrangements can take from the underlying normative justifications that animate forms of nationalism. Such analytical sloppiness collapses the nation into the state as a unified, singular, and ahistorical entity.

Finally, a fourth and related characteristic of this literature is the utter failure to theorize resources and the nation. Both are taken as pre-given, discreet entities that exist independently of the historically constituted social relations of which they are a part. Wilson (2015, 400), for instance, attributes the spatial unevenness of global resource production and consumption not to patterns of uneven capitalist development but to the "arbitrary spread of minerals and energy around the globe." In this view, natural resources are discreet things somewhere "out there," waiting to be discovered—a perspective that reinforces calls for liberalizing trade and investment, as well as warnings of the threat posed to capital by resource nationalism.

Toward a critical-geographical understanding of resource nationalism

In contrast to realist analyses that view resource nationalism as located solely in the realm of economics, I argue that resource nationalism is more properly understood as constructed historically through the interplay of two distinct but related spheres of social life: political economy and cultural politics. While there is no doubt that economic factors are of primary importance in understanding the geographies and histories of resource nationalism, I would argue that analytical focus should also be trained on extra-economic factors such as symbolic meanings, historically constituted power relations, and collective memory. Moreover, critical analyses of resource nationalism demand the recognition that economic relations are inseparable from political relations—the unity of production and power under capitalism—signaled by political economy. In this view, resources are materially produced through the social relations of production and consumption which bring them into being and assign them value (Harvey 1974). This is more than just a material, economic process. Indeed, in a Gramscian sense, the production of resources and resource imaginaries is at its heart an ideological project that serves either to bolster or challenge the hegemonic power of ruling classes (Marston and Perreault 2017). The conjoining of political economy and cultural meaning is evident in the patriotic murals and statuary found in mining districts the world over. Such is the case with the monument to the miner in Oruro Bolivia (see Figure 11.1). Shirtless and masculine, the figure in the monument holds aloft a rifle—symbolic of the miners' role in defeating the Bolivian military in the 1952 Social Revolution. From his wrists, hang broken chains symbolizing the workers' liberation from exploitation of (mostly) foreign capitalists. Similar monuments, with nearly identical iconography, can be found in mining communities throughout the Bolivian Andes.

Such "banal" or everyday examples of nationalist sentiment serve at once to produce a form of political subjectivity and bind it to particular resource regimes. This articulation (Hall 1996) is evident in the pithy slogan "país minero" or "mining country," adopted at various times in Bolivia, Chile, and Peru. There is nothing inevitable about the conjoining of national identity and natural resources; such an articulation must be forged and continually (re)constructed. This ideological work is evident in the 1972 poster, Cobre Chileno ("Chilean Copper"), produced by the brothers Vicente and Antonio Larrea to celebrate Chile's national "Día de la Dignidad" ("Day of Dignity"), which marked the one-year anniversary of the nationalization of the country's copper industry, a move taken by then-President Salvador Allende (and which served as part of the justification for the US-backed coup in 1973 that brought Augusto Pinochet to power; see Figure 11.2).

The colorful illustration represents various sectors of Chilean society (a miner, a doctor, a gaucho, a Mapuche indigenous person, etc.), seated atop a platform made of copper ingots.

Figure 11.1 Monument to the miner, Oruro, Bolivia. This monument was built in 1962, on the 10th anniversary of Bolivia's Social Revolution. The figure holds aloft a rifle, symbolizing the miners' pivotal role in defeating the Bolivian army. Broken chains dangle from his wrists, symbolizing liberation from the oppressive control of foreign capitalists. Photo by author.

In this image, copper is literally the foundation upon which society rests. Above the figures is the text adapted from a poem by Chilean Nobel laureate Pablo Neruda:

> CHILEAN COPPER
> you are the fatherland [*la patria*], plain and people
> sand, clay, school, house
> resurrection, fist, offensive
> order, parade, attack, wheat
> struggle, grandeur, resistance.

In this rendering, copper is far more than just a commodity. Instead, it serves symbolically to link diverse identities with a common national project. Indeed, the poem evokes nation-building

COBRE CHILENO

tu eres la patria, pampa y pueblo.
arena, arcilla, escuela, casa
resurreccion. puño. ofensiva
orden, desfile, ataque, trigo
lucha, grandeza, resistencia.

Figure 11.2   *Cobre Chileno* (Chilean Copper), by graphic artists Vicente Larrea, Antonio Larrea, and Luis Albornoz. Poem by Pablo Neruda. This poster was produced in 1972 to commemorate Salvador Allende's nationalization of copper and was one of a series of nationalist posters produced at the time, which celebrated the nationalization of copper and Allende's socialist government. Reproduced with permission of the artists. Image courtesy of the Center for the Study of Political Graphics (Culver City, California).

in a literal, material sense: sand, clay, school, and struggle are the bedrocks upon which a common sense of belonging is built.

Similar examples are not hard to find. Podeh (2011) details the symbolism of Iraq's "nationalization day" celebrations of the 1970s, which commemorated the nationalization of the country's oil industry—an event represented at the time as central to national pride and power. Not surprisingly, given its geopolitical importance, oil inspires considerable nationalist sentiment in countries as diverse as Niger (Hicks 2015) and Ecuador (Rosales 2017). In this regard, Venezuela is archetypal. With the world's largest proven petroleum reserves and a large national oil company (PDVSA: Petróleos de Venezuela, SA), oil has long been associated with

Venezuelan national identity and development. Under both center-right and more recently left-wing governments, oil has functioned to bind together people, state, and nation (Schiller 2011). In his analysis of Venezuelan petropolitics, Fernando Coronil (1997) views Venezuela as an "oil nation" composed of two bodies, a "political body" made up of citizens and a "natural body" comprising the subsoil: "By condensing within itself the multiple powers dispersed throughout the nation's two bodies, the state appeared as a single agent endowed with the magical powers to remake the nation" (Coronil 1997, 4). For decades, Venezuelan oil endowed the state with political legitimacy, ushering in an era of modernity and vast wealth. Under the leftist government of Hugo Chávez, which coincided with a period of relatively high oil prices, oil was the central motor of the so-called Bolivarian Revolution, providing hard currency and thus a source of state largesse and political power.

Something similar was at work in Bolivia following the election of President Evo Morales, who sought to strengthen state control of natural gas, his government's single largest source of hard currency, and a potent symbol of national pride. Elected in December 2005 after five years of near constant protest (Perreault 2006), Morales was swept to power on a wave of resource nationalism centered on calls to nationalize the country's natural gas industry. Shortly after assuming office, in January 2006, he released a presidential decree calling for the "nationalization" of the country's natural gas industry (to be accomplished through the forced renegotiation of contracts with international hydrocarbons firms). The decree's name, "Heroes of the Chaco" (*Héroes del Chaco*), references the calamitous Chaco War that Bolivia fought against Paraguay in the 1930s, in the eastern scrublands where the country's gas fields are located. Bolivia suffered some 65,000 casualties and lost much of its national territory, but its forces managed to stop the Paraguayan advance and protect the country's hydrocarbons resources. Thus, in referencing this war—fought in a region far distant from the Andean west where Morales' political base was located—the "Heroes of the Chaco" decree discursively articulates Bolivia's natural resource endowments to a spatial imaginary embodied in the heroic defense of national territory (Kohl and Farthing 2012; Perreault 2013).

Such narratives are not unique to resource-exporting countries of the global South, and a version of it is found in the United States, which is both a major producer and importer of resources. The US coal–producing region of southern Appalachia figures prominently in nationalist narratives of energy independence, produced by rugged (mostly white) men (see Fabricant, Chapter 26 this volume). For example, the West Virginia coal executive (and onetime congressional candidate) Don Blankenship was known to make speeches dressed in matching star-spangled shirt and baseball cap, standing before an enormous American flag. Coal was also central to Donald Trump's presidential campaign in 2016, in which the declining fortunes of the US coal industry figured prominently in a campaign fueled largely by nationalist nostalgia and the politics of white grievance. These narratives rest on a sense of collective vulnerability to forces beyond people's control: hostile foreign powers and insecurity arising from imperial overreach coupled with cultural and demographic changes at home (Koch and Perreault 2018). In this context, for a particular segment of the US population, coal represents an atavistic longing for an imagined past of prosperity, secure from cultural change, and the incursions of unknowable others. Aspirational narratives of "energy independence" rest upon (and in turn help reinforce) geographical imaginaries in which US lifestyles and energy consumption patterns are held hostage to unseen global forces that can only be defeated through relentless efforts to "drill baby drill." Resource nationalism, in this sense, provides an ideological frame that binds together the political economies of resource consumption with the cultural politics of national belonging, all rooted in a geographical imaginary that posits the United States as vulnerable to the whims of global energy markets and broad demographic and cultural change. While the power of Venezuelan oil and West Virginia coal are surely rooted in their economic might, their political potency extends well beyond questions of economic policy. To the extent that resources such as oil, coal, and minerals are imbued with cultural meaning, they become signifiers of the nation: symbols of national belonging and nationalist sentiment (Perreault and Valdivia 2010). As with any national imaginary, resource nationalism also delineates an inside and an outside of the nation. In these cases, resources such as oil and coal (or natural gas, copper, etc.) are imagined to belong to, and should be used to benefit, particular social groups and not others.

If resource nationalism is an expression of resource anxiety, then the source of that anxiety largely depends on a country's position along a given resource commodity chain. For resource-*importing* countries such as the United States, Japan, and most western European countries, anxieties stem from concern over the possible disruptions in the flow of oil and gas imports at the hands of hostile foreign states or nonstate actors (Bridge and Le Billon 2013). As Huber (2013) has demonstrated, for instance, the specter of oil sheikhs posing a mortal threat to the American way of life proliferated in the United States following the 1973 OPEC oil embargo. Meanwhile, for resource-*exporting* states, particularly those in the global South, resource anxieties are largely rooted in concerns about foreign control over national resources. In a striking example, Valdivia (2008; see also Sawyer 2002) shows how in Ecuador, those politicians who facilitated foreign involvement in the country's oil sector were labeled *vendepatrias*—literally "sellers of the fatherland." Throughout Latin America, resource endowments are commonly referred to as *patrimonio nacional*—national patrimony—the same term used for architectural and other cultural treasures of historic importance. It is worth noting that *patria* (fatherland) and *patrimonio* (patrimony) both stem from the Latin word *pater* or father. That resources such as oil, gas, and minerals occupy the same discursive slot as cultural history tells us that these resources are far more than economic in their importance. Politically as well as etymologically, then, resources and nation take root in the same conceptual soil (Perreault and Valdivia 2010).

Conclusion

Nationalism takes many forms. During the last 20 years, as part of the broader rise of anti-neoliberal, anti-globalization critiques from both the left and the right, latent nationalist sentiments have erupted to the surface in forms both bellicose and banal. One particularly visible and widespread form is resource nationalism. While resource nationalism is nothing new (Young 2017), it is nevertheless of enduring relevance to understanding the uneasy relationship among resources, the nation, and the state. It is, as such, of central importance to the field of critical resource geography, as well as to critical nature-society scholarship more broadly.

I have argued here that a critical-geographical approach to resource nationalism may be contrasted with mainstream, realist analyses along four broad axes. First, realist accounts are distinguished by their economistic reductionism that views resource nationalism solely in terms of particular sets of economic policies and analyzes the state solely in terms of its role as policy maker. Moreover, these works ignore cultural factors such as popular mobilization, the role of historical memory, and the histories of (neo)colonialism that have shaped views of resource extraction in both the global South and global North. By contrast, critical accounts of resource nationalism are grounded in political economy and rigorous analyses of power (e.g., Kaup and Gellert 2017). Moreover, such analyses of resource nationalism account for extra-economic factors such as historical memory in shaping the production of (contested) meanings of resources and the nation (Perreault 2018) and are therefore attentive to the role of ideology in shaping the political economies of natural resource production and consumption.

Second, mainstream, realist accounts of resource nationalism take the nation and nationalism as unproblematic. These works do not account for the ways that the nation is historically constituted and contested, or the political work done by various forms of nationalism, including resource nationalism. Further, mainstream and realist accounts tend to conflate nation with state, sidestepping altogether the thorny issues of power and the constructed nature of the nation in resource nationalism. By contrast, critical geography (like critical scholarship more generally) has long viewed the nation as a historical construction—even a persistent and pernicious fiction (Koch and Perreault 2018; Smith 1986; Sutherland 2012)—and as such takes seriously the power of nationalism as a political, ideological force.

Third, and relatedly, mainstream, realist authors focus solely on the central state as the locus of resource nationalism. To be sure, the central state is the primary actor in driving most forms of resource nationalism, particularly as it concerns trade, investment, and other forms international relations. Nevertheless, these analyses invariably ignore other sources and forms of resource nationalism, including substate and nonstate actors, which can play an outsized role in shaping resource policies and politics. By contrast, critical-geographical analyses of resource nationalism account for the diversity of actors and spatial scales involved in the production of resource nationalism, including regional elites, social movements, and labor unions (Perreault and Valdivia 2010).

Fourth, mainstream and realist approaches unfailingly adopt an Aristotelian view of resources as "out there in the world, waiting to be discovered." Wilson's (2015, 400) comically naïve claim that the spatial unevenness of resource production and consumption are a result of the "arbitrary spread" of resources across the globe is paradigmatic of this view. Little attention is given in these accounts to actual political economies of natural resources, let alone their historical contingency (Harvey 1974). Critical resource geographers, in contrast, take as axiomatic that resources exist only in relation to the mode of production that produces them, gives them meaning, and imbues them with value, a view that demands attention to historically and geographically specific relations of production and consumption.

A critical-geographical approach to resource nationalism, then, places greater attention on the specificities and contingencies of history and geography than has been the case with mainstream, realist approaches. Further, such an approach centers relations of power even as it decenters the central state as the primary locus of nationalist politics. Finally, it understands resources and the nation as constructed through the frictions of social relations and, further, as co-constituted as effects of social and socio-natural ordering (Bridge 2014). In these ways, a critical-geographical approach to resource nationalism allows for multi-scalar analyses that integrate the political economic with the cultural political, thus permitting a more supple and far-reaching understanding of the relationship between resources and the nation. Resource nationalism is sure to remain a potent feature of national and international politics for the foreseeable future. Critical resource geographers are well poised to trace the contours of these dynamics and to provide a corrective to the conceptually anemic analyses of realist authors.

Note

1 Thanks to Matt Himley, Gaby Valdivia, and Elizabeth Havice for their guidance and patience. Special thanks to Vicente Larrea who graciously allowed me to reproduce the poster *Cobre Chileno*.

References

Anthias, Penelope. 2018. *Limits to Decolonization: Indigeneity, Territory, and Hydrocarbon Politics in the Bolivian Chaco*. Ithaca, New York: Cornell University Press.

Andreasson, Stefan. 2015. "Varieties of Resource Nationalism in Sub-Saharan Africa's Energy and Minerals Market." *The Extractive Industries and Society* 2: 310–319.

Anderson, Benedict. 1991. *Imagined Communities*. London: Verso.

Bebbington, Anthony, and Denise Humphreys Bebbington. 2011. "An Andean Avatar: Post-Neoliberal and Neoliberal Strategies for Securing the Unobtainable." *New Political Economy* 16 (1): 131–145. doi: 10.1080/13563461003789803.

Biermann, Christine. 2016. "Securing Forests from the Scourge of Chestnut Blight: The Biopolitics of Nature and Nation." *Geoforum* 75: 210–219. doi: 10.1016/j.geoforum.2016.07.007.

Billig, Michael. 1995. *Banal Nationalism*. London: Sage.

Breuilly, John. 2008. "Introduction." *Nations and Nationalism*. 2nd ed., by Ernest Gellner, xiii–liii. Ithaca, NY: Cornell University Press.

Bremmer, Ian, and Robert Johnston.. 2009. "The Rise and Fall of Resource Nationalism." *Survival* 51 (2): 149–158. doi: 10.1080/00396330902860884.

Bridge, Gavin.. 2009. "Material Worlds: Natural Resources, Resource Geography and the Material Economy." *Geography Compass* 3 (3): 1217–1244. doi: 10.1111/j.1749-8198.2009.00233.x.

Bridge, Gavin. 2010. "Resource Geographies I: Making Carbon Economies, Old and New." *Progress in Human Geography* 35 (6): 820–834. doi: 10.1177/0309132510385524.

Bridge, Gavin. 2014. "Resource Geographies II: The Resource-State Nexus." *Progress in Human Geography* 38 (1): 118–130. doi: 10.1177/0309132513493379.

Bridge, Gavin, and Philippe Le Billon. 2013. *Oil*. Cambridge, UK: Polity.

Bustos, Beatriz. 2010. "Geographies of Knowledge Production in a Neoliberal Setting: The Case of Los Lagos Region, Chile." PhD dissertation, Syracuse University.

Childs, John. 2016. "Geography and Resource Nationalism: A Critical Review and Reframing." *The Extractive Industries and Society*, 3: 539–546. doi: 10.1016/j.exis.2016.02.006.

Coronil, Fernando. 1997. *The Magical State: Nature, Money and Modernity in Venezuela*. Chicago, IL: University of Chicago Press.

Davidov, Veronica. 2015. "Beyond Formal Environmentalism: Eco-Nationalism and the 'Ringing Cedars' of Russia." *Culture, Agriculture, Food and Environment* 37 (1): 2–13. doi: 10.1111/cuag.12043.

Ganbold, Misheelt, and Saleem H. Ali. 2017. "The Peril and Promise of Resource Nationalism: A Case Analysis of Mongolia's Mining Development." *Resources Policy* 53: 1–11. doi: 10.1016/j.resourpol.2017.05.006.

Grimley, Scott. 2015. Resource Nationalism. *AusIMM Bulletin*, April, 22–24.

Hall, Stuart. 1996. "On Postmodernism and Articulation: An Interview With Stuart Hall." In *Stuart Hall: Critical Dialogues in Cultural Studies*, edited by David Morley, and Kuan-Hsing Chen, 131–150. London: Routledge.

Harvey, David. 1974. "Population, Resources and the Ideology of Science." *Economic Geography* 50 (3): 256–277. doi: 10.2307/142863.

Hicks, Celeste. 2015. *Africa's New Oil: Power, Pipelines and Future Fortunes*. London: Zed.

Himley, Matthew. 2014. "Mining History: Mobilizing the Past in Struggles Over Mineral Extraction in Peru." *Geographical Review* 104 (2): 174–191. doi: 10.1111/j.1931-0846.2014.12016.x.

Hirsch, Robert L. 2008. "Mitigation of Maximum World Oil Production: Shortage Scenarios." *Energy Policy* 36 (2): 881–889. doi: 10.1016/j.enpol.2007.11.009.

Hobsbawm, Eric J. 1990. *Nations and Nationalism: Programme, Myth, Reality*. Cambridge, UK: Cambridge University Press.

Hobsbawm, Eric J., and Terence Ranger, eds. 1983. *The Invention of Tradition*. Cambridge, UK: Cambridge University Press.

Huber, Matt. 2013. *Lifeblood: Oil, Freedom and the Forces of Capital*. Minneapolis, MN: University of Minnesota Press.

Joffé, George, Paul Stevens, Tony George, Jonathan Lux, and Carol Searle. 2009. "Expropriation of Oil and Gas Investments: Historical, Legal and Economic Perspectives in a New Age of Resource Nationalism." *Journal of World Energy Law & Business* 2 (1): 3–7. doi: 10.1093/jwelb/jwn022.

Kaplan, Robert. 2013. *The Revenge of Geography: What the Map Tells Us About Coming Conflicts and the Battle Against Fate*. New York, NY: Random House.

Kaup, Brent Z., and Paul K. Gellert. 2017. "Cycles of Resource Nationalism: Hegemonic Struggles and the Incorporation of Bolivia and Indonesia." *International Journal of Comparative Sociology* 58 (4): 275–303.

Klare, Michael. 2002. *Resource Wars: The New Landscape of Global Conflict.* New York, NY: Henry Holt.

Klare, Michael. 2004. *Blood and Oil: The Dangers and Consequences of America's Growing Dependency on Imported Petroleum.* New York, NY: Henry Holt.

Koch, Natalie, and Tom Perreault. 2018. "Resource Nationalism." *Progress in Human Geography* 43 (4): 611–631. doi: 10.1177/0309132518781497.

Kohl, Ben, and Linda Farthing. 2012. "Material Constraints to Popular Imaginaries: The Extractive Economy and Resource Nationalism in Bolivia." *Political Geography* 31 (4): 225–235. doi: 10.1016/j.polgeo.2012.03.002.

Kosek, Jake. 2006. *Understories: The Political Life of Forests.* Durham: Duke University Press.

Le Billon, Philippe. 2013. "Resources." In *The Ashgate Research Companion to Critical Geopolitics*, edited by Klaus Dodds, Merje Kuusand, and Joanne Sharp, 282–303. New York, NY: Routledge.

Marston, Andrea, and Tom Perreault. 2017. "Consent, Coercion and *Cooperativismo*: Mining Cooperatives and Resource Regimes in Bolivia." *Environment and Planning A* 49 (2): 252–272. doi: 10.1177/0308518X16674008.

Perreault, Tom. 2006. "From the Guerra del Agua to the Guerra del Gas: Resource Governance, Neoliberalism and Popular Protest in Bolivia." *Antipode* 38 (1): 150–172. doi: 10.1111/j.0066-4812.2006.00569.x.

Perreault, Tom. 2013. "Nature and Nation: Hydrocarbons Governance and the Territorial Logics of 'Resource Nationalism' in Bolivia." In *Subterranean Struggles: New Geographies of Extractive Industries in Latin America*, edited by Anthony Bebbington, and Jeff Bury, 67–90. Austin: University of Texas Press.

Perreault, Tom. 2018. "Mining, Meaning and Memory in the Andes." *The Geographical Journal* 184 (3): 229–241. doi: 10.1111/geoj.12239.

Perreault, Tom, and Gabriela Valdivia. 2010. "Hydrocarbons, Popular Protest and National Imaginaries: Ecuador and Bolivia in Comparative Context." *Geoforum* 41 (5): 689–699. doi: 10.1016/j.geoforum.2010.04.004.

Podeh, Elie. 2011. *The Politics of National Celebrations in the Arab Middle East.* Cambridge: Cambridge University Press.

Radcliffe, Sarah, and Sallie Westwood. 1996. *Remaking the Nation: Place, Identity and Politics in Latin America.* New York, NY: Routledge.

Riofrancos, Thea. 2020. *Resource Radicals: From Petro-Nationalism to Post-Extractivism in Ecuador.* Durham: Duke University Press.

Romero, Simon. 2013. "Argentina Falls from Its Throne as King of Beef." *New York Times*, 13 June. http://www.nytimes.com/2013/06/14/world/americas/argentina-falls-from-its-throne-as-king-of-beef.html.

Rosales, Antulio. 2017. "Contentious Nationalization and the Embrace of the Developmental Ideals: Resource Nationalism in the 1970s in Ecuador." *The Extractive Industries and Society* 4: 102–110. doi: 10.1016/j.exis.2016.12.007.

Ross, Michael. 2012. *The Oil Curse: How Petroleum Wealth Shapes the Development of Nations.* Princeton, NJ: Princeton University Press.

Ross, Michael. 2015. "What Have We Learned About the Resource Curse?" *Annual Review of Political Science* 18: 239–259. doi: 10.1146/annurev-polisci-052213-040359.

Sawyer, Suzana. 2002. "Bobbittizing Texaco: Dis-Membering Corporate Capital and Re-Membering the Nation in Ecuador." *Cultural Anthropology* 17 (2): 150–180.

Schiller, Naomi. 2011. "'Now That the Petroleum Is Ours': Community Media, State Spectacle and Oil Nationalism in Venezuela." In *Crude Domination: An Anthropology of Oil*, edited by Andrea Behrends, Stephen P Reyna, and Günther Schlee, 190–219. New York, NY: Berghahn Books.

Smith, Anthony D. 1986. *The Ethnic Origins of Nations.* Oxford: Basil Blackwell.

Sutherland, Claire. 2012. *Nationalism in the Twenty-First Century: Challenges and Responses.* New York, NY: Palgrave Macmillan.

Swyngedouw, Erik. 1999. "Modernity and Hybridity: Nature, *Regeneracionismo*, and the Production of the Spanish Waterscape." *Annals of the Association of American Geographers* 89 (3): 443–465. doi: 10.1111/0004-5608.00157.

Swyngedouw, Erik. 2015. *Liquid Power: Contested Hydro-Modernities in 20th Century Spain.* Cambridge, MA: MIT Press.

Valdivia, Gabriela. 2008. "Governing the Relations between People and Things: Citizenship, Territory and the Political Economy of Petroleum in Ecuador." *Political Geography* 27 (4): 456–477. doi: 10.1016/j.polgeo.2008.03.007.

Vandergeest, Peter, and Nancy Lee Peluso. 1995. "Territorialization and State Power in Thailand." *Theory and Society* 24 (3): 385–426.

Vivoda, Vlado.. 2009. "Resource Nationalism, Bargaining and International Oil Companies: Challenges and Change in the New Millennium." *New Political Economy* 14 (4): 517–534. doi: 10.1080/13563460903287322.

Watts, Michael. 2004. "Antinomies of Community: Some Thoughts on Geography, Resources and Empire." *Transactions of the Institute of British Geographers* 29: 195–216. doi: 10.1111/j.0020-2754.2004.00125.x.

Williams, Colin, and Anthony D. Smith. 1983. "The National Construction of Social Space." *Progress in Human Geography* 7: 502–518. doi: 10.1177/030913258300700402.

Wilson, Jeffrey D. 2015. "Understanding Resource Nationalism: Economic Dynamics and Political Institutions." *Contemporary Politics* 21 (4): 399–416. doi: 10.1080/13569775.2015.1013293.

Young, Kevin A. 2017. *Blood of the Earth: Resource Nationalism, Revolution and Empire in Bolivia.* Austin: The University of Texas Press.

Zimmermann, Erich W. 1933. *World Resources and Industries: A Functional Appraisal of the Availability of Agricultural and Industrial Materials.* New York, NY: Harper.

12

Resources in a world of borders, boundaries, and barriers
Dividing, circumscribing, confining

Kathryn Furlong, Martine Verdy, and Alejandra Uribe-Albornoz

Introduction

In this chapter, we explore the role of borders, boundaries, and barriers in the (re)making of resources and their effects. These seemingly disparate sites—the frontiers at which resources are ostensibly blocked, corralled, circumscribed, and channeled—are likewise important loci of crossing, fluidity, and porosity where power over resources and their effects can be contested, redefined, and obfuscated. Thus, when considered together, borders, boundaries, and barriers can prove generative of new ways of thinking about key themes in critical resource geographies (CRG). Specifically, we engage with political borders and their work in simultaneously dividing and integrating territories via projects of resource exploitation; ecological boundaries used to separate complex, contested, and shifting ecosystems from a range of activities at their socio-technically defined limits; and the barriers of urban infrastructure intended to canalize and contain urban resources but leaving us with only partial stories of resource urbanization. In dialoguing across these distinct cases, the discourses, tools, and artifacts of scientific and technical knowledge emerge as integral to the mobilization of borders, boundaries, and barriers in the (re)making of resource worlds.

The relevance of borders, boundaries, and barriers in forging resource geographies is evidenced by the attention they have been given in relation to key CRG themes like exclusion and enclosure, natural scales and conservation, and resource metabolism. With respect to transboundary resources, scholars have shown how borders create differential spaces of resource governance that inhibit indigenous resource practices (Norman and Bakker 2009), while empowering colonial resource control (Norman 2014). Here, political and epistemological barriers act together to reinforce transnational techno-scientific resource perspectives over more localized forms of knowledge (Fox and Sneddon 2019). Indeed, borders and boundaries are defined through the confluence of biophysical and social claims (Fall 2005). Similarly, work on ecological boundaries has examined how these represent approximations of ecological divisions that reflect the objectives and discourses of distant policy makers over the lived realities of local resource use and management (Robbins 2001). Such objectives are often justified through the discourses of science and technology. Which scientific approach will dominate, however, is bound up with other forms of knowledge-power. In work that explores watersheds as the "natural scale" of resource management, for example, the delimitation of a given watershed is shown to be an effect of politics, history, and neoliberal science as opposed to a strict result of hydrological analysis (Cohen 2012; Graefe 2011; Molle 2009). In terms of resource urbanization, scholarship has focused on infrastructure as a key instrument of resource metabolism. Here infrastructure acts as a conduit and a barrier, channeling resources to specific spaces and groups, thereby producing multiple inequalities within cities, at sites of resource extraction, as well as across the hinterlands that are bypassed along the way (Gandy 2002; Swyngedouw 2004).

Taking these explorations as starting points, we bring questions of stability and instability, immobility, and mobility into reflections on resource-making at the sites of confinement, crossing, and transgression that borders, boundaries, and barriers provide. For Bridge (2009), what constitutes a resource, how and for whom is fluid, relational, and contested, resting on flows of valuation and knowledge, the relative importance of which is subject to relations of power. This makes instability a key feature of resources and of how we need to think about them. An engagement with borders, boundaries, and barriers helps to bring this instability into focus as they tend to produce arbitrarily divergent effects on either side of an imagined or physical line, one whose effects are constituted through relations of power (Simon 2010). Moreover, as relations of power shift or particular resources become visible and valued, the function and location of a border, boundary, or barrier can change. This can engender major shifts in resource use. The restriction of activities in militarized border zones, for example, can have the effect of pushing residents from extractive to conservation economies—transforming trees from the resource "timber" into the resource "habitat" (Furlong 2006). Thus, while borders, boundaries, and barriers often act to confine resource flows, restricting how and by whom they are accessed—be it through the work of a political border, a conservation area, or a piped water network (Kooy and Bakker 2008; Norman 2014; Ojeda 2012)—they can also wield great power through their instability and porosity, the resource mobilities they allow in addition to those they restrict. The importance of resource mobility in resource-making is made clear in the extensive transportation and energy infrastructures needed to transform remote territories into repositories of resource wealth and economic development (Peyton 2017). But, as we shall see in the sections below, (in)mobility is also important in understanding the mutually constituting effects of resources and the borders, boundaries, and barriers with which they interact.

By bringing borders, boundaries, and barriers into dialogue as multiple loci of resource-making, we hope to bring these insights into focus. First, by analyzing hydroelectric development as a border-crossing project, an opportunity is created to complicate the relationship between resource development and nation. Here, technical discourses are mobilized through legal processes, converting the border from a line between two territories into an effective barrier. Stabilizing the border enables the imposition of conditions for resource mobility. These conditions serve to concentrate the benefits of resource exploitation on one side of the border, and the costs on the other. The nationalist discourses that result mask the interdependence of the territories in question. Second, by examining the work of GIS software in defining ecosystem boundaries, we show how such boundaries—although ostensibly for conservation—also serve to define spaces of extraction. This is produced through the interaction of the ecosystem's own complexity, its spatiotemporal instability, competing and powerful interests, and the software itself. Together these make the boundary mobile and its resource effects unstable. Third, in terms of resource urbanization, infrastructure has been approached as a barrier between resources and cities, containing resources for urban consumption. Ironically, however, what has been confined is our perception of urban resources. Understanding infrastructure as a barrier has led us to neglect the importance of its porosity and limited scope, obfuscating the diverse ways that resources are urbanized, their multiple mobilities, interactions, and effects. This chapter treats each of our three cases in turn. We conclude with some considerations emanating from our triptych, focusing in particular on how thinking across borders, boundaries, and barriers in relation to resources and techno-scientific discourses draws attention to the centrality of mobility—or immobility—in the (re)making of resources and the distribution of their benefits and harms.

Activating borders: resource mobility and nation-building

Resources matter to the formation of nation and nationalism. The state's different attempts to access and control resources are seen as integral to the production of nation and socionatural relations, forming a basis for resource nationalism (Huber 2019; see also Perreault, Chapter 11 this volume). In this nexus of resource-making and state-making, scientific and political practices, and discourses play a central role (Bridge 2014). Large hydroelectric developments have been important in this regard. Here, scholars have tied the construction of large dams to nation-building around the world through the concentration of knowledge, power, and capital, the exclusion and displacement of other resource and territorial claims, and the mobilization of discourses of modernization based on the techno-scientific mastery of nature (Desbiens 2013; Evenden 2009; Mitchell 2002).

Studies of resource nationalism tend to focus within the borders of a given state or subnational entity. As such, there are increasing calls to take resource geographies beyond the state, to understand them in a more relational way (Bridge 2009, 2014; Le Billon 2007). A perspective that explores the ways in which resource- and state-making projects mobilize and work through borders can shed new light on resource geographies. Our study of the Churchill Falls hydroelectric development, which links the province of Quebec with Newfoundland and Labrador (Canada), upends the accepted history of Quebec's resource nationalism. Where it is commonly accepted that the province's nationalist discourse of "maîtres chez-nous" (masters at home) was made possible through the massive damming of its own rivers (e.g., Desbiens 2013), our work shows that it was the construction of a 5200-MW dam in Newfoundland's northern territory of Labrador that generated the wealth and the training for the construction of the family of Quebec dams that came to symbolize that province's modernity and national independence (Verdy 2018). To become "maîtres chez nous," it was necessary to become "maîtres chez-eux" (masters abroad).

Here, the relationship between nationalism and resource exploitation can only be fully understood relationally, between territories. The border is a key instrument in this process. It comes into being through the mobilization of power, science, and the law and serves to divide the benefits of resource extraction from its costs, drawing on and reinforcing nationalist and exclusionary discourses in the process. The Churchill Falls hydroelectric development, completed in 1976, remains among the largest dams in the world. Located in Labrador, it is cut off from the rest of Newfoundland by the Strait of Belle Isle and bordered by the North Atlantic to the east and the province of Quebec to the south and west (see Figure 12.1). Thus, in order for the resource to reach markets and consumers—and thus "become" (Zimmerman 1933)—it had to be transmitted via infrastructure across the territory of Quebec and across the theretofore dormant interprovincial border, a border that had up to then demarcated a sparsely populated territory considered remote, barren, and of little economic or political interest.

Large resource development projects like hydropower tend to reproduce the inequalities that they were promised to resolve. Planned and conceived in contexts of unequal power, they aggravate social and economic divisions, concentrating as opposed to redistributing wealth and power (Mitchell 2002; Worster 1985). In a transboundary context, this pattern is repeated with the border acting as a key instrument through which to redefine national identities

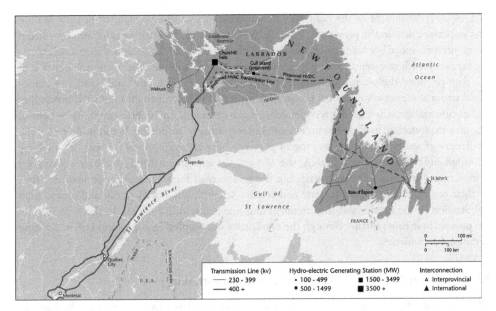

Figure 12.1 The Churchill Falls hydroelectric project in Northern Canada (Froschauer 1999, 110).

Source: Government of Canada, Department of Energy, Mines and Resources, Energy Sector, The National Atlas of Canada, 5th edition, Electricity Generation and Transmission, 1983, MCR4069 and MCR4144; Government of Newfoundland, "New Churchill River Developments Plus HVDC Infeed to Island," 1998. Use of the Atlas data is subject to the Open Government Licence—Canada.

(Anderson 1983; Paasi 1996). Once resources were "discovered" in Labrador, the Quebec government began to mobilize the border as an instrument of territorial division. By refusing to otherwise allow the electricity to cross the border—effectively immobilizing the resource—it was able to compel the Newfoundland government to sell 90 percent of the dam's electricity at a low and static price to Hydro-Québec for a period of 65 years (until 2041). Thus, on one side of the border, the revenues of the electricity sales that would fund Quebec's subsequent dams were concentrated, while the responsibility and costs of maintaining and operating the dam accumulated on the other.

Subsequent governments of Newfoundland and Labrador looked to the legal system to gain access to a greater portion of the electricity to which the province was entitled by law. Yet, the courts continually favored the position of Hydro-Québec, which mobilized techno-scientific discourses to argue that it would not be "economically and technically feasible" to build the infrastructure necessary for the electricity to bypass Quebec. As Jepson (2012) demonstrates, in legal processes over resource access, the object of dispute is often modified through a discursive transformation from something legal to something technical, supported by scientific claims. Such claims, while referred to frequently in settling legal disputes, remain "highly contested, contingent on particular localized circumstances, and freighted with buried presumptions about the social world in which they are deployed" (Jasanoff 1995, xiv). Through legal process, apparently objective legal issues, given their actual ambiguity, are transformed into technical ones allowing contentious political questions to be ignored. Here, the law and its relationship to science are integral in the production of space, and in the effects of the border upon it (Paddison 1983).

Despite lying outside of the politically defined territory of Quebec, Churchill Falls was the first important dam in the province's hydroelectricity network. Resource exploitation mediated by an interprovincial border, science, and the law gave Quebec de facto control over a neighboring territory, with one-tenth of its population and economic output. The economic benefits of the dam provided the wealth for Quebec to become "masters at home" and build a resource-based national identity. A relational perspective shows how the construction of such hierarchies and exclusions through a border is dependent on the economic and political capacity to control resource exploitation, making resources and borders mutually producing. It also underscores the challenges of resource mobility and the importance of overcoming them for resource-making. Through differential power relations, the border was instrumentalized as a barrier between resources and markets, effectively stripping the resources of their economic potential and thus of their resource status. Here, the law and techno-scientific discourses justify and enable such stabilization of the border and the attendant and unequal resource effects. In the next section, we pursue these issues further through the confluence of complex ecologies and GIS in defining ecological boundaries.

Mobile boundaries: GIS, scale, and resource-making

There is a vast literature on ecological boundary-making and its effects on resource-livelihoods. Scholars, for example, have examined how such acts of delimitation—supported by colonial and postcolonial logics (Goldman 2005)—can result in restricted access to or even expulsion from territories newly defined for conservation (Adams and Hutton 2007; see also Lunstrum and Massé, Chapter 30 this volume). In other cases, the boundaries create stark divisions between resource practices on either side of the imagined line where ecological transitions and livelihood strategies are actually gradual, shifting, and interwoven (Robbins 2001; Simon 2010). As a result, processes of boundary-making and stabilization shape relationships between nature and society (Simon 2010), reifying the apparent division between them (Fall 2002). Delineation is thus inherently political and rife with complexity. Boundaries are established to define distribution, access, and control over ecosystems and the resources within them (Graybill 2014).

As such, research tends to focus on the contested nature of what is to be permitted and what is to be prohibited within the newly defined ecological zones. In the case of Colombian paramos, contestation over the boundary is as much about the activities that will be authorized at the boundary's edge as it is about the processes that will be fostered and protected within it. Here, extraction and conservation are bound together within the same process of ecological boundary-making. Paramos, found mainly in the high-altitude plateaus of the Andes, are defined as ecosystems located above the tree line and below the permanent ice cap. At 37, Colombia is home to nearly half of all the paramos in the world. These high-elevation neotropical ecosystems are characterized by their unique varieties of endemic flora and fauna. They are important sources of water and rich in metals and minerals (Rivera and Rodrigues 2014). Paramos are the main source of water for rural communities living in their proximity and provide up to 70 percent of the water for cities like Bogotá and Bucaramanga (Prieto Rozo 2017). Paramos are also valued as sites of resource development, with mining and agribusiness heavily engaged in their delimitation. The particular concern of these economic actors is not with conservation but rather with ensuring that certain lands and resources will be located beyond the paramo's ecological perimeter, where resource exploitation will be permitted.

In Colombia, the government-led processes to define the boundaries of the country's paramos are shaped by a confluence of social, economic, political, and environmental factors that are interpreted and territorialized through GIS science. GIS translates data into simplified, coherent, and accessible representations of reality, encoding particular visions and ideas about how the world is and ought to be organized. On the one hand, it enables the classification of information by establishing areas of interest, generating compact, comparable, and transferable units of knowledge that represent the location of an ecosystem or the presence of a resource through a socially assigned value (Kirsch 1995). The work also facilitates the mapping of greater territorial extents than were previously possible, particularly in areas with low physical accessibility, reducing costs through the virtual elimination of fieldwork (Sheppard 1995). On the other hand, by producing knowledge that is transferable and easy to replicate, GIS generates broad definitions of nature. It can therefore result in oversimplification as it homogenizes data through the elimination of nuance and particularities (Sheppard 1995). In these processes, GIS tends to reinforce dominant ideas by overlooking the lack of consensus within the scientific community as well as perspectives and studies beyond the academy (Kitchin, Perkins, and Dodge 2011). GIS—like mapping more generally—can thus become a tool for representing powerful political and economic agendas in the development of strategies for resource management (Monmonier 1991).

Despite being unique to the Andes—and thus geographically limited—there is no universal definition that encapsulates all of the nuances and distinctive elements of a paramo. Thus, defining the ecological boundaries of paramos is an exercise in ambiguity requiring choices about simplification in the context of shifting, complex, and contested ecologies. This uncertainty is heightened by the multiple scales at which paramo definition is shaped, contested, and transformed. The result has been instability, whereby boundaries shift in relation to the evolving contest over conservation versus extraction, what one weekly news magazine summarized as "Water or gold?" (Revista Semana, July 12, 2019). In the case of Santurban, a paramo adjacent to the city of Bucaramanga, the boundary—within which land-use restrictions apply—has been the subject of multiple contestations over whether or not it should encompass two municipalities that are sites of multinational mining activities comprising 5–8 percent of the paramo in question.

In this dispute, GIS mapping has been central, with the scale of visualization mobilized as a key instrument of boundary-(re)making. In 2012, the Ministry of Environment and Sustainable Development (MinAmbiente) commissioned a study from the Alexander von Humboldt Biological Resources Research Institute of Colombia. This study was conducted at a scale of 1:100,000. It resulted in a contiguous paramo covering 142,608 ha within which the contested municipalities were included. MinAmbiente rejected this delimitation. In 2014, it commissioned a new study using a finer scale of 1:25,000 and an expanded set of landscape categories. These changes modified the modeling of the boundary, yielding a paramo of 99,994 ha (30-percent smaller), the exclusion of the two municipalities, and a cottage cheese of different spatial categories as opposed to a contiguous paramo (Figure 12.2). This new definition was immediately and successfully challenged in the constitutional court by environmental groups. A revised boundary has yet to be determined. Instead, MinAmbiente has been granted numerous extensions to commission additional studies in an effort to produce adjacent yet disconnected spaces where resource exploitation can proceed.

In the context of Santurban, it is greater precision rather than simplification in GIS science that is used to disaggregate the paramo into a patchwork of apparently distinct ecosystem qualities, making more space available for resource exploitation. Producing maps at a finer scale and with greater precision enabled the Ministry to make "visible" spaces within the paramo that could be redefined as another type of ecosystem and thus reopened for other types economic and extractive activities. A paramo boundary, in its mobility, is thus a marker of the shifting power dynamics between conservation and extraction, where scientific precision as opposed to simplification creates ambiguity and thus acts as a tool of power. In the next section, we examine

LOCATION OF THE SANTURBAN PARAMO

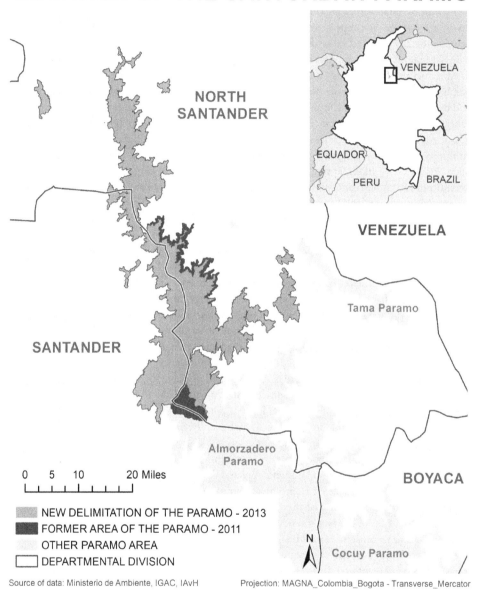

Figure 12.2 Current and former extent of the Santurban Paramo, located in the departments of Santander and North Santander in Colombia.

Source: The data are taken from the *Instituto Geográfico Agustin Codazzi* (IGAC) and the *Investigación de Recursos Biológicos Alexander von Humboldt* (IAvH). The "new delimitation of the paramo" layer comes from a map published by the Ministry of Environment and Sustainable Development in 2014 (MinAmbiente 2014).

how the reification of infrastructure in scholarship has led to our own blindness toward the multiple complexities of resource urbanization.

Porous barriers: resource urbanization beyond infrastructure

Recently, scholars working from the perspective of southern urbanism have multiplied calls to liberate resource urbanization from the conceptual confines of infrastructure. While the view of infrastructure as the mediating agent between nature and the city was central to the development of scholarship on resource urbanization (Gandy 2005; Swyngedouw, Kaïka, and Castro 2002), in recent years, scholars have questioned its applicability to the realities of southern cities (Jaglin 2014; Lawhon, Ernstson, and Silver 2014). In most cities, resource materially exceeds infrastructure in numerous ways. For water alone, these include a variety of nonregulated and nonnetworked water sources, contaminated flows, urban runoff, streams, stagnant pools, and flooding. This urbanization of water beyond infrastructure thus has key implications for human health and well-being.

In this section, rather than a border or boundary connected to a particular territory or ecosystem, we explore a physical barrier cum conceptual barrier that has served to confine our thinking about resource urbanization. Even if it is the experiences of southern cities that have been used to expose the approach's limitations, the issues are not confined to them. The Canadian city of Montreal and its suburbs, for example, regularly experience severe flooding as winter snowmelt swells rivers and lakes overwhelming dikes and turning streets into rivers navigable by canoe. The city has also been compelled to "flush" flows of runoff and sewage directly into the St. Lawrence River as they exceed its infrastructural capacity. These episodes are qualified as emergencies or as exceptional, whereas they are actually relatively recurrent and common. Yet, the belief that infrastructure will be able to contain these resource (over)flows—to perform its work as an impenetrable barrier between nature and the city—holds fast. Plans are made and remade for infrastructure to be renewed, reinforced, and expanded, while discussions about approaching these flows differently—through such things as land-use change or green infrastructures—are sidelined. Here, the myth of infrastructure's capacity to contain and control resources appears more stable than any suburban dike.

In upholding a view of urban nature contained and corralled by infrastructure, scholarship on resource urbanization tends to reduce infrastructure to a conduit for resource flows, a heuristic device for connecting nature and the city. This approach was central in dismantling the binary of urban/artificial-rural/natural in the 1990s by demonstrating the interdependence of urban and rural spaces as well as cities as natural environments dependent on, and vibrating with, flows of resources. Using a Marxist perspective, scholars reworked approaches to resource metabolism, unearthing the power relationships involved in and inequalities produced through resource urbanization. Here, infrastructure was the chief instrument of resource metabolism, connecting cities to their hinterlands. The water coursing through such infrastructure provided a lens through which to understand not only the circulation of resources but also of power in the city (Swyngedouw, Kaïka, and Castro 2002). Infrastructure was seen as a kind of "exoskeleton" lulling urban dwellers into a technical unconscious, unaware of the "invisible" work of infrastructure in containing and channeling the resources necessary to sustain urban life (Gandy 2005), consciousness waking only momentarily upon breakdown (Graham 2010).

The perspective, while important, has been subject to three interrelated critiques, each in some way calling for a "decentering" of infrastructure—that is, an examination of resource flows beyond the barriers of any conduit. The first called for engagement with the lessons of science and technology studies, in which the effects of any technical artifact—including infrastructure—are unstable and cannot be determined a priori by policy, technical design, or structural factors (Coutard and Guy 2007). The second emphasized the fact that the focus on centralized infrastructure networks reduced questions of inequality to the binary of connection/disconnection when relationships to water in southern cities are more complex. In particular, the binary masks the lack of standardization within the formal system whereby even those with connections rely on a diversity of water sources (Gopakumar 2014). Thus, instead of developing a deeper understanding of the diverse realities and creative potential of water's urbanization in southern cities (Boland 2007), the infrastructural focus was said to lead to a cataloguing of "failed examples" (Lawhon, Ernstson, and Silver 2014, 501).

The third critique calls for engagement with the innumerable ways in which resources exceed infrastructure in cities, which are obfuscated by the assumption that infrastructure acts as a barrier, distancing urban residents from ecological processes (Furlong and Kooy 2017). Despite the focus on water in studies of resource urbanization, scholars found that even generalized issues like wastewater, system losses, and drainage were "hardly ever addressed" (Zimmer 2010, 345), that little is said about water's biophysical properties and their potential "influence [on] the socio-spatial development of cities" (Braun 2005, 646), and that little attention is paid to the vulnerability of tens of millions of urban residents to ecological processes (Vollmer and Grêt-Regamey 2013). These ecological connections are everywhere in the production of "uneven urbanization" (Batubara, Kooy, and Zwarteveen 2018). In Colombia, residents suffer uneven exposure to episodes of drought where stream and rainwater collection are important water provision strategies and to flash floods and landslides where drainage infrastructure is poor or absent (Acevedo Guerrero 2018). In a recent article with Michelle Kooy, we addressed these issues through water's urbanization in Jakarta, emphasizing how ecological connections between water flows link "seemingly disjointed spaces of 'wealth and impoverishment'" through groundwater extraction and contaminant flows that engender land subsidence, saltwater intrusion, flooding, and degradation of the waters upon which low-income households depend, increasing ecological and socioeconomic fragmentation across the city (Furlong and Kooy 2017, 894). In all of these ways, urban resources and their mobilities exceed the presumed barrier of urban infrastructure and our ideas about it.

Conclusions

In this chapter, we explored interactions among borders, boundaries, and barriers and the discourses, tools, and artifacts of techno-science in the production of uneven resource geographies. The cases brought together here show that for different borders, boundaries, and barriers, the conditions of crossing are subject to power relations in ways that reconfigure the socio-spatial relations around the resources and the territories in question. In the first case, we show how the border is mobilized through unequal power relations and techno-scientific discourses to successfully block access to electricity markets for Labrador while assuring a constant flow of electricity and profits to Quebec. The border between Quebec and Labrador may exist on a map, but it is largely a conceptual artifact, the meanings and effects of which shift with evolving struggles for resource control. The second case of ecosystems delimited by GIS shows that the boundary must be made mobile to release resources from protected ecosystems and render them exploitable. Using a finer scale, ecological areas like paramos are segmented into different ecosystem types in order to create spaces of exploitation within the spaces initially drawn for conservation. The third case shows how the idea that infrastructure would contain urbanized resources led to the neglect of a diversity of urban resource flows. This conceptual barrier led to a misreading of the physical barrier, overlooking its porosity and limited reach.

This engagement with borders, boundaries, and barriers helps push debates in the CRG literature forward by encouraging new ways of thinking about resource-making and the power relations that shape resource access. Such perspectives combine issues of exclusion and enclosure, natural scales and conservation, and resource metabolism with questions of resource stability and instability, immobility, and mobility. Given that resources are located in space, they must often traverse space if they are to "become," that is to reach sites of transformation, exchange, and consumption. This or its prevention requires the transgression or stabilization of borders, boundaries, and barriers of all kinds. Often this engages the discourses, tools, and artifacts of techno-science and implies some kind of infrastructure and some kind of containment. The type of containment, moreover, can have significant implications for the transformation of space, markets, and the resources themselves (Cronon 1991). For O'Neill (2013), the particularity of infrastructure is not simply its function as a conduit of flows, but its inherent right to transgress the borders of private property, expropriating tracts of land (with the backing of government) as it goes. In this sense, thinking through (in)stability and (in)mobility in relation to borders, boundaries, and barriers, while acknowledging the porosity and limits of infrastructure, can be important for picking apart the shifting complexities of evolving and contested resource worlds.

References

Acevedo Guerrero, Tatiana. 2018. "Water, Arroyos, and Blackouts: Exploring Political Ecologies of Water and the State in Barranquilla." PhD diss., Université de Montréal.

Adams, William, and Jon Hutton. 2007. "People, Parks and Poverty: Political Ecology and Biodiversity Conservation." *Conservation and Society* 5 (2): 147–183.

Anderson, Benedict. 1983. *Imagined Communities: Reflections on the Origin and Spread of Nationalism.* London: Verso.

Batubara, Bosman, Michelle Kooy, and Margreet Zwarteveen. 2018. "Uneven Urbanisation: Connecting Flows of Water to Flows of Labour and Capital Through Jakarta's Flood Infrastructure." *Antipode* 50 (5): 1186–1205. doi: 10.1111/anti.12401.

Boland, Alana. 2007. "The Trickle-Down Effect: Ideology and the Development of Premium Water Networks in China's Cities." *International Journal of Urban and Regional Research* 31 (1): 21–40. doi: 10.1111/j.1468-2427.2007.00702.x.

Braun, Bruce. 2005. "Environmental Issues: Writing a More-than-Human Urban Geography." *Progress in Human Geography* 29 (5): 635–650.

Bridge, Gavin. 2009. "Material Worlds: Natural Resources, Resource Geography and the Material Economy." *Geography Compass* 3 (3): 1217–1244. doi: 10.1111/j.1749-8198.2009.00233.x.

Bridge, Gavin. 2014. "Resource Geographies II: The Resource-State Nexus." *Progress in Human Geography* 38 (1): 118–130.

Cohen, Alice. 2012. "Rescaling Environmental Governance: Watersheds as Boundary Objects at the Intersection of Science, Neoliberalism, and Participation." *Environment and Planning A* 44 (9): 2207–2224.

Coutard, Olivier, and Simon Guy. 2007. "STS and the City: Politics and Practices of Hope." *Science Technology and Human Values* 32 (6): 713–734.

Cronon, William. 1991. *Nature's Metropolis: Chicago and the Great West.* New York, NY: W.W. Norton & Company.

Desbiens, Caroline. 2013. *Power Form the North: Territory, Identity, and the Culture of Hydroelectricity in Quebec.* Vancouver: University of British Columbia Press.

Evenden, Matthew. 2009. "Mobilizing Rivers: Hydro-Electricity, the State, and World War II in Canada." *Annals of the Association of American Geographers* 99 (5): 845–855. doi: 10.1080/00045600903245847.

Fall, Juliet. 2002. "Divide and Rule: Constructing Human Boundaries in 'Boundless Nature.'" *GeoJournal* 58 (4): 243–251.

Fall, Juliet J. 2005. *Drawing the Line: Nature, Hybridity and Politics in Transboundary Spaces.* Aldershot, UK: Ashgate Publishing Limited.

Fox, Coleen A., and Christopher S. Sneddon. 2019. "Political Borders, Epistemological Boundaries, and Contested Knowledges: Constructing Dams and Narratives in the Mekong River Basin." *Water* 11 (3): 413.

Froschauer, Karl. 1999. *White Gold: Hydroelectric Power in Canada.* Vancouver: University of British Columbia Press.

Furlong, Kathryn. 2006. "Unexpected Narratives in Conservation: Discourses of Identity and Place in Šumava National Park, Czech Republic." *Space and Polity* 10 (1): 47–65.

Furlong, Kathryn, and Michelle Kooy. 2017. "Worlding Water Supply: Thinking Beyond the Network in Jakarta." *International Journal of Urban and Regional Research* 41 (6): 888–903.

Gandy, Matthew. 2002. *Concrete and Clay: Reworking Nature in New York City.* Cambridge, MA: MIT Press.

Gandy, Matthew. 2005. "Cyborg Urbanization: Complexity and Monstrosity in the Contemporary City." *International Journal of Urban and Regional Research* 29 (1): 26–49.

Goldman, Michael. 2005. *Imperial Nature: The World Bank and the Making of Green Neoliberalism.* New Haven, CT: Yale University Press.

Gopakumar, Govind. 2014. "Experiments and Counter-Experiments in the Urban Laboratory of Water-Supply Partnerships in India." *International Journal of Urban and Regional Research* 38 (2): 393–412. doi: 10.1111/1468-2427.12076.

Graefe, Olivier. 2011. "River Basins as New Environmental Regions? The Depolitization of Water Management." *Procedia Social and Behavioral Sciences* 14: 24–27.

Graham, Stephen. 2010. "When Infrastructure Fails." In *Disrupted Cities: When Infrastructure Fails*, edited by Stephen Graham, 1–26. New York, NY: Routledge.

Graybill, Andrew R. 2014. "Boundless Nature Borders and the Environment in North America and Beyond." In *The Oxford Handbook of Environmental History*, edited by Andrew C. Isenberg, 668–687. New York, NY: Oxford University Press.

Huber, Matt. 2019. "Resource Geographies II: What Makes Resources Political?" *Progress in Human Geography* 43 (3): 553–564.

Jaglin, Sylvy. 2014. "Regulating Service Delivery in Southern Cities: Rethinking Urban Heterogeneity." In *The Routledge Handbook on Cities of the Global South*, edited by Susan Parnell, and Sophie Oldfield, 434–447. London: Routledge.

Jasanoff, Sheila. 1995. *Science at the Bar: Law, Science, and Technology in America.* Cambridge, MA: Harvard University Press.

Jepson, Wendy. 2012. "Claiming Space, Claiming Water: Contested Legal Geographies of Water in South Texas." *Annals of the Association of American Geographers* 102 (3): 614–631. doi: 10.1080/00045608.2011.641897.

Kirsch, Scott. 1995. "The Incredible Shrinking World? Technology and the Production of Space." *Environment and Planning D: Society and Space* 13 (5): 529–555. doi: 10.1068/d130529.

Kitchin, Rob, Chris Perkins, and Martin Dodge. 2011. "Thinking About Maps." In *Rethinking Maps: New Frontiers in Cartographic Theory*, edited by Martin Dodge, Rob Kitchin, and Chris Perkins, 1–25. London: Routledge.

Kooy, Michelle, and Karen Bakker. 2008. "Splintered Networks: The Colonial and Contemporary Waters of Jakarta." *Geoforum* 39 (6): 1843–1858.

Lawhon, Mary, Henrik Ernstson, and Jonathan Silver. 2014. "Provincializing Urban Political Ecology: Towards a Situated UPE Through African Urbanism." *Antipode* 46 (2): 497–516. doi: 10.1111/anti.12051.

Le Billon, Philippe. 2007. "Geographies of War: Perspectives on 'Resource Wars'." *Geography Compass* 1 (2): 163–182.

MinAmbiente. 2014. "Gestión Integral del Territorio para la Conservación del Complejo de Páramos Jurisdicciones-Santurbán-Berlín." Ministerio de Ambiente y Desarrollo Sostenible. Accessed July 5, 2020. https://www.minambiente.gov.co/images/sala-de-prensa/Documentos/2014/diciembre/191214_mapa_delimitacion_santurban.pdf.

Mitchell, Timothy. 2002. *Rule of Experts: Egypt, Techno-Politics and Modernity.* Berkeley, CA: University of California Press.

Molle, François. 2009. "River-Basin Planning and Management: The Social Life of a Concept." *Geoforum* 40 (3): 484–494. doi: 10.1016/j.geoforum.2009.03.004.

Monmonier, Mark. 1991. *How to Lie with Maps*. Chicago, IL: University of Chicago Press.

Norman, Emma S. 2014. *Governing Transboundary Waters: Canada, the United States, and Indigenous Communities*. London: Routledge.

Norman, Emma S., and Karen Bakker. 2009. "Transgressing Scales: Water Governance Across the Canada-US Borderland." *Annals of the Association of American Geographers* 99 (1): 99–117. doi: 10.1080/00045600802317218.

Ojeda, Diana. 2012. "Green Pretexts: Ecotourism, Neoliberal Conservation and Land Grabbing in Tayrona National Natural Park, Colombia." *The Journal of Peasant Studies* 39 (2): 357–375. doi: 10.1080/03066150.2012.658777.

O'Neill, Phillip M. 2013. "The Financialisation of Infrastructure: The Role of Categorisation and Property Relations." *Cambridge Journal of Regions, Economy and Society* 6 (3): 441–454. doi: 10.1093/cjres/rst017.

Paasi, Anssi. 1996. *Territories, Boundaries and Consciousness: The Changing Geographies of the Finnish-Russian Border*. Chichester, UK: John Wiley and Sons.

Paddison, Ronan. 1983. *The Fragmented State: The Political Geography of Power*. New York, NY: St. Martin's Press.

Peyton, Jonathan. 2017. *Unbuilt Environments: Tracing Postwar Development in Northwest British Columbia*. Vancouver: University of British Columbia Press.

Prieto Rozo, Andrea. 2017. *Conflictos Socioambientales en los Páramos de la Sabana de Bogotá*. Bogotá, Colombia: Associación Ambiente y Sociedad.

Rivera, David, and Camilo Rodrigues. 2014. "Guía Divulgativa de Criterios Para la Delimitación de Páramos de Colombia." Ministerio de Medio Ambiemnte, Vivienda y Desarrollo Territorial y Instituto de Investigación de Recursos Biológicos Alexander von Humboldt.

Robbins, Paul. 2001. "Fixed Categories in a Portable Landscape: The Causes and Consequences of Land-Cover Categorization." *Environment and Planning A: Economy and Space* 33 (1): 161–179. doi: 10.1068/a3379.

Revista Semana. July 12, 2019. "Santurbán sigue en limbo." Accessed September 20, 2019. https://sostenibilidad.semana.com/impacto/articulo/santurban-sigue-en-el-limbo/44976.

Sheppard, Eric. 1995. "GIS and Society: Towards a Research Agenda." *Cartography and Geographic Information Systems* 22 (1): 5–16. doi: 10.1559/152304095782540555.

Simon, Gregory L. 2010. "The 100th Meridian, Ecological Boundaries, and the Problem of Reification." *Society & Natural Resources* 24 (1): 95–101. doi: 10.1080/08941920903284374.

Swyngedouw, Erik. 2004. *Social Power and the Urbanization of Water: Flows of Power*. Oxford: Oxford University Press.

Swyngedouw, Erik, Maria Kaïka, and Esteban Castro. 2002. "Urban Water: A Political Ecology Perspective." *Built Environment* 28 (2): 124–137.

Verdy, Martine. 2018. "Relations Interterritoriales, Hydroélectricité et Pouvoir: Le Cas du Fleuve Churchill au Labrador." PhD diss., Université de Montréal.

Vollmer, Derek, and Adrienne Grêt-Regamey. 2013. "Rivers as Municipal Infrastructure: Demand for Environmental Services in Informal Settlements Along an Indonesian River." *Global Environmental Change* 23 (6): 1542–1555. doi: http://dx.doi.org/10.1016/j.gloenvcha.2013.10.001.

Worster, Donald. 1985. *Rivers of Empire: Water, Aridity, and the Growth of the American West*. New York, NY: Pantheon Books.

Zimmer, Anna. 2010. "Urban Political Ecology: Theoretical Concepts, Challenges, and Suggested Future Directions." *Erdkunde* 64 (4): 343–354. doi: 10.2307/25822107.

Zimmerman, Erich Walter. 1933. *World Resources and Industries*. New York, NY: Harper and Brothers.

13

Pets or meat

A resource geography of dogs in China, from Chairman Mao (1949–1976) to the Pet Fair Asia Fashion Show (2015–2020)

Heidi J. Nast

Introduction

In 1993, Wolfgang Natter and J.P. Jones published an article (with the same implicative title as this one) analyzing capitalism's relation to labor through the lens of Michael Moore's "docu-comedy," *Roger and Me* (1989; Natter and Jones 1993).[1] Set in Flint, Michigan, the film charts the devastations that followed GM's decision to relocate its automotive plants to northern Mexico, just when GM was making record profits. For the authors, insights into why this happened can be seen in the survivalist operations of one of Moore's interviewees, Rhonda Britton. Unemployed, Rhonda has taken to breeding and selling rabbits, which she has packed so tightly into cages that they have started cannibalizing one another. Some customers buy the rabbits as pets, while she slaughters and sells (or uses) the unsold pets as meat. This distinction, the authors aver, is analogous to how capitalism treats labor: as either "pets" (well-fed commoditized labor power on a leash) or "meat" (rendered "redundant" and socially dead). This binary maps somewhat comfortably onto another one, that between exchange and use value, a division that invites us to think about Rhonda in terms of her place (domicile) (on production of nature and labor, see also Schlosser, Chapter 2 this volume). In butchery, it is a theater where Rhonda stages a return to "nature," taking what she needs from the freely reproducing planetary life around her, especially the rabbit. In merchandising the rabbits as pets or meat, by contrast, her domicile becomes a site of retail, a point of purchase. The imaginary and symbolic flexibility of the rabbit brings to light two different geopolitical economies operating in the singularity of Rhonda's domicile and person: the agrarian and the service oriented.

This chapter explores the contradictions of the variously nested pet-meat distinction through the body of the dog in China, where an inordinate amount of "western" capital previously withdrawn through deindustrialization of the "west" was rerouted as foreign direct investment (FDI). Most of this was funneled into China's highly subsidized coastal Special Economic Zones (SEZs), which over time proliferated inland toward China's international borders. The wealth generated there was so great that within decades, China emerged from poverty to become the world's second wealthiest nation with the largest middle class. The dog has been an enduring witness to these epic changes, its own interests naturally buoyed by its evolutionary connection to the human.

This chapter broadens the theoretical utility of the pet-meat distinction, by centering the canine-human *relation* within the great geopolitical economic shifts that came to characterize twentieth-century China. In so doing, it shows the complicated ways in which use value and exchange value are imagined through species and space (on the liveliness of use and exchange value see Naslund and McKeithen, Chapter 6 this volume). The chapter begins by tracing out how the dog was often called upon to rescue China's first communist leader, Chairman Mao Tsetung (1949–1976), from the profound economic and political consequences of his narcissism. It ends by exploring how Mao's successor, Paramount Leader Deng Xiaoping (1978–1992), likewise kept the dog in his employ, this time to address the anomie and loneliness that came from neoliberal reforms and a one-child policy at odds with the economic need to cultivate domestic markets. Here, urban elites drew upon the dog's remarkable phenotypical range to disappear the rural "use value" dog into the pedigreed cosmopolitan world of appearances and exchange.

Mao's narcissism and canine sacrifice

The Korean War, 1950–1953

In the spring of 1957, William Kinmond, a writer with Toronto's *Globe and Mail*, went to "Red China" to report on what communist life looked like there, within months of his return compiling his daily entries into a book, *No Dogs in China* (Kinmond 1957). His visit came at the end of the Hundred Flowers Movement (HFM), a nearly year-long period of openness to the west that also encouraged artists and intellectuals to question the progress of a communism that Mao Tsetung had put into play (Reporters without Borders [RSF] 2009). Kinmond's book title refers to an entry where he talks about how, in the three weeks he had been in China, he had not seen a single dog. Puzzled, Kinmond (163) queries his interpreter who responds that:

> [t]here are no dogs in any of our cities. We killed them all….They were all killed when the US started germ warfare in Korea. We found *the dogs were carriers of the germs* so we had to destroy them. It was a difficult decision to make because we Chinese like dogs (author emphasis).

Finding these statements preposterous, Kinmond seeks additional counsel at the Chinese Foreign Office where he meets with a young Chinese official to whom he poses the same question, receiving a similar reply: "It happened in 1952 when the US aggressors spread germ warfare to northeast China" (164). Kinmond eventually visits China's Minister of Health. She concurs and elaborates:

> In 1952, to protect the country against the bacteriological war then being waged, a great nation-wide patriotic health movement was launched. The people everywhere destroyed dogs, rats, mosquitoes, and other pests, along with their breeding places. In factories, offices, and military units, everyone turned out periodically for thorough cleanups, and this is still being done.
>
> (Kinmond 1957, 164)

Leitenberg's (2016) work provides clues as to what Kinmond's interlocutors were referring to and why Kinmond was confused. Working with government documents from the USSR, China, North Korea, and the United States, and the recollections of the Chinese virologist, Wu Zhili, Leitenberg pieces together a bit of Cold War intrigue related to the Korean War (1950–1953;

see Jian 1998). In early 1951, the Chinese state accused the United States of using bacteriological, biological, and chemical weapons in the Korean War, including ticks and fleas, insects associated with dogs. In early 1952, it claimed that US aircraft were carrying out thousands of germ-warfare sorties over North Korean and Chinese trenches, though it had been unable to down any aircraft or identify the technologies used. When the WHO or the International Committee of the Red Cross mobilized to verify the claims, China and North Korea refused them access, opting to work, instead, with investigators tied to the Soviets. International scientists ridiculed the longest of the resultant reports (669 pages) for its lack of field study and rigor, e.g., identifying insects as vectors that could not have survived under the subzero conditions in which they were found. Wu Zhili was called in to independently investigate the matter. The Soviet Union and North Korea commissioned similar studies. None of the teams found evidence of germ warfare, their findings largely ignored by China and certain influential North Korea actors. Mao's decision to eliminate all of the dogs in China was thus accomplished to assert truths that would gain him local and international support for his revolutionary leadership and the war.

Still, the enormity of canine sacrifice does not explain why no dogs existed in 1957. The dogs involved were members of China's rural canine landraces, the population of which would have rebounded as dogs crossed China's extensive border and reproduced.

Clues into why their absence continued and why officialdom continued to fabricate the germ warfare fiction can be found by looking in a different viral direction, namely, rabies in humans (Zhang et al. 2005). The year Kinmond visited China was also when dog-transmitted cases of rabies in humans peaked. For Mao, the HFM, which saw thousands of international visitors coming to "see" and report on what communist in China looked like, was not the time to have a rabies crisis. By sacrificing *all* dogs in China, Mao disappeared the problem. Whereas in the North Korean context, the dog was sacrificed to demonstrate China's higher moral ground in relation to the United States, during the HFM, it helped Mao save international face. During the Great Famine (GF) that soon followed, Mao rendered all of rural China a sacrifice zone, the evolutionary ties between human and dog giving both parties equal standing.

The Great Famine (1959–1961)

The GF began only months after Mao launched the Great Leap Forward (GLF; 1958–1962) and led to roughly 36 million mostly rural deaths and the disappearance of the rural dog (Meng, Qian, and Yared 2015; Yang 2008; Shapiro 2001).[2] The GLF was *supposed* to have shown the world, including the Soviets (Khrushchev did not support it), that the Chinese peasantry was not only capable of growing enough grain to feed China's massive population but could also produce enough to pay off Soviet loans ahead of time *and* reenergize the nation's flailing heavy industry by building and operating several million backyard smelters. Mao estimated that in 15 years, peasant productivity would rival that of industrial workers in the Soviet Union, Great Britain, and United States (Li 2015; Chang and Halliday 2005; Yang 2008).

The elimination of rural dogs had to do with how Mao saw *himself* as embodying a perfect communism perennially in peril. For Mao, real and potential antagonists existed everywhere and were bent on defeating the revolution and *him*, a signature personalism that he used to call the citizenry *personally* to his defense. His persistent calls for help in the face of interminable danger reinforced the superordinate position he gave *loyalty* in comparison to other forms of care, like attentiveness, recognition, play, gentleness, and nurturance. Loyalty to *him* mattered most. His need for security weaponized and redirected the tenderness of interpersonal life toward punitive defensive ends; of seeking and finding fault in one another where there was none. The punishments meted out to the disloyal were brutal, public, and often fatal.

Figuring out how best to demonstrate loyalty became a source of anxiety for everyone and led to citizen turning against citizen, on the one hand, and administrative cadres turning against peasants, on the other. Cadres were especially important in that it was they who reported on provincial agricultural and industrial output. Well aware of what might befall them if productivity levels did not conform with those Mao had calculated unreasonably, the cadres securitized their futures by citing amounts in *excess* of the already impossible. This demonstration of loyalty through excess proved calamitous for farmers in that the cadre reports determined the quantity of product that the state requisitioned for use in urban areas and for export. Inflated production figures effectively increased the amount requisitioned. To give face to the lie, the cadres paid for the additional grain they needed by seizing the grain allotments of peasants, leaving them with little on which to live. As hunger grew, some of those caught trying to sneak into the (commandeered) grain stores were brutally punished, including children. Rural folk consequently sought out things to eat that were otherwise foreign to their sensibilities: rodents, dogs, tree bark, cotton unraveled from clothing and shoes, sorghum husks, sorghum stalks, insects, and, eventually, the deceased human. Between 1 and 3 million peasants committed suicide. Women not uncommonly took their children with them. One mother accused of having "stolen" food to feed her two children tied them to her body and jumped into a river, knowing that if the authorities killed her, her children would also die (Dikötter 2011, 305; see also Zhi-Sui 1996).

Despite its physical death, the rural dog continued on in language, used as a powerful signifier within communist discourse to describe one's political enemies, those whom one despised, or sadistic political sentiments and acts. Its use speaks not only to a vaulting of the human, but to the cynicism of using a companionate species that had evolved to become the most humanly devoted. Its cynicism was likewise sexed in that village dogs were treated as mongrels, products of bitches unfettered by the disciplinary processes of pedigreed race-making and "breed." This wilding association of rurality and mongrel-bitch is not unlike the sexed and quasi-racialized distinction made between peasant and city-dweller. To wit, Confucian moral distinctions made between the rarefied *junzi*, persons dedicated to living cultured lives of the mind, and the *xiaoren*, that majority ruled by common bodily desires for comfort, food, and "partisan" belonging (Liang 2015, 23; Xiucheng 2015).

The human dog, the children of dogs, and the Red Guard's Cultural Revolution (1966–1968)

[T]ruckloads of Red Guards [traveled]… to Qinghua [University] campus [in Beijing], where they beat the administrators and professors. After … their blood stained the ground, [s]omeone marked a circle around the blood and wrote "dog blood." … [On] June 11–12, 1966, some students organized a "dog-beating team" to beat those who had been accused of being "members of the black gang" or other enemies. They insisted that those people were "dogs" and not human … Students from "bad family backgrounds" were called "children of dogs" (狗崽子) and many of them were physically humiliated or tortured … At Beijing First Middle School, there was … a "team of the children of dogs" … about two hundred students from families with "problems." They were separated from other students and were forced to "work for self-reform."

Red Guards from …the Girls Middle School attached to Beijing Teachers University, held a "struggle meeting" against their classmates… [T]en were Red Guards and sat on chairs, ten were from "bad families" and stood in front of the classroom, while the rest of the class sat on the floor … A huge slogan "Down with the children of dogs" … was pasted on the wall. A long rope went around the necks of the ten students who were "struggled against" … [T]he accused were ordered to "confess" their "reactionary thoughts" and their parents' "crimes," then repeat "I am a child of a dog. I am a rotten egg. I deserve to die."

(Wang 2001, 11, 15–28, 32–33; excerpts recount Red Guard activities and are based on oral interviews with survivors from 95 Beijing schools)

On May 16, 1966, the Central Committee of the Chinese Communist Party proclaimed the beginning of a Great Proletarian Cultural Revolution (CR), a revolution of the mind that would cleanse the Chinese people of staid, nonrevolutionary thinking and of bureaucracies that were as elaborate as they were corrupt. Despite the initiative's reference to the proletariat, the first two years were about students, Mao cultivating the zeal of millions of "Red" school-age children to rid the Party of his growing number of critics. Only the children whose *parents* were in one of Mao's Five *Red* Categories could join, that is, those who at the time of the revolution were poor or lower middle peasants, workers, revolutionary soldiers in the Peoples Liberation Army (PLA), the revolutionary cadres, and revolutionary martyrs (Yi 2019, 19; Zhang and Wright 2018; Song 2011). At a superficial level, Mao saw these children as best positioned in age, temperament, and fidelity to wage a revolution against the staid hierarchies of the Confucian past, what Mao called the Four Olds: old things; old ideas; old customs; old habits. While the students referred to themselves as the Red Guard (RG) and considered themselves a revolutionary vanguard, the movement was inorganic. At the crudest level, the Cultural Revolutionary Group (CRG) that Mao installed would use student energies against the educational system to up-end the Party base, which Mao felt no longer supported him.

Months before the CR launch, the CRG had curated RG leaders from the ranks of Beijing's Red students and faculty. It also endorsed the first "big character poster" that a Beijing University student displayed on campus on May 25, 1966, accusing the university president of being an anti-revolutionary and asking the Party to dismiss him. Students across Beijing followed suit, creating their own posters to decry and humiliate their teachers. Posters bearing revolutionary slogans or defamations were plastered on the walls of school buildings and faculty offices as well as on public buildings and streets. Still others were displayed on the homes and offices of class enemies or made to hang around the necks of the accused, as were chalkboards (Powell and Wong 1997). These were of a piece with posters the state manufactured to extol Mao's virtues (Figure 13.1; cf. Schrift 2001).

As the opening passages of this section indicate, the RG drew on the "dog" to vilify and discipline the *children* of parents Mao had placed in one of his "Black" categories. Just as opposition to Mao had grown with each overreach and failure, so did Mao's categories, from Four to Five (during the 1957 Anti-Rightist Campaign) to *Nine* during the CR. These included landlords, rich peasants, counterrevolutionaries, bad elements, and rightists, as well as four new sorts of persons—capitalists, capitalist roaders, traitors, and spies. During the early months of the CR, being Red or Black was less about choice and more about genetics, blood meanings holding additional species significance in that persons in a Black category were seen as more akin to dogs than humans (Figure 13.1; Yi 2019; Zhang and Wright 2018; Durdin 1971; below).

To ensure the movement became national, RG members were provided free access to trains, with the directive to create a national revolutionary network. Restraints removed, the RG spilled out of Beijing campuses and into the city, wreaking considerable "revolutionary" violence during a month known as Red August that would inspire similar RG violence across China. RG members conscripted their enemies to dismantle public monuments and religious structures. They vandalized and ransacked homes of those believed to be "bourgeois" (including teachers) and destroyed artwork, books, music, jewelry, religious items, and, even, potted plants. Pet cats were bludgeoned to death and thrown into the streets. While most pet dogs in Beijing

Figure 13.1 Smash the dog head of whoever opposes Chairman Mao! A Chinese Cultural Revolution propaganda poster from 1967. Courtesy of Private Collection© The Chambers Gallery, London/Bridgeman Images. See Powell and Wong (1997, 783) for how the same "dog" poster was tailor made to attack Mao's enemies by name.

had been largely eliminated, those that remained were exterminated, as were dogs in other cities (Dikötter 2016, 86; Smith 2001; Tyler 1996; Heaslet 1972).

With student freedoms, the scale and tenor of RG support for Mao swelled; *millions* of students from outside Beijing arrived by train for eight pro-Mao rallies in Tiananmen Square, with Mao there to greet them. Yet, the reason that Mao grew the RG was ultimately not cultural but political. For Mao, the CR represented an opportunity to assemble millions of new, energetic, and politically inexperienced Red allies who would protect and serve him against all odds. The political risks of enfranchising them handsomely paid off. Mao revitalized his base, the sheer numbers of RG supporters stabilizing Mao's position and giving opponents considerable pause. The politics in which Mao engaged was not just about orchestrating a *show* of force but overthrowing the Party base.

While Maoist China had long used dog metaphors to denigrate enemies (running dogs of imperialism), the CR was the first time they were deployed against children. Their power is particular, taken from a metaphor subset that depicts the dog as a species having little value and thus, in comparison with the human, degenerate (Lu 2020; Wu and Zhao 2018; Hatalova 2007). It is this *degeneracy* that made the "dog" so punishing to children. The making of this degeneracy would not, however, have been disinterested. It would have been tied to the dogs involved and those with whom they were associated. For millennia, this would have been China's landrace of rural dogs that guarded farming households, barking at, and often (famously) biting, strangers. Thus, the negative valence needs to be seen in relation to the positionality of the speaker. Presumably, these would have included intruders (non-rural folk) who dominated the countryside and had negative (class- and race-inflected) understandings of the *peasantry*.

The origins of the CR lie in the narcissistic paranoia that followed from the One Hundred Flowers Movement (OHF; 1956–1957), which Mao and the CCP initiated as a means for allowing citizens to voice ideas about the revolution for future conversations that could collectively "resolve contradictions" (Grunfeld 2019, 467; Bing 2018). The Party was inundated with critical responses, and the OHF was quickly shut down. In its place, Mao launched the Anti-Rightist Campaign (1957), which identified some 500,000 class enemies who would be reeducated, tortured, or executed, including leading intellectuals, scholars, and artists.

That Mao felt the need for such a Campaign and for the CR that followed speaks to the *geographical* nature of the anxieties the OHF provoked: until the OHF, Mao had seen the expulsion of western imperialist-capitalists and his severing of relations with the noncommunist world as having made China's revolution secure. Yet, the scale of resistance channeled by the OHF told him that the same "bourgeois elements" the communists had ousted were now residing *in* the Party itself (Harding 1997). Ultimately, Mao's relegation of dissent to an avowed lesser species speaks to his profound fears of difference, which lies at the root of all narcissism.

Deng's dogs

There were two historic initiatives that Deng Xiaoping (1979–1992) enacted as Paramount Leader that allowed the "bourgeois" pedigree dog to make its urban return. First, he opened up the country to FDI, which he directed toward a proliferating array of coastal SEZs. This he did not only to build up China's productivity and technological know-how but to generate employment. Secondly, he introduced the urban one-child policy (1979–2015), the logic being that Chinese standards of living would rise more quickly if wages and salaries were distributed across smaller households.

With access to China's extraordinarily cheap labor, foreign manufacturers made China into the pet-commodity capital of the world, dog-centered consumption driving the global pet market overall.[3] Most pet goods were exported to the United States and other wealthy nation-states. With the sustained oil prices that followed the 1973 crisis and the deindustrialization and global recession that ensued, the United States demand for commodity dog goods mushroomed as owners shifted emotional *and* financial investments away from children and family and toward pets. The pet dog body, like that of the human, became a dynamic *relational* locus of business and love where investments could be intensified and extended across a life course. The pet industry developed "humanization" as its main marketing technique, which allowed the dog to be imaginarily recast as the ideal companion across very different consumer segments. In the United States, humanization made the dog central to the nation's economic recovery and its shift to services (Euromonitor International 2014; Dayan 2014; Bainbridge 2008; Epley, Waytz, and Cacioppo 2007; Nast 2006).

In urban China, similar markets emerged toward the end of the 1990s for different reasons, their existence predicated on Deng's second round of reforms (1992–2001), on which President Jiang Zemin (1993–2003) elaborated. Deng's influence was pivotal: in 1992, in the wake of the Tiananmen massacre and growing Party criticism of the SEZs, he staged an internationally publicized tour of the southern SEZs to show foreign investors just how committed he was to their expansion. His PR strategy worked, and FDI grew by orders of magnitude. Whereas in 1980 90 percent of the population lived in poverty (<$2/day), by 2000 that number had fallen to 40 percent. The impoverished would be replaced by a "low-income class" ($2–$10/day) that would grow from roughly 10 percent of the population in 1981 to 55 percent in 2000. By the new millennium, a strong lower and upper "middle class" materialized, growing from 2 percent to 39 percent of the population between 1999 and 2013. It was their demand that would make domestic markets possible (China Power Team 2019; Chen 2018; Degen 2009; Tomba 2004).[4]

One of the most compelling demands to emerge during Zemin's presidency centered on dog ownership and the *reproductive* contradictions that Deng's reforms had put into play. The one-child policy had worn away at the extended familial roots of Chinese society and generated considerable loneliness, while the number, variety, and size of SEZs had grown at an astronomical pace, generating widespread alienation and new emotional needs (China Power Team 2019). Simultaneously, the capitalist elites that the state had studiously assembled were growing in number and force, broadening and intensifying class demands in unanticipated racialized political ways (Tomba 2004).

Pet dog ownership and the black market

One of the biggest demands was for pet dogs, especially pedigrees, despite the Party's objection to them (Winn 2014; Wei 2008). Similar demands were registering in privileged nation-states worldwide, a measure of how quickly capital's circulation through the SEZs brought larger cosmopolitan needs, sensibilities, and aesthetics to bear. Chinese domestic and international travel deepened these connectivities, as did state investments in universities and growing public access to the Internet. By the new millennium, China was its own cosmopolitan force (China Power Team 2019).

By the early 1990s, the desire for dog ownership had become widespread across Chinese cities, a remarkable development given that purchasing a dog was a clandestine activity that involved considerable risk. Those best able to procure pedigrees came from the ranks of those elite Han Chinese who had benefited the most from neoliberal reforms and had the social and political capital needed to access and navigate the black market, of which there were several kinds. The first market emerged after the fall of the Soviet Union and was controlled by Russian smugglers who either stole the pedigreed dogs of elite individual owners or kennels or were approached by these same parties interested in cashing in on the exceptionally high prices that Chinese elites were known to pay. These smugglers reached their Chinese markets through the trans-Siberian railway, drugging and placing their pedigreed produce in suitcases for the journey. Rural Hong Kong provided another source, city entrepreneurs enlisting rural dwellers to run pedigree puppy mills (Blass 1993).[5] Mainland peasants provided additional, if counterfeit stock. They cosmetically altered pups to make them *look* pedigreed, capitalizing on the aspirational nature of elites while ironically pointing to the fleetingness of appearances.[6] The pedigrees initially in greatest demand were the toy breeds that Chinese royalty and wealthy urbanites had favored in the nineteenth and early-twentieth centuries, especially the Pekingese.[7] Their associative exclusivity re-invoked the racialized distinctions between pedigreed and landrace canines that had obtained in pre-communist eras and that had likewise shaped perceptions of the urbanite and farmer.

Black market dogs circulated through several vendor channels in the 1990s, including regular city markets, like Beijing's Shuizhuizi Market. Blass (1993) records the sale of a purebred miniature pinscher for $2635 one Sunday there in 1993. (The national GDP per capita, then, was less than $700.) At the same time, the especially prestigious Pekingese and other small breeds like the Shih Tzu, Chin, and Shar-pei could sell for about $5000. Vendors additionally sold *rural* dogs as pets for less than $8, expanding the demographic basis of pet dog ownership. Those that pined to own a dog but had little means for doing so also had options. They could visit the Divine Land of Beloved Dogs (or Divine Land Dog Lovers' Paradise), an enclosed commercial area about an hour-long bus ride north of Beijing. Upon paying a small gate fee, visitors entered grounds where they could mingle with the dogs. For an additional 35 cents to 1 dollar, a visitor could rent a dog for an hour, sometimes with a leash (Saywell 1996; Southerl 1994; Blass 1993). Just as with finance capitalism's ascent in the United States, then, the dog provided common relief to groups otherwise disconnected.

In 1989 and 1990, in the lead-up to the 1990 Asian Games in Beijing, municipal police cracked down on dog ownership, ostensibly fearing a potential rabies outbreak, which would cause China to lose face. The founder of one informal Beijing shelter recalled how the police raided her establishment and clubbed all ten of her dogs to death along with the pet dogs of her neighbors (Tyler 1996).[8] Fears of being found out by the authorities compelled many dog owners to keep their pets inside and walk them at night, some taking the extra precaution of cutting their vocal cords.[9]

The reason authorities were so harsh had not only to do with the communist government being largely averse to pet dogs, on principle, but also with the fact that cities were growing at unprecedented rates and were completely unprepared. They had no infrastructure for integrating pet dogs into city life (dog parks, small animal veterinary care); no institutions to issue licenses, assess public health risks, or deal with rescue, care, and control; and no civic means for educating a public about pet dog-keeping. While eliminating pet dogs in Beijing before the Asian Games was partly about public health risks, it also had had to do with staving off a potential PR debacle of visitors snapping photos of stray dogs and ungodly amounts of dog excrement on Beijing's streets.

Cashing in: commodity humanization and the commodity dog in the Chinese megalopolis

Despite the difficulties that owners faced in securing an actual pet dog body and keeping it safe from the authorities, pet dog ownership became ever-more popular. In the 1990s, some megalopolitan areas, like Shanghai, Xiamen, and Beijing, rolled back their anti-canine proscriptions, not because communist officials had a change of heart, but because of the *political* charge that commodity dog ownership now held.

Pedigree dog ownership had become a mark of privilege among the elite that the state had cultivated and wanted to keep, while many other urban dwellers had found that canine companionship kept their loneliness at bay. Dog ownership was becoming integral to China's social *stability* (Wei 2008; Tomba 2004). By the late 1990s, many municipalities had relented and started implementing fee-driven licensing and regulatory systems that interpellated the dog into the city and nation-state. By then, city dog populations were enormous. When Beijing unveiled its licensing program in 1994, 400,000 dogs were estimated to live there, a number that climbed to 600,096 in 2006 and 1 million by 2012 (Winn 2014; Wei 2008).

The daunting practical challenges of accommodating these numbers led some municipal authorities to declare a maximum *size* for pet dogs: no taller than 35 cm at the shoulder. Later, cities like Beijing (2006), Chengdu (2009), Shanghai (2011), Changzhou (2015), and Qingdao (2017) implemented one-*dog* policies that gave owners one of two choices: send their additional dog(s) to live in the countryside or have the authorities take them to a shelter. Changzhou dog owners protested this last directive so strongly that the municipality soon overturned it.

This rise in commodity dog ownership and affection has to be seen as globally exceptional. For half a century prior, Mao had framed the pet dog as antithetical to the "people," a proprietary distraction, an opiate of the aristocrat and the wealthy. As a repository of affection, it took elites away from questions of power and the impoverished lives around them. Drawing on ancient beliefs of the dog as a lesser species than the human, Mao had regularly sacrificed dogs to hide the devastating effects of his narcissism, save face, and garner sympathy. The impetus behind the profound and summary turn toward urban pet ownership in the 1990s could therefore not have come only from a citizenry realizing the emotional benefits of canine companionism. It also came from the state recognizing that the class-striated nature of pet ownership demand could be leveraged toward another one of the state's neoliberalizing objectives, namely, to cultivate local markets as means for recirculating worker earnings and reenergizing capital investments. Pet humanization became a critical means of shoring up the reproduction-oriented spending that Deng's engineered decline in population growth rates had weakened. Its play on care converted the libidinal energies no longer expended in child-centered biological-social activities into "reproductive savings" that could be funneled into homologous human care-related goods and services (Nast 2017).

In 1997 and 1998, just prior to and after the Asian financial crisis, the state supported two massive pet trade shows in especially successful coastal SEZs with ready access to high-performing local and international markets. Both became annual events. The first was the Chinese International Pet Show (CIPS), held in Guangzhou, adjacent to Shenzhen and across from Hong Kong. CIPS (held every other year in Shanghai), initially involved mostly *domestic* wholesalers, distributors, retailers, and budding pet industry professionals, including veterinary. Even so, CIPS worked to cultivate international linkages—and did so successfully[10]: in 2000, there were 142 exhibitors (nine "foreign") and 3806 industry professionals, whereas in 2017, there were 1293 exhibitors (285 foreign) and 60,000 industry professionals.

Pet Fair Asia (PFA) first launched in Shanghai in 1998 where it has been held ever since. PFA was an initiative of VNU Exhibition Asia, one of China's first joint ventures exhibition companies and an affiliate of Royal Dutch Jaarbeurs. Shanghai is the nation's wealthiest and largest city as well as a global financial hub with the second-most capitalized stock market in the world. In 2019, the five-day PFA event covered nearly 2 *million* square feet of floor space within which 1591 exhibitors showcased 20,000 brands and engaged with 67,163 industry professionals. PFA aggressively pursued market diversification, expansion, and rationalization. Most innovatively, it has operated as an institutional clubhouse for an accelerating diversity of pet industry professionals. In 2010, PFA sponsored the first annual International Pet Industry Summit (IPIS); in 2013, it hosted the annual Pet Hospital Management meeting; and in 2015, it launched its first Pet Fashion Show, featuring designers from across Europe, Asia, and China (Figure 13.2). Most recently, it reached out to consumers, permitting them to attend one of its five days, the same day as the fashion show which, in 2019, nearly 204,000 consumers attended.

Figure 13.2 Promotional poster for Pet Fair Asia's fourth annual Pet Fashion Show (2019), now a key part of PFA's annual pet trade show event that launched in Shanghai in 1998 where it is still held, today. The Pet Fashion Show is meant to attract a key consumer demographic, educated and wealthy youth in China, which is also being encouraged to break into international Pet Fashion design. Courtesy of Pet Fair Asia—VNU Exhibitions Asia.

PFA's cultivation of consumers represents a departure from pet industry practice and shows the degree to which *reproductive* interests—for love, companionship, family, children, loyalty, stability—fuel this market, the "aliveness" and companionate nature of the dog pulling productivity and markets together in globally epic proportions.

Within a decade, dozens of shows like the CIPS and PFA were being organized at local and regional levels. As in other global markets, they have provided fodder for a vast dog-related service industry (competitive designer grooming, veterinary subspecialties, schooling and agility training, retail).[11] In seeing the choices available globally, some elite Chinese tastes have shifted away from traditional toy breeds and toward medium-size and large dogs associated with the west—the state censuring of which remains hotly contested.

Conclusions

This entry has examined the changing fortunes of the dog in China to demonstrate the nested and fungible nature of the use-exchange value distinction, historically and across species, in the service of power. For Mao, the dog's greatest (use) value lay in its negation, in its sacrifice. Such sacrifices were realized during the Korean War when China accused the United States of engaging in germ warfare; in 1956, when a rabies epidemic threatened to upend Mao's efforts to showcase communism internationally; and during the GF, when Mao created such famishment that peasants were compelled to eat their own dogs and, eventually, their own kind. The dog's unparalleled loyalty, heightened through species difference, provoked in Mao a special animus, despite loyalty being the trait he most desired. This animus is inexplicable as simply a regionalized cultural phenomenon. A cultural ranking of species cannot explain the intensity with which its lesser-ness was leveraged in metaphor and practice for widespread sadistic (political use value) ends. To wit, the youthful RG derogating Mao's "class" enemies as dogs and submitting them to harm strangely considered dog appropriate.

Mao realized a different use value in his negation (sacrifice) of the urban pedigreed "toy" dog, dogs bred for the pleasures of royalty and other urban elites. Their value came from their market-mediated diminutive looks (aesthetics) and docility, which allowed owners to contain them on their lap and in their sleeve. The toy dog was a conspicuous, living, breathing reminder of how class privilege operates by redirecting care to what one owns. In diverting the gaze away from the relational other, the commodity-dog helped make class injuries less significant. For Mao, sacrificing *this* dog was about destroying the ownership-intimacy relation. Its use value as sacrifice lay in prying the owner away from the owned, inflicting a didactic pain that would free the world from possession.

Deng Xiaoping's policies brought this same urban bourgeois commodity-dog back, the contradictions of FDI and the one-child policy rekindling the ownership-intimacy relation at a globally unprecedented scale. The sociospatial and bodily alienations and desires that transpired, fueled, and suckled one another were changing the dog into the single most important living site of nonhuman emotional and financial investment in urban China, in keeping with trends in other privileged worlds.

Unlike the rabbit, the dog evolved to become the most loyal of human companions, it dyadic companionism making plain that it is *relationality* (which is always materially and spatially mediated) that renders the pet-meat distinction possible. The theoretical importance of the relational shows further how the exchange-use value distinction is *reproductively* circumscribed by power. Mao slaughtered the "bourgeois" fetish dog as a means of disappearing exchange value and dissolving the intimacy-ownership connection, while Deng and his successors used the exchange value of the commodity-dog for new geopolitical economic ends.

The rural dog, like the peasant, has been expended and expendable. The pet-meat distinction is shown to be more than a metaphorical device for describing the status of labor under capitalism. It tells us about—it *is* about—the singularity of *re*production as power.

Notes

1 Many thanks to Professor Xing Lu for helping me think more exactingly about how and why dog metaphors were used during the Cultural Revolution; Ian Liujia Tian for productive theoretical discussions about the rural-urban divide; Gaby Valdivia for her editorial fortitude and patience; and Temi Famodu and Elizabeth Havice for their assistance in preparing the final manuscript.

2 I use Yang's (2008) estimate. He worked for decades for the state newspaper, *Xinhua*, and had privileged access to primary documents that were not part of those which Deng Xiaoping declassified after Mao's death.

3 Walmart began offering dog-related items manufactured in China in the mid-1980s, such as heated and unheated dog houses, squeaky plush toys, dog beds, and dog food. The 1980s was also when PetCo (1980) and PetsMart (1986; later, PetSmart), opened the world's first "big box" pet stores with many goods manufactured in China.

4 The professional "classes" are high-income earners that include lawyers and businessmen, investors, and insurers, bankers, and developers, professors, and doctors, as well as scientists, planners, and architects. It was to them that markets would begin to cater towards the end of the 1990s, especially pet-dogs markets.

5 At the time, Hong Kong was mostly rural (Spollen 2011).

6 This was common in Victorian England; elite reactions against it leading to the creation of the British Kennel Club (Nast 2020).

7 Starting with the Second Opium War, the British stole or arranged illicit deals with palace officials to secure specimens they could bring home for pedigree breeding purposes (Cheang 2006).

8 Lu Di (b. 1927) was a former university professor who began sheltering dogs during the Cultural Revolution. At the time of her interview, she was searching for alternative ways to keep alive the Association of Small Animal Protection, which she founded in 1992 (Wei 2008; Tyler 1996). Beijing's bicycle-riding police were initially the only municipal workforce with the authority to bludgeon dogs. The well-known Beijing filmmaker, Ning Ying, refers to this in her darkly comedic cinema verité film, *On the Beat* (Maslin 1996).

9 Devocalization has long been practiced on humans, while canine debarking has a long history in the pet dog world. A 1966 issue of *Good Housekeeping* (a US weekly) noted its currency in the United States and that it was falling out of favor.

10 The Pet Food and Articles Inspection and Quarantine Branch, China Entry-Exit Inspection and Quarantine Association (CIQA) is under the Ministry of Civil Affairs.

11 A third major commercial pet event launched in 2014, the China Pet Expo, lies outside this work's purview.

References

Bainbridge, Jane. 2008. "Sector Insight: Pet Food and Petcare Retailing – Pampered Pets Buoy Sales." *Marketing*, December 3, 28. https://www.questia.com/magazine/1G1-189989756/sector-insight-pet-food-and-petcare-retailing-pampered.

Bing, Wang, dir. 2018. *Dead Souls. Mandarin with English Subtitles. Les Films d'ici*. Paris, France: Distributed by Icarus Films, released at Cannes Film Festival, France. Film.

Blass, Anthony. 1993. "Traveller's Tales: The Return of the Pekingese." *Far Eastern Economic Review* 156 (27): 30.

Chang, Jung, and Jon Halliday. 2005. *Mao: The Unknown Story*. New York, NY: Random House Books.

Cheang, Sarah. 2006. "Women, Pets, and Imperialism: The British Pekingese Dog and Nostalgia for Old China." *Journal of British Studies* 45 (2): 359–387.

Chen, Chunlai. 2018. "The Liberalisation of FDI Policies and the Impacts of FDI on China's Economic Development." In *China's 40 Years of Reform and Development, 1978–2018*, edited by Ross Garnaut, Ligang Song, and Cai Fang, 595–617. Acton: Australian National University Press.

China Power Team. 2019. "How Well-off is China's Middle Class?." *China Power*. Last modified May 29, 2019. https://chinapower.csis.org/china-middle-class/.

Dayan, Colin. 2014. "Dogs Are Not People." *Boston Review*, January 23. http://bostonreview.net/books-ideas/colin-dayan-dogs-are-not-people-humanity.

Degen, Ronald Jean. 2009. "Opportunity for Luxury Brands in China." *IUP Journal of Brand Management* 6 (3/4): 75–85.

Dikötter, Frank. 2016. *The Cultural Revolution: A People's History 1962–1976*. New York, NY: Bloomsbury Publishing.

Dikötter, Frank. 2011. *Mao's Great Famine*. London: Bloomsbury Publishing.

Durdin, Tillman. 1971. "China Transformed by Elimination of 'Four Olds.'" *New York Times*, May 19. https://www.nytimes.com/1971/05/19/archives/china-transformed-by-elimination-of-four-olds.html.

Epley, Nicholas, Adam Waytz, and John T. Cacioppo. 2007. "On Seeing Human: A Three-Factor Theory of Anthropomorphism." *Psychological Review* 114 (4): 864–886.

Euromonitor International. 2014. Pet Humanisation: The Trend and its Strategic Impact on Global Pet Care Markets. https://www.euromonitor.com/pet-humanisation-the-trend-and-its-strategic-impact-on-global-pet-care-markets/report.

Grunfeld, Thomas A. 2019. "Review of Dead Souls, directed by Wang Bing." *Critical Asian Studies* 51 (3): 467–469.

Harding, Harry. 1997. "The Chinese State in Crisis, 1966–69." In *The Politics of China: The Eras of Mao and Deng*, edited by Roderick MacFarquhar, 148–247. Cambridge: Cambridge University Press.

Hatalova, Henrieta. 2007. "The Dog as Metaphor or Symbol in Chinese Popular Phraseology." *Asian and African Studies* 16 (2): 160–185.

Heaslet, Juliana Pennington. 1972. "The Red Guards: Instruments of Destruction in the Cultural Revolution." *Asian Survey* 12 (12): 1032–1047.

Jian, Chen. 1998. "Not Yet a Revolution: Reviewing China's 'New Cold War Documentation.'" Paper presented at the Cold War International History Conference, National Archives at College Park, MD, September 25–26.

Kinmond, William. 1957. *No Dogs in China*. New York, NY: Thomas Nelson & Sons.

Leitenberg, Milton. 2016. "A Chinese Admission of False Korean War Allegations of Biological Weapon Use by the United States." *Asian Perspective* 40 (1): 131–146.

Li, Lifeng. 2015. "Rural Mobilization in the Chinese Communist Revolution." *Journal of Modern Chinese History* 9 (1): 95–116.

Liang, Kai. 2015. "The Making of the *"Junzi"*: A Modern Perspective on the Pedagogy of *Junzi* in the *Analects*." MA /MSc Thesis, Ohio State University, Graduate Program in East Asian Languages and Literatures.

Lu, Xing. 2020. *Rhetoric of the Chinese Cultural Revolution*. Columbia, SC: University of South Carolina Press.

Maslin, Janet. 1996. "The Lot of Beijing's Finest is Not Happy or Heroic." *New York Times*, April 6. https://www.nytimes.com/1996/04/06/movies/film-festival-review-the-lot-of-beijing-s-finest-is-not-happy-or-heroic.html.

Meng, Xin, Nancy Qian, and Pierre Yared. 2015. "The Institutional Cause of China's Great Famine, 1959–1961." *Review of Economic Studies* 82: 1586–1611.

Nast, Heidi J. 2020. "D is for Dog." In *Animalia: An Anti-Imperial Bestiary of Our Times*, edited by Antoinette Burton, and Renisa Mawani, 45–63. Durham, NC: Duke University Press.

Nast, Heidi J. 2017. "Queering the Maternal?: Unhinging Supremacist Geographies of the Machine, Markets, and Recreational Pleasure." *Society & Space* [online journal] www.societyandspace.org/articles/queering-the-maternal-unhinging-supremacist-geographies-of-the-machine-markets-and-recreational-pleasure.

Nast, Heidi J. 2006. "Loving….Whatever: Loving….Whatever: Alienation, Neoliberalism and Pet-Love in the Twenty-First Century." *ACME* 5 (2): 300–327.

Natter, Wolfgang, and John Paul Jones III. 1993. "Pets or Meat: Class, Ideology, and Space in *Roger and Me*." *Antipode* 25 (2): 140–158.

Powell, Patricia, and Joseph Wong. 1997. "Propaganda Posters from the Chinese Cultural Revolution." *Historian* 59 (4): 777–793.

Reporters without Borders (RSF). 2009. "Sixty Years of News Media and Censorship." *Reporters without Borders*. Last modified January 2016. https://rsf.org/en/reports/sixty-years-news-media-and-censorship.

Saywell, Trish. 1996. "It's a Dog's Life." *Far Eastern Economic Review* 159 (46): 62.

Schrift, Melissa. 2001. *Biography of a Chairman Mao Badge*. New Brunswick, NJ: Rutgers University Press.

Shapiro, Judith. 2001. *Mao's War Against Nature: Politics and the Environment in Revolutionary China*. Cambridge: Cambridge University Press.

Smith, Craig. S. 2001. "Peixian Journal; Local Treat Angers World Pet Lovers." *New York Times*, July 7. https://www.nytimes.com/2001/07/07/world/peixian-journal-local-treat-angers-world-pet-lovers.html.

Song, Yongyi. 2011. "Chronology of Mass Killings During the Chinese Cultural Revolution (1966–1976)." Mass Violence and Resistance – Research Network, *SciencesPo*. https://www.sciencespo.fr/mass-violence-war-massacre-resistance/en/document/chronology-mass-killings-during-chinese-cultural-revolution-1966-1976.html.

Southerl, Daniel. 1994. "Beijing is Going to the Pekingese." *Washington Post*, February 7. https://www.washingtonpost.com/archive/politics/1994/02/07/beijing-is-going-to-the-pekingese/6b66e743-a1dc-4232-8bfe-be3fe0e7367f/.

Spollen, Jonathan. 2011. "Hong Kong's Forgotten Villages." *New York Times*, January 18. https://www.nytimes.com/2011/01/19/world/asia/19villages.html.

Tomba, Luigi. 2004. "Creating an Urban Middle Class: Social Engineering in Beijing." *The China Journal* 51: 1–26.

Tyler, Patrick E. 1996. "For Dogs Pursued Pitilessly, Sanctuary." *New York Times*, November 29. https://www.nytimes.com/1996/11/30/world/for-dogs-pursued-pitilessly-sanctuary.html.

Wang, Youqin. 2001. "Student Attacks Against Teachers: The Revolution of 1966." *Issues and Studies* 37 (2): 1–56.

Wei, Michael. 2008. "Dog Owners Maddened by Beijing Canine Restrictions." *Reuters*, July 16. https://www.reuters.com/article/us-olympics-china-pets/dog-owners-maddened-by-beijing-canine-restrictions-idUSPEK26979420080717.

Wu, Ruixue, and Zuri Zhao. 2018. "A Comparative Study on Conceptual Metaphors Between English and Chinese Dog Idioms." *International Journal of Arts and Commerce* 7 (3): 125–134.

Winn, Patrick. 2014. "China Denounces Pet Dogs as Filthy Imports from the West." *Public Radio International*, August 13. https://www.pri.org/stories/2014-08-13/china-denounces-pet-dogs-filthy-imports-west.

Xiucheng, Pang. 2015. "Unifying vs. Diversifying Approaches and Relevant Reflection on Translation of Cultural Keywords: Based on the Case Analysis of *Junzi* and its Counterpart *Xiaoren*." *Linguistics and Literature Studies* 3(4): 169–178.

Yang, Jisheng. 2008. *Tombstone: The Great Chinese Famine, 1958–62*. New York, NY: Farrar, Straus, and Giroux.

Yi, Xiaocuo. 2019. "Blood Lineage." In *Afterlives of Chinese Communism: Political Concepts from Mao to Xi*, edited by Christian Sorace, Ivan Franceschini, and Nicholas Loubere, 17–22. Canberra: Australian National University Press and Verso Books.

Zhang, Yong-Zhen, Cheng-Long Xiong, Dong-Lou Xiao, Ren-Jie Jiang, Zhao-Xiao Wang, Ling-Zhu Zhang, and Zhen F. Fu. 2005. "Human Rabies in China," letter to the editor. *Emerging Infectious Diseases* 11 (12): 1983–1984.

Zhang, Joshua, and James D. Wright. 2018. *Violence, Periodization and Definition of the Cultural Revolution*. Leiden: Brill.

Zhi-Sui, Li. 1996. *The Private Life of Chairman Mao*. New York, NY: Random House.

14

The social production of resources

A Marxist approach

Matthew T. Huber

Introduction

Resource geography is often a project of tracing the social relations of resource use back to the source (Emel, Bridge, and Krueger 2002). Yet, before a resource becomes an "input," it must first be classified as a "resource" in the first place. Erich Zimmermann's (1951, 15) truism, "resources are not; they become," suggests resources only become seen as resources in specific historical conditions. Zimmermann's maxim makes the process of becoming appear as a complex and multifaceted process including culture, knowledge, technology, politics, and more, but under capitalism there is actually only one factor that matters: whether or not it is profitable to produce a particular resource.[1] Of course, cultural, political, and scientific factors are all very important in shaping what is seen as "profitable" in the first place—and profitability itself must often be produced. But, for the resource producer under capitalism, all these other factors are subordinate to the one question of whether resource production will produce monetary returns to investors. Just because a resource is "known" to exist; even if we have the technological machines and skills to produce useful properties from a forested landscape or a mine; even if there exists a culture of risk-taking; if enough investors with money capital do not think it worth the investment, these resources will not be produced. Indeed, it is this overarching concern with money—itself an abstraction of value—that allows all more concrete concerns such as technology, labor processes, scientific knowledge, and the like, to be subordinated to the quantitative logic of monetary accumulation. I argue here that while this profit imperative is often taken-for-granted, it is an extremely restrictive constraint. I suggest that moving beyond capitalism—from a Marxist view this would mean toward some kind of ecosocialist form of production—would mean opening up the process of resource becoming to a much wider set of criteria, including the social and cultural construction of social and ecological needs.

In this chapter, I will trace the importance of Marxist political economy to contemporary resource geography. A Marxist analysis can help us understand the specific social forms through which resources are seen as "useful." In other words, the ways in which resources become resources has to do with what mode of production predominates at a given historical moment. This is important to contemporary resource geography because it allows us to identify the structural logics that underlie resource production of all kinds under capitalism and not focus entirely on the contingent or "concrete" aspects of individual resources, such as their specific materialities or even injustices wrapped up in their extraction (as important as these all are). Rather than get bogged down in the "modes of production" debates (slave, feudal, Asiatic, capitalist, etc.; see Foster-Carter 1978), I draw from Neil Smith's (1984) "production of nature" thesis to explain three specific forms through which resources are socially produced: production for use, production for exchange, and production for accumulation or profit. I close by speculating on what an ecosocialist resource production system would look like.

The production of resources

In this chapter, I will focus on "natural resources"—or those aspects of the nonhuman environment that get classified as resources. Bridge (2009, 648) suggests a resource is a "product of biological, ecological, or geological processes…that satisfies human wants." Yet, these products don't just arrive to society ready to satisfy these wants. They must also be produced through human labor as a combination of manual effort and systems of knowledge (see Kama, Chapter 5 this volume). Even foraging societies organized forms of labor to produce their means of subsistence. Therefore, a historical materialist perspective places production—in the form of human interchange with nature—at the core of its analytical framework. Marx and Engels ([1845] 1970, 42) explain in The German Ideology:

> The first premise of all human history is, of course, the existence of living human individuals. Thus the first fact to be established is the physical organization of these individuals and their consequent relation to the rest of nature…What they are, therefore, coincides with their production, both with what they produce and with how they produce. The nature of individuals thus depends on the material conditions determining their production.

Thus, on the one hand, production to "satisfy human wants" is universal to all human societies, but on the other hand, we can acknowledge the culturally differentiated ways through which "human wants" are negotiated and imagined. The very category of resource itself in Western contexts denotes a discrete thing to be extracted from the environment, but this is not the only way through which human wants are imagined in relation to the nonhuman world (more on this below).

In Neil Smith's (1984, 55) theory of the production of nature, he makes clear that any analysis of socionatural production must first grasp a notion of "production in general": "By producing the means to satisfy their needs, human beings collectively produce their own material life, and in the process produce new human needs whose satisfaction require further productive activity." Thus, production, or the labor required for production, is what Marx ([1867] 1976, 290) called the "the everlasting nature-imposed condition of human existence."

This transhistorical insight is only so helpful. The key utility of the historical materialist framework is to understand how production changes throughout history. To understand the historically specific forms of production, Marx coined the term "mode of production," and most famously began Capital as an analysis of societies "in which the capitalist mode of production prevails" ([1867] 1976, 125). This has led to a schematic and sometimes mechanical conception of different historical "modes of production," such as the capitalist, slave, feudal, and awkwardly named Asiatic modes, respectively. Historians and Marxists alike have hotly debated how much or little these "modes" apply to actually existing societies (Foster-Carter 1978). These controversies led to the so-called mode of production debates of the 1970s leading many to understand societies—or "social formations"—as complex articulations of many different modes of production at once (see Hall 2003).

A historical materialist resource geography would seek to historicize the particular social relations of resource production. Rather than differentiating between distinct modes, analyzing the differential goals of production can help us understand the contrasting ways in which resources become resources in particular contexts. Following Smith (1984), I explain three particular forms of resource production: production for use, production for exchange, and production for profit (or capitalist production).

Production for use

For the vast majority of human history, production was almost always for direct use—either the direct use of a person, household, or community. In production for use, resources are evaluated in terms of their potential to fulfill human needs. In this sense, Zimmermann's aphorism must pay special attention to the culturally specific ways local communities construct their own versions of needs above the "bare minimum" of subsistence. Subsistence production also characterizes thousands of still-existing peasant and indigenous cultures who orient their production systems toward local needs (see Bennholdt-Thomsen and Mies 1999). For many of these cultures, Zimmermann's saying would better be described simply as "Resources are not…" The very concept of "resource" is itself a highly instrumental vision of the natural world as a discrete field of atomized "things" that can be extracted—or separated from the environment—for human use. Where I live in Central New York, the tribal leader Oren Lyons (2006) of the Onondaga Nation—one of several tribes making up the Haudenosaunee Confederacy—argued:

> Where our white brother will talk about water and trees and animals and fish as resources we talk about them as relatives. That's a whole different perspective. If you think that they're relatives and you understand that then you're going to treat them differently.

Even under capitalism, we continue to run households and other social spheres largely in terms of provisioning human needs—and this process is shaped by the heavily gendered work of social reproduction (Federici 2012). Making dinner is a process of defining what a particular household group needs. A bag of potatoes is a resource that has already gone through significant social and economic processes before reaching the home, but it only truly becomes a useful resource when household members cook them with the definite goal of making digestible (i.e., useful) food.

Yet, we shouldn't assume production for use or need is always innocuous or without power and hierarchy. Even in massively unequal agricultural civilizations elites forced slaves and to produce use values for their own enjoyment—from pyramids to exotic spices and gold. Perhaps the use value of labor itself—the capacity to perform bodily work—was the most critical resource in societies powered by muscle power (Williams 1944; Wrigley 2010). Whether through slavery or serfdom, that resource of "labor" was obtained through coercion.

One thing Marx makes clear about production for use is that there were material limits to how many use values would make sense to accumulate by a specific class of people. Because every use value serves a qualitatively specific need, even the richest person will grow tired of accumulating palaces, sugar, or spices. Ancient empires would literally have spatial limits to how many use values could be stockpiled. At the extreme level, nomadic cultures—whether pastoralist or hunter-gatherer—could only accumulate as many use values as they could carry (see, e.g., Gowdy 1998).

169

Production for exchange

The material limits to a system based on production for use sooner or later can transcend those limits through the development of a surplus. Smith (1984, 59) suggests, "The production of surplus is a necessary if not sufficient condition for the regular exchange of use-values to occur. With production for exchange, the relation with nature is not long exclusively a use value relation…" As soon as communities start producing for exchange, it is also possible for an increasingly complex division of labor to arise based on resource specialization—wherein some individuals or communities devote most or all of their time to producing one specific resource. With such an increasingly complex division often come even more stark and hierarchical class systems of control and coercion of labor.

Production for exchange immediately leads to a different social relationship with resource production. Instead of culturally evaluating resources in terms of specific social needs, those in power start to view resources as an instrumental means to an end. David Ricardo's (1821, 307) theory of "comparative advantage" argues regions that specialize in one commodity will gain from trade in other commodities from other regions. His classic example is wine and cloth. In this exchange economy, wine producers view their vineyard not purely as a source of wine and festiveness, but as a production system that will allow them to acquire something else entirely, cloth. It will also force resource producers to start abstracting from the specific use value of the wine to compare how much wine can be exchanged for the specific quantities of cloth desired. In systems based on production for exchange, a quantitative logic of commensurability takes hold and resources are produced in specific units—bushels of wheat, for example—that can be clearly divided and measured.

Marx explains "production for exchange" is subsumed in a specific logic of circulation. In the case of direct exchange—wine for cloth—the logic of circulation is straightforward: commodity for commodity, or C-C. Yet, it is not always the case where producers of specific use values can find an exchange partner with the specific use value they desire. Thus, money emerges with one of its most important functions: to facilitate exchange. The logic of circulation in this kind of exchange commodity is C-M-C. In other words, an individual produces a commodity to exchange for money and uses the money to acquire another use value. This, of course, makes the quantitative logic of an exchange economy even starker. Instead of measuring the commensurability of two different use values, exchange-producers can evaluate a resource in a pure quantity of money—say $50 per barrel of wine or ton of cloth.

Marx points out, however, that the end goal of this circulation process is still serving needs or use values. "The repetition or renewal of the act of selling in order to buy finds its measure and its goal (as does the process itself) in a final purpose which lies outside it, namely consumption, the satisfaction of definite needs" (Marx [1867] 1976, 252). Although resources are viewed through an abstract and instrumental logic of money, the end goal of resource production is to acquire other useful things.

Historically, such trading systems vary in form and content. For example, while there are important historical debates on when capitalism began and what the principle differences are between mercantilism, commercial capitalism, and industrial capitalism, the mercantilist trading system can be seen as a particular kind of exchange economy. In his classic book *Capitalism and Slavery*, Eric Williams (1944, 51) suggests, "The seventeenth and eighteenth centuries were the centuries of trade, as the nineteenth century was the century of production." He explains how the "triangular trade" in the 1600s—focused on slaves, sugar, manufactured goods, and other commodities—was oriented around the exchange of specific use values desired by elites: "In this triangular trade England—France and Colonial America equally—supplied the exports and the ships; Africa the human merchandise; the plantations the colonial raw materials" (Williams 1944, 51). In this brutally exploitative system, the goals of "exchange" of specific goods predominated over money making in general. (Although it should be said merchants were enriched through this trading system.) It was the plantation owners' need for slave labor, elite desire for sugar, and dispersed demands for cheap manufactured goods that drove the triangular trading system.

Thus, just like production for use, there are still definite material limits to a resource production system based off exchange. Production is limited by the needs for material items of consumption. This is why Marx is at pains to point out that the mere existence of exchange and money does not make a mode of production capitalist. What makes capitalist resource production different than production for exchange?

Capitalist production: resources for profit

As Williams suggests, the nineteenth century was characterized by a new form of production that transformed global economic relations. The industrial revolution revolutionized the productive forces through the implementation of fossil fuel powered automatic machinery (Wrigley 2010). Apart from the materiality of machinery, however, industrial production also required significant up-front capital investment for machines, factory space, and other increasingly large-scale aspects to production (see, e.g., Cronon 1991, 85, on the case of railroad investment). This all meant the need for huge sums of money capital. The fact that industrial producers had to secure money investment first meant their productive orientation must also be directed narrowly to money making for profit. This is not simply a product of greed, but also the institutional constraints of capitalism which force producers to pay interest to bankers and rent to landlords before they take a profit for themselves.

As Marx painstakingly shows in Capital, the formula for capital takes on a much different orientation than production for exchange. Rather than a producer of a specific use value seeking to acquire other use values (C-M-C), production is now guided by an investor seeking a money return on investment. That is, money invested in commodities and coming out with more money at the end of the process (M-C-M'). Thus, if previous systems focused ultimately on "use values," the ways through which resources "became" resources was subject to manifold cultural, economic, and political processes as complex and different as use values themselves. With capitalism, the only question that really matters in whether or not a resource is extracted or produced becomes: is it profitable to do so? One could have the technological know-how, the cultural demand for the product, and the state contract to extract in a specific property, but if the monetary costs of production outweigh the revenues from selling that resource, production won't happen. Of course, for resources to become profitable, it relies on these other factors such as cultural desire, technology, and the like, but capitalism is a system where the profit imperative dominates the underlying production system.

Since M-C-M' is ultimately guided by money, another problem with resource-based capitalist production is the fact that how much M can be generated through extraction is subject to chaotic booms and busts and volatility (Carter, Rausser, and Smith 2011). This means that while it may be profitable for a resource to "become" a resource in a certain historical moment, this can change rapidly. For example, consider the mammoth unconventional oil reserves known as the "tar sands" of Alberta. A historical summary of commercial interest in the region asserts that as early as 1883, Canadian geologists reported that the Athabasca oil sands represented, "the most extensive petroleum field in America, if not the world...it is probable this great petroleum field will assume enormous value in the near future" (Hein 2000, 2–3). Yet, despite this knowledge about the viability of the tar sands as a resource, production did not even start until 1967. And, even then, production was slow. Although there was much investment during the 1970s oil price spike, a lot of those plans were curtailed with the subsequent oil price collapse of the 1980s. In 1998, Alberta produced only 653,200 barrels per day in total crude bitumen (Alberta Energy Regulator 2019). Yet, by 2018, it was producing just over 3 million barrels per day (Alberta Energy Regulator 2019). What changed? The spike in the price of oil made investment in tar sands production viable and profitable. In fact, prior to around 2004, the tar sands weren't even included in calculations of Canada's total "oil reserves." In other words, they weren't even seen as resources at all. To be sure, we cannot assume "oil prices" are themselves apolitical products of purely market forces. The high prices that made tar sand oil profitable are subject to larger power struggles within OPEC to constrain production, wars in the Middle East taking production "offline" (Bichler and Nitzan 2004), financial futures market speculation (see, Labban 2010), and other specific policies meant to provide the conditions for accumulation for oil capital. The conditions of profitability themselves must be produced (Huber 2011).

Nevertheless, much of the discussion of resource extraction avoids this fundamental constraint of a capitalist society: resources don't become resources unless they are profitable. Recently there has been much discourses on the "ethical" dimensions of resource extraction. Certain diamonds are sold as "conflict-free" (Le Billon 2008), food supply chains labeled "sustainable" (Friedberg 2017), and the ecologically destructive tar sands oil is somehow constructed as "ethical oil" because it comes from a "friendly" nation (even as its form of extraction is extremely unfriendly and unethical to the ecosystems and first nation communities nearby) (Levant 2011). These efforts come with mixed results, but they often elide how resources can only be ethically provisioned insofar as they are profitable. This often leads to ethical products coming with a price premium—thus only appealing to relatively privileged consumers (Finn 2017). Without the profit constraint, the realm of ethics could apply much more widely to resource production.

Similarly, resource extraction is also shaped by discourses of "geopolitics" that assume resource control and access is determined exclusively through "political" or state strategic considerations of "national security" and interimperial rivalries (Le Billon 2005). Some narratives of resource geopolitics often assume it is states—not private capitalist firms—who ultimately are the ones who control resource extraction (Huber 2012).[2] This obscures the fact that it is often states aligned with private extractive capital based outside the territory of the state who do the resource production (Emel, Huber, and Makene 2011). States receive rents for such arrangements, but it is not always the case that these rents benefit the "nation" or "society" in which the resources are extracted. It is also equally clear that private capital uses geopolitical discourses to justify extractive regimes—even when their actions prove otherwise. For example, oil and gas capital within the United States constantly proclaimed the national security benefits of hydraulic fracturing—producing domestic oil and gas for domestic consumption—only to aggressively lobby for repeals of export restrictions once the fracking boom created a glut and price collapses (Sica and Huber 2017). Profits required export markets regardless of national security concerns.

We don't often consider how constraining the profitability requirement is over resource production. It overrides all ethical questions of human rights and sustainability. In fact, it is more often the case that the less a resource production system abides by these moral concerns the more profitable they can be. Surely there must be a more just way to organize resource production?

Ecosocialist resources?

Socialism is often defined as "production for social need" (Pepper 2002, 51). In other words, we would call this production for use. Yet, we have already reviewed this form of production. Is socialism then a "backward" mode of production that seeks to resurrect preindustrial agrarian or local subsistence production systems? Not really. Precapitalist production for use was for the use of private individuals, family groups, or, at most, larger regional communities. As we know, capitalism develops through the forcible expropriation of these scattered forms of land-based private and communal production systems (including peasant and indigenous modes of subsistence). This is what Marx ([1867] 1976, 873–940) called "primitive accumulation," and it is important to emphasize that this process is always ongoing and unfinished.

In *Capital*, Marx ([1867] 1976) painstakingly shows how the specifically capitalist mode of production possesses structural tendencies that push individual capitalists to develop increasingly social forms of production. He first covers the complex—but mainly people-driven—forms of cooperation and divisions of labor in the system of "manufacture." As Adam Smith ([1776] 1902) recognized, these social divisions of labor created gains in productivity summed up by what Marx ([1867] 1976, 458) called the "collective worker." For Marx, however, the system of manufacture only socialized production based on old "artisanal" methods into the single space of the factory. What Marx ([1867] 1976, 646) calls the "specifically capitalist mode of production" is based on massive technological revolutions of the mode of production itself. In Marx's time, this was of course the "industrial revolution" and the development of automatic machinery as a replacement for the muscle-driven "collective" forces of human labor. This wasn't only social because of the cooperative arrangement of living workers in a factory but mainly because of the complex social systems of knowledge—engineering, thermodynamics, chemistry, etc.—that become integrated directly into production. As Marx ([1858] 1973, 694) puts it in his famous "fragment on machines" in the *Grundrisse*, this leads to a situation where "[t]he accumulation of knowledge and of skill, of the general productive forces of the social brain, is thus absorbed into capital, as opposed to labour, and hence appears as an attribute of capital."

This is the core contradiction. As capital develops increasingly socialized production systems—think of the global supply chains that provision much of your life—it maintains a private system of appropriation based on profit. Engels ([1880] 1918, 105) summarizes it best in *Socialism: Utopian and Scientific* as, "The contradiction between socialized production and capitalistic appropriation." For Marx ([1867] 1976, 772–781), this contradiction is intensified as more and more capital is concentrated and centralized into fewer and fewer hands (see Campling, Chapter 16 this volume). The only logical resolution for this contradiction is the mass expropriation of private capital in order to take social control over what is already a socialized production system.

This is the standard Marxist story of the progression toward a socialist mode of production. But, for resource geography, we need a properly ecosocialist politics of resource production. The fact is—and much of the history of actually existing socialist regimes proves this—production for social need does not guarantee a larger commitment to socioecological needs. (That is, producing in a way that sustains the ecological relations on which human society—and nonhuman life—depends.) As reviewed above, we should contemplate the problem with the very category of "resource" and "use" specifically. Ecosocialism would need to revise production as less a purely instrumental relationship and more of a cooperative relationship with ecological systems.

Yet, again, ecosocialism—in Marxist terms at least—can't be seen as a return to local or precapitalist communal systems of production for use (although we certainly can learn immensely from such systems).[3] For one reason, the planetary crisis of climate change and ecological breakdown requires social control and coordination at a global level (see Mann and Wainwright 2018). We need a global resource geography to tackle climate change. Ecosocialism must be about harnessing the more-than-local socialized systems of capitalist production to make them fully social in the sense of socially and democratically controlled. For Marx ([1894] 1981, 911), socialism required overcoming the private property relations that monopolized all the control and benefits from socialized production:

> From the standpoint of a higher socioeconomic formation, the private property of particular individuals in the earth will appear just as absurd as the private property of one man in other men. Even an entire society, a nation, or all simultaneously existing societies taken together, are not the owners of the earth. They are simply its possessors, its beneficiaries, and have to bequeath it in an improved state to succeeding generations, as boni patres familias [good heads of the household].

Right now, those living under capitalist social relations access resources through the capitalist market. This has proven a highly unstable and volatile way to organize resource production: subject to booms, busts, overproduction, and colossal waste. The promise of ecosocialism is that we could actually consciously plan and direct our social relationship to nature and resource production. Marx believed if this social production were under social—that is, transparent and democratic—control, the gains from production could be distributed in a way to allow for what he called "disposable time" for all members of society (Marx [1858] 1973, 708). This is another reason for not wanting a return to an agrarian local economy: it would take a lot of work/labor, which always raises questions of social justice in terms of who will do this labor (see Huber 2019). Socialism is about a historical opportunity that only capitalism presents to abolish poverty and create free time for all.

For Marx, freedom itself is only possible through more direct control of our relationship to nature and resource production. As Marx ([1894] 1981, 959) explains in his famous text on the "realm of freedom," "Freedom in this field can only consist in socialized man, the associated producers, rationally regulating their interchange with Nature, bringing it under their common control, instead of being ruled by it as by the blind forces of Nature." Under capitalism, our interchange with nature is regulated by the "blind forces" of the market. We thus can only see "solutions" in terms of those that would somehow internalize ecological problems or "costs" into market prices and market forces—but this project has proven unsuccessful (Aronoff 2017). With the planet literally on fire, the only option is to take social control over resources themselves and subject resource production to different criteria other than private profit.

Conclusion

If resources are not, they become, we need deeper critical understanding of how constrained the becoming of resources is under capitalism. In this chapter, I've applied a Marxist approach to historicize how resources become under different logics of production. Only capitalism subjects resource production to the unlimited imperatives of monetary accumulation and market competition—generating global resource complexes where the control and benefits flow to a tiny international elite of investors and resource-owning states. Moreover, unlike precapitalist systems, more and more humans rely not on land or direct resource production for use but money and commodities for survival. Both capitalists and workers alike under capitalism are fundamentally alienated from the resource-basis of their lives. This toxic cocktail of private profit and mass alienation from nature has created a global resource system bound for collapse. As Marx ([1867] 1976, 381) said, "Capital … takes no account of the health and the length of life of the worker, unless society forces it to do so." Capital also takes no account of life at all and is taking the planet to the brink. We, as "society," need to force it to stop.

Notes

1   There are, of course, some caveats here. Public provision of certain resources (e.g., water) precludes the profitability criterion. Some oil states provision oil beneath the cost of production. Even *private* resource capital will keep producing "unprofitable" resources to recoup already sunk costs and debts (yet, in this case, it is the profitability imperative itself that drives this seemingly irrational production).

2   I should admit that there are significant *state-owned enterprises* involved in resource extraction all across the world—especially in the world of oil and gas, such as the cases of Saudi Aramco, the Chinese state-owned oil firms, and Russian firms (see, Bridge and Le Billon 2017).Yet, even these state-owned firms often produce for the global market and treat profitability as the ultimate arbiter of whether or not resources get produced or not.

3   A classical socialist perspective—drawing from Lenin (1914)—also strongly argues for the self-determination of all nations and peoples.This should include a politics that supports the self-determination of indigenous nations and other peasant-based movements for land and food sovereignty.

References

Alberta Energy Regulator. 2019. "Figure S3.1 Alberta Crude Bitumen Production." *ST98: Alberta Energy Outlook 2019*. Available at: aer.ca/providing-information/data-and-reports/statistical-reports/st98/statistics-and-data.

Aronoff, Kate. 2017. "No Third Way for the Planet." *Jacobin*, May 10th.

Bennholdt-Thomsen, Veronika, and Maria Mies. 1999. *The Subsistence Perspective: Beyond the Globalized Economy*. London: Zed.

Bichler, Shimshon, and Jonathan Nitzan. 2004. "Differential Accumulation and Middle East Wars: Beyond Neo-Liberalism." In *Global Regulation. Managing Crises after the Imperial Turn*, edited by Kees van der Pijl, Libby Assassi, and Duncan Wigan, 43–60. London: Palgrave Macmillan.

Bridge, Gavin. 2009. "Resource." In *The Dictionary of Human Geography*, edited by Derek Gregory, Ron Johnston, Geraldine Pratt, Michael Watts, and Sarah Whatmore, 648–649. Oxford: Wiley.

Bridge, Gavin, and Phillipe Le Billon. 2017. *Oil*. Malden, MA: Polity Press.

Carter, Colin A., Gordon C. Rausser, and Aaron Smith. 2011. "Commodity Booms and Busts." *Annual Review of Resource Economics* 3: 87–118.

Cronon, William. 1991. *Nature's Metropolis: Chicago and the Great West*. New York, NY: W.W. Norton.

Emel, Jody, Gavin Bridge, and Rob Krueger. 2002. "The Earth as Input: Resources." In *Geographies of Global Change: Remapping the World in the Late Twentieth Century*, edited by Peter J. Taylor, Michael Watts, and Ron J. Johnston, 377–390. Oxford: Blackwell.

Emel, Jody, Matthew T. Huber, and Madoshi H. Makene. 2011. "Extracting Sovereignty: Capital, Territory and Gold Mining in Tanzania." *Political Geography* 30 (2): 70–79.

Engels, Frederick. [1880] 1918. *Socialism: Utopian and Scientific*. Chicago, IL: Charles Kerr.

Federici, Silvia. 2012. *Revolution at Point Zero: Housework, Reproduction, and Feminist Struggle*. Oakland, CA: PM Press.

Finn, S. Margot. 2017. *Discriminating Taste: How Class Anxiety Created the American Food Revolution*. New Brunswick, NJ: Rutgers University Press.

Foster-Carter, Aidan. 1978. "The Modes of Production Controversy." *New Left Review* I/107 (January–February): 47–77.

Friedberg, Susanne. 2017. "Big Food and Little Data: The Slow Harvest of Corporate Food Supply Chain Sustainability Initiatives." *Annals of the American Association of Geographers* 107 (6): 1389–1406.

Gowdy, John, ed. 1998. *Limited Wants, Unlimited Means: A Reader on Hunter-Gatherer Economics and The Environment*. Washington, DC: Island Press.

Hall, Stuart. 2003. "Marx's Notes on Method: A 'Reading' of the '1857 Introduction'." *Cultural Studies* 17(2): 113–149.

Hein, Frances J. 2000. "Historical Overview of the Fort McMurray Area and Oil Sands Industry in Northeast Alberta." Edmonton: Alberta Energy and Utilities Board. Albert Geological Survey.

Huber, Matthew T. 2011. "Enforcing Scarcity: Oil, Violence, and the Making of the Market." *Annals of the Association of American Geographers* 101 (4): 816–826.

Huber, Matthew T. 2012. "Energy, Environment and the Geopolitical Imagination." *Political Geography* 31 (6): 402–403.

Huber, Matthew T. 2019. "Ecosocialism: Dystopian and Scientific." *Socialist Forum*. Available at: https://socialistforum.dsausa.org/issues/winter-2019/ecosocialism-dystopian-and-scientific/.

Labban, Mazen. 2010. "Oil in Parallax: Scarcity, Markets, and the Financialization of Accumulation." *Geoforum* 41 (4): 541–552.

Le Billon, Phillipe. 2008. "Diamond Wars? Conflict Diamonds and Geographies of Resource Wars." *Annals of the Association of American Geographers* 98 (2): 345–372.

Le Billon, Phillipe. 2005. *The Geopolitics of Resource Wars*. London: Routledge.

Lenin, Vladimir I. 1914. *The Right of Nations to Self-Determination*. Available at: https://www.marxists.org/archive/lenin/works/1914/self-det/.

Levant, Ezra. 2011. *Ethical Oil: The Case for Canada's Oil Sands*. Toronto: McClelland and Stewart.

Lyons, Oren. 2006. "Oren Lyons: On the Indigenous View of the World." By Leila Connors. Available at: https://ratical.org/many_worlds/6Nations/OrenLyons-IndigenousWorldView.html.

Mann, Geoff, and Joel Wainwright. 2018. *Climate Leviathan: A Political Theory of Our Planetary Future*. London: Verso.

Marx, Karl, and Frederick Engels. [1845] 1970. *The German Ideology*. London: Lawrence and Wishart.

Marx, Karl. [1858] 1973. *Grundrisse*. London: Penguin.

Marx, Karl. [1867] 1976. *Capital Volume 1*. London: Penguin.

Marx, Karl. [1894] 1981. *Capital Volume 3*. London: Penguin.

Pepper, David. 2002. *Eco-Socialism: From Deep Ecology to Social Justice*. London: Routledge.

Ricardo, David. 1821. *On the Principles of Political Economy and Taxation*. London: John Murray Albemarle-Street.

Sica, Carlo, and Matthew T. Huber. 2017. "'We Can't Be Dependent on Anybody': The Rhetoric of 'Energy Independence' and the Legitimation of Fracking in Pennsylvania." *Extractive Industries and Society* 4 (2): 337–343.

Smith, Adam. [1776] 1902. *The Wealth of Nations*. New York, NY: P.F. Collier and Sons.

Smith, Neil. 1984. *Uneven Development: Nature, Capital, and the Production of Space*. Athens, GA: University of Georgia Press.

Williams, Eric. 1944. *Capitalism and Slavery*. Chapel Hill, NC: University of North Carolina Press.

Wrigley, E.A. 2010. *Energy and the English Industrial Revolution*. New York, NY: Cambridge University Press.

Zimmermann, Erich W. 1951. *World Resources and Industries: A Functional Appraisal of the Availability of Agricultural and Industrial Materials*. New York, NY: Harper.

15

World-systems theory, nature, and resources

Paul S. Ciccantell

Introduction

World-systems theory often paid only limited attention to issues of space, nature, and resources. Classic works by Wallerstein (1974) and Arrighi (1994, 2007) often examined resources as key sectors in long-term social change, but these sectors were largely undertheorized. Chase-Dunn and Hall (1997) pay more explicit attention to resources, especially food, as central elements of bulk goods networks and the role of space and location in long-term social evolution. In recent years, world-systems theory and research focused increasing attention on these issues. Three strands are particularly important: Ecological Unequal Exchange (Jorgenson 2006; Jorgenson and Rice 2012), world-ecology (Moore 2015; Patel and Moore 2017), and raw materialist lengthened global commodity chains (Ciccantell and Smith 2009; Sowers, Ciccantell, and Smith 2014, 2017).

In this chapter, I describe each of these theoretical approaches and their utility for understanding resources and the broader relationship between society and nature. Following the broader themes of this volume, I will focus this analysis on issues of space, the relationship between "nature" and society, and how to understand "resource-making" as a long-term sociohistorical process. This discussion will also highlight the overlaps with and divergences from the Global Production Network/Global Value Chain approaches in geography and other social sciences.

The chapter concludes with an illustration utilizing the raw materialist lengthened global commodity chains approach to examine a newly made resource, natural gas extracted from shale via hydraulic fracturing, and its evolution over time in a comparative historical framework, focusing on how this material became a critical resource and the effects of the creation and evolution of this new relationship between society and nature.

World-systems theory and nature

World-systems theory in its early years paid little attention to the environment, although Hopkins and Wallerstein's (1986) formulation of commodity chains emphasized the role of the natural environment as the starting point of many commodity chains. World-systems analyses sometimes noted the natural and environmental characteristics of particular places and industries (e.g., Tomich 1990), but these characteristics were largely analyzed as external to the structures and mechanisms of the world-system. However, by the late 1980s, a number of world-systems scholars began to incorporate the environment more explicitly (e.g., Bunker 1985; Barham, Bunker, and O'Hearn 1990; Ciccantell 1994; Dunaway 1996). World-systems analyses now often focus a great deal of attention on the natural environment and seek to incorporate the environment into the theoretical framework (see, e.g., Moore 2015; Ciplet, Roberts, and Khan 2015).

In the broadest sense, world-systems analysis is well-suited to analyze space, because its organizing principle is explicitly spatial, with three zones making up the capitalist world-economy. This spatiality is often metaphorical rather than geographic, and the physical geographic boundaries of these zones are often unclear and/or topics of intense debate and disagreement. Similarly, world-systems theory's focus on the longue durée provides a window to examine long-term sociohistorical processes such as "resource-making," or how natures of various kinds become resources and are incorporated into systems of appropriation, valuation, transformation, and disposal (see also Kama, Chapter 5 this volume). Moreover, world-systems analysis provides a long-term approach to examining the relationship between society and nature and how this relationship has changed over time.

Ecological Unequal Exchange

One of the most influential streams of this environmentally conscious world-systems analysis is Ecological Unequal Exchange (EUE); this chapter will turn now to outlining its key tenets and analytical framework and its utility for examining issues of space, nature and society, and resource-making.

The EUE approach uses sophisticated quantitative analytic methods to address world-systems and environmental sociology research questions. Four major tenets of EUE are as follows:

1. EUE focuses analytic attention on the appropriation, use, and flows of resources and the resulting waste, as well as the myriad environmental impacts of this extraction, processing, and consumption (Jorgenson and Rice 2012).
2. Unequal material exchange relations, ecological interdependencies, and unequal power across the zones of the world-economy create and reproduce multiple forms of inequality (Hornborg 1998; Roberts and Parks 2007; Jorgenson and Rice 2012).
3. Global inequalities create an apparent contradiction: overconsumption of resources but with relatively less environmental degradation in core countries, while underconsumption of resources in the periphery leaves most residents with poor living and health standards, inadequate incomes, highly polluted shantytowns, and degraded ecosystems (Hornborg 2001; Jorgenson 2003; Jorgenson and Rice 2012).
4. Extractive regions suffer tremendous negative environmental and social consequences from EUE (Bunker 1985; Hornborg 2001; Jorgenson 2003; Bunker and Ciccantell 2005).

Topics examined in the EUE literature include, for example, climate change (Roberts and Parks 2007), the role of primary products exports in EUE, the impact of the military on the environment (Hooks and Smith 2005; Clark, Jorgenson, and Kentor 2010), the disposal of various forms of waste in the periphery (Frey 1995, 1998), deforestation (Burns, Kick, and Davis 2006), the growth of slums in the periphery (Jorgenson, Rice, and Clark 2010), and coal consumption (Clark, Jorgenson, and Auerbach 2012). In summary, EUE offers a world-systems theory–based model for understanding the intrinsic interdependence between the capitalist world-economy and the global environment as well as methodological tools for examining this interdependence.

In EUE, space is typically seen as a variable of location, especially of location within particular national boundaries, in contrast to the focus in geography on space as spatial relationships and particular places with distinct topographies, ecosystems, communities, connections to transport systems, etc. The availability of data limits EUE analyses to recent decades, meaning that these analyses do not examine truly long-term processes of sociohistorical change or the evolving relationship between nature and society. Interestingly, EUE focuses not on resource-making but instead on the loss or destruction of resources via pollution; the extraction and export of resources from peripheries to core nations; the environmental and health costs of resource exports; and other damages from resource extraction, trade, and consumption. In a sense, the focus is on resource un-making as a critical factor in the capitalist world-economy. In this framework, capitalist extraction, processing, and consumption destroy clean water and air; turn landscapes into unproductive, poisoned wastelands; and threaten the health of populations in both extractive and consuming regions.

World-ecology

The world-ecology model is grounded within the world-systems tradition but seeks a fundamentally different approach to understanding long-term social and environmental change that overcomes the Cartesian dualism that separates people and nature into distinct realms of analysis and of reality. The world-ecology approach "unfolds from a rich mosaic of relational thinking about capitalism, nature, power, and history…[and] says that the relationality of nature implies a new method that grasps humanity-in-nature as a world-historical process" (Moore 2015, 3). From this perspective, "capitalism is a world-ecology, joining the accumulation of capital, the pursuit of power, and the co-production of nature in dialectical unity" (Moore 2015, 3). Moore's work has generated a rapidly growing new field of world-ecology scholarship, including an annual international conference and a rapidly growing body of academic research.

Fundamentally, Moore's world-ecology seeks to build on Marx' and Wallerstein's understandings of capitalism but without accepting any sort of philosophical or analytical separation between society and "nature." The goal is to build an analytic model and language that overcome this dualism in order to formulate an understanding of the current geopolitical, socioeconomic, and environmental crises that offers an organizing principle that reveals the inseparability of what are often portrayed as separate social and environmental issues that threaten the very survival of humanity and the "web of life" (Moore 2015) of which humanity is a part. As Moore (2015, 3) puts it, "humans make environments and environments make humans—and human organization." The central overarching analytic questions become: "First, how is humanity *unified* with the rest of nature within the web of life? Second, how is human history a *co-produced* history, through which humans have put nature to work—including other humans—in accumulating wealth and power?" (9). Notably, the analytical model that Moore develops parallels work in geography that similarly seeks new approaches to overcome nature-society dualism (see, e.g., Swyngedouw 1999; Bakker and Bridge 2006).

The world-ecology model places a great deal of focus on space, in part because Moore was trained as a geographer. Moore is particularly concerned to emphasize that the world-ecology model builds on two conceptions that "every effort to accelerate turnover time implies a simultaneous restructuring of space…[and] the accumulation of capital is the production of space" (Moore 2015, 10). Within this framework, capital accumulation operates over time and through space and also depends fundamentally on appropriating from nature the "'Four Cheaps' of labor-power, food, energy, and raw materials" (17). Patel and Moore (2017) expand this list to "Seven Cheaps" of cheap nature, cheap money, cheap work, cheap care, cheap food, cheap energy, and cheap lives, but the fundamental approach remains rooted in the inseparability and coproduction of "nature" and "society" as well as the dependence of capital accumulation and long-term socioeconomic and environmental change on the availability of these "cheaps" and the crises that result when these essential inputs are not available.

Moore's world-ecology framework is particularly useful in explaining the role of biological nature in the capitalist world-economy. Moore's (2009) analysis of the role of the island of Madeira as a sugar commodity frontier in shaping colonialism in Latin America, supporting capital accumulation in Europe, and the impacts of this process on people and the environment is extremely insightful and compelling. Commodity frontiers are a key concept for examining the role of resources in the capitalist world-ecology. Europe, Latin America, and the capitalist world-economy would not have developed and underdeveloped in the ways that they have without this sugar commodity frontier as Patel and Moore (2017) emphasize. Similarly, Moore's (2010a, 2010b, 2010c) analysis of the role of timber supplies to the Dutch Republic from Norway brings into sharp focus how economic ascent in Holland literally and figuratively rested on the extraction of natural resources from its extractive periphery in Norway. Dutch innovations in shipbuilding that made it a global power (Bunker and Ciccantell 2005) could not have occurred in the absence of two inseparable processes: Dutch social action to develop shipbuilding technologies and shipping organizations and the extraction of Norwegian wood to supply Dutch shipyards with their essential raw materials. As Patel and Moore (2017) argue, capitalism simply could not have developed in the ways that it has without access to these cheap resources; "resource-making" is an essential element of capital accumulation and capitalism itself.

Moore (2010c, 2015) and Patel and Moore (2017) utilize this analytic framework to examine commodity frontiers for mineral extraction as well, particularly the role of Potosí in South America as the most important source of silver to fund the Spanish Empire. Marley (2016) similarly examines coal in Appalachia as a commodity frontier. Again, this framework provides important insights into the role of these commodity frontiers as key locations and time periods in which particular commodity frontiers shaped the capitalist world-economy and world-ecology. However, the framework is less directly useful for examining geological nature, including mineral resources and landscapes. Sugar and timber are biological resources that are produced and reproduced in human-scale timescales (measured in years and decades) and are clearly the products of the web of life that includes both "nature" and humans. Patel and Moore's (2017) chapter on energy contains interesting insights about peat and wood as resources, but this chapter does not provide much insight into coal and oil, resources that are produced on a very different timescale that have been essential to capital accumulation for a long period of time but are not coproduced in any meaningful sense with humans and society, which is an issue that a number of scholars have sought to address (see, e.g., Boyd, Prudham, and Schurman 2001; Boyd and Prudham 2017). Resource-making, extraction, transport, processing, and consumption of these and other mineral resources are certainly contributing to the environmental problems such as climate change that motivate Moore (2015) and Patel and Moore (2017), but their analysis focuses on the need for these resources to be cheap rather than how their materiality shapes and constrains capitalist development since the Industrial Revolution. Coal was historically and remains in the twenty-first century a critical ingredient in all cases of transformative economic ascent (Bunker and Ciccantell 2005, 2007; Ciccantell and Gellert 2018), but its extraction and consumption by humans are shaped by much longer term material and sociohistorical characteristics, although its waste products also drive the environmental crisis of climate change that motivates much of Moore's (2015) and Patel and Moore's (2017) analytic efforts focused on the web of life.

Raw materialist lengthened global commodity chains

The raw materialist lengthened global commodity chains theoretical and methodological model brings together the global commodity chains model (Hopkins and Wallerstein 1986; Gereffi and Korzeniewicz 1994; Bair 2005, 2009) and new historical materialism (Bunker and Ciccantell 2005, 2007). The goal is to move past the dualisms between the global and the local and between nature and society, a goal shared with the world-ecology approach that, if achieved, could make it difficult to maintain the analytic separation between humans and the environment, which allows policy debates to be framed in dualistic terms such as "jobs versus the environment."

The raw materialist model (Bunker and Ciccantell 2005, 2007) begins from a focus on the material process of economic ascent in the capitalist world-economy. The key problem for rapidly growing economies over the last five centuries has been obtaining raw materials in large and increasing volumes to supply their continued economic development in the context of economic and geopolitical cooperation and conflict with the existing hegemon and other rising economies. Economies of scale in resource extraction, processing, and transport offer opportunities to reduce costs and create competitive advantages relative to the existing hegemon and other rising economies, but raw materials depletion and increasing distance create diseconomies of space (increasing costs due to the need to bring raw materials from ever more distant extractive peripheries to the consuming regions) that make finding economic, technological, and sociopolitical fixes to sustaining economic ascent via increasing economies of scale difficult to achieve, maintain, and eventually reconstruct on an even larger scale.

Successfully resolving this contradiction relies on the creation of generative sectors: industries that create backward and forward linkages; create patterns of relations between firms, sectors, and states; stimulate a range of technical skills and learning and social institutions to fund and promote them; and kindle the creation of a financial system to meet complex and costly capital needs across borders (Bunker and Ciccantell 2005, 2007). In short, generative sectors drive economic ascent of states and economies. Building these generative sectors is a highly contentious and tenuous process that must be maintained in dynamic tension; it is far more common for efforts in rising economies to create and maintain these sectors to fail than to succeed. While the unit of analysis is the world-system as a whole, this analytic approach focuses attention on the role of states, firms, and the relationships between states and firms in shaping long-term competition and ascent in the world-system.

These processes of economic ascent and economic and geopolitical competition with existing hegemons have driven long-term change in the capitalist world-economy over the last five centuries (Bunker and Ciccantell 2005). The most dramatic and rapid processes of economic ascent restructure national economies and the world economy in support of national economic ascent. The competitive advantages created by organizational and technological innovations in generative sectors and by subsidies in the form of low-cost raw materials from peripheries lead to global trade dominance. Economic and political competition from the existing hegemon and other ascending economies shapes and constrains long-term success, making economic ascent and challenges to existing hegemons extremely difficult. The most successful cases of ascent restructure and progressively globalize the world economy, incorporating and reshaping economies, ecosystems, and space. The historical sequence of rapidly ascending economies from

Holland to Great Britain to the United States to Japan led to dramatic increases in the scale of production and trade, building generative sectors in iron and steel, petroleum, railroads, ocean shipping, and other raw materials and transport industries that drove the economic ascent of their economies and states and impoverished their raw materials peripheries that supply key inputs to ascendant economies (Bunker and Ciccantell 2005, 2007).

The raw materialist lengthened global commodity chains (GCCs) model (Ciccantell and Smith 2009; Sowers, Ciccantell, and Smith 2014, 2017) begins analysis of any commodity chain by focusing on raw materials extraction and processing and on the transport and communications technologies that link the multiple nodes of the commodity chain from its raw materials' sources through industrial processing to consumption and eventually waste disposal. This approach contrasts sharply with most work in the GCCs tradition that focuses on industrial production and consumption and pays little attention to the upstream parts of commodity chains (Ciccantell and Smith 2009). This materially and spatially grounded approach allows analysis of the economic, social, and environmental dimensions of these chains at each node.

Equally important, this approach provides a lens to examine spatially based disarticulations— that is, the marginalization or outright elimination of particular nodes from a GCC (Bair and Werner 2011)—and contestations over extraction, processing, transport, consumption, and waste disposal across these chains. This grounded analysis can examine development trajectories and the sociopolitical conflicts over the division of costs and benefits in particular nodes and across these commodity chains. This approach highlights the role of contestation and resistance to the construction and reproduction of a particular commodity chain in particular places, for example, labor movements and social movement organizations seek to achieve their goals despite resistance from firms and states that oppose these goals, for instance, indigenous groups resisting the redefinition of Amazonian ecosystems into reservoirs for hydroelectric dams (Ciccantell 1994) or the Sierra Club's efforts to end coal use (Ciccantell and Smith 2009; Sowers, Ciccantell, and Smith 2014, 2017; Ciccantell and Gellert 2018).

This model, thus, emphasizes long-term historical change in the world-system as a whole and in particular places and times, and it allows world-systemic comparative analysis that makes nested comparisons within the broader world-system and over time comparisons across commodity chains. Space appears in a variety of critical roles, including as location of extraction, processing, consumption, and waste disposal; as a barrier that must be overcome via transport systems; and as the naturally determined locations of mineral deposits, forests, farmland, and other natural products that social action via technological and organization innovations, geopolitics, and firm and state strategies turn into resources for the capitalist world-economy. The relationship between society and nature plays a central role in shaping and constraining the evolution of the capitalist world-economy via these issues of space and location, the process of resource-making, the criticality of social processes that discover new technological and organizational means to utilize resources, and the centrality of this nature-society relationship in long-term processes of socioeconomic and geopolitical change.

Raw materialist lengthened GCCs and critical resource geography

The raw materialist lengthened GCCs approach (Ciccantell and Smith 2009) was developed precisely because of the same issue that Bridge (2008) identified in the economic geography Global Value Chain (GVC) literature: a lack of attention to the upstream end of commodity chains, parts of the chain that are profoundly different from "footloose" labor-intensive factories that were the focus of most of the literature. Bridge (2008) reconstructs the global value chain approach by emphasizing the issues of materiality and territoriality that are so salient at the extractive stage of production chains. Bridge's (2008) analysis is also concerned with the long-standing debate over the existence of a "resource curse" that dooms resource-rich nations to developmental failure.

Bridge's (2008) framework shares several key elements with the raw materialist lengthened GCCs approach: (1) an emphasis on materiality; (2) recognition of the need to examine the entire commodity chain; (3) a focus on issues of space, territoriality, and the role of the state because of the fixity of resources in space/location/territory; (4) the examination of the developmental impacts of extraction; and (5) a desire to develop a methodology for comparing commodity chains. Ironically, the Ciccantell and Smith (2009) article and our conference presentations in earlier years were developed completely unaware of Bridge's (2008) parallel efforts in geography, highlighting the need to build the interdisciplinary dialogue that motivates the current edited volume. To Bridge's (2008, 407) credit, he is aware of and incorporates aspects of the foundations of the raw materialist approach, particularly the emphasis on economies of scale in Bunker and Ciccantell (2005).

Despite these extensive shared interests and emphases, there are several important distinctions between Bridge's (2008) approach and the raw materialist lengthened GCC model. Most prominently, the raw materialist lengthened GCC model has a much longer term analytic focus, emphasizing processes of socioeconomic change at the national and global levels over decades and centuries, including the central role of processes of national economic ascent as transforming the broader capitalist world-economy. A second difference is the explicit focus on the geopolitical dimensions of commodity chains. Third, our approach emphasizes the role of labor and resistance as integral factors in shaping GCCs and the articulation and disarticulation of particular nodes (Sowers, Ciccantell, and Smith 2014, 2017).

Recent work in the global value chain framework similarly seeks to integrate analytically the upstream end of commodity chains and the material characteristics of resources (and their extraction, processing, and consumption) as essential factors that shape these chains. Havice and Campling (2017) and Campling and Havice (2019) use the tuna global value chain as a case study of how to rework this approach to take seriously "nature" and materiality. Their papers bring resources into a central place in analyzing value chains, arguing that understanding of chain governance needs to expand beyond efforts to create typologies of chains to include an explicit emphasis on environmental conditions of production and environmental problems (Havice and Campling 2017). The careful analysis of the materiality of tuna and of the firm strategies and governance issues across the tuna value chain and contestation between actors provides the sort of rich, grounded understanding of a chain that begins with a "natural resource" that is fundamentally shaped by the relationship between nature and society, materiality, and space as location and territory (Havice and Campling 2017; Campling and Havice 2019; see also Campling, Chapter 16 this volume) that the raw materialist lengthened GCC approach seeks to build. This reformulated GVC approach further shares our methodological emphasis on developing deep familiarity with the chains under study via combining interviews, observations, and quantitative data. This approach also seeks to extend the analytic focus to a much longer term than is typical for GVC analysis and begins to dialogue with the world-systems-based world-ecology framework for understanding and overcoming the dualistic understanding of the relationship between nature and society, for instance, via the concept of commodity frontiers (Campling and Havice 2019). While the raw materialist lengthened GCCs approach still has an explicit longer term and geopolitical focus, there is a great deal of opportunity for dialogue with this framework in critical resource geography.

Conclusion: illustration utilizing the raw materialist lengthened GCCs framework

How can this framework be used to analyze key issues of resource-making, nature-society relationships, and space? A recent case of resource-making and creating a growing global commodity chain, natural gas extracted from shale via hydraulic fracturing, presents perhaps the most obvious and recent case of resource-making on a large and potentially globally transformative scale. Although these deposits of gas contained in shale rock formed in geological processes beginning millions of years ago, and geological knowledge about their existence has existed for decades, these deposits were not resources even 30 years ago. Technological innovations in directional drilling, hydraulic fracturing, and exploration turned these previously socially useless natural formations into socially useful resources. The excess capacity and low prices of natural gas in North America spurred a huge rush of potentially hundreds of billions of dollars in investment in pipelines, port and processing facilities, and specialized ships to transport shale gas as liquefied natural gas (LNG) thousands of miles to markets in Japan, South Korea, China, and Europe, rapidly accelerating the growth of a new GCC in LNG. This new resource and new GCC demonstrate the utility of the raw materialist lengthened GCC approach for analyzing the inseparability of natural and social processes. Without naturally produced gas deposits in shale formations, the potential for social action to build the LNG GCC would have been very limited, while, without social action to develop technologies to explore, extract, and transform this gas into LNG, these gas deposits would have remained undisturbed and particular ecosystems and communities would not have faced air and water pollution, earthquakes from wastewater injection wells, and myriad other environmental and social impacts from fracking and LNG infrastructure (Ciccantell 2020).

In the political realm, natural gas exports as LNG have recently taken on a new economic and geopolitical identity as what the Trump Administration labeled "freedom gas." In a fundamental sense, this represents a key dimension of the "America First" policy orientation of the Trump Administration that rejected traditional Republican allegiance to globalization and free trade. This policy orientation guided US efforts to delink from or at least restructure key elements of the US role in the world economy and geopolitics, including by pulling back from free trade agreements in favor of economic nationalist polices, reducing dependence on traditional alliances, and delinking from some global commodity chains, most notably the Middle East–based oil industry commodity chain. This effort to create what the Trump Administration describes as energy dominance rests fundamentally on a dual redefinition of nature: making shale gas into a resource via fracking and turning natural gas into a globally tradable commodity, LNG. This dual redefinition rests on the social redefinition of large rural areas of the United States into natural gas extracting peripheries, many of which had been agricultural peripheries prior to the shale boom. The integration of these areas into the national natural gas commodity chain and now the global LNG commodity chain has disrupted local socioeconomic and environmental systems in the service of capital accumulation by core firms and investors and to provide energy to rapidly ascending China. The raw materialist lengthened GCC approach emphasizes the inseparability of both nature and society and of the global and local, with fracking peripheries in Appalachia, North Dakota, and West Texas now helping fuel China's economic ascent (Ciccantell 2020).

The United States is returning to its historical role of energy exporter, as it was in the early decades of the oil industry. From the raw materialist lengthened GCC perspective, the United States is moving back to being an extractive periphery, in this case for ascendant China, as it was for Europe and especially Great Britain well into the nineteenth century. This re-peripheralization is exploiting US natural resources in support of China's economic ascent while leaving large areas of the United States with depleted resources, damaged ecosystems, and disrupted communities from fracking (Ciccantell 2020).

In short, the raw materialist lengthened GCCs framework offers an analytic method to examine central issues of critical resource geography. This framework incorporates analysis of space, processes of long-term change, and the role of raw materials extraction and processing in the political economy of the capitalist world-economy, offering fresh insights into issues of central concern to critical resource geography. Given these shared concerns, there are clearly significant opportunities for scholars in sociology, geography, and other social sciences who employ these frameworks to share findings and work jointly to examine pressing issues such as climate change and other products of resource use, consumption, and waste, and the impacts of processes of making and remaking of resources that are, for example, turning the Amazon rainforest into cattle pasture and soybean farms to supply China's economic ascent.

References

Arrighi, Giovanni. 1994. *The Long Twentieth Century: Money, Power, and the Origins of Our Times*. London: Verso.

Arrighi, Giovanni. 2007. *Adam Smith in Beijing: Lineages of the 21st Century*. New York, NY: Verso.

Bair, Jennifer. 2005. "Global Capitalism and Commodity Chains: Looking Back, Going Forward." *Competition and Change* 9 (2): 153–180.

Bair, Jennifer, ed. 2009. *Frontiers of Commodity Chain Research*. Stanford, CA: Stanford University Press.

Bair, Jennifer, and Marion Werner. 2011. "The Place of Disarticulations: Global Commodity Production in La Laguna, Mexico." *Environment and Planning A* 43: 998–1015.

Bakker, Karen, and Gavin Bridge. 2006. "Material Worlds? Resource Geographies and the 'Matter of Nature." *Progress in Human Geography* 30 (1): 5–27.

Barham, Bradford, Stephen G. Bunker, and Denis O'Hearn, eds. 1990. *States, Firms, and Raw Materials: The World Economy and Ecology of Aluminum*. Madison, WI: University of Wisconsin Press.

Boyd, William, and W Scott Prudham. 2017. "On the Themed Collection, 'The Formal and Real Subsumption of Nature'." *Society & Natural Resources* 30 (7): 877–884.

Boyd, William, W. Scott Prudham, and Rachel A. Schurman. 2001. "Industrial Dynamics and the Problem of Nature." *Society & Natural Resources* 14 (7): 555–570.

Bridge, Gavin. 2008. "Global Production Networks and the Extractive Sector: Governing Resource-Based Development." *Journal of Economic Geography* 8 (3): 389–419.

Bunker, Stephen G. 1985. *Underdeveloping the Amazon*. Chicago, IL: University of Chicago Press.

Bunker, Stephen G., and Paul S. Ciccantell. 2005. *Globalization and the Race for Resources*. Baltimore, MD: Johns Hopkins University Press.

Bunker, Stephen G., and Paul S. Ciccantell. 2007. *An East Asian World Economy: Japan's Ascent, With Implications for China*. Baltimore, MD: Johns Hopkins University Press.

Burns, Thomas, Edward Kick, and Byron Davis. 2006. "A Quantitative Cross-National Study of Deforestation in the Late 20th Century: A Case of Recursive Exploitation." In *Globalization and the Environment*, edited by Andrew Jorgenson, and Edward Kick, 37–60. Leiden, NL: Brill.

Campling, Liam, and Elizabeth Havice. 2019. "Bringing the Environment into GVC Analysis: Antecedents and Advances." In *Handbook on Global Value Chains*, edited by Stefano Ponte, Gary Gereffi, and Gale Raj-Reichert, 214–227. Cheltenham: Edward Elgar.

Chase-Dunn, Christopher, and Thomas Hall. 1997. *Rise and Demise: Comparing World Systems*. New York, NY: Routledge.

Ciccantell, Paul S. 1994. "Raw Materials, States and Firms in the Capitalist World Economy: Aluminum and Hydroelectricity in Brazil and Venezuela." PhD diss., University of Wisconsin-Madison.

Ciccantell, Paul S. 2020. "Liquefied Natural Gas: Redefining Nature, Restructuring Geopolitics, Returning to the Periphery?" *The American Journal of Economics and Sociology* 79 (1): 265–300.

Ciccantell, Paul S., and Paul Gellert. 2018. "Raw Materialism and Socio-Economic Change in the Coal Industry." In *Oxford Handbook of Energy and Society*, edited by Debra Davidson, and Matthias Gross, 113–136. Oxford: Oxford University Press.

Ciccantell, Paul S., and David A. Smith. 2009. "Rethinking Global Commodity Chains: Integrating Extraction, Transport, and Manufacturing." *International Journal of Comparative Sociology* 50 (3–4): 361–384.

Ciplet, David, J. Timmons Roberts, and Mizan Khan. 2015. *Power in a Warming World: The New Global Politics of Climate Change and the Remaking of Environmental Inequality*. Cambridge, MA: MIT Press.

Clark, Brett, Andrew Jorgenson, and Daniel Auerbach. 2012. "Up in Smoke: The Human Ecology and Political Economy of Coal Consumption." *Organization and Environment* 25 (4): 452–569.

Clark, Brett, Andrew Jorgenson, and Jeffrey Kentor. 2010. "Militarization and Energy Consumption: A Test of Treadmill of Destruction Theory in Comparative Perspective." *International Journal of Sociology* 40 (2): 23–43.

Dunaway, Wilma. 1996. *The First American Frontier: Transition to Capitalism in Southern Appalachia, 1700-1860*. Chapel Hill, NC: University of North Carolina Press.

Frey, R. Scott. 1995. "The International Traffic in Pesticides." *Technological Forecasting and Social Change* 50 (2): 151–169.

Frey, R. Scott. 1998. "The Hazardous Waste Stream in the World-System." In *Space and Transport in the World-System*, edited by Paul S. Ciccantell, and Stephen G. Bunker, 84–103. Westport, CT: Greenwood Press.

Gereffi, Gary, and Miguel Korzeniewicz, eds. 1994. *Commodity Chains and Global Capitalism*. Westport, CT: Praeger.

Havice, Elizabeth, and Liam Campling. 2017. "Where Chain Governance and Environmental Governance Meet: Interfirm Strategies in the Canned Tuna Global Value Chain." *Economic Geography* 93 (3): 292–313.

Hooks, Gregory, and Chad Smith. 2005. "Treadmills of Production and Destruction: Threats to the Environment Posed by Militarism." *Organization and Environment* 18 (1): 19–37.

Hopkins, Terence K., and Immanuel Wallerstein. 1986. "Commodity Chains in the World-Economy Prior to 1800." *Review (Fernand Braudel Center)* 10 (1): 157–170.

Hornborg, Alf. 1998. "Ecosystems and World Systems: Accumulation as an Ecological Process." *Journal of World-Systems Research* 4 (2): 169–177.

Hornborg, Alf. 2001. *The Power of the Machine: Global Inequalities of Economy, Technology, and Environment*. New York, NY: AltaMira Press.

Jorgenson, Andrew. 2003. "Consumption and Environmental Degradation: A Theoretical Proposition and Cross-National Study of the Ecological Footprint." *Social Problems* 50 (3): 374–394.

Jorgenson, Andrew. 2006. "Unequal Ecological Exchange and Environmental Degradation: A Theoretical Proposition and Cross-National Study of Deforestation, 1990–2000." *Rural Sociology* 71 (4): 685–712.

Jorgenson, Andrew, and James Rice. 2012. "The Sociology of Ecologically Unequal Exchange in Comparative Perspective." In *Routledge Handbook of World-Systems Analysis*, edited by Salvatore Babones and Christopher Chase-Dunn, 431–439. New York, NY: Routledge.

Jorgenson, Andrew, James Rice, and Brett Clark. 2010. "Cities, Slums, and Energy Consumption in Less-Developed Countries, 1990–2005." *Organization and Environment* 23 (2): 189–204.

Marley, Benjamin. 2016. "The Coal Crisis in Appalachia: Agrarian Transformation, Commodity Frontiers and the Geographies of Capital." *Journal of Agrarian Change* 16 (2): 225–254.

Moore, Jason. 2009. "Madeira, Sugar, and the Conquest of Nature in the 'First' Sixteenth Century: Part I: From 'Island of Timber' to Sugar Revolution, 1420-1506." *Review of the Fernand Braudel Center* 32 (4): 345–390.

Moore, Jason. 2010a. "Amsterdam Is Standing on Norway Part 1: The Alchemy of Capital, Empire, and Nature in the Diaspora of Silver, 1545–1648." *Journal of Agrarian Change* 10 (1): 35–71.

Moore, Jason. 2010b. "Amsterdam Is Standing on Norway Part 2: The Global North Atlantic in the Ecological Revolution of the Long Seventeenth Century." *Journal of Agrarian Change* 10 (2): 188–227.

Moore, Jason. 2010c. "This Lofty Mountain of Silver Could Conquer the Whole World': Potosi and the Political Ecology of Underdevelopment, 1545–1800." *The Journal of Philosophical Economics* 4 (1): 58–103.

Moore, Jason. 2015. *Capitalism in the Web of Life: Ecology and the Accumulation of Capital*. London: Verso.

Patel, Raj, and Jason Moore. 2017. *A History of the World in Seven Cheap Things: A Guide to Capitalism, Nature, and the Future of the Planet*. Oakland, CA: University of California Press.

Roberts, J. Timmons, and Bradley Parks. 2007. *A Climate of Injustice: Global Inequality, North-South Politics, and Climate Policy*. Cambridge, MA: MIT Press.

Sowers, Elizabeth, Paul S. Ciccantell, and David A. Smith. 2014. "Comparing Critical Capitalist Commodity Chains in the Early Twenty-First Century: Opportunities for and Constraints on Labor and Political Movements." *Journal of World-Systems Research* 20 (1): 112–139.

Sowers, Elizabeth, Paul S. Ciccantell, and David Smith. 2017. "Are Transport and Raw Materials Nodes in Global Commodity Chains Potential Places for Worker/Movement Organization?" *Labor and Society* 20 (2): 185–205.

Swyngedouw, Erik. 1999. "Modernity and Hybridity: Nature, *Regeneracionismo*, and the Production of the Spanish Waterscape, 1890–1930." *Annals of the Association of American Geographers* 89 (3): 443–465.

Tomich, Dale. 1990. *Slavery in the Circuit of Sugar: Martinique in the World Economy, 1830–1848*. Baltimore, MD: Johns Hopkins University Press.

Wallerstein, Immanuel. 1974. *The Modern World-System*. New York, NY: Academic Press.

16

The corporation and resource geography

Liam Campling

Why should critical resource geographers care about the corporation?

Whether we like it or not, corporations are the dominant economic agents of the world economy and are leading political actors (Crouch 2011; Forsgren 2017; van Apeldoorn and de Graaff 2017; Koch 2020).[1] The corporate form is the most global and ubiquitous capitalist entity. It is one with which we all coexist and upon which most of us depend in our daily lives. Corporations control vast proportions of the means and conditions of production and reproduction of contemporary capitalism across much of the planet including extractive natures, and this control is increasingly centralized. For example, the last 20 years have seen the world's top 2000 firms increase their proportion of total sales or revenue and capture an even higher rate of profit in the process (UNCTAD 2018). At the start of the 2020s, the idea that corporations *inhibit* social progress is increasingly widespread (Talbot 2020).

Despite a quarter of a century of research on firms through the lens of global value chains (GVCs) and global production networks (GPNs) since Gereffi and Korzeniewicz's seminal (1994) edited collection, the difference that nature makes in the study of multinational corporations (MNCs) and interfirm relations remains almost entirely excluded. But as readers of this *Handbook* will know, the MNCs that control and integrate complex networks of global production in the early-twenty-first century *depend* on extractive natures and natural resource industries (Bridge 2008). Indeed, the history of extractive natures and resource geographies is central to the emergence of "the corporation": from the British and Dutch East India companies (Brandon 2017) to the first MNCs which developed in industries based on the direct appropriation of natural resources including minerals, oil, food, and fibers (Jones 2005). More broadly, the emergence of an imperialist world-market in the "long sixteenth century" was linked to colonial silver mining in Latin America and, in turn, large-scale forestry and coal and iron extraction in Western Europe for ship-building and steel-making that supported modernizing war and transport industries (Moore 2015). Cotton and food production spurred the capitalist transformation of agriculture on a world-scale based upon colonial, gendered, and racialized divisions of labor while providing raw material for industrial machinery and nourishment for workers in the metropoles (Mintz 1986; Baptist 2016). Likewise, firms were central to generating the waterpower and coal and oil extraction that fueled the trajectory of capitalist production from the industrial revolution to Fordism and beyond (Malm 2016). Today, contemporary networks of corporate control are held together by "systems integrators"—otherwise referred to as "lead firms" (Gibbon and Ponte 2005)—with enormous procurement expenditure at the apex of global supply chains (Nolan, Zhang, and Liu 2008); systems that are always-already underpinned and reproduced by extractive natures. Plainly, the history of capitalism is a history of enterprises appropriating and transforming nature at ever-increasing scale and pace.

This chapter sets out to explore two questions. In what ways might the study of the corporation be useful to critical resource geography (CRG)? And what can CRG contribute to our understanding of the political economy of the corporation? It does so in three subsequent sections. The first sets out some contours of "the corporation"—and the business enterprise more broadly—as a powerful political-economic actor pulled by the tendency to the concentration of capital and argues for the need to think beyond populist tropes when studying the corporation and its role in capitalism and resource industries. The second situates the corporation in capitalism and points to Marx's method as a way to do so. The third returns to the corporation as an actor in the hierarchical context of interfirm relations in GVCs and GPNs. My underlying argument is that in taking the corporation more seriously as a differentiated object of study and as political-economic agents—as "classes of capital"—CRG will be able to better understand capitalist strategies in the transformation of nature into commodities—"extractive natures"—and, in turn, the (re)production of natural resource industries and the landscapes which they inhabit. Additionally, those interested in the critical study of the corporation can learn a great deal from CRG, including its nonessentialist understanding of "resources" and its emphasis on landscapes, materiality, and place.

Pinning down "the corporation"

Much critical scholarship on the firm has focused on legal status. (For extended treatments of the corporation, law, and capitalism, see Baars [2019] and Barkan [2013].) Legal form has real-world manifestations in terms of corporate governance, public regulation, and tax systems (e.g., Ireland 2010; IGLP Law and Global Production Working Group 2016; Hager and Baines 2020). Legal form will always be context-specific, whether an enterprise is *incorporated* as a legal "person" separate from its owners, a state-owned enterprise (SOE), a "partnership" in US law or "limited liability company" in UK law (where owners are responsible for debt), or a "firm" or "enterprise" that may not have any legal standing but represent, for example, an informal family business, a household or an individual. The latter point is important in accounting for the informal economy given the ongoing significance of informal economic activity, such as "peasant" farming and artisanal and small-scale fishing and mining throughout the world especially in the global South. The rest of this chapter uses the terms "enterprise," "firm," and "corporation" interchangeably to indicate this ambiguity with the crucial caveat that a corporation proper – i.e., when treated by a legal system as a "legal person" – has distinctive and important rights which give it potential power and thus *does* require different treatment (Robé 2020). For resource geographers, what may matter more than legal form is the political-economic agency of the enterprise or firm as "an *organization* performing an economic activity" (Robé 2020, 195). Here I intend to emphasize the different *strategies* that enterprises use to make money (see the next section).

What most often comes to mind in imagining the corporation is the MNC—aka transnational corporation (TNC) or multinational enterprise (MNE)—a broad-brush definition of which is provided by the OECD:

> [MNCs] usually comprise companies or other entities established in more than one country and so linked that they may co-ordinate their operations in various ways. While one or more of these entities may be able to exercise a significant influence over the activities of others, their degree of autonomy within the enterprise may vary widely from one multinational enterprise to another. Ownership may be private, state, or mixed.
>
> (OECD 2008, 12)

This definition is useful because it gets to the heart of the MNC as an organizational form—that it represents the internationalization of capital (Jenkins 1987)—*and* that there is very wide scope of what this means in practice with some MNCs exerting considerable control over other firms (see below on GVCs and GPNs) and some being private or owned by the state. As I will go on to argue, it is important to differentiate *between* firms as they often deploy quite different strategies, even in the same segment of the same industry. The OECD's definition also makes clear that separate divisions of an MNC may have more or less autonomy from each other, including even from corporate headquarters. This means that it is also at times important to differentiate *within* firm; for example, one division may be in a jurisdiction that implements strong environmental and labor standards while another division of the same firm may be based in a jurisdiction where environmental and labor laws are weak and/or not implemented and corporate managers decide to not voluntarily introduce standards applied elsewhere.

In principle, the academic field of International Business (IB) should provide a starting point for CRG to think about the corporation. While mainstream IB does involve study of MNCs, it does so almost entirely from the perspective of *exchange* relations and has nothing to offer to the analysis of global production (Ietto-Gillies 2007). The leading school of thought in IB is "internalization" theory. Like transaction cost economics, this approach recognizes the existence of market imperfections but claims that these are the result of "natural" transaction costs (Williamson 1981). For example, Oliver Williamson (1995) famously uses the example of diamond trading techniques to argue that De Beers' practices are about efficiency maximization in a market typified by high transaction costs, *not* the exertion of monopoly power. In internalization theory, the "natural" costs of doing business in the market *justify* the hierarchical organization of the MNE—i.e., it is more efficient for firm A to internalize firm B than to transact on the market (Rugman 1986; Hennart 2001; Buckley and Casson 2009). Mainstream IB suffers from profound analytical blind spots—issues which are often central concerns in CRG. Aside from its complete rejection of power-based explanations for the internationalization of the firm, today's IB literature simply fails to consider the relationship between MNCs and conflict as well as inequality and uneven development (Eunni and Post 2006; Rygh 2019). This is despite that the first theorist of foreign direct investment—Stephen Hymer—was deeply interested in MNCs as manifestations of international *production* and market power. In contrast to internalization theory, Hymer (1970, 1972) argued that MNCs' market power allows them to *create* oligopolistic market imperfections, that foreign direct investment is a mechanism of corporate *control* of markets, and that these processes often reproduce uneven development in the global South, including in extractive industries.

Surface-level business school readings of Adam Smith the world over mirror internalization theory's tendency to brush over "real competition" (see below)—a reading that idealizes decentralized market competition through the "invisible hand." But Smith himself—like

Hymer—made very clear the strategy of capitalists to "widen the market and to narrow the competition.... [T]o levy, for their own benefit, an absurd tax upon the rest of their fellow-citizens" (1991, 219–220). Almost 250 years since Smith's foundational writing on political economy, the dynamics of competitive accumulation are ongoing, as is Hymer's (1975) observation—following Marx—that the tendency to corporate concentration and centralization reproduces uneven development. Ever-bigger firms own or control a greater proportion of global production and, in their role as systems integrators, are able to appropriate an increasing share of wealth through their ability to squeeze suppliers, intensify the exploitation of labor, and pass on costs (Nolan and Zhang 2010; Starrs 2013). While this is accompanied with countertendencies such as new businesses emerging in new product markets or to capture the monopoly profits made by concentrated MNCs, big business lobbies, and state managers are increasingly bypassing this competitive leveler, including through the rollback of competition policy since the 1980s (Crouch 2011; Christophers 2016).

While mainstream IB ignores the politics of the corporation and its centrality to global production and thus resource geographies, CRG and cognate fields tend to deploy the category of the corporation in an untheorized and ahistorical way. Most often, "the corporation" is set up in analytical—and political—contradistinction to "the community." Connected, there is a distinct tendency in CRG for research design to be focused on case study analyses in particular places, often centered around a community scale. There are two major problems with the counter-pointing of "corporation" (bad) and "community" (good). First, it side-steps important sources of differentiation among the diversity of "firms" that make up the continuum of resource appropriation in any particular geography: for instance, from globally dominant agribusiness and mining MNC at one end of the continuum to artisanal/small-scale farmers and miners on the other. These different enterprises all have the *potential* to accumulate capital through the exploitation of labor and appropriation of nature. For example, we might want to think through whether artisanal/small-scale production embedded in a community is capitalist or not. One category for doing so is that of "petty commodity producers." While not legally incorporated, artisanal/small-scale farmers or miners own the means of production and might thus be thought of as (often informal) "enterprises." However, they may either hold the class position of labor or that of nascent capitalists if they are "net buyers of labor power," engaged in simple capitalist reproduction (Pérez Niño 2016; Pattenden 2018). Determining which class position a petty commodity producer falls into is thus an empirical question, but it starts to open up the analytical question of differentiating among "classes of capital" (Baglioni 2015).[2] Another axis in the differentiation of firms engaged in extractive natures is that they are always subject to distinct *contexts*: for instance, the negotiation of resource access (customary rights, illegal squatting, state concessions); relations with buyers (from local traders to global markets); access to finance and its often short-termism disciplines demanding returns (money lenders to private equity and shareholders); and the purchase of means of production (from the rudimentary technology used in placer mining to the underground mining of deep deposits).

Second and connected, the counterpointing of "corporation" and "community" reproduces populist framings of "us" versus "them" without carefully teasing apart the distinctive social relations that may or may not articulate solidarities and frame political strategy among social groups, such as between those who depend on selling their labor for their material survival and those who live from returns on assets (Borras 2020; Barca 2019; Johnson, Chapter 24 this volume). The risk here is that researchers end up defining "the community" as "capital's other"—as Bernstein (2014) puts it—but without carefully setting out the material and sociopolitical axes that provide the bases for solidarity among the "other." Political rhetoric is an important tool of mobilization and organizing, but it is less useful in analyzing particular social relations under capitalism and the distinctive *landscapes*—"the phenomenal form of the social processes and practices of production, consumption, and exchange" (Mitchell 2003, 240)—within which people live, work, and die. The point here is that the boundaries between "community" and "the corporation" can often be blurred—and are often linked and mediated via the transformation of nature into commodities—and that we need to more fully theorize capitalism to sharpen analytically our take on these morphing boundaries.

Situating the corporation in capitalism

In this section, I draw on Marx's analytical categories used in his critique of political economy to suggest ways to differentiate the economic and political "interests" of firms involved in any resource geography. This seeks to move beyond simplistic notions of the corporation to understand the systemic tendencies that shape business strategies by highlighting the fundamental antagonism in capitalist social relations of production between the asset-owning ruling class and the laboring classes who must sell their labor-power to survive. A note of caution: this approach should not involve reducing the study of resource geographies to antagonistic class relations alone—whether in the "community," nation-state, and/or world-system—as under capitalism these relations are *always* articulated with other sources of social domination and subordination, such as gender and race, and they are far from the sole orientation of political struggle (Hall 1980; Campling et al. 2016).

Both liberal (in the European, not North American, sense) and populist (both left and right) approaches to the corporation in academic literature and political discourse tend to divorce their analyses from a broader theorization of the corporation *in* capitalism. The analytical and, in turn, political consequence of this tendency is that solutions to the "excesses" of corporate power tend to veer variously towards: reformist tinkering through liberal, market-based governance (CSR, codes of conduct); the fetishizing of "national" capitals on the populist or fascist right; criticism of vague notions of unfettered "corporate power" on the populist left, often delineated by "national" economic projects focused on social and/or environmental justice; or of state intervention to "pick winners" in order to counter structural unevenness in the world economy (e.g., the "developmental state").

By contrast, the Marxist analysis suggested here seeks to get at the essence of capitalist competition. It approaches the corporation as a legal and organizational manifestation of capitals-in-competition over the capture and distribution of surplus value and the varied forms of exploitation upon which the production of this value depends. Each individual enterprise *may* seek to be more or less "ethical" or "sustainable" and they may even be relatively successful in doing so. But at the level of the system as a whole—the totality of enterprises—the "dull compulsion" of the capitalist imperative demands that enterprises "accumulate or die" (Fine 1984, 36). As noted earlier, MNCs and other types of firms use distinct *strategies*—they have relative political-economic agency—and these accumulation strategies *do* differ, often considerably. Strategies include "classic" profit based on the direct capture of surplus value from unpaid work in the labor process (e.g., in the factories and fields), which is shaped by cost-cutting strategies (wages, increasing working-time, technological, and/or organizational change); rent captured through asset ownership (land, brands, intellectual property); interest accrued from loans; value appropriated by a "lead firm" from another firm in a value chain; and/or financialized strategies based on the doctrine of shareholder value such as share buybacks or private equity engaged in asset stripping (Andersson and Haslam 2012; Lazonick 2014; Quentin and Campling 2018). Regardless of the particular strategy used, an underlying dynamic driving *real* competition is the "gravitational pull" of the capital-relation—that is the shared, structural objective of accumulating capital through the exploitation of labor and appropriation of nature—and that this is "antagonistic by nature and turbulent in operation," which is as different from the idealized neoliberal imaginary of competition as equilibrium and efficiency "as war is from ballet" (Shaikh 2016, 259).

But who or what is "capital"? A direct but potentially obscure answer is that it is a *social relation*—not a group of individuals or a set of things. But then who are "the capitalists"? Asset owners? Executive managers? Fat cats? "Capitalists" include those who buy labor-power and extract unpaid labor in the form of "surplus value" (or profit); those who own or control assets that are based on the active capture of surplus value from others in the circuit of capital (e.g., banking and finance, owners of land or intellectual property); and those for whom the majority of their remuneration is *tied to* the enhancement of the capture of value (e.g., corporate executives). Given this heterogeneous definition of those who make up the class of capitalists and the differentiated strategies that they might use to capture value through the legal and organizational form of the corporation, it is once again useful to refer to them as "classes of capital" (Baglioni 2015).[3]

Yet even if the "gravitational pull" of competitive accumulation and its social relations of production and general forms of domination (e.g., wage-labor) may capture the "essence" of global capitalism, these abstract relations manifest in infinitely diverse forms in particular times, spaces, and places. Here it is useful to distinguish between (a) *capital as a process* and living-breathing, eating-shitting *historical capitalism* (Parenti 2015), as this allows the researcher to be clear that the abstract theoretical logics of capital accumulation manifest very differently in messy reality. There is a second cross-cutting dimension between (b) capital *in general* and capital *in particular*—or between the social totality of enterprises and the individual enterprise (Banaji 2010). These two cross-cutting dimensions matter, because where the "empirical" is generally observed through historical capitalism *in particular*—as is the case in most research in CRG—it can only be understood if situated within capital as process/capital in general. Put differently, if we want to be able to explain the relationship between the corporation and resource geographies beyond discrete dynamics in particular places and to extend the analysis to articulations within *and* between entire landscapes, we need to get beyond populist conflations of capital in particular (e.g., a "bad apple") with capital in general.

A Marxist approach *starts from* a general theorization of capitalism and the social relations of production that typify it, which are based on diverse forms of exploitation including, but far from being limited to, the wage-relation (i.e., unpaid labor captured by capitals in the form of "surplus value"). This does not mean that all such labor is based on wages per se as historical capitalism works "through a multiplicity of forms of exploitation *based on* wage-labor" (Banaji 2010, 145). There is a tricky but important distinction here between wages (or a salary) as a simple, descriptive term and *wage-labor* as a concrete category of "capital-positing, capital-creating labor" (Marx 1993, 463). For example, a wage-labor *relation* can be "disguised" as self-employment in agriculture, construction, and the "gig" economy (e.g., Bernstein 2010; Pattenden 2016; Moore and Newsome 2018). And as a recent flowering of research debating social reproduction argues, exploitation is shaped and regenerated by highly gendered labor segmentation in the workplace, the community, and the household (Baglioni 2018; Mezzadri 2019). More broadly, labor can be more or less "free" or "unfree" under historical capitalism; slavery, debt bondage, etc. have always coexisted with salaried-work. The existence of the former does not make these forms of exploitation noncapitalist, including in agriculture, fisheries, and other resource sectors (Lerche 2007; Vandergeest and Marschke 2020). Further, these forms of exploitation are institutionally mediated in various ways at different scales, including diverse management regimes, households, and/or debt relations (Hanlon 2015; Mezzadri 2017).

While this approach starts abstractly with the "general forms of domination" under capitalism—i.e., the relations of production in the circuit of capital at the level of the social totality of enterprises (Banaji 2010)—as already suggested, it *demands* careful analysis of particular places and spaces in all of their infinity variety.

Marx described his method as "rising from the abstract to the concrete," where the "concrete" "is the concentration of many determinations" (Marx 1993, 101). The identification of "abstract" and "concrete" does not denote "theory" versus "empirical." It signifies, rather, the importance of utilizing *general* concepts and categories ("capitalism," "class," "surplus value") to identify and analyze *particular* historical-geographical social forms (for example, the corporation, processes of local class formation, and specific nation-states). Put differently, the "concrete" does not mean "the empirical" but a greater level of conceptual specification that seeks to better proximate the diverse phenomenal forms of social relations, histories and geographies—"many determinations." These "forms" matter. They are not simply "functions" of the capital relation; for example, while commodity prices are a phenomenal representation of value, they have very real effects (Campling et al. 2016; Capps 2016).

In sum, by using Marxist method we can begin to splice apart the different roles and strategies of firms in capitalism. In contrast to more populist takes that might pit the corporation against the community or the small-scale farmer against corporate agribusiness, we can start to splice apart the class dynamics of accumulation and how they manifest in particular resource geographies. A central axis in doing so in CRG might be the organization of global production and the distribution of value from extractive natures, to which we turn.

The corporation, the geographical distribution of value, and extractive natures

Networks of firms involved in a sequence of commodity-producing activities linking workers, households, and states have been a central feature of global production since the emergence of the new international division of labor from the 1970s with the rise of manufacturing in the global South. GVC, GPN, and related but less-known frameworks such as systems of provision (Fine and Leopold 1993) have sought to understand this phenomenon. Commodity chains have, of course, underpinned the capitalist world economy since its emergence in the long sixteenth century. However, outsourcing and offshoring associated with, but not limited to, the neoliberal era saw the unfolding of new layers of complexity as to how GVCs are organized spatially, the increasing technical fragmentation of global production, and, crucially, the power relations shaping linkages and relations among firms. The latter—"chain governance"—is the principal conceptual innovation of GVC analysis and where the focus is commonly on the power relation of "lead firms," which are normally MNCs (Bair 2009).

Extractive natures are shaped profoundly by power relations in GVCs and vice versa (see also, Ciccantell, Chapter 15 this volume). Certain firms "lead" because they are the (temporary) winners of capitalist competition in one or more nodes of a GVC. These dynamics of competitive accumulation are often mirrored through GVCs—generating what Nolan, Zhang, and Liu (2008) call the "cascade effect" where the tendency to centralization among "lead" firms is mimicked by their suppliers, including because lead firms seek to directly articulate with a smaller number of suppliers to more effectively "appropriate value [and] pass on risk and costs" (Havice and Campling 2017, 294). Giant contractors, such as Foxconn with its 1 million workers and Yue Yuen with its global dominance of footwear manufacturing, neatly illustrate this trend (Chan, Pun, and Selden 2013; Kumar 2020). But even critical analyses of GVCs rarely trace the components of commodities back to their origins in extractive natures (see Campling and Havice 2019)—and as suggested below, this is a major area where CRG can contribute by studying differentiated firms in contemporary capitalism.

Critical political economy approaches to the study of the corporation in global production seek to understand the geographical transfer of value (Hadjimichalis 1984) and the articulations of the law and GVCs in this process (IGLP Law and Global Production Working Group 2016). Those working in critical accounting, for example, have shown how the market power of lead firms has resulted in highly uneven patterns of distribution of value in GVCs, emphasizing the complexities of regulating the "post-national" corporation in the contemporary world economy (Haslam 2013; Froud et al. 2014). Today's so-called superstar firms are capturing an entirely unequal share of the global economic pie through the mechanisms of GVCs and the law, most notably through the control of intangible assets that act as mechanisms of value *capture* (UNC-TAD 2017, 2018; Durand and Milberg 2018; Baglioni, Campling, and Hanlon 2020). Yet what often gets lost in thinking about intangibles is that they themselves *depend* upon very tangible industries (from servers to undersea cables and from energy production to laboring bodies), and, further, the initial creation of the value that "superstar firms" capture is *necessarily* underpinned by extractive natures (Baglioni and Campling 2017; Quentin and Campling 2018).

GPN analysis advances the conceptual category of "territorial embeddedness" to refer to the degree of "anchoring" of a firm in different places, which of course includes embedded biological and geophysical natures. The idea here is that foreign direct investment may see a corporation "become embedded" by absorbing "the economic activities and social dynamics that already exist" there (Henderson et al. 2002, 452). This anchoring can work in several ways, but of particular importance to us here is how an incoming firm may be attracted by extractive natures such as land (soil and laboring bodies), mineral deposits, a fish population, etc., in combination with state concessions or other policies, which can affect the developmental prospects of a location and change social relations there. Context matters as firms behave differently in different spaces and places and adjust their business strategies and day-to-day operations in relation to national (and in larger countries, local/municipal) governments, labor regimes, and other non-state actors, such as nongovernmental organizations. This is of special importance in extractive industries, because as Gavin Bridge has made clear, they "face some clear limitations to spatial flexibility," which "exerts a powerful influence on the location of competition in the production network, the form that competition takes and on relations of dependency between holders and seekers of resources" (Bridge 2008; also, Ciccantell and Smith 2009). Put directly, the *state* can play a particular role in (re)producing natural resource geographies, including in shaping the accumulation strategies of firms (Emel, Huber, and Makene 2011; Havice and Reed 2012; Parenti 2015).

At the heart of CRG—often through case study analysis—is an understanding of the contexts in which natural resources are situated, which means that however "footloose" capital might be—as Hymer taught us—it must *always* be rooted in particular places. There are fruitful connections here between CRG and contemporary frameworks for the study of firms such as the GPN category of "territorial embeddedness." CRG recognizes that natural resources are sociopolitical constructions because what is a "resource" is inherently dynamic as they are made and unmade according to different times and places—and under capitalism, the gravitational pull of *profit* is what matters most to the organization of economic life (see also Huber, Chapter 14 this volume). And as we have noted, some aspects of the materiality of natural resources confer on these static properties (Baglioni and Campling 2017). Whether nonrenewable or renewable-but-exhaustible, they compel the production process to conform to their geophysical/ biological characteristics. Compare, for example, relational social-environmental conditions of resource extraction in terms of mining and the geophysical (Capps 2012), forestry and topology or terrain (Prudham 2005), farming and soil/climate (Mann and Dickinson 1978), and fisheries and biology—i.e., fish move but within spatial-biological limits (Campling 2012). Each of these couplets suggests an opening to think about the corporation as the dynamic articulation of nature and capitalist relations rather than focusing only on the impact of the corporation *on* nature. Further, within each of these extractive natures, we will find differentiated classes of capital, from petty commodity producers to MNCs. And, in turn, by using the lens of CRG to understand extractive natures, we might better understand the corporation because nature imposes disciplines on capitalist accumulation as a result of geophysical/biological characteristics and therefore contributes to differentiation among classes of capital and shapes the conditions for real competition. CRG combined with a political economy of the corporation can then help to break down the separation of nature and society through the lens of capitalist relations. Or to rephrase, political economy can draw on CRG to bring nature into the theorization of capitalism, the analysis of differentiated corporate strategy, and "classes of capital."

Conclusion

In the context of the climate crisis and rapid biodiversity loss, it is now obvious that social science needs to find new ways of seeing, thinking, and *acting*: we need to put power, value, and nature in relationship to each other. After all, capitalism exists *through* the environment (Moore 2015). This is what CRG does and, I argue, can do even better if it engages more with the insights of Marxist political economy and the study of the corporation. But this should not involve the collapsing of the social and the environmental as separate categories in so far as *political* responses to the role of capitalist relations of production in planetary decline must be forged and set in motion with struggles *against* exploitation in its various forms and *for* the redistribution of wealth (Malm 2018). This may not be news to those working in CRG; it is certainly an area where resource geographers are able to offer pivotal and potentially profound insights to other cognate fields.

Resource geography's theorizing of corporations should be sensitive to their differentiated economic, organizational, and political strategies, which can only be comprehended through careful empirical study. Expressed more broadly, CRG could contribute to theorizing the relationship between political-economic processes and ecological ones, without which we risk "that global environmental problems (and thus their proposed 'solutions') become separated from their political-economic foundations, and *vice versa*" (Katz-Rosene and Patterson 2018, 4). For example, despite a quarter of a century of political economy research on GVCs and GPNs, the appropriation of nature in general and natural resource industries in particular remain marginal both theoretically and empirically (Baglioni and Campling 2017). There are advances here, including work on how chain governance and environmental governance interact, with the latter being used as a tool by lead firms to accumulate more value and pass on costs (Havice and Campling 2017; Ponte 2019), and studies of how ecological contradictions produced in the capitalist drive to accumulate value manifest in disease and socioeconomic crisis, in turn reconfiguring resource geographies and production networks (de la Cruz and Jansen 2018; Irarrázaval and Bustos-Gallardo 2019). The particular differences that nature makes in confronting corporations as they seek to appropriate and transform nature spur continuous geographical, organizational, and technological change, which has implications far wider than the specific moment of production, therefore opening-up the sphere of circulation as a crucial area of study in CRG (Arboleda 2020). In sum, individual corporations are a manifestation of historical capitalism, but to understand the *particular* forms they may take in extractive natures, we need to theorize capital as a process and the social totality of enterprises ("real competition"). The differentiation of classes of capital therein opens up rich avenues for CRG to systematically critique corporate capitalism and may even contribute in small, but important, ways to changing it.

Notes

1 I wish to thank Elizabeth Havice and Matthew Himley for inviting me to contribute to this handbook and for their generous and generative editorial comments and suggestions on various drafts of this chapter. I also thank Jeremy Anderson and Elena Baglioni for important comments on the final draft. All of which resulted in improvements, but any remaining clumsiness and errors are my own.
2 The notion of "classes of capital" is used as an alternative to the more abstract and often loaded term "fractions of capital," in part because the former encourages a greater focus on "concrete" social relations (see below) and in part to side-step often sectarian debates on the capitalist state.
3 I am indebted here to Gavin Capps who is developing a parallel project on "capitalist classes," and I would be remiss to not acknowledge our long conversations on this theme.

References

Andersson, Tord, and Colin Haslam. 2012. "The Private Equity Business Model: On Terra Firma or Shifting Sands?" *Accounting Forum* 36 (1): 27–37.

Arboleda, Martin. 2020. "Towards an Agrarian Question of Circulation: Walmart's Expansion in Chile and the Agrarian Political Economy of Supply Chain Capitalism." *Journal of Agrarian Change* 20 (3): 345–363. doi: 10.1111/joac.12356.

Baars, Grietje. 2019. *The Corporation, Law and Capitalism: A Radical Perspective on the Role of Law in the Global Political Economy*. Leiden, UK: Brill.

Baglioni, Elena. 2015. "Straddling Contract and Estate Farming: Accumulation Strategies of Senegalese Horticultural Exporters." *Journal of Agrarian Change* 15 (1): 17–42.

Baglioni, Elena. 2018. "Labour Control and the Labour Question in Global Production Networks: Exploitation and Disciplining in Senegalese Export Horticulture." *Journal of Economic Geography* 18 (1): 111–137.

Baglioni, Elena, and Liam Campling. 2017. "Natural Resource Industries as Global Value Chains: Frontiers, Fetishism, Labour and the State." *Environment and Planning A: Economy and Space* 49 (11): 2437–2456.

Baglioni, Elena, Liam Campling, and Gerard Hanlon. 2020. "Global Value Chains As Entrepreneurial Capture: Insights from Management Theory." *Review of International Political Economy* 27 (4): 903–925. doi: 10.1080/09692290.2019.1657479.

Bair, Jennifer, ed. 2009. "Global Commodity Chains." In *Frontiers of Commodity Chain Research*. Stanford, CA: Stanford University Press.

Banaji, Jairus. 2010. *Theory as History: Essays on Modes of Production and Exploitation*. Leiden, UK: Brill.

Baptist, Edward E. 2016. *The Half Has Never Been Told: Slavery and the Making of American Capitalism*. New York, NY: Basic Books.

Barca, Stefania. 2019. "The Labor(s) of Degrowth." *Capitalism Nature Socialism* 30 (2): 207–216.

Barkan, Joshua. 2013. *Corporate Sovereignty: Law and Government Under Capitalism*. Minneapolis, MN: University of Minnesota Press.

Bernstein, Henry. 2010. *Class Dynamics of Agrarian Change*. Halifax, NS: Fernwood.

Bernstein, Henry. 2014. "Food Sovereignty via the 'Peasant Way': A Skeptical View." *Journal of Peasant Studies* 41 (6): 1031–1063.

Borras, Saturnino M. 2020. "Agrarian Social Movements: The Absurdly Difficult But Not Impossible Agenda of Defeating Right-Wing Populism and Exploring a Socialist Future." *Journal of Agrarian Change* 20 (1): 3–36.

Brandon, Pepijn. 2017. "Between Company and State: The Dutch East and West India Companies as Brokers Between War and Profit." In *The Corporation: A Critical, Multidisciplinary Handbook*, edited by Grietje Baars, and André Spicer, 215–225. Cambridge: Cambridge University Press.

Bridge, Gavin. 2008. "Global Production Networks and the Extractive Sector: Governing Resource-Based Development." *Journal of Economic Geography* 8 (3): 389–419.

Buckley, Peter, and Mark Casson. 2009. "The Internalisation Theory of the Multinational Enterprise: A Review of the Progress of a Research Agenda after 30 Years." *Journal of International Business Studies* 40 (9): 1563–1580.

Campling, Liam. 2012. "The Tuna 'Commodity Frontier': Business Strategies and Environment in the Industrial Tuna Fisheries of the Western Indian Ocean." *Journal of Agrarian Change* 12 (2–3): 252–278.

Campling, Liam, Satoshi Miyamura, Jonathan Pattenden, and Benjamin Selwyn. 2016. "Class Dynamics of Development: A Methodological Note." *Third World Quarterly* 37 (10): 1745–1767.

Campling, Liam, and Elizabeth Havice. 2019. "Bringing the Environment into GVC Analysis: Antecedents and Advances." In *Handbook on Global Value Chains*, edited by Stefano Ponte, Gary Gereffi, and Gale Raj-Reichert. Cheltenham, UK: Edward Elgar.

Capps, Gavin. 2012. "Victim of Its Own Success? The Platinum Mining Industry and the Apartheid Mineral Property System in South Africa's Political Transition." *Review of African Political Economy* 39 (131): 63–84.

Capps, Gavin. 2016. "Tribal-Landed Property: The Value of the Chieftaincy in Contemporary Africa." *Journal of Agrarian Change* 16 (3): 452–477.

Chan, Jenny, Ngai Pun, and Mark Selden. 2013. "The Politics of Global Production: Apple, Foxconn and China's New Working Class." *New Technology, Work and Employment* 28 (2): 100–115.

Christophers, Brett. 2016. *The Great Leveler: Capitalism and Competition in the Court of Law*. Harvard: Harvard University Press.

Ciccantell, Paul, and David A. Smith. 2009. "Rethinking Global Commodity Chains: Integrating Extraction, Transport, and Manufacturing." *International Journal of Comparative Sociology* 50 (3–4): 361–384.

Crouch, Colin. 2011. *The Strange Non-Death of Neoliberalism*. Cambridge: Polity.

de la Cruz, Jaye, and Kees Jansen. 2018. "Panama Disease and Contract Farming in the Philippines: Towards a Political Ecology of Risk." *Journal of Agrarian Change* 18 (2): 249–266.

Durand, Cedric, and William Milberg. 2018. "Intellectual Monopoly in Global Value Chains." Working Paper, Department of Economics, The New School for Social Research, New York.

Emel, Jody, Matthew T. Huber, and Madoshi H. Makene. 2011. "Extracting Sovereignty: Capital, Territory, and Gold Mining in Tanzania." *Political Geography* 30 (2): 70–79.

Eunni, Rangamohan V., and James E. Post. 2006. "What Matters Most? A Review of MNE Literature, 1990–2000." *The Multinational Business Review* 14 (3): 1–28.

Fine, Ben. 1984. *Marx's Capital*. London: Macmillan.

Fine, Ben, and Ellen Leopold. 1993. *The World of Consumption: The Material and Capital Revisited*. 2nd ed. London: Routledge.

Forsgren, Mats. 2017. *Theories of the Multinational Firm: A Multidimensional Creature in the Global Economy*. Cheltenham, UK: Edward Elgar.

Froud, Julie, Sukhdev Johal, Adam Leaver, and Karel Williams. 2014. "Financialization Across the Pacific: Manufacturing Cost Ratios, Supply Chains, and Power." *Critical Perspectives on Accounting* 25 (1): 46–57.

Gereffi, Gary, and Miguel Korzeniewicz, eds. 1994. *Commodity Chains and Global Capitalism*. Westport, CT: Praeger.

Gibbon, Peter, and Stefano Ponte. 2005. *Trading Down: Africa, Value Chains, and the Global Economy*. Philadelphia, PA: Temple University Press.

Hadjimichalis, Costis. 1984. "The Geographical Transfer of Value: Notes on the Spatiality of Capitalism." *Environment and Planning D: Society and Space* 2 (3): 329–345.

Hager, Sandy Brian, and Joseph Baines. 2020. "The Tax Advantage of Big Business: How the Structure of Corporate Taxation Fuels Concentration and Inequality." *Politics & Society* 48 (2): 275–305.

Hall, Stuart. 1980. "Race, Articulation and Societies Structured in Dominance." In *Sociological Theories: Race and Colonialism, 305–45*. Poole: UNESCO.

Hanlon, Gerard. 2015. *The Dark Side of Management: A Secret History of Management Theory*. London: Routledge.

Haslam, Colin, ed. 2013. "The Apple Business Model." *Accounting Forum* 37 (4): 245–248.

Havice, Elizabeth, and Liam Campling. 2017. "Where Chain Governance and Environmental Govern-ance Meet: Interfirm Strategies in the Canned Tuna Global Value Chain." *Economic Geography* 93 (3): 292–313.

Havice, Elizabeth, and Kristin Reed. 2012. "Fishing for Development? Tuna Resource Access and Industrial Change in Papua New Guinea." *Journal of Agrarian Change* 12 (2–3): 413–435.

Henderson, Jeffrey, Peter Dicken, Martin Hess, Neil Coe, and Henry Wai-Chung Yeung. 2002. "Global Production Networks and the Analysis of Economic Development." *Review of International Political Economy* 9 (3): 436–464.

Hennart, Jean-Francois. 2001. "Theories of the Multinational Enterprise." In *The Oxford Handbook of International Business*, edited by Alan Rugman, and Thomas L. Brewer, 127–148. London: Oxford University Press.

Hymer, Stephen. 1970. "The Efficiency (Contradictions) of Multinational Corporations." *The American Economic Review* 60 (2): 441–448.

Hymer, Stephen. 1972. "The Internationalization of Capital." *Journal of Economic Issues* 6 (1): 91–111.

Hymer, Stephen. 1975. "The Multinational Corporation and the Law of Uneven Development." In *International Firms and Modern Imperialism*, edited by Hugo Radice, 128–152. Harmondsworth: Penguin.

IGLP Law and Global Production Working Group. 2016. "Recognising the Constitutive Role of Law in Global Value Chains: A Research Manifesto." *London Review of International Law* 4 (1): 57–79.

Ietto-Gillies, Grazia. 2007. "Theories of International Production: A Critical Perspective." *Critical Perspectives on International Business* 3 (3): 196–210.

Irarrázaval, Felipe, and Beatriz Bustos-Gallardo. 2019. "Global Salmon Networks: Unpacking Ecological Contradictions at the Production Stage." *Economic Geography* 95 (2): 159–178.

Ireland, Paddy. 2010. "Limited Liability, Rights of Control and the Problem of Corporate Irresponsibility." *Cambridge Journal of Economics* 34 (5): 837–856.

Jenkins, Rhys. 1987. *Transnational Corporations and Uneven Development: The Internationalization of Capital and the Third World*. London: Routledge.

Jones, Geoffrey. 2005. *Multinationals and Global Capitalism: From the Nineteenth to the Twenty-First Century*. Oxford: Oxford University Press.

Katz-Rosene, Ryan, and Matthew Patterson. 2018. *Thinking Ecologically About the Global Political Economy*. London: Routledge.

Koch, Natalie. 2020. "The Corporate Production of Nationalism." *Antipode* 52 (1): 185–205.

Kumar, Ashok. 2020. *Monopsony Capitalism: Power and Production in the Twilight of the Sweatshop Age*. Cambridge: Cambridge University Press.

Lazonick, William. 2014. "Profits Without Prosperity." *Harvard Business Review*, Sept.

Lerche, Jens. 2007. "A Global Alliance Against Forced Labour? Unfree Labour, Neo-Liberal Globalization and the International Labor Organization." *Journal of Agrarian Change* 7 (4): 424–452.

Malm, Andreas. 2016. *Fossil Capital: The Rise of Steam Power and the Roots of Global Warming*. London: Verso.

Malm, Andreas. 2018. *The Progress of This Storm: Nature and Society in a Warming World*. London: Verso.

Mann, Susan, and James M. Dickinson. 1978. "Obstacles to the Establishment of Capitalist Agriculture." *Journal of Peasant Studies* 5 (4): 466–481.

Marx, Karl. 1993. *Grundrisse der Kritik de politischen Ökonomie [Fundamentals of Political Economy Criticism]*, translated by Martin Nicolaus. Harmondsworth: Penguin.

Mezzadri, Alessandra. 2017. *The Sweatshop Regime: Labouring Bodies, Exploitation, and Garments Made in India*. Cambridge: Cambridge University Press.

Mezzadri, Alessandra. 2019. 'On the Value of Social Reproduction." *Radical Philosophy* 204: 33–41.

Mintz, Sidney. 1986. *Sweetness and Power: The Place of Sugar in Modern History*. London: Penguin.

Mitchell, Don. 2003. "Dead Labour and the Political Economy of Landscape – California Living, California Dying." In *Handbook of Cultural Geography*, edited by Kay Anderson, Mona Domosh, Steve Pile, and Nigel Thrift. London: SAGE.

Moore, Jason W. 2015. *Capitalism in the Web of Life: Ecology and the Accumulation of Capital*. London: Verso.

Moore, Sian, and Kirsty Newsome. 2018. "Paying for Free Delivery: Dependent Self-Employment as a Measure of Precarity in Parcel Delivery." *Work, Employment and Society* 32 (3): 475–492.

Nolan, Peter, Jin Zhang, and Chunhang Liu. 2008. "The Global Business Revolution, the Cascade Effect, and the Challenge for Firms From Developing Countries." *Cambridge Journal of Economics* 32 (1): 29–47.

Nolan, Peter, and Jin Zhang. 2010. Global Competition After the Financial Crisis. *New Left Review II* 64 (Jul.–Aug.): 97–108.

OECD. 2008. *Guidelines for Multinational Enterprises*. Paris: OECD.

Parenti, Christian. 2015. "The Environment Making State: Territory, Nature, and Value." *Antipode* 47 (4): 829–848.

Pattenden, Jonathan. 2016. "Working at the Margins of Global Production Networks: Local Labour Production Regimes and Rural-Based Laborers in South India." *Third World Quarterly* 37 (10): 1809–1833.

Pattenden, Jonathan. 2018. "The Politics of Classes of Labour: Fragmentation, Reproduction Zones, and Collective Action in Karnataka, India." *The Journal of Peasant Studies* 45 (5–6): 1039–1059.

Pérez Niño, Helena. 2016. "Class Dynamics in Contract Farming: The Case of Tobacco Production in Mozambique." *Third World Quarterly* 37 (10): 1787–1808.

Ponte, Stefano. 2019. "Green Capital Accumulation: Business and Sustainability Management in a World of Global Value Chains." *New Political Economy* 25 (1): 72–84. doi: 10.1080/13563467.2019.1581152.

Prudham, Scott. 2005. *Knock on Wood: Nature as Commodity in Douglas-Fir Country*. New York, NY: Routledge.

Quentin, David, and Liam Campling. 2018. "Global Inequality Chains: Integrating Mechanisms of Value Distribution into Analyses of Global Production." *Global Networks* 18 (1): 33–56.

Robé, Jean-Philippe. 2020. *Property, Power and Politics*. Bristol: Bristol University Press.

Rugman, Alan. 1986. "New Theories of the Multinational Enterprise: An Assessment of Internalization Theory." *Bulletin of Economic Research* 38 (2): 101–118.

Rygh, Asmund. 2019. "Multinational Enterprises and Economic Inequality: A Review and International Business Research Agenda." *Critical Perspectives on International Business*. doi: 10.1108/cpoib-09-2019-0068.

Shaikh, Anwar. 2016. *Capitalism: Competition, Conflict, Crises*. Oxford: Oxford University Press.

Smith, Adam. 1991. *Wealth of Nations*. Amherst, NY: Prometheus Books.

Starrs, Sean. 2013. "American Economic Power Hasn't Declined—It Globalized! Summoning the Data and Taking Globalization Seriously." *International Studies Quarterly* 57 (4): 817–830.

Talbot, Lorraine. 2020. "Why Corporations Inhibit Social Progress: A Brief Review of Corporations from Chapter 6 'Markets, Finance and Corporations. Does Capitalism Have a Future?'." *Review of Social Economy* 78 (2): 128–138.

UNCTAD. 2017. *Trade and Development Report 2017—Beyond Austerity: Towards a Global New Deal*. Geneva: UNCTAD.

UNCTAD. 2018. *Trade and Development Report 2018: Power, Platforms, and the Free Trade Delusion*. Geneva: UNCTAD.

van Apeldoorn, Bastiaan, and Naná de Graaff. 2017. "The Corporation in Political Science." In *The Corporation*, edited by Grietje Baars, and André Spicer, 134–159. Cambridge, UK: Cambridge University Press.

Vandergeest, Peter, and Melissa Marschke. 2020. "Modern Slavery and Freedom: Exploring Contradictions Through Labour Scandals in Thai Fisheries." *Antipode* 52 (1): 291–315.

Williamson, Oliver E. 1981. "The Modern Corporation: Origins, Evolution, Attributes." *Journal of Economic Literature* 19 (4): 1537–1568.

Williamson, Oliver E. 1995. "Hierarchies, Markets and Power in the Economy: An Economic Perspective." *Industrial and Corporate Change* 4 (1): 21–49.

Section III
Doing critical resource geography
Methods, advocacy, and teaching

17

Life with oil palm

Incorporating ethnographic sensibilities in critical resource geography

Eloisa Berman-Arévalo

Introduction

Critical resource studies are concerned with the relations of power that configure and underlie the production of resources. For example, scholars have studied how nature becomes "resources" through its commodification and the creation of capitalist value (Huber 2018; Moore 2015), destabilized the boundaries between natural and social worlds (Castree 2003; Whatmore 1999), and highlighted the colonial origins and assumptions that underlie dominant schemes of resource use and extraction (Braun 2002; Willems-Braun 1997). Feminist, post-structural, and postcolonial critiques pave the way for scholars in critical resource geographies to extend the critical scrutiny of resources to the research practices and epistemic frames through which resources are known, characterized, and defined. This chapter focuses on ethnography as a mode of knowledge making that enables a critical reflection on the intersubjective, ethical, material, and political aspects of research relationships and that can provide nuanced accounts of the everyday politics and geographies of resource production.

Specifically, the chapter draws on feminist geography to develop the concept of "ethnographic sensibilities" in studies of resource-making in extractive spaces. Anthropologist Carol McGranahan defines ethnographic sensibility as a "sense of the ethnographic as the lived expectations, complexities, contradictions, possibilities and grounds of any given cultural group" (2018, 1). Here, I expand on this definition, taking "ethnographic sensibilities" as a mode of engagement with people, places, and field-based knowledge-making premised on the researcher's physical and epistemological exposure to the places and relations through which research unfolds.

Ethnographic sensibilities make a critical intervention in resource geographies by granting epistemic and political importance to the "minor" stories of ordinary life and instigating an opening in public imagination to the world-making that takes place among those who inhabit spaces of extraction. At the same time, by situating resource production within the broader web of material, affective and social relations of "life-with-resources" (Valdivia 2017) and drawing on the researcher's embodied participation therein, ethnographic sensibilities can provide nuanced understandings of the ways in which people live with resource extraction and grapple with its effects.

First, the chapter offers an overview of ethnography's critical trajectories in human geography and situates the concept of ethnographic sensibilities within the contributions of feminist geographers to ethnography's critical turn. Next, it contextualizes my own research on oil palm in Colombia in order to suggest practical ways in which researchers of resource production may incorporate ethnographic sensibilities on the field. These include a sensitive openness to the broader dynamics of life-with-resources (beyond the resource itself), an embodied exposure to place and relations, and a reflexive and multilevel writing practice. The conclusion discusses the spatial entanglements between the community's everyday life and resource extraction as an example of the theoretical potential of ethnographic sensibilities.

Ethnography and its critical trajectories in human geography

Ethnography extends well beyond a set of research techniques applied through long-term immersion in local life (McGranahan 2018, 2014; Ortner 2006). As a field-based method based on the researcher's embodied participation in a given group's everyday life, ethnography constitutes a vantage point from which to perceive, understand, and write about social worlds. Ethnography is itself a *theory* in its analytic orientation toward the often unpredictable, ambiguous, and contradictory rhythms and logics of social life, through which sociocultural groups collectively make sense of the world (McGranahan 2018, 1). It is, furthermore, a theoretical proposition on translation, which is premised on the recognition of the researcher's subjectivity in deciding whether and how to make visible particular aspects of social life and categories of analysis (Biehl and McKay 2012). In this sense, ethnography recognizes the impossibility of social research to arrive at fixed "truths" about the always uncertain dynamics of everyday life. Finally, ethnography is also a personal mode of narration that foregrounds the ethnographers' own lived and learning experiences, makes the ethnographer herself part of the object of study (see, for instance, Tsing [2005]), and allows for creative dialogue between the researcher's voice and the conceptualizations of informants (Biehl and McKay 2012).

Ethnography in geography

Ethnography's attention to people's perceptions and experiences, as well as its traditionally place-based immersion among a social group, has long made it a useful method in qualitative geographic research. Its use in geography can generate fine-grained accounts of the "connection between the life-world of a social group and the geographic world they construct" (Herbert 2000, 551) and help geographers develop a deep understanding of socio-spatial relations from the perspective of both meaning and materiality.

The relationship between ethnography and geographic research has shifted over time. In recent decades, geographers have not only *used* ethnography but also contributed to its critical potential. In the first half of the twentieth century, geographers within the cultural ecology tradition conducted long-term, field-based ethnographic fieldwork in order to study particular societies' knowledge, beliefs, and material practices regarding the biophysical world (St Martin and Pavlovskaya 2009). Later, in the 1970s, humanistic geography incorporated ethnographic approaches by focusing on the mundane and the everyday and by addressing the complexity and particularities of subjects' spatial experiences (Crang and Cook 1995). Ethnography added thickness to humanistic geographers' observations of space, place, and culture and provided a subject- and place-based approach to a field then dominated by positivist epistemologies, universalizing claims, and quantitative methods (Ley and Samuels 1978). Yet, as geographers used ethnography, they also reproduced a view that cultures were transparent, homogenous, and uncontested and available for objective study by a detached researcher (Crang and Cook 1995, 5). In the 1990s, geography's cultural turn livened geographic theorizations on space, place, and the politics of academic production, leading, for instance, to relational and processual understandings of space and to the recognition of the interscalar configuration of purportedly "local" geographies situating the researcher within power-laden multi-scalar spatialities. These developments resonated with contemporary critiques to single site ethnographies and led geographers to interrogate the power relations and spatial practices involved in ethnographic fieldwork (Cook 2019).

Feminist geography, in particular, addressed important critiques to the colonial and patriarchal foundations of social science research, thereby participating in the making of a reflexive, positioned, and embodied ethnography committed to subverting different forms of power hierarchies. In response to the assumption that social phenomena could be observed through a "detached gaze" (Haraway 1988), feminist geographers stressed the importance of considering how the researchers' race, class, and gendered positionalities conditioned knowledge production (McDowell 1992; Rose 1997; Staeheli and Lawson 1995). They insisted on the ethical and political imperative of reflexive accounts of research relationships. Influential works by Cindy Katz (1994, 2001) in global political economy or Juanita Sundberg (2003, 2015) in political ecology, for example, helped blur boundaries between "desk" and "field," questioning the fictitious making of a "field" as a passive space separate from the centers of knowledge-making from which the researcher normally operates. These authors further explored the imperial underpinnings of traditional geographic fieldwork and called upon scholars to integrate knowledge-making with activism for social change.

Additionally, research on the embodied and emotional aspects of spatial relations and engagement with the ordinary and the intimate as spheres of analysis (Sharp 2007; on spatial and corporeal dimensions of precarity see Mollett, Chapter 8 this volume) informed feminist geographers' methodological stances on how these dimensions influence research relations. Therefore, ethnographic relations themselves were conceived as embodied, intersubjective, intimate, and emotional. Finally, recent feminist scholarship in geography has critically addressed question of translation and ethnographic rendering. Works by Richa Nagar, for instance, engage the political potential of translation through an ethic of "radical vulnerability" committed to intimate and longstanding relationships of co-creation between researcher and the collectives she accompanies (Nagar 2018). Here, research involves all participants embracing their incommensurabilities between worlds and creating knowledges that are always fluid and unfinished. By recognizing and incorporating the epistemic agency of collaborators and sensing the multiple "epistemic energies" that unfold in the field (Muppidi 2015), this approach to research enables researchers to account for multiple and complex forms of creating knowledge and exercising politics.

Ethnographic sensibilities for critical resource geographies

Critical ethnographic approaches inform recent geographic research in development and nature-society relations. This research includes multi-scalar approaches (Hart 2004; Tsing 2005) as well as "thick" ethnographic accounts of particular populations and places (Bobrow-Strain 2007; Sultana 2011; Ybarra 2017). Both localized and multi-scalar works have produced nuanced conceptions of the spatialities, subjectivities, and temporalities through which resource-making occurs on the ground, shedding light on the multiple meanings and reappropriations of resources as they become part of peoples' everyday lives (Bobrow-Strain 2007; Sultana 2011) and on the contestations around their commodification and use.

Ethnographic sensibilities draw from the feminist contributions outlined above and broaden the epistemological registers through which resources are studied. Specifically, an ethnographic sensibility involves a willingness to listen, feel, recognize, and interact with people and places in their full complexity, attending to the ordinary and unpredictable logics and practices of social groups (McGranahan 2018). It requires an opening of the researcher's sensitive registers in order to become acutely aware and account for esthetic, affective, and mundane aspects of life and place. In the words of Richa Nagar, these are "the tones and textures, memories and feelings, logics and poetics, traumas and hauntings—of people, places, and times as well as the seemingly mundane truths of life that remain distant or insignificant in the imagination of mainstream academia" (2018, 22). It is these ungraspable "atmospherics resonant in a scene" (Berlant and Stewart 2019, 34) that constitute the ethnographic grounds in which ethnographers dwell—both materially and analytically.

Ethnographic sensibilities further constitute a mode of engagement in which the researcher *exposes* herself to the material conditions of the field and the intimacy of field relations and makes her knowledge frames vulnerable through epistemic dialogue with those with whom she studies. In turn, writing (or any other form of representation) becomes a reflexive and positioned exercise that accounts for the dilemmas and tensions that inform methodological and analytical decisions.

Incorporating ethnographic sensibilities to the study of resources involves situating resources within the broader set of practices and relations of "life-with resources," understood as the entanglements between resource production, everyday life, and politics (Valdivia 2017; Valdivia and Lu 2020). In turn, engaging life-with-resources as ethnographic grounds requires researchers to attend to the affective, ordinary, and mundane aspects of resource relations and to interrogate the everyday ways in which resource politics, resource spaces, and resources are constantly reconfigured and redefined (see also Bryan, Chapter 37 this volume). This perspective has important spatial implications, as it blurs the conceptual distinction between spaces of extraction and spaces of community life and focuses on the multiple ways in which resource production and everyday life overlap and interweave.

Ethnographic sensibilities enable a shared exposure to the material conditions of extractive spaces, a condition that has ethical and analytic implications for the study of resource relations. By experiencing the materiality of life amidst extraction researchers can understand how the environmental and social effects of extractivism merge into the everyday and become an ordinary matter. Relations of intimacy with people and place may lead researchers to share the ordinariness of extraction and thereby have a deeper understanding of the ambivalences and contradictions of political reactions toward socio-ecological harm. At the same time, sustained engagement with a given community constitutes a precondition for witnessing the intermittent and unpredictable instances in which resource extraction becomes an extraordinary matter and generates different forms of organized resistance.

Shared vulnerability further informs the translation of field experiences into ethnographic narratives, generating an ethical commitment to reflexive and positioned writing practices of the researcher's knowledge frames. Based on my research on the everyday politics of palm oil in the municipality of Marialabaja in Colombia's Caribbean, the next section suggests practical ways in which ethnographic sensibilities may enhance critical resource geography.

Researching life with oil palm through ethnographic sensibilities

My research studies the everyday politics of palm oil in the municipality of Marialabaja in Colombia's Caribbean (on the spatial configurations of land relations, see also Mollett, Chapter 8 this volume). Palm oil plantations in Marialabaja expanded at an alarming rate over the last two decades, growing from 570 ha in 2001 to 8310 ha in 2012 (Secretaría de Agricultura 2011), and reaching 11,022 hectares by 2015. This expansion was enabled by piecemeal dispossession in the context of paramilitary violence, a history that haunts collective imaginaries around the resource (Berman-Arévalo 2021). Palm oil introduced new forms of agrarian production, characterized by precarious wage labor, impersonal agrarian relations, and high levels of water and pesticide use.

In Marialabaja, palm oil is inseparable from everyday life. Plantations encroach upon crops, homes, soccer fields, and even elementary schools (Figure 17.1). They merge with spaces of everyday use and circulation and become part of the inhabited landscape. Furthermore, palm oil labor is one of the few sources of income for young men, without which many families could not survive. Most middle-aged men have, at least once, worked as an oil palm laborer. Anything, from beers and billiards to school uniforms and cooking oil, can be acquired with palm oil money. Despite palm oil's "ordinariness" and in the midst of the apparently stable rhythms of life with oil palm, there are sudden (and unpredictable) instances when it is collectively recognized as a threat that mobilizes different forms of resistance.

My ethnographic fieldwork lasted over 15 months and took place intermittently over the course of three years. After two month-long stays in 2013 and 2014 respectively, many phone conversations, and communications via social media, I spent one year living in the region and staying in a particular village, Paloaltico, for periods of three to four weeks, interrupted by week-long stays in nearby cities. Throughout my fieldwork, I became immersed in the many stories, practices, and events that were seemingly unrelated to oil palm, constantly blurring the conceptual boundaries that separate "resource politics" and everyday life. Practical, ethical, and

Figure 17.1 Women telling stories by the water well. Photo by author.

analytical decisions were informed by ethnographic sensibilities. Three aspects proved meaningful for theorizing resource geographies: a sensitive openness to the broader dynamics of life-with-oil palm, an embodied exposure to the material conditions of place, and a reflexive and multilevel writing practice.

Sensitive openness

A "sensitive openness" requires the researcher to make herself vulnerable by (temporarily) letting go of her research agendas and conceptual frames in order to interact with people and places in ways that are receptive to the multiple voices and events that emerge spontaneously in social life. Practicing this "radical vulnerability" (Nagar 2018; Nagar and Shirazi 2019) involves listening, observing, participating, and letting ourselves be "touched" by conversations and practices that may appear minor and ordinary. It requires granting importance to the matters that are important to research participants, individually and collectively, even though they may seem unrelated to our research interests. We are receptive to people's stories, interests, desires, and practices; we *empathize*, letting go of preconceived narratives about others and opening to relations of intimacy and mutual vulnerability. Finally, a sensitive openness requires time: remaining present, keeping in touch, and coming back. With time, unexpected events may unfold that force researchers to reconsider previous judgments and analyses and to remain open to surprises. The following excerpt from my field diary illustrates the potential of this openness in revealing the textures of resource politics:

> This morning, Eloisa, the 59-year-old grandmother, lost her temper when Danielito, the neighbor, complained that her grandchildren were "stealing" his tamarind. I could not believe it when the sweet 59-year grandmother yelled and banged pots and let out the most improper curse words in regional slang. "How could he be so stingy?" she wondered out loud. "Since when does fruit on a tree belong to anyone?" It was Danielito's tamarind tree, but the fruit fell on Eloisa's yard. She was furious. She told me later that she had woken up with her cables crossed that morning. She had spent the night thinking about how her daughter had called her to asked her for 100,000 thousand pesos (approximately UDS$35) for tuition, and she had no money to give her. It broke her heart.
> (Personal field diary, Paloaltico, January 31, 2015)

I was able to witness the above event by spending time in Eloisa's back yard, a place where the extended family got together daily to drink coffee and gossip on recent events. I hardly missed a coffee session. The type of quarrel we saw that day was quite uncommon. Everybody was surprised at Eloisa's reaction and talked about the incident for the rest of the day. I too participated in the later conversations. I listened, commented, and "stuck around." Without me asking, Eloisa eventually opened to me about how she was feeling. I later reflected on this event through several iterations of writing. I realized its connection with an increasingly cash-dependent economy in which money scarcity is the norm and common resources become private. My openness to Eloisa's frustrations further allowed me to realize the emotional and gendered aspects of such dynamics: the rage, the sorrow, the frustration with the loss of collective practices, and the uneven effects of new agro-capitalist economies on mothers—expected to provide but hardly accessing a wage.

These phenomena were contained in an ordinary event to which I remained open. Given my interest in the politics of oil palm, I could have chosen to limit my observations to the direct effects of palm expansion on community life, to actively search for resistance or dispossession. However, events like the one narrated above showed me that such concepts alone could not capture the complex operations of politics, subjectivity, and material conditions of everyday life. In Paloaltico, suffering and vitality, passivity and sudden rage, the conscious realization of injustice, all ran through the fabric of social life. As I was invited into the realm of ordinary life, it became clear that oil palm was entangled with a broader spectrum of practices, desires, preoccupations, power relations, and spatialities, which often provincialized palm oil production.

Embodied exposure

An embodied exposure involves experiencing place and relations in their full material and emotional dimensions. During my time in Paloaltico, I went swimming with the children in the nearby reservoir; I participated in the school marathon along the plantation's irrigation canals; I drank the water; I helped cook despite my clumsiness with the wood stove; I danced in the village cantinas. Women picked my head for lice, sitting in the patio where everybody else practiced similar grooming activities. This was not only participant observation but also an act of physical exposure and subversion of my embodied privilege as an urban, white-mestiza woman. It involved vulnerability but also generated social bonding and broke down boundaries of race and class, albeit ephemerally. Shared vulnerability to the materiality of extractive spaces provided a unique perspective to understand how the environmental and social effects of extractive agro-capitalism become entangled with everyday practices and "descend" into the realm of the ordinary (Das 2015, 71). As the ethnographic vignette below suggests, this immersion made my own body a site where the contested relations of resource-making became a concrete reality:

> The parasite coming out of my body at the university library was a crass reminder of how entangled my body had been with the material conditions of field. My body literally contained residues of water most likely contaminated with agro-industrial fertilizers, decomposing fish, and animal feces.
>
> As I reflected on the politics of contamination, I remembered Leticia, the name the community gave to the village water well. Surrounded by oil palm and colonized by its roots, Leticia was also a place of enjoyment and storytelling. During the course of my fieldwork, Leticia would eventually disappear, its water completely depleted. Leticia's disappearance would interrupt its taken-for-grantedness as well as the ordinariness of its contaminated materiality. For many, it became a symbol of the tragic effects of palm oil encroachment on village commons, a reminder of the threats posed by oil palm to physical survival. For others, it was just another reminder of the material precarity to which they were increasingly getting accustomed to.
>
> For me too, drinking contaminated water had become an ordinary matter, an ordinariness interrupted by the shared witnessing of events, and now by the realization of parasites in my body. As I shared everyday life in the community, I found the privilege of bottled water to be ethically and politically inconsistent with everyday conversations on racialized privilege and the devaluation of black bodies. Yet this embodiment of my politics had also contributed to the naturalization of oil palm's impacts.

Exposing my body to the material conditions of oil palm space was a practical decision driven by ethical and political motives. Making myself vulnerable by experiencing place "like everybody else" allowed me to see the politics of resource contamination in a different light: I realized that political reactions to the harms of oil palm could not be separated from the fact of living amidst extraction and becoming used to it. It was not a question of acquiescence versus resistance but rather of the unpredictable instances when parasites, dried wells, or dead fish rendered oil palm an extraordinary matter.

Writing ethnographic sensibilities

As revealed in the above vignette, ethnography did not only happen on the field. It also took shape through the reflexive and creative process of accounting for field experiences through ethnographic writing. In this sense, the potential of ethnographic sensibilities on the field was realized in retrospect, as writing generated critical awareness of the risks and potentials of exposure and opened analytic avenues for understanding the entanglements between politics, materiality, and the body of the researcher herself.

Reflexive and iterative writing practices enable a mode of analysis in which conceptual and theoretical propositions emerge from a personal, affective attempt to grapple with the complex, fragmented, unpredictable nature of social life, while seeing its particularities in light of a broader set of multi-scalar processes and relations. Particular narrative forms further help the researcher embrace the vulnerability of her own knowledge frames and materialize an ethical commitment to allow the ordinary to become an epistemological register and a site of politics.

I developed a writing practice at three levels: field notes; a field diary; and ethnographic narratives. While the boundaries between these three categories are often blurred (Taussig 2011, 5), each one lies at a distinct place in the continuum between recording, reflection, and analysis.

Fieldnotes *recorded* fragments of experience and retained them for later elaboration. I spontaneously wrote names, numbers, ideas, maps, perceptions, or details of events or conversations. The apparently mechanical practice of writing notes enhances the ethnographer's perception of what is happening around them. As Michael Taussig (2001) suggests, notes are like a "new organ" that "incorporates other worlds into one's own" (5). While random, fragmented, and lacking coherence, field notes also lay the grounds for analysis, allowing researchers to access the latent "worlds of expressivity" of the ground by remembering its details (Berlant and Stewart 2019, 34).

A field *diary* constituted a second level of writing. During my fieldwork, regular diary entries helped me process the day's experiences and elaborate intuitive thoughts. Written by hand and from the field, the texts evoked sensations and affective qualities of place and events, while ordering them in a narrative flow and pointing toward particular analytic directions. The diary became a creative outlet that helped me express a personal voice in an emerging academic narrative and was the basis for later developing ethnographic text.

During my visits to regional capitals and later in a North American university, writing *ethnographic narratives* became at once a mode of analysis and a personal and creative exercise of coming to terms with analytic distance while materializing an ethical commitment to recognize and incorporate participants' knowledge frames. Ethnographic narratives connected concrete elements of the field experience with conceptual and theoretical abstraction. I used stories and ethnographic vignettes to open my chapters or articles in order to connect both reader and writer with the textures of social life. This practice offered a sensible register for readers to be touched by the social realities presented and at the same time enabled me to be openly partial and process my political-ethical commitments.

A sensitive openness, an embodied presence, and reflexive writing are just three practical elements of what I call ethnographic sensibilities. The following section brings these three elements together in an analysis of the entanglements between the spaces of resource extraction and the spaces of community life (Figure 17.2):

> *Life-with-oil palm: the entangled spaces of palm and community*
> In Marialabaja palm grows into schools, soccer fields, water wells and homes. In turn, people, animals, memories, and demons inhabit palm oil plantations by walking, resting, feeding, and even making love in the shade of palm oil trees.

Figure 17.2 Children playing soccer in an oil palm plantation. Photo by author.

Although they overlap, the spatialities of palm and people (and animals and others) are also in tension and sometimes in opposition. Palm oil production limits the land available for traditional plots, and is accompanied by permanent—yet incomplete—efforts to territorialize through enclosure and boundary setting. An instance of palm's violent territorialities unfolded one morning while enjoying morning in the patio of my host family. In the middle of the conversation, we saw a dying pig make its way back towards the house. The skin on her forehead was split in half, enough to expose her bare skull. We knew what had happened: the guards in the neighboring palm oil plantation had hurt the pig to keep her away. It was a symbolic demonstration of power and a direct threat to local livelihoods. "Animals can't roam freely anymore, the palm people don't want them around," explained 42-year old Mercedes. This was a violent encounter between securitized plantation space and the mobile spatialities of animal circulation, which materialized the spatial conflict between private, capitalist resource-making and the possibility for other kinds of non-human elements to become resources for community-members.

The ethnographic vignette above illustrates how particular fieldwork practices and observations informed my theorization on resource geographies and how both experience and analysis come together in ethnographic text. As the narrative reveals, the geographies of oil palm and of everyday life exist in relationship to each other. They are sometimes opposing, sometimes entangled, and often unpredictable. They both contain and enable resource-making practices in a constant negotiation between ago-capitalist interests and securing everyday life. As resource space is co-inhabited by women, men, children, security guards, capitalists, water wells, and animals, they together become active agents in the making of the lived geographies of oil palm. Although palm oil production threatens livelihoods, the spaces of extraction and the spaces community also overlap, interweave, and mutually transform each other in a dynamic process of space-making that always involves conflict and articulations (Hart 2004). In this process, the landscape of resource production is imbued with sociality, thereby becoming a social landscape constituted not only by palm oil production but by the stories, practices, and relations of individuals and collectives (Tsing 2005, ix). Ethnographic sensibilities enhance our exploration of these material, social, and symbolic reappropriations of resource space.

Important political implications arise out of such ethnographically informed analyses of resource spatialities. Common-held geographic imaginaries envision "community territories" that are spatially, culturally distinct from the spatialities of extraction (Zibechi 2012). An ethnographic analysis of resource spatialities in Marialabaja reveals unstable, relational territorialities characterized by a permanent play between boundary setting and border-crossing, where community reappropriation and exclusion coexist, and where local political subjectivities and practices emerge both *in opposition to* and *entangled with* capitalist resource-making.

Conclusion

This chapter developed the concept of ethnographic sensibilities and argued that they constitute a mode of engagement with people, place, and knowledge making that can add theoretical and epistemological nuance to critical resource geographies. Ethnographic sensibilities situate resources within the affective, social, and material dynamics of life-with-resources, generating a commitment to attend to the ordinary aspects of resource relations and further engage the ordinary as an epistemological register to understand the ways in which resource geographies and politics unfold on the ground.

A sensitive openness to the ethnographic grounds and an embodied exposure to the materiality of extractive space can generate a shared vulnerability between the researcher and those with whom she studies. This condition informs nuanced understanding of the ways in which people live with extraction and grapple with its effects, and it creates an ethical commitment to account for the epistemological and political importance of the minor stories of ordinary life. Provincializing resources as part of a broader web of everyday relations invites researchers to conceptualize the geographies of resource production in ways that attend to the overlappings and entanglements with the spaces and spatial practices of community life.

Ethnography is not free of challenges and limitations. Boundaries of race, class, or geographic origin, though they may be temporarily blurred, always condition our ethnographic narratives and perspectives. After all, life-with-resources is, for the researcher, a temporary experience from which she can exit. Moreover, despite an intention to grant political importance to ordinary life and incorporate collaborator's perceptions and understandings into our own conceptualizations, ethnography will always be an uneven practice of academic representation and translation controlled by the researcher and conditioned by prevalent academic standards. Despite these limitations, as this chapter suggests, a critical ethnographic practice is a somewhat utopian attempt to escape these power imbalances, create alternative relationships on the field, and offer sensitive openings for readers to connect with others' life worlds as they make life happen amidst the uneven relations of resource extraction.

References

Berlant, Lauren, and Kathleen Stewart. 2019. *The Hundreds*. Durham, NC: Duke University Press.

Berman-Arévalo, Eloisa. 2021. "Mapping Violent Land Orders: Armed Conflict, Moral Economies, and the Trajectories of Land Occupation and Dispossession in the Colombian Caribbean." *Journal of Peasant Studies*. 48 (2): 349-367. doi: org/10.1080/03066150.2019.1655640.

Biehl, Joao, and Ramah McKay. 2012. "Ethnography as Political Critique." *Anthropological Quarterly* 85 (4): 1209–1228.

Bobrow-Strain, Aaron. 2007. *Intimate Enemies: Landowners, Power, and Violence in Chiapas*. Durham, NC: Duke University Press.

Braun, Bruce. 2002. *The Intemperate Rainforest Nature, Culture, and Power on Canada's West Coast*. Minneapolis, MN: University of Minnesota Press.

Castree, Noel. 2003. "Commodifying What Nature?" *Progress in Human Geography* 27 (3): 273–297.

Cook, Ian. 2019. "Ethnography in Human Geography." In *SAGE Research Methods Foundations*, edited by Paul Atkinson, Sara Delamont, Alexandru Cernat, Joseph Sakshaug and Richard A. Williams. *SAGE Research Methods Foundations*. https://www.doi.org/10.4135/9781526421036861343

Crang, Michael, and Ian Cook. 1995. *Doing Ethnographies*. Norwich, UK: Geobooks.

Das, Veena. 2015. "What Does an Ordinary Ethics look Like?" In *Four Lectures on Ethics. Anthropological Perspectives*, edited by Michael Lambek, Veena Das, Didier Fassin, and Webb Keane, 53–125. Chicago, IL: HAU Books.

Haraway, Donna. 1988. "Situated Knowledges: The Science Question in Feminism and the Privilege of Partial Perspective." *Feminist Studies* 14 (3): 575–599.

Hart, Gillian. 2004. "Geography and Development: Critical Ethnographies." *Progress in Human Geography* 28 (1): 91–100.

Herbert, Steve. 2000. "For Ethnography." *Progress in Human Geography* 24 (4): 550–568.

Huber, Mathew T. 2018. "Resource Geographies I: Valuing Nature (or Not)." *Progress in Human Geography* 42 (1): 148–159.

Katz, Cindy. 1994. "Playing the Field: Questions of Fieldwork in Geography." *The Professional Geographer* 46 (1): 67–72.

Katz, Cindy. 2001. "On the Grounds of Globalization. A Topography for Feminist Engagement." *Signs: Journal of Women and Culture in Society* 26 (4): 1213–1235.

Ley, David, and Marwyn Samuels. 1978. "Introduction: Contexts of Modern Humanism in Geography". In *Humanistic Geography: Prospects and Problems*, edited by David Ley and Marwyn Samuels. Chicago, IL: Marooufa Press.

McDowell, Linda. 1992. "Doing Gender: Feminism, Feminists and Research Methods in Human Geography." *Transactions of the Institute of British Geographers* 17 (4): 399–416.

McGranahan, Carole. 2014. "What Is Ethnography? Teaching Ethnographic Sensibilities Without Fieldwork." *Teaching Anthropology* 4: 23–26.

McGranahan, Carole. 2018. "Ethnography Beyond Method. The Importance of an Ethnographic Sensibility." *SITES: New Series* 15 (1): 1–10.

Moore, Jason. 2015. *Capitalism in the Web of Life. Ecology and the Accumulation of Capital*. London: Verso.

Muppidi, Himadeep. 2015. *Politics in Emotion: The Song of Telangana*. London: Routledge.

Nagar, Richa. 2018. "Hungry Translations. The World Through Radical Vulnerability." *Antipode* 51 (1): 3–24.

Nagar, Richa, and Roozbeh Shirazi. 2019. "Radical Vulnerability." In *Antipode Keywords*, edited by the Antipode Editorial Collective, 236–242. Hoboken, NJ: Wiley-Blackwell.

Ortner, Sherry. 2006. *Anthropology and Social Theory: Culture, Power and the Acting Subject*. Durham, NC: Duke University Press.

Rose, Gillian. 1997. "Situating Knowledges: Positionality, Reflexivities and Other Tactics." *Progress in Human Geography* 21: 305–320.

Secretaría de Agricultura. 2011. *Informe de los Consensos Agropecuarios del Departamento de Bolívar, Periodo 2000–2011*. Cartagena: Colombia.

Sharp, Joanne. 2007. "Geography and Gender: Finding Feminist Political Geographies." *Progress in Human Geography* 31 (3): 381–387.

St Martin, Kevin, and Marianna Pavlovskaya. 2009. "Ethnography." In *A Companion to Environmental Geography*, edited by Noel Castree, David Demeritt, Diana Liverman, and Bruce Rhoads, 370–383. Hoboken, NJ: Wiley-Blackwell.

Staeheli, Lynn, and Victoria Lawson. 1995. "Feminism, Praxis, and Human Geography." *Geographical Analysis* 27: 321–338.

Sultana, Farhana. 2011. "Suffering for Water, Suffering from Water: Emotional Geographies of Resource Access, Control and Conflict." *Geoforum* 42: 163–72.

Sundberg, Juanita. 2003. "Masculinist Epistemologies and the Politics of Fieldwork in Latin Americanist Geography" *The Professional Geographer* 55 (2): 181–191.

Sundberg, Juanita. 2015. "Ethics, Entanglement, and Political Ecology." In *Routledge Handbook of Political Ecology*, edited by Gavin Bridge, and James McCarthy, 117–126. London: Routledge.

Taussig, Michael. 2011. *Fieldwork Notebooks*. 100 Notes, 100 Thoughts: Documenta Series 001. Accessed January 1. https://d13.documenta.de/#/participants/participants/michael-taussig/

Tsing, Anna. 2005. *Friction: An Ethnography of Global Connection*. Princeton, NJ: Princeton University Press.

Valdivia, Gabriela. 2017. "'Wagering Life' in the Petro-City: Embodied Ecologies of Oil Flow, Capitalism, and Justice in Esmeraldas, Ecuador." *Annals of the American Association of Geographers* 108 (2): 549–557.

Valdivia, Gabriela, and Flora Lu. 2020. "Extractive entanglements: Environmental justice and the realpolitik of life-with-oil." In *Extraction, Entanglements, and (Im)materialities: Reflections on the Methods and Methodologies of Natural Resource Industries Fieldwork*, edited by Adrienne Johnson and Anna Zalik. Environment and Planning E: Nature and Space. doi: 10.1177/2514848620907470.

Whatmore, Sarah. 1999. "Hybrid Geographies: Rethinking the Human in Human Geography." In *Human Geography Today*, edited by Doreen Massey, John Allen, and Phillip Sarre, 24–39. Cambridge, UK: Polity Press.

Willems-Braun, Bruce. 1997. "Buried Epistemologies: the Politics of Nature in (Post) Colonial British Columbia." *Annals of the Association of American Geographers* 87 (1): 3–31.

Ybarra, Meghan. 2017. *Green Wars: Conservation and Decolonization in the Maya Forest*. Oakland, CA: University of California Press.

Zibechi, Raúl. 2012. *Territories in Resistance: A Cartography of Latin American Social Movements*. Oakland, CA: AK Press.

18

Institutional ethnography

A feminist methodological approach to studying institutions of resource governance

Emily Billo

Introduction: institutions and critical resource geographies

Critical resource geographers have an abiding interest in institutions. Yet, conceptual and methodological approaches to the study of institutions have varied and have been shaped by developments in subfields including economic geography, political ecology, and feminist geography. For instance, the institutional turn in economic geography has shaped studies of resource governance, defining institutions as rules and norms that undergird the "economic landscape." Critical resource geographers guided by urban regime theory have tended to focus on environmental policy-making, analyzing state and market actors that govern "geographies of urban expansion" (Himley 2008, 437). Those influenced by regulation theory have focused on the state, as well as subnational scales of government, as institutional sites of analysis. From a regulationist perspective, analysis begins with "historically and geographically specific" resource conflicts "codified as the institutions and social practices" in which resource extraction is rooted (Bridge and Jonas 2002, 759). Political ecologists have defined institutions as "rules-in-use" (Watts 2000, 40) to understand resource access, use, and conservation (Himley 2008). They have paid particular attention to the role of the state in property rights and the impacts of privatization, colonialism, and imperialism on shared, communal rights (Himley 2008). Political ecologists have mobilized the concepts of "enclosure" and "accumulation by dispossession" (Harvey 2003) to analyze state and capital arrangements that dispossess traditional resource users and often erase traditional, customary social institutions that govern access to resources (Bridge and Perreault 2009; Himley 2008).

This chapter builds on these discussions of institutions in economic geography and political ecology, complementing them with insights from post-structuralism and feminist geography in order to reveal how institutions of resource governance are materially and discursively constructed and with what effects. Institutions do not emerge as fully formed rules and norms but are constructed via social relationships and are "actively produced through daily social contexts" (Billo and Mountz 2016, 202). That is, institutional boundaries are permeable and constituted through subjectivity and daily life (Billo and Mountz 2016; Rutherford 2007). This approach to institutions advances critical resource geographers' attention to the relationship between subject formation and the rules and norms of institutions of resource governance. In practice, methodologies attuned to the microscale, intimate relationships of resource governance, including subject formation, help reveal the role of institutions in facilitating resource extraction. Moreover, methodological focus on "the everyday" can draw attention to difference and inequality and the lived experiences of differently situated subjects (Bee, Rice, and Trauger 2015). Within this frame, the methodological approach of institutional ethnography (IE) offers critical resource geographers a powerful tool for revealing how institutions of resource governance operate in and through everyday, embodied relationships.

In this chapter, I discuss and reflect on my experience undertaking an IE of one particular institution of resource governance: corporate social responsibility (CSR) programs in the Ecuadorian oil sector (for more on ethnography in critical resources geography, see Berman-Arévalo, Chapter 17 this volume; for more on feminist methods, see Johnson, Chapter 24 this volume). CSR programs emerged within neoliberalism to alleviate the negative socio-ecological impacts of extraction. They typically couple extraction with project-based local economic and social development opportunities. I use this case to illustrate the potential of IE to reveal new insights into the institution as a site of power that facilitates resource extraction as well as how researcher and research participant subjectivities are produced through everyday experiences with institutions of resource governance. In the next section, I discuss the feminist epistemological foundations of IE and review how scholars have conducted IE. In the penultimate section, I examine CSR in an indigenous community in Ecuador at the site of oil extraction. I discuss how my own gender subjectivities as a researcher were shaped in relation to power dynamics associated with CSR, which attuned me to the intersection of gender and indigenous subjectivities around the implementation of CSR projects. In the conclusion, I discuss the broader significance of IE for critical resource geography.

Conducting an institutional ethnography

Methodologically, scholars studying institutions commonly aim to map them onto landscapes or to identify the material realities and effects of institutions (Del Casino et al. 2000). In these approaches, the interviewer often "penetrates" the institution rather than beginning with their embodied relationship to the institution. This approach tends to produce a fixed notion of the institution, both socially and spatially, and institutions become "repressive" and "autonomous" with the power to "affect social relations" (Billo and Mountz 2016, 202). Further, this approach tends to conceive of institutions as bounded, with clearly defined populations, policies, and cultures, and focuses on uncovering how discourses may flow outward from them (Bebbington et al. 2004, 2006; Goldman 2004; King 2009; Perreault 2003a, 2003b).

In contrast, IE, a feminist methodology, begins with the researcher's own relationship to the institution under study (Smith 1987). Founded on critical reflexivity, this approach attunes researchers to positionality (how we are situated on the basis of identities we hold, e.g., gender, race, class, and sexuality) and subjectivity (sense of self), which are inscribed on bodies (Hiemstra and Billo 2017). Critical reflexivity is fundamental to feminist epistemologies that challenge the notion of objectivity or the pursuit of one truth in knowledge production (England 1994; Katz 1994; Kobayashi 2009; Nagar 2002). With critical reflexivity as a starting point, ethnographers utilize interviews, participant observation, and archival research and analysis to make sense of "what people do, as well as what they say" (Herbert 2000, 552).

Following a feminist epistemology and drawing on Dorothy Smith, I begin with the notion that institutions themselves are constructed through material and discursive social relations, which in turn, bring new sorts of relations (including identities) into existence (Billo and Mountz 2016). This enables me to examine how institutions materialize through embodied relationships, where "power moves through dis/embodiments" (Mountz 2004, 328). To do so, I conceive of institutions as dispersed and embedded in daily life, with the division between the inside and outside of the institution blurred. I focus on subjectivity and daily life, rather than only the formal structure of the institution (Nelson 1999; Sawyer 2004). IE allows me to uncover institutional spaces often overlooked in studies of resource governance by carrying out embodied analyses of institutions that produce understandings of the everyday and intimate.

Below, I draw on my own research to discuss multiple data collection methods that are useful in IE but ultimately highlight participant observation as a method particularly attentive to embodiment and subject formation because it enables the researcher to see and experience the daily life of the institution, or the "thick description" outlined by Clifford Geertz (1973). When coupled with attention to researcher positionality, participant observation reveals the workings of institutions, particularly how embodiments and encounters are gendered, raced, classed, and sexualized, and how categories of identification exercise power in the everyday life of the institution (Billo and Mountz 2016). How people embody systems and structures of power is relevant to forms of knowledge production; the focus on embodiment emphasizes the physical and social situatedness of bodies in relation to institutions. Moreover, participant observation can reveal emotions, subjectivities, and power struggles that other methods (e.g., interviews) cannot always as readily access. In sum, as a key method of IE, participant observation offers a valuable opportunity for researchers to learn how power operates across daily socio-spatial productions of the institution.

Background: CSR programs in the oil sector in Ecuador

CSR programs expanded in the neoliberal era and have been lauded for their ability to "go beyond" the legal requirements of states in social and environmental regulation of extractive industries. Where the state is "rolled back" under neoliberal policies and corporate-centric approaches to governing are "rolled-out," this kind of corporate "self-governing" with little government oversight has become common practice in spaces of extraction (Hilson 2012; Himley 2008; McCarthy and Prudham 2004; Peck and Tickell 2002). CSR programs are often implemented as infrastructural or development projects—medical facilities, transportation, schools, microcredit programs, and so on—and have been shown to shape and be shaped by relationships of power that emerge through social relations in places where the programs are implemented (Billo 2015, 2019; Li 2015; Pearson et al. 2019; Shever 2012).

CSR programs in Ecuador play out in a historical-political context in which resources, nation, and citizenship are shaped via hydrocarbons governance (Perreault and Valdivia 2010; Valdivia 2008). In the 1970s, commercial extraction of oil began in Ecuador and became tightly linked to ideas of modernization and progress. However, hydrocarbons governance, including CSR programs, have perpetuated neocolonial, patriarchal relations of capitalist systems of production (Pearson et al. 2019; Valdivia 2008). This is not least because oil extraction takes place in indigenous territories, where resource struggles revolve around social and environmental impacts of the industry (Sawyer 2004). Government elites have argued that indigenous peoples are expected to become "modern citizens," including through participation in the oil complex's neocolonial systems of governance.

By the 1980s, neoliberal policies underpinned the development of the oil industry in Ecuador, leading to increased presence of private, multinational oil companies in the Amazon region and cuts in public spending (Sawyer 2004; Valdivia 2008). Social relationships shifted accordingly. Private companies introduced new institutions of governance, including state-mandated CSR programs in impacted communities. As nation-wide protests by indigenous peoples through the

1990s and 2000s resulted in indigenous rights to territory and representation in congress, companies expanded their use of CSR programs to gain permission to enter indigenous territory and continue extractive operations (Sawyer 2004). Today the programs are no longer mandated by the Ecuadorian state, but, as statements and actions by companies like Repsol—a Spanish-owned multinational oil firm—illustrate, they are considered by industry actors to be essential to operations in the region (Billo 2019; Pearson et al. 2019).

CSR programs are thoroughly embedded in everyday life in zones of extraction today, producing particular subjectivities and identities (Billo 2015, 2019; Li 2015; Pearson et al. 2019; Velásquez 2012; Warnaars 2012). Companies that introduce CSR programs often present a coherent, monolithic discourse that the program will alleviate the negative socio-ecological impacts of resource extraction and provide economic development opportunities related to extractive activities. Critical ethnographies of CSR programs have revealed how companies garner support for their operations through everyday relationships with people living in the communities where they operate (Pearson et al. 2019). IE complements this work, revealing how CSR programs produce particular kinds of subjects. How are CSR programs intersecting with and reformulating identities and social relations in indigenous territories? And what strengths does IE have for understanding the microscale, intimate relations of this particular institution? My IE of CSR programs in one indigenous community in Ecuador revealed that the programs construct particular gender subjectivities; it also revealed *how* they do so. In turn, my examination of processes of subject formation emerging within, through, and beyond CSR programs enhanced understanding of institutions by demonstrating that the institution is not uniform across time and place.

Knowledge production through IE

In my field work, the spatiality and geographies of oil extraction in Pompeya, an indigenous Kichwa community, shaped my own researcher subject formation, and my relationships to the practices of CSR revealed insights into the functioning of the oil complex. In this section, I examine how IE enabled me to understand the institution of CSR as coproduced through gender subjectivities, including my own. In particular, I demonstrate how my understanding of CSR took form in two ways. First, institutional power operated on me, contributing to my own gender subjectivities and subject formation. My gender subjectivities circumscribed my research process, at once limiting my access to some parts of the institution but attuning me to parts of the institution that I had not previously considered in my study and leading me to reformulate my methods. Second, my own gendered subject formation produced in relation to CSR programs attuned me to the production of gendered subjectivities as a site of institutional power. As a result, I added to my research methods, complementing my planned interviews with participant observation to enhance my understanding of identities and subjectivities, including my own, produced through CSR programs. This method revealed the institution to be not only defined by rules and norms but also comprising everyday, embodied social relationships.

Researcher subjectivities and coproduction of the institution

My project was designed to understand CSR programs through indigenous peoples' everyday lived experiences. During preliminary research in Ecuador in 2007, I approached several private, multinational oil companies about studying their CSR programs. Repsol was the only company to agree to my research plan, because they were proud of their operations in the region. Other companies were wary of critique that might result from my research. Repsol gave me "permission" to work at their port of entry in Pompeya; travel into Repsol's oil block along an access road that began in Pompeya required additional company permits. Repsol and Petroecuador, the Ecuadorian state oil company operating on Pompeya's territory, prohibited any photos of their operations. Careful monitoring of who and what information comes and goes from the region is an example of one corporate strategy for holding power over others that seek to engage with them.

My plan to live and work in Pompeya required only permission of Pompeyans, granted via the community president. I also established a collaborative, logistical relationship with an Ecuadorian nonprofit organization, *Fondo Ecuatoriano Populorum Progressio* (FEPP), with a long history of work in the region and in Pompeya specifically. FEPP personnel introduced me to community members to help me begin interviews. They initially provided my transportation into Pompeya, located in Ecuador's northern Amazon, and I lived in a house FEPP maintained in the community. FEPP also received funding from Repsol to implement CSR programs in communities along the Napo River.

I began my research in Pompeya in 2008, flying in and out of the region every two to three weeks, between the regional capital, Coca, and Quito, the capital of Ecuador. I also conducted interviews in Repsol's offices in Quito. On one of these trips, I was on the same flight as a Repsol executive and his colleague from an environmental consulting company, a recipient of a CSR contract. As I described my research design, the employee of the consulting company asked me, "CSR is just a discourse, how can you study it?" The Repsol executive proudly responded that CSR is also evident in the many infrastructural projects in the region. Together the employee and executive's comment suggested that CSR is both discourse and material practice that shapes everyday lives. I knew that IE, with its focus on embodied, everyday relationships, was an especially useful methodology for understanding how institutions like CSR are constructed by a combination of dynamic discourses and practices that shape material relations that govern human behavior. However, as I went about my research in Pompeya, I soon learned via my Pompeyan research assistant that residents were wary of speaking with me about Repsol. My foreignness, race, and gender led residents to assume that Repsol employed me; they were hesitant to respond to my questions about CSR. I recalibrated my interviews to ask about community histories and cultures in relation to the arrival of companies. I conducted these interviews with community elders, all men, who had signed contracts with companies to implement CSR programs as community leaders. I did not record interviews to alleviate any nervousness on the part of participants. It was much harder to speak with women, who often deferred to their husbands. I did not know whether Pompeyans desired my presence, their opinions on my research questions, or really anything about their thoughts on the role of Repsol in their community.

Given these challenges in Pompeya, I began to expand my research to local and regional government, nonprofit, and indigenous organization offices. Some of these nonprofits and indigenous organizations benefitted from lucrative financial contracts with Repsol to implement CSR programs, while local state officials helped to facilitate the programs in communities (see Billo 2015; Pearson et al. 2019). I noticed that these interviews with mostly elite actors who were often far removed from the daily realities of Pompeya tended to narrate a static, disembodied story of the operation of CSR programs in the community.

My own uncertainty and confusion about where and how to find an entry point for studying lived experiences of CSR was heightened through everyday research relationships. Indeed, my IE of CSR began to reveal as much about me, as it did CSR programs (compare with Mountz 2007). The everyday realities of oil extraction weighed on my daily life and welfare, as I sought to situate myself in the research project and to understand forms of power that shaped everyday lives. I realized that when I designed my research, I had underestimated my own embodied needs in my field research site in Pompeya. I did not have access to potable water and worried about my safety as a single woman working in a masculine space of extraction. I had not fully considered prior to research, and nor could I have known, the overwhelming loneliness and the constant, unwanted advances of male oil workers during long periods of field research in a remote location (Billo and Hiemstra 2013).

My efforts to attend to my own physical and emotional needs eventually led me to restructure the research project, in the process revealing to me that I myself had become a subject of the institution. I made shorter trips into the region, renting an apartment in Quito. When I did arrive in Pompeya, I stayed with nuns in the Catholic mission (thereby linking me to another colonizing institution in the region). This placed me on the opposite side of the Napo River from the center of Pompeya, where I had to cross in oil company boats each day to meet with residents and conduct interviews. The last boat crossing was at 4:30 p.m., just as residents returned home after working on their farm plots. I had no other way of returning to the mission and routinely missed the chance to interview residents. However, living with the nuns provided access to drinking water and a private room I could return to in the evening (Billo and Hiemstra 2013).

Throughout all this, I often sensed people fixating on my embodied presence as I moved through the masculine spaces of extraction. I became acutely attuned to my gender subjectivities as I moved through spaces of CSR (Billo and Hiemstra 2013). In Pompeya, I found that waiting for the company-operated boats made me more vulnerable to male oil worker advances. I tried to arrive only a few minutes before river crossings to avoid interactions with company employees. I scheduled interviews with residents to ensure I had somewhere to be, rather than hoping to encounter residents at home by chance. If I did not have an interview, I waited at the nuns' house, secluded from the outside environment. If a scheduled interview did not materialize, I hoped that the few women I had gotten to know in Pompeya might be home and willing to chat with me.

Production of my own gender subjectivities through CSR programs and the broader oil complex made it evident to me that institutional dynamics shape social relationships, identities, power, and agency, including my own. Restructuring my research to attend to my own embodiment shifted my access to the everyday spaces I hoped to study (Billo and Hiemstra 2013), just as it embedded me more fully in the social relations of the institutions that governed extraction. As a researcher, I worked to constantly negotiate "difference, identity, and change" to live in the borders or in-between spaces of research, rather than (re)produce boundaries and differences in an effort to soldier on and collect data (Katz 1994; Katz 1996, 182; Kobayashi 1994).

Longstanding company presence in Pompeya led to a distinct set of power relationships between companies and indigenous peoples and to which I was not immune. My daily life in Pompeya was shaped by these power dynamics. For instance, the pervasive masculinity/machismo culture underpinned all aspects of life and influenced—and limited—when I decided to go outside to conduct interviews and my travel into and out of the community. Confined by these power relations, data that I collected in interviews with corporate, state, and indigenous actors led to a static or fixed vision of the historical and ongoing implementation of CSR programs that was defined via corporate and state rules and norms. Yet these stories contradicted my own observations of everyday life with CSR in Pompeya.

Gendered indigenous subject formation through CSR programs

With the realization of my own alignment to CSR programs, I sought to find ways to observe the institution in action. Because my own gendered subject formation attuned me to the paternalism linked to CSR programs, to expand my understanding of the institution, I began to focus on how women narrated their own participation in the programs in CSR meetings. I turned to participant observation to observe the meetings and roundtables in which Pompeyans and CSR representatives negotiate the terms and practices of the institution (CSR programs). This shift in research design allowed me to understand how Repsol's CSR programs produce gendered indigenous subjects in relation to oil extraction. I was able to see how women negotiated their gendered identities in spaces of CSR, something that interviews did not readily highlight. I examine these practices of negotiation through one example that focused on "women's empowerment": a CSR women's microcredit program.

Repsol entices Kichwa women to participate in microcredit projects with the promise of having their own economic income and independence, an approach rooted in an ideal of creating liberal subjects. Repsol's microcredit program was designed to create opportunities for women to earn their own money through raising and selling agricultural products in local markets. In interviews, Repsol executives said that the company's goal in this microcredit program included intervening in "machista" (male-chauvinist) Kichwa culture; my observations suggested it did little in the way of this. Through IE, I found that participation in the microcredit program perpetuates masculinity and patriarchal relations (Billo 2019).

The meeting spaces revealed power relationships between indigenous men and women in the context of the microcredit CSR program. I observed that Pompeyan men often dominated meeting spaces that were meant to be focused on a women's microcredit program, ensuring that the project continued. Men attended the microcredit meetings because they often participated in the labor of *mingas* (community workdays) that directed the microcredit funded agricultural projects. During meetings, men would often blame the microcredit group's lack of organization on women who were also present at the meeting. I observed that meetings offered opportunity for men to dominate and control community space and, more broadly, the "opportunity" for local people that comes from oil via resources from CSR programs (Billo 2019).

In interviews, Respol executives framed the microcredit project around what indigenous women lacked, or their gendered deficiencies. In practice, women faced a double burden, where their embodied intersecting gender and indigenous subjectivities subjected them to neocolonial systems intent on bringing them into corporate projects that sustained Repsol's presence in the region. In meetings about the microcredit project, women said they were tired of CSR programs that encumbered their already busy lives but were encouraged to keep participating by male residents. For example, only after her husband left the meeting, one female resident said, "My husband is always saying that I should do things with the organization. But, I also have to wash the clothes and do things in the house." Thus, CSR programs create new gender inequalities. Despite Repsol's aim to intervene in a male dominated Kichwa culture, the social relationships in the meeting suggested that CSR programs reproduced patriarchal community relationships (Billo 2019).

IE enabled me to see that intimate, microscale social relationships that form within CSR programs circumscribe gendered indigenous subjects. Companies present these programs as designed to challenge indigenous traditions, patriarchal structures, and oppression of indigenous women linked to work in the home. These practices produce particular material outcomes that structure everyday lives, but that can be at odds with the programs' stated aims. Furthermore, because women perceive CSR programs as a burden, the projects themselves are not directed by the needs of women, but instead by Repsol-imposed norms and rules, that ultimately aim to sustain oil extraction in the region (Billo 2019). Yet, CSR programs are more than just rules or norms that affect social relations. The programs operate through strategic embodiments, in this case through construction of gender identities and women's roles in sustaining CSR programs that facilitate the extractive economy. IE locates the power of the institution through embodiment, disrupting institutional coherence.

Conclusions: IE in studies of subject formation in critical resource geographies

Critical resource geographers have focused on subject formation in the context of institutional relationships that structure access to territory and resources. IE enables a focus on construction of identities via institutional discourses and practices to reveal where and how people are located in the everyday materialities of the extractive landscape. Studying institutions through embodied relationships, including the researcher's, reveals the specific histories and geographies that situate people in particular extractive contexts. Embodied methods of knowledge production that attend to subjectivities, such as participant observation, hold the potential to uncover institutional spaces often overlooked or inaccessible through other research methods, like interviews. IE can expand critical resource geographers' attention to everyday, intimate scales and sites of analysis.

IE invites critical resource geographers to see how the discourse of an institution and the material outcomes of its activities may not align in a coherent and fixed way in daily lives. The approach advanced in this chapter emphasizes how institutions contribute to strategic embodiments, determining people's relationships to each other, and to institutions and power. In this approach, institutions are understood as the products of social relations that, as they become embedded within everyday life, produce the on-the-ground socio-spatial practices (e.g., rules and norms) that govern extractive spaces. Focusing on how institutions produce particular kinds of subjects thus reveals the operation and construction of institutional power. IE, as methodology, can enable the researcher to see and reveal dominant discourses and practices that structure resource use, access, and control.

References

Bebbington, Anthony, Scott Guggenheim, Elizabeth Olson, and Michael Woolcock. 2004. "Exploring Social Capital Debates at the World Bank." *Journal of Development Studies* 40 (5): 33–64.

Bebbington, Anthony, Michael Woolcock, Scott Guggenheim, and Elizabeth Olson, eds. 2006. *The Search for Empowerment: Social Capital as Idea and Practice at the World Bank.* Boulder, CO: Kumarian Press.

Bee, Beth, Jennifer Rice, and Amy Trauger. 2015. "A Feminist Approach to Climate Change Governance: Everyday and Intimate Politics." *Geography Compass* 9 (6): 1–12.

Billo, Emily. 2015. "Sovereignty and Subterranean Resources: An Institutional Ethnography of Repsol's Corporate Social Responsibility Programs in Ecuador." *Geoforum* 59: 268–277.

Billo, Emily. 2019. "Gendering Indigenous Subjects: An Institutional Ethnography of Corporate Social Responsibility in Ecuador." *Gender, Place & Culture.* doi: 10.1080/0966369X.2019.1650723.

Billo, Emily, and Nancy Hiemstra. 2013. "Mediating Messiness: Expanding Ideas of Flexibility, Reflexivity, and Embodiment in Fieldwork." *Gender, Place & Culture* 20 (3): 313–328.

Billo, Emily, and Alison Mountz. 2016. "For Institutional Ethnography: Geographical Approaches to Institutions and the Everyday." *Progress in Human Geography* 40 (2): 199–220.

Bridge, Gavin, and Andrew E. G. Jonas. 2002. "Governing Nature: The Reregulation of Resource Access, Production, and Consumption." *Environment and Planning A* 34: 759–766.

Bridge, Gavin, and Tom Perreault. 2009. "Environmental Governance." In *A Companion to Environmental Geography,* edited by Noel Castree, David Demerrit, Diana Liverman, and Bruce Rhoads, 475–497. Oxford, UK: Blackwell.

Del Casino, Vincent J. Jr, Andrew J. Grimes, Stephen P. Hanna, and John Paul Jones III. 2000. "Methodological Frameworks for the Geography of Organizations." *Geoforum* 31 (4): 523–538.

England, Kim V. L. 1994. "Getting Personal: Reflexivity, Positionality, and Feminist Research." *The Professional Geographer* 46 (1): 80–89.

Geertz, Clifford. 1973. *The Interpretation of Cultures: Selected Essays.* New York, NY: Basic Books.

Goldman, Michael. 2004. "Ecogovernmentality and Other Transnational Practices of a 'Green' World Bank." In *Liberation Ecologies: Environment, Development, Social Movements*, edited by Richard Peet, and Michael Watts, 153–177. London: Routledge.

Harvey, David. 2003. *The New Imperialism*. Oxford: Oxford University Press.

Herbert, Steve. 2000. "For Ethnography." *Progress in Human Geography* 24 (4): 550–568.

Hiemstra, Nancy, and Emily Billo. 2017. "Introduction to Focus Section: Feminist Research and Knowledge Production in Geography." *The Professional Geographer* 69 (2): 284–290.

Hilson, Gavin. 2012. "Corporate Social Responsibility in the Extractive Industries: Experiences from Developing Countries." *Resources Policy* 37 (2): 131–137.

Himley, Matthew. 2008. "Geographies of Environmental Governance: The Nexus of Nature and Neoliberalism." *Geography Compass* 2 (2): 433–451.

Katz, Cindi. 1994. "Playing the Field: Questions of Fieldwork in Geography." *The Professional Geographer* 46 (1): 67–72.

Katz, Cindi. 1996. "Expeditions of Conjurors: Ethnography, Power, and Pretense." In *Feminist Dilemmas in Fieldwork*, edited by Diane L. Wolf, 170–184. Nashville, TN: Westview Press.

King, Brian. 2009. "Commercializing Conservation in South Africa." *Environment and Planning A: Economy and Space* 41 (2): 407–424.

Kobayashi, Audrey. 1994. "Coloring the Field: Gender, "Race," and the Politics of Fieldwork." *The Professional Geographer* 46 (1): 73–80.

Kobayashi, Audrey. 2009. "Situated Knowledge, Reflexivity." In *International Encyclopedia of Human Geography*, edited by Rob Kitchin, and Nigel Thrift, 138–143. Oxford: Elsevier.

Li, Fabiana. 2015. *Unearthing Conflict: Corporate Mining, Activism, and Expertise in Peru*. Durham, NC: Duke University Press.

McCarthy, James, and Scott Prudham. 2004. "Neoliberal Nature and the Nature of Neoliberalism." *Geoforum* 35: 275–283.

Mountz, Alison. 2004. "Embodying the Nation-State: Canada's Response to Human Smuggling." *Political Geography* 23 (3): 323–345.

Mountz, Alison. 2007. "Smoke and Mirrors: An Ethnography of the State." In *Politics and Practice in Economic Geography*, edited by Adam Tickell, Eric Sheppard, Jamie Peck, and Trevor Barnes, 38–48. Thousand Oaks, CA: SAGE Publications.

Nagar, Richa. 2002. "Footloose Researchers, 'Traveling' Theories, and the Politics of Transnational Feminist Praxis." *Gender, Place & Culture: A Journal of Feminist Geography* 9 (2): 179–186.

Nelson, Diane M. 1999. *A Finger in the Wound: Body Politics in Quincentennial Guatemala*. Berkeley, CA: University of California Press.

Pearson, Zoe, Sara Ellingrod, Emily Billo, and Kendra McSweeney. 2019. "Corporate Social Responsibility and the Reproduction of (Neo)Colonialism in the Ecuadorian Amazon." *Extractive Industries and Society* 6 (3): 881–888.

Peck, Jamie, and Adam Tickell. 2002. "Neoliberalizing Space." *Antipode* 34 (3): 380–404.

Perreault, Tom. 2003a. "A People with Our Own Identity: Toward a Cultural Politics of Development in Ecuadorian Amazonia." *Environment and Planning D: Society and Space* 21 (5): 583–606.

Perreault, Tom. 2003b. "Changing Places: Transnational Networks, Ethnic Politics, and Community Development in the Ecuadorian Amazon." *Political Geography* 22 (1): 61–88.

Perreault, Tom, and Gabriela Valdivia. 2010. "Hydrocarbons, Popular Protest and National Imaginaries: Ecuador and Bolivia in Comparative Context." *Geoforum* 41 (5): 689–699.

Rutherford, Stephanie. 2007. "Green Governmentality: Insights and Opportunities in the Study of Nature's Rule." *Progress in Human Geography* 31 (3): 291–307.

Sawyer, Suzana. 2004. *Crude Chronicles: Indigenous Politics, Multinational Oil, and Neoliberalism in Ecuador*. Durham, NC: Duke University Press.

Shever, Elana. 2012. *Resources for Reform: Oil and Neoliberalism in Argentina*. Palo Alto, CA: Stanford University Press.

Smith, Dorothy E. 1987. *The Everyday World as Problematic: A Feminist Sociology*. Toronto: University of Toronto Press.

Valdivia, Gabriela. 2008. "Governing Relations Between People and Things: Citizenship, Territory, and the Political Economy of Petroleum in Ecuador." *Political Geography* 27 (4): 456–477.

Velásquez, Teresa A. 2012. "The Science of Corporate Social Responsibility (CSR): Contamination and Conflict in a Mining Project in the Southern Ecuadorian Andes." *Resources Policy* 37 (2): 233–240.

Warnaars, Ximena S. 2012. "Why Be Poor When We Can Be Rich? Constructing Responsible Mining in El Pangui, Ecuador." *Resources Policy* 37 (2): 223–232.

Watts, Michael J. 2000. "Contested Communities, Malignant Markets, and Gilded Governance: Justice, Resource Extraction, and Conservation in the Tropics." In *People, Plants and Justice: The Politics of Nature Conservation*, edited by Charles Zerner, 21–51. New York, NY: Columbia University Press.

19

Critical physical geography

In pursuit of integrative and transformative approaches to resource dynamics

Christine Biermann, Stuart N. Lane, and Rebecca Lave

Introduction

As an answer to the perennial calls for integrative geographic scholarship, critical physical geography (CPG) brings together deep knowledge of biophysical systems and attention to social dynamics and structural relations of power (Lave et al. 2014; Lave, Biermann, and Lane 2018). In bridging the human-physical divide, CPG aims to not only produce fresh geographic insights but also to transform the social and ecological worlds we study and inhabit and the scientific cultures within which we work. As such it presents a novel framework and unique methodological opportunities for the study of resource geography. More specifically, a CPG approach can provide more expansive answers to typical resource geography questions about resource-making at multiple scales; the knowledge politics of resource management; and the eco-social impacts of commodification, circulation, consumption, and disposal of resources.

In this chapter, we introduce the intellectual foundations of CPG, reflecting on their relevance to issues of resources. In studying natural processes, knowledge politics, and structural power relations together, CPG expands upon an idea that has long circulated within political ecology and critical geographic approaches to resources: that the matter of nature matters (Bakker and Bridge 2006). Indeed, our vision for CPG shares much in common with political ecology, leading some to question the need for CPG as a distinct intellectual project and subfield (Rochner et al. 2019). We contend that there is indeed an important distinction. While political ecology takes seriously the materiality of nature and a handful of political ecologists even work with biophysical data, the integration of social and biophysical research is not central to its mission. In other words, it may be widely recognized with political ecology that resources themselves are co-constituted through a suite of interacting biogeophysical and social processes, but it is far less common for research to actually integrate methods from physical and human geography to examine the dynamics of these interacting processes as they shape resources and resource-making (compare to Mark et al. 2010; McClintock 2015). Complementing existing work in political ecology and resource geography, CPG encourages an integrative, iterative mode of analysis and provides a useful model for geographers who seek to develop approaches and insights that span the human/physical divide. To bring this model to life, we intersperse

CPG research examples that draw upon qualitative and quantitative methods from across the natural sciences, social sciences, and humanities throughout the chapter.

But the potential value of this framework goes beyond an expansive methodological toolkit. As we envision it, CPG offers a disciplinary home for integrative work that grapples with values and transformative ambitions. Effectively, CPG research argues that "if we cannot avoid values we should engage them" (Biermann, Lane, and Lave 2018, 567). As such this approach calls for research that "[fosters] the causes of justice alongside those of sustainability" (Rochner et al. 2019), "internalize[s] and expand[s] rigorous self-critique," and "swing[s] the doors open to diverse publics and ways of knowing" (Biermann, Lane, and Lave 2018, 564).

Intellectual foundations

We turn first to the three core tenets that provide an intellectual foundation for CPG research. Rather than being solely the domain of CPG, these ideas have emerged out of decades of research within political ecology, science and technology studies (STS), and cognate fields. We repeat these tenets here and reflect on their relevance to issues of resources. Investigating physical processes, social power relations, and knowledge politics *together,* we argue, expands the intellectual and practical value of critical resource geography scholarship.

> **Core tenet 1:** Most landscapes are now deeply shaped by human actions and structural inequalities around race, gender, and class. These power relations are not social drivers, external to nature and shaping it from the outside. Rather, structural power relations incorporate and draw on the materiality of nature, creating inextricably eco-social systems.
>
> (Lave, Biermann, and Lane 2018, 5)

This idea recognizes that the worlds we live in and study are products of intertwined social and biophysical forces. A soil scientist, for example, might understand soil classification as determined by geological, biological, and chemical processes, which are at times disrupted by human activity. A social scientist, on the other hand, might note how particular social dynamics lead to soil degradation but might not view soils themselves as central to "the social." CPG research on agricultural systems in Hungary (Engel-DiMauro 2018) and dryland West Africa (Turner 2018) demonstrates, by contrast, that even the chemical properties of soils (e.g., soil pH, composition, nutrient and moisture content, etc.) are products of biophysical and social factors that cannot be easily disentangled. This notion applies similarly to water (Arce-Nazario 2018), wildlife (Goldman 2018), streams (Ashmore 2018), and numerous other research objects.

To critical resource geographers, this insistence on an eco-social material world is likely to be familiar. Studies of resources have long shown that so-called natural resources are not naturally resources (Hudson 2001; quoted in Bridge 2010): they do not arrive fresh from nature but *become* resources through particular structural power relations and ways of valuing and organizing nature. In the process, the materiality of the resource itself is reshaped. As a familiar corollary, "matter *matters*": the ecological dynamics and biophysical particularities of a resource also shape broader social dynamics and ways of organizing nature (Bakker and Bridge 2006). Turning back to the example of soils, a CPG approach reflects the idea that resources such as soils are not only produced (in part) through social relations but also are themselves productive of social relations.

> **Core tenet 2:** The same power relations that shape the landscapes we study also shape who studies them and how we study them. Both natural and social science are inextricably

imbricated in social, cultural, and political-economic relations that affect the questions we ask (or ignore), the way we conduct our research, and even our findings.

(Lave, Biermann, and Lane 2018, 5)

This second tenet of CPG draws on insights from STS, and feminist postcolonial science studies in particular, to foreground the *knowledge politics* at play in environmental scholarship. Both biophysical and social science methods yield knowledge that is partial, power-laden, and situated (Haraway 1988; Nightingale 2003). Even seemingly detached and neutral observations—of primate behavior (Haraway 1989) or Antarctic ice (O'Reilly 2016), for example—have been demonstrated to be bound up in social dynamics. As such, we recognize that our own projects and research communities, as well as taken-for-granted concepts and frameworks, are inevitably shaped by social, cultural, and political-economic relations. This is certainly relevant for resource geography, which was imbricated in a variety of imperial, colonial, and nationalist interests as it took shape in early- to mid-twentieth century Europe and North America (Zimmerer 2010). Traces of these histories remain salient in conservation and resource management today (e.g., Biermann 2016; Davis 2018; Duvall, Butt, and Neely 2018; Sayre 2018); if we are blind to these traces, we risk reproducing injustices alongside flawed accounts of the world.

The politics of knowledge production, however, need not be nefarious or elaborate to be influential. An example from the first author's research can help to illustrate some of the seemingly mundane, everyday knowledge politics at work in the field of tree-ring science (Biermann and Grissino-Mayer 2018). A survey of tree-ring researchers found that the disciplinary boundaries between ecology, tree physiology, and climatology inhibit communication and collaboration around controversial concepts, methods, and theories related to physiological responses of trees to climatic changes. While these disciplinary divisions are engrained in our understanding of academia, they are indeed a form of politics that shape what knowledge is produced, by whom, and through what methods. Furthermore, disciplinary divisions work to uphold hegemonic representations of resources and narrowly circumscribe the types of methods and data that are deemed appropriate and trustworthy. For example, the discipline of forestry science has played a crucial role in constructing timber as a resource and defining how we measure concepts like forest health, productivity, and resilience.

In the same survey discussed earlier, several respondents also indicated that a gendered and hierarchical culture within tree-ring science has long shaped how ideas and techniques circulate. In the 1930s, for example, Tennessee Valley Authority (TVA) scientist Florence Hawley faced sexism from leading male scholars that prevented the publication of her research on tree growth and precipitation variability in the Tennessee River basin (Smith 2019). More recently, a survey respondent described how new statistical techniques from outside the tree-ring community have been slow to trickle in and influence research practices, attributing much of the resistance to persistent sexism and elitism. These politics of knowledge production have important consequences for thinking about and managing resources, as tree-ring studies provide historical information on climate and ecological variability to enhance resource management and planning (e.g., water managers use tree-ring data to understand historical droughts and thus better prepare for future droughts.)

Opening the black box of science allows us to consider how the choices of scientists, the methods they pursue, and the claims produced are all mediated by social relationships (King and Tadaki 2018). This form of reflexivity, we believe, needs not weaken science's capacity to produce understandings of the world and to inform decision-making. Indeed, the claim that excluding social relationships from the practice of science is a necessary component of scientific authority is itself a political statement because of the implied transfer of power to scientists

(Lane 2017). Rather than undermining the ability of science to innovate, reflexivity about the politics of knowledge production can help scientists to generate new questions about the world around us and potentially even launch entirely new areas of inquiry (Stengers 2013).

> **Core tenet 3.** The knowledge we produce has deep impacts on the people and [systems] we study. The myth of the ivory tower is just that: a myth. Our research has unavoidably political consequences; our choice is thus not between being political or apolitical, but among different possible political commitments.
>
> (Lave, Biermann, and Lane 2018, 5)

CPG's third foundational idea flags that we do not do research in a vacuum. Just as knowledge production is shaped *by* politics, so too does knowledge *constitute* politics. In short, the knowledge we produce—and the processes by which it is produced—have real material and political socioecological impacts. While control over the impacts of one's research is elusive, we as researchers do have some choice in how our research impacts material environments and social relations, as well as how our research is framed and applied, by whom, and to what ends (Law 2018). When we decide who to include in framing the problem, where to perform fieldwork and with what methods, and even how and where to disseminate our findings, we are engaging in political decisions. Recognizing them as such is central to the mission of CPG.

Two examples from CPG research illustrate both the political commitments and consequences of research on resource-making. In the first example, geographer Nathan McClintock (2015) designed empirical research to assess and explain the socio-spatial patterning of lead (Pb) contamination in urban soils in Oakland, California. Rather than adopting a politically neutral approach, McClintock aligned his work with environmental and social justice activist movements, hopeful that a CPG approach would articulate with their radical and material politics of place—a politics that extends beyond a single contaminated site. Beginning with quantitative geochemical and geospatial analysis of soils, McClintock found that soil Pb concentrations were indeed significantly higher in the low-income flatlands of West Oakland than in the more affluent and predominantly white Oakland Hills, and that soil Pb levels were strongly linked to the density of pre-1940s housing, suggesting that deteriorating housing stock is a key source of contamination. Using these findings as a starting point, McClintock then turned to the archives and to urban political ecology theoretical concepts (e.g., social metabolism) to trace the history of urbanization and the subsequent devaluation of the flatlands of West Oakland. The resultant CPG places the inequitable distribution of soil Pb in a broader historical context and empirically demonstrates "how the ebbs and flows of capital have resulted in disproportionate concentrations of various forms of Pb in the soils of the flatlands" (McClintock 2015, 70). By bringing to light soil contamination, linking it to longer term social processes, and sharing data with activist groups, McClintock's research aims to further both justice and sustainability. More broadly, this example illustrates how CPG can be used to document the ecological and social afterlives of resource-making, which in this case include the commodification, circulation, consumption, and disposal of Pb.

A second example deals with an increasingly prevalent form of resource-making: market-based environmental management. Advocates of this form of environmental management argue that pricing and establishing markets for nature will help to "save" it. Political ecologists and critical nature-society geographers have launched numerous critiques of market-based environmental management, arguing that it may worsen ecological outcomes while also contributing to economic injustices. In their study, Lave et al. (2018) investigated the material effects of stream mitigation banking, a form of market-based neoliberal environmental management, on streams in North Carolina. Drawing on critical nature-society geography, they hypothesized that, in following economic logic to maximize the number of credits produced by a project ("credit chasing") while minimizing restoration costs, stream mitigation banking would ultimately "produce substantially worse outcomes for fluvial systems than were produced previously through non-market approaches" (Lave et al. 2018, 448). Their data did not support these hypotheses; it indicated instead that the effects of stream mitigation banking are not all that different from nonmarket forms of stream restoration. In other words, this particular form of neoliberal, market-based environmental management appears to be neither significantly better than nonmarket approaches (as advocates argue) nor significantly worse (as critics suggest). This project—from its framing to its hypotheses to its findings—is wrapped up in the politics of environmental management. Instead of imagining that their empirical research was taking place outside this political realm, the researchers explicitly sought to engage and contribute new perspectives and data to political debates about market-based approaches.

Integrative and iterative methods

But how does one actually *do* the integrative and iterative research that CPG envisions? First, one needs research questions that require both biophysical and social analysis to answer. Certainly not all research agendas necessitate integrated questions; much social research need not incorporate methods originating in physical geography, and much physical geography research need not engage with social theory or methods, which are the foci of critical human geography. But given that resources themselves are widely understood to be simultaneously social and natural, we argue that there remains a pressing need for research on resources that examines biophysical systems, structural power relations, and knowledge politics *together*.

Because the range of issues that CPG researchers address is so broad—from soils to water to fracking to forests (and beyond)—CPG embraces an expansive methodological toolkit. There is no single set of CPG methods (as remote sensing and agent-based modeling dominate in land change science, for example); instead, the methods that CPG researchers employ are chosen in relation to the specific problem at hand. CPG research is defined by a "reach across traditional ideas of what are admissible methods, whether in the natural or social sciences" (Lave, Biermann, and Lane 2018, 10). CPG is explicitly interdisciplinary in its embrace of mixed methods, but it is also implicitly interdisciplinary in its insistence on a reflexive approach to knowledge production, integrating insights from STS (and the broader social sciences and humanities) with physical science methods.

Figure 19.1 presents a framework for thinking about the diversity of methods available for environmental and resource scholarship. We distinguish between natural science and social science methods as well as quantitative and qualitative methods, recognizing that there may be slippage and overlap across these distinctions. CPG research typically spans multiple boxes, allowing triangulation among sometimes quite disparate data sources. This mixing of methods is both an epistemological and a political strategy, a direct outgrowth of the core tenets discussed in the previous section. For resource geographers, incorporating methods from different epistemological traditions can highlight the partiality and power-laden nature of knowledge, echoing geographer Andrea Nightingale's call to use different methods to "challenge the hegemony of positivist science within mainstream academic and policy circles" (2003, 77).

Despite recent emphasis within the academy on interdisciplinarity, transdisciplinarity, and mixed-methods research, there remain real barriers to designing and performing integrative research that stretches across the squares in Figure 19.1. It is our hope that CPG can provide additional justification to support and protect those who would like to conduct research and

	Quantitative Methods	*Qualitative Methods*
Biophysical Science	dendrochronology isotope geochemistry climate modeling geospatial analysis	soil classification landscape observation species distribution mapping
Social Science	surveys econometrics demographic analysis geospatial analysis	interviews archival/historical discourse analysis participant observation

Figure 19.1 A methods four-square for thinking about the diversity of methods and ways of knowing in the biophysical and social sciences.

teach in ways that bring together methods, tools, and ideas from disparate traditions: remote sensing and ethnography, dendrochronology and discourse analysis, and water quality analysis and oral histories, to name a few examples. Reaching across these methodological divisions, we believe, can strengthen the explanatory power of our research, as well as increase its practical and intellectual significance. In the sections that follow, we draw on existing scholarship to explore how CPG works in practice and how researchers grapple with the political and normative dimensions of their research.

Mixing methods to research resource-making: an example from Sulawesi, Indonesia

Let's turn to another research example to consider what this integrative impulse looks like in practice, and how mixing methods can yield new insights about land as a resource; its role in global agricultural commodity chains; and patterns of its enclosure, commodification, and conversion. Geographer Lisa Kelley investigates environmental change in the Sulawesi province of Indonesia, where over the last few decades many forest and agro-forest ecosystems have been replaced by monocultures. The landscapes of Sulawesi are very clearly eco-social hybrids, impacted as much by biophysical processes as by human actions and structural power relations (see CPG's first core tenet). To better understand how social dynamics structure the expansion of commodity crops (and cacao in particular), Kelley draws on a methodological toolkit that includes remote sensing, land use and cover change (LUCC) mapping, oral histories, interviews, household surveys, and analysis of secondary data on agricultural production, soils, topography, and migration (Kelley, Evans, and Potts 2017; Kelley 2018).

The benefits of mixed methods for Kelley's research are not merely additive. Instead, this carefully selected combination of methods allows her to address some of the shortcomings of existing approaches to land change science. Remote sensing, Kelley notes, has yielded plenty of information on the macroscale patterns and processes of land use change in Indonesia. It has demonstrated that smallholder cacao plantings tend to accelerate forest clearance, as it quickly becomes more profitable for smallholders to clear additional forest than to replant existing sites (in line with the theory of forest rents). But where remote sensing has fallen short is in capturing and interpreting the *social and ecological variability* within these macroscale patterns, particularly regarding the socio-political and historical dynamics operating on a local scale. For example, how do the social contexts in which cacao is planted shape LUCC trajectories? In what contexts has cacao expansion hastened deforestation, and in what contexts has it not? What accounts for the regional differences in cacao-related deforestation? Without addressing these questions, a reliance on remote sensing data alone might "reproduce simplistic tropes surrounding environmental change and/or misinterpret underlying causal dynamics" (Kelley 2018, 24), even as it appears to be an objective and impartial *view from nowhere* (Haraway 1988).

Qualitative social science methods, on the other hand, are able to document in detail localscale social dynamics but have less to say about landscape-scale patterns of environmental change in the study area and throughout Sulawesi. By linking remote sensing and LUCC mapping with in-depth field-intensive qualitative methods, Kelley investigates a wider set of processes that shape commodity crop expansion and ultimately produces a richer account of material environmental changes and the social relations structuring them: in short, the world of resource-making in Sulawesi. In contrast to existing LUCC science research on commodity crop expansion, Kelley's CPG approach pushes back against the notion that regional, national, and global scales of analysis (which are common scales of analysis in remote sensing) are inherently associated with greater generalizability of findings, while local-scale social dynamics are seen as complex and ungeneralizable. Through an emphasis on heterogeneity rather than generalizability, this approach addresses the politics of scientific knowledge production (see CPG's second core tenet), using methods from different intellectual traditions to question the preferential treatment that large-scale, remote sensing–based analyses often receive from within the policy sphere.

What results does this mixed-methods approach yield? Most broadly, Kelley corroborates the established link between cacao expansion and forest cover loss in Sulawesi. Yet she also finds that "much of what has looked like deforestation for cacao has had as much or more to do with the politics of access to land and capital" (Kelley 2018, 32). In other words, the rapid cacao expansion that occurred in certain areas was not inevitable but was enabled by local and regional land policies and unequal distributions of wealth that produced an "engine of expansion" (31). Demonstrating the variability within macroscale patterns of forest cover change, Kelley's methods reveal both forest cover *gain* due to cacao plantings in former grasslands or swidden fields, and significant forest cover *loss* unrelated to the cacao economy.

This type of integrative analysis was pursued with policy implications in mind, illustrating the third core tenet of CPG: that our research has real impacts on policies, people, and landscapes. Kelley (2018) notes that the existing gap between macroscale and local-scale research has produced a set of assumptions that unnecessarily constrain potential policy responses. By examining the variability within commodity crop expansion and land change in Sulawesi, Kelley (2018, 23) hopes to "[shift] persistent assumptions guiding policy," which, when applied to highly variable places and communities, might further exacerbate existing inequalities while also failing to protect forests. Thinking about resources more generally, one take-home message is that a mixed-methods, CPG approach may help to challenge overly simplistic, mono-causal assumptions that continue to inform many resource management policies and practices. Notably, this is the same goal that motivated much of the early work in political ecology and the political economy of resources (e.g., Blaikie and Brookfield 1987).

Normative research and critical resource geography

While CPG emphasizes the potential benefits of integrating methods from across the natural and social sciences and humanities, we note that it is far from the only field with an eye toward interdisciplinarity (see, for example, land change science, socioecological systems, agroecology, and sustainability science, among others). However, CPG diverges from many other integrative approaches in a crucial way: rather than implicitly or explicitly pursuing value-neutral research, it encourages researchers to openly grapple with the normative and political dimensions of their research.

Values and politics are entangled in resource scholarship in a variety of ways: in the normative frameworks within which research is situated (e.g., sustainability, energy efficiency, environmental justice, carrying capacity, just transitions, and ecological integrity); in the values, perspectives, and positionality of the researcher; in the power dynamics that affect knowledge production; and in the ways research findings come to justify or challenge particular policies, practices, or relations. It is not possible—and perhaps not even desirable, we contend—to achieve a value-neutral resource geography (see also, Bebbington et al., Chapter 21 this volume; Shapiro-Garza et al., Chapter 20 this volume). Instead, CPG offers an intellectual home for research that is reflexive about its own normative assumptions and openly grapples with questions about how to transform social and environmental relations and cultures and politics of science.

Certainly not all research that uses the CPG label or framework shares the same values or ambitions, and even within commonly shared values (e.g., justice), there exist many shades of gray. Still, we can identify three broad normative goals that emerge from the current body of CPG research. These are:

1. To foster the cause of justice and challenge inequalities and injustices through environmental research and inquiry.
2. To reveal environmental management and policy failures and expand the range of possible solutions available for environmental problem-solving.
3. To transform the scientific process and culture from the inside: to make science more just, humble, slow, creative, inclusive, and socially engaged (Biermann, Lane, and Lave 2018; Lane 2017; Mountz et al. 2015).

Much research within critical resource geography, as we understand it, shares many of these values—particularly around fostering social and environmental justice, revealing unseen processes of resource-making, contesting resource management failures, and countering dominant frameworks that constrain available solutions. Given this overlap, CPG might offer an inviting pathway for justice-minded resource geographers to work across the divide between critical approaches and technical/physical and problem-oriented approaches to the study of resources (Calvert 2016).

There are real obstacles to working across these traditions, however. Some scholars, for example, may hesitate to engage in technical/physical problem-oriented research, fearing that such approaches are inevitably "narrow, instrumental, and formulaic," leading to "vast oversimplification of conclusions, management lessons, and landscapes" (Gillett et al. 2018, 521), or worse that they "merely serve to patch up the present system, aid the legitimation of the state, and bolster the forces of capitalism... to create inequality" (Pacione 2004, 31). On the other hand, resource scientists engaged in technical/physical research may see social relations and power dynamics as outside the scope of their research or expertise. Further, many applied projects are expected to proceed at a relatively fast pace, producing clear-cut answers and solutions that do not square well with the context-rich knowledge that an iterative and integrative critical approach might yield.

Yet despite these obstacles, it is worth considering: can integrating the critical social and technical/physical dimensions of resources further the goals of critical resource geography?

Might CPG offer a useful framework for critical resource geographers to engage in integrative work directed at specific problems? To consider these questions, let's turn to an example of problem-oriented CPG research: the University of Massachusetts' Department of Geoscience RiverSmart Communities program. Combining social and river science, institutional and policy research, and outreach, the program aims to address river flooding in New England and help communities become "river-smart": "managing rivers and riverside landscapes, as well as… [human] actions and expectations, so people and communities are more resilient to river floods" (University of Massachusetts Amherst 2019). The project leaders describe RiverSmart as an applied project "buttressed by a CPG ontology and epistemology," which leads researchers to "ask interconnected questions, seek feedback, follow up with further questions, and learn iteratively" (Gillett et al. 2018, 521).

Thus far, the RiverSmart program has produced not only academic research publications but also: interactive community meetings and activities; landowner and community factsheets on flooding, fluvial geomorphology, and mitigation; profiles of institutions that are successfully helping communities become river-smart; a policy recommendation report; and, a database of fluvial geomorphic assessment techniques. These products reflect multiple transformative ambitions: to increase awareness of and resilience to floods among the public, to foster interdisciplinarity within and beyond the academy, and to empower a variety of publics (landowners, community leaders, grassroots activists, etc.) to creatively devise and advocate for solutions that work for their communities (Gillett et al. 2018).

This example illustrates how a specific material problem or challenge can facilitate conversations that reach across disciplines and engage diverse publics outside the academy. An applied problem can also make it easier to understand and empirically address resources as "diverse and contingent socio-natural hybrids, variable and coevolving, messy in ways that belie straightforward measurement or theorization" (Gillett et al. 2018, 518). In a sense, then, we raise the possibility that CPG—with its attention to knowledge politics, material natures, and social dynamics—might open new avenues for critical resource geographers (and beyond) to explore the possibilities of problem-oriented research while remaining attuned to some of the trenchant critiques of such work.

What might we expect for the future of CPG as a framework, method, and normative approach? First, we recognize that CPG is not yet as grounded as it could be. As a new subfield, thus far much of the published work has been primarily conceptual rather than empirical in nature. It has been stronger at arguing for what is needed than actually embarking upon projects in practice. Conceptual, field-building work is both strategic and necessary. By establishing CPG's main intellectual foundations, we aim to loosen the constraints posed by the existing political economy of the academy, giving students (and others) a formal subfield within which to pursue questions of power, justice, knowledge production, and physical geography. Yet we now need to "walk our talk" (Rochner et al. 2019, 211): to develop bodies of empirical research that exemplify the utility of CPG's integrative, reflexive, and transformative ambitions while also building upon work in other fields that shares these core values. This is where we, as CPGers, meet fellow travelers in critical resource geography, and we look forward to seeing what will emerge at this intersection.

References

Arce-Nazario, Javier. 2018. "The Science and Politics of Water Quality." In *The Palgrave Handbook of Critical Physical Geography*, edited by Rebecca Lave, Christine Biermann, and Stuart Lane, 464–484. London: Palgrave Macmillan.

Ashmore, Peter. 2018. "Transforming Toronto's Rivers: A Socio-Geomorphic Perspective." In *The Palgrave Handbook of Critical Physical Geography*, edited by Rebecca Lave, Christine Biermann, and Stuart Lane, 485–511. London: Palgrave Macmillan.

Bakker, Karen, and Gavin Bridge. 2006. "Material Worlds? Resource Geographies and the 'Matter of Nature.'" *Progress in Human Geography* 30 (1): 5–27. doi: 10.1191/0309132506ph588oa.

Biermann, Christine. 2016. "Securing Forests from the Scourge of Chestnut Blight: The Biopolitics of Nature and Nation." *Geoforum* 75: 210–219. doi: 10.1016/j.geoforum.2016.07.007.

Biermann, Christine, and Henri Grissino-Mayer. 2018. "Shifting Climate Sensitivities, Shifting Paradigms: Tree-Ring Science in a Dynamic World." In *The Palgrave Handbook of Critical Physical Geography*, edited by Rebecca Lave, Christine Biermann, and Stuart Lane, 201–225. London: Palgrave Macmillan.

Biermann, Christine, Stuart Lane, and Rebecca Lave. 2018. "Critical Reflections on a Field in the Making." In *The Palgrave Handbook of Critical Physical Geography*, edited by Rebecca Lave, Christine Biermann, and Stuart Lane, 559–573. London: Palgrave Macmillan.

Blaikie, Piers, and Harold Brookfield. 1987. *Land Degradation and Society*. London: Methuen.

Bridge, Gavin. 2010. "Resource Geographies I: Making Carbon Economies, Old and New." *Progress in Human Geography* 35 (6): 820–834. doi: 10.1177/0309132510385524.

Calvert, Kirby. 2016. "From 'Energy Geography' to 'Energy Geographies': Perspectives on a Fertile Academic Borderland." *Progress in Human Geography* 40 (1): 105–125. doi: 10.1177/0309132514566343.

Davis, Diana. 2018. "Between Sand and Sea: Constructing Mediterranean Plant Ecology." In *The Palgrave Handbook of Critical Physical Geography*, edited by Rebecca Lave, Christine Biermann, and Stuart Lane, 129–151. London: Palgrave Macmillan.

Duvall, Chris, Bilal Butt, and Abigail Neely. 2018. "The Trouble with Savanna and Other Environmental Categories, Especially in Africa." In *The Palgrave Handbook of Critical Physical Geography*, edited by Rebecca Lave, Christine Biermann, and Stuart Lane, 107–127. London: Palgrave Macmillan.

Engel-DiMauro, Salvatore. 2018. "Soils in Ecosocial Context: Soil pH and Social Relations of Power in a Northern Drava Floodplain Agricultural Area." In *The Palgrave Handbook of Critical Physical Geography*, edited by Rebecca Lave, Christine Biermann, and Stuart Lane, 393–419. London: Palgrave Macmillan.

Gillett, Nicole, Eve Vogel, Noah Slovin, and Christine Hatch. 2018. "Proliferating a New Generation of Critical Physical Geographers: Graduate Education in UMass's RiverSmart Communities Project." In *The Palgrave Handbook of Critical Physical Geography*, edited by Rebecca Lave, Christine Biermann, and Stuart Lane, 515–536. London: Palgrave Macmillan.

Goldman, Mara. 2018. "Circulating Wildlife: Capturing the Complexity of Wildlife Movements in the Tarangire Ecosystem in Northern Tanzania from a Mixed Method, Multiply Situated Perspective." In *The Palgrave Handbook of Critical Physical Geography*, edited by Rebecca Lave, Christine Biermann, and Stuart Lane, 319–338. London: Palgrave Macmillan.

Haraway, Donna. 1988. "Situated Knowledges: The Science Question in Feminism and the Privilege of Partial Perspective." *Feminist Studies* 14 (3): 575–599. doi: 10.2307/3178066.

Haraway, Donna. 1989. *Primate Visions: Gender, Race, and Nature in the World of Modern Science*. New York, NY: Routledge.

Hudson, Ray. 2001. *Producing Places*. New York, NY: Guilford Press.

Kelley, Lisa C. 2018. "The Politics of Uneven Smallholder Cacao Expansion: A Critical Physical Geography of Agricultural Transformation in Southeast Sulawesi, Indonesia." *Geoforum* 97: 22–34. doi: 10.1016/j.geoforum.2018.10.006.

Kelley, Lisa C., Samuel G. Evans, and Matthew D. Potts. 2017. "Richer Histories for More Relevant Policies: 42 Years of Tree Cover Loss and Gain in Southeast Sulawesi, Indonesia." *Global Change Biology* 23 (2): 830–839. doi: 10.1111/gcb.13434.

King, Leonora, and Marc Tadaki. 2018. "A Framework for Understanding the Politics of Science." In *The Palgrave Handbook of Critical Physical Geography*, edited by Rebecca Lave, Christine Biermann, and Stuart Lane, 67–88. London: Palgrave Macmillan.

Lane, Stuart. 2017. "Slow Science, the Geographical Expedition, and Critical Physical Geography." *The Canadian Geographer/Le Géographe Canadien* 61 (1): 84–101. doi: 10.1111/cag.12329.

Lave, Rebecca, Matthew W. Wilson, Elizabeth S. Barron, Christine Biermann, Mark Carey, Chris Duvall, Leigh Johnson, et al. 2014. "Intervention: Critical Physical Geography." *The Canadian Geographer/Le Géographe Canadien* 58 (1): 1–10. doi: 10.1111/cag.12061.

Lave, Rebecca, Christine Biermann, and Stuart Lane. 2018. "Introducing Critical Physical Geography." In *The Palgrave Handbook of Critical Physical Geography*, edited by Rebecca Lave, Christine Biermann, and Stuart Lane, 3–21. London: Palgrave Macmillan.

Lave, Rebecca, Martin Doyle, Morgan Robertson, and Jai Singh. 2018. "Commodifying Streams: A Critical Physical Geography Approach to Stream Mitigation Banking in the USA." In *The Palgrave Handbook of Critical Physical Geography*, edited by Rebecca Lave, Christine Biermann, and Stuart Lane, 443–463. London: Palgrave Macmillan.

Law, Justine. 2018. "The Impacts of Doing Environmental Research." In *The Palgrave Handbook of Critical Physical Geography*, edited by Rebecca Lave, Christine Biermann, and Stuart Lane, 89–103. London: Palgrave Macmillan.

Mark, Bryan, Jeffrey Bury, Jeffrey M. McKenzie, Adam French, and Michel Baraer. 2010. "Climate Change and Tropical Andean Glacier Recession: Evaluating Hydrologic Changes and Livelihood Vulnerability in the Cordillera Blanca, Peru." *Annals of the Association of American Geographers* 100 (4): 794–805. doi: 10.1080/00045608.2010.497369.

McClintock, Nathan. 2015. "A Critical Physical Geography of Urban Soil Contamination." *Geoforum* 65: 69–85. doi: 10.1016/j.geoforum.2015.07.010.

Mountz, Alison, Anne Bonds, Becky Mansfield, Jenna Loyd, Jennifer Hyndman, Margaret Walton-Roberts, Ranu Basu, et al. 2015. "For Slow Scholarship: A Feminist Politics of Resistance through Collective Action in the Neoliberal University." *ACME: An International Journal for Critical Geographies* 14 (4): 1235–1259.

Nightingale, Andrea. 2003. "A Feminist in the Forest: Situated Knowledges and Mixing Methods in Natural Resource Management." *ACME: An International Journal for Critical Geographies* 2 (1): 77–90.

O'Reilly, Jessica. 2016. "Sensing the Ice: Field Science, Models, and Expert Intimacy with Knowledge." *Journal of the Royal Anthropological Institute* 22: 27–45. doi: 10.1111/1467-9655.12392.

Pacione, Michael. 2004. "The Principles and Practices of Applied Geography." In *Applied Geography: A World Perspective*, edited by Antoine Bailly, and Lay Gibson, 23–45. Boston, MA: Kluwer Academic Publishers.

Rochner, Maegen, Becky Mansfield, M. Jahi Chappell, Erica Smithwick, Adam Romero, Stuart Lane, Rebecca Lave, and Christine Biermann. 2019. "The Palgrave Handbook of Critical Physical Geography." *The AAG Review of Books* 7 (3): 203–213. doi: 10.1080/2325548X.2019.1615327.

Sayre, Nathan. 2018. "Race, Nature, Nation, and Property in the Origins of Range Science." In *The Palgrave Handbook of Critical Physical Geography*, edited by Rebecca Lave, Christine Biermann, and Stuart Lane, 339–355. London: Palgrave Macmillan.

Smith, Laura. 2019. "Reevaluating the Florence Hawley Collection and Development of Hydroclimate Reconstructions in the Tennessee Valley Region." Paper presented at the Annual Meeting of the American Association of Geographers, Washington, DC, April 6.

Stengers, Isabelle. 2013. *Une Autre Science est Possible! Manifeste pour un ralentissement des sciences*. Paris: La Découverte.

Turner, Matthew. 2018. "Questions of Imbalance: Agronomic Science and Sustainability Assessment in Dryland West Africa." In *The Palgrave Handbook of Critical Physical Geography*, edited by Rebecca Lave, Christine Biermann, and Stuart Lane, 421–441. London: Palgrave Macmillan.

University of Massachusetts Amherst. 2019. "RiverSmart Communities." Accessed 3 February 2019. https://extension.umass.edu/riversmart/.

Zimmerer, Karl S. 2010. "Retrospective on Nature-Society Geography: Tracing Trajectories (1911–2010) and Reflecting on Translations." *Annals of the Association of American Geographers* 100 (5): 1076–1094. doi: 10.1080/00045608.2010.523343.

20

Praxis in resource geography

Tensions between engagement and critique in the (un)making of ecosystem services

Elizabeth Shapiro-Garza, Vijay Kolinjivadi,
Gert Van Hecken, Catherine Windey, and Jennifer J. Casolo

Introduction

The phrase "engaged critical scholarship" can seem an oxymoron. Some critical scholars claim that the role of the critic is only to challenge the normative assumptions underlying practice. They claim that to engage with the "subjects" of research in ways meant to support social change is to serve as "handmaidens" to pragmatic needs (Brenner 2009, 201), too focused on the immediate to take on more profound, structural challenges (Mohan 2006). On the other hand, practitioners often challenge purely critical scholars for what they portray as undue abstraction and "expert" elitism, divorced from the perspectives and needs of the communities and movements with whom they claim solidarity (Refstie 2018), leading to what McNay (2014, 4) has called "social weightlessness." These tensions between critique and engagement are acute in resource geography, where the material and social inevitably intermix, leading to struggles over how nature is defined, valued, and controlled, as well as to disempowerment and dispossession. This chapter explores these tensions through the lens of four cases drawn from our differing experiences as critical scholars but holding in common direct engagement with marginalized communities and a focus on a particular conceptualization and approach to natural resource management: payments for ecosystem services (PES).

Some critical scholarship has itself been critiqued for taking an overly deductive, "strong theory" approach. Detractors claim that by presupposing an "essentialist, usually structural vision of what is" (Gibson-Graham 2008, 618), these scholars can fail to recognize alternative theories or practices, thereby reinforcing the hegemony of the very structures they critique. Feminist and postcolonial/decolonial epistemologies offer the basis for a counter approach: critical engaged scholarship that is inductive, grounded, contextualized, and collaborative. A key aspect of this type of scholarship is the concept of *praxis*, a combination of "reflection and action directed at the structures to be transformed" (Freire 1970, 126). It assumes that everyone is able to apply critical perspectives to assess existing practice, pushing professional scholars to therefore engage with "the excluded and oppressed," to generate grounded theory through collective observation and reflection (see also Fabricant, Chapter 26 this volume). Following Dussel, praxis generates critique from a position outside of hegemonic structures and discourses, beyond even the Euro-centrism that informs much critical scholarship, in order to create new narratives of what was, is, and "can be" that "demand explanation" (2011, 21). Praxis therefore provides a conceptual framework and approach for engaged critical scholars to align themselves with those who are socially, economically, and/or politically marginalized and to amplify their critical discursivity if it emerges in order to challenge hegemonic assumptions and discourses.

However, there is risk that this approach can further oppress those it purports to serve if it perpetuates a "White Saviour Complex" among researchers (e.g., Straubhaar 2015) or if it ignores power inequities in access to material resources and information (Harney and Moten 2013). Scholars must also contend with and be explicit about their own limitations, acknowledging the ways in which their agency is embedded in structural and historical power relations and hierarchical and colonial forms of academic knowledge creation that influence actionable outcomes from engaged research. Acknowledging these hazards, Harney and Moten (2013) call for engaged critical scholars to be reflexive and explicit about these constraints and, if desiring to build upon theory to enact transformative change, be willing to contest and defy the obligations of their institutional affiliations, academic expectations, and their sense of professional accomplishment within a Western educational model.

In this chapter, we collectively reflect on engagements within one area of critical scholarship in resource geography: PES policies and initiatives. The PES approach provides financial incentives to landowners, through market or "market-like" transactions, to manage ecosystems in ways thought to increase the production of specific "services," such as greenhouse gas sequestration, biodiversity conservation, or cleaner or greater quantities of water for downstream communities (Shapiro-Garza et al. 2020). Starting in the late 1980s, ecologists and economists joined forces in developing the concept of ecosystem services and in promoting PES as a mechanism to ensure their provision. The former were interested in making explicit the value of "nature" to humans, while the latter hoped to increase the efficiency and effectiveness of environmental interventions by making the economic value of healthy ecosystems both legible within the economy and recognized by capitalist markets.

PES has since been put into practice on a global scale, with over 500 existing programs and an estimated US$30–50 billion in annual transactional value (Salzman et al. 2018). As a relatively recent attempt to conceptualize and promote the conditions under which an entirely new natural resource is created and recognized by capitalist economies (Huber 2018), it is possible to observe the process of creation and the varied outcomes of these attempts, offering insights into many of the central questions in critical resource geography. Because PES initiatives are often implemented within communities rich in natural resources but marginalized from capitalist economies and political power, they are also key sites for engagement by critical scholars.

In this chapter, we explore the tensions inherent in engaged critical scholarship through reflections on our own research focused on PES but based in a wide variety of geographies and contexts: the Democratic Republic of Congo (DRC), Nicaragua, Canada, and Mexico. In presenting these reflections, we understand the tensions we identify as complex and ever-evolving, requiring constant, reflexive vigilance to address, noting that we continue to navigate these tensions wherever we are on our paths as engaged critical scholars: from advanced doctoral student to researchers with over twenty-five years of experience. In order to practice the reflexivity we advocate, we start from the position that all knowledge is "situated" in the norms, conditions, and prevalent ontologies of particular times and geographies, and we recognize that our own positionalities are influenced by the dynamics, concerns raised, and outcomes of each case.

Green carbon economy as modernity in the Democratic Republic of Congo (DRC)—reflections by Catherine Windey

Trained as an anthropologist, I am finishing a PhD in Development Studies based on a poststructuralist, decolonial, and feminist understanding of knowledge formation and plural epistemological approaches. I have conducted extensive field research in the DRC on the implementation of the Reduced Emissions from Deforestation and Degradation (REDD+) framework through which (sub)national actors in developing countries are paid from multinational funds to reduce deforestation. I analyze REDD+ rationale in DRC as a capitalist eco-modernization project in which the framing of carbon as a scarce economic resource allows the reproduction of market logics and the expansion of economistic rationality while erasing other values of forests (Gómez-Baggethun and Ruiz-Pérez 2011). For me, as a scholar committed to highlighting a "pluriverse" of ecologies and ways of knowing and being with forests, engagement means thinking about economic alternatives together with Congolese civil society organizations (CSOs), researchers, and farmers in the communities targeted by REDD+. Since 2018, I have participated in a collaborative action-research project that aims to better integrate the voices and needs of communities near Kisangani, an isolated and economically marginalized region of the DRC, into REDD+ forest policies. When I initiated my research, I assumed that local voices would be critical of the modernist market focus of REDD+. As I explore here, however, engagement often requires revisiting our assumptions.

The framing of the DRC's REDD+ strategy positions "local communities" as homogeneous "poor shifting cultivators" who are unproductive in terms of both economic growth and forest conservation, obscuring the full complexity of social identities and of people's lives. Justified by this narrative, the policy reorganizes trees and forests into "integrated landscapes" of privately held zones of capitalist productivity: areas for extractive timber and mineral production, intensive agricultural plantations, and conserved carbon sinks that can supposedly yield profit for these "local forest farmer communities" through the sale of carbon credits or as plantation laborers or petty commodity producers (Windey and Van Hecken 2019; Windey 2020). The local, predominantly male farmers, Congolese environmental CSOs and local scholars from this region with whom I engage rarely appear to resist this utilitarian and seemingly disempowering framing. At times, they even reproduce it. The postcolonial and post-conflict context of this region of Kisangani is characterized by the (quasi-)absence of industries and infrastructures (accelerated since the 1980s), a lack of investments that have isolated the region, and high political instability, with a constant reshuffling of ministries that has led to very low levels of institutional trust. In such a context, nostalgic references to "colonial modernity"—associated with stability and prosperity and with the ability to transform trees or "soil into something that becomes money," as one local Banjwade farmer said—were common. For many of the forest farmers in the REDD+ intervention zones of this region, carbon, despite being perceived as an invisible and uncertain resource, appeared to be an opportunity to improve material conditions and achieve a productive transformation. While recognizing the hardships of labor conditions under Western domination, these farmers also often voiced their regret for the disappearance of international agribusiness companies and the paid labor opportunities they provided and saw their potential return as a pathway to development. While forests have symbolic, social, and spiritual values for these forest farmers, they also viewed the conversion of forests into "productive lands" for agriculture or carbon as a pathway to modernization and social and environmental justice for the DRC.

How should I respond when I encounter the seeming reproduction of a monocultural perspective instead of the alternatives and resistance to REDD+ discourses I expected? As a white

Belgian scholar, how should I engage with the aspirations of local collaborators which reproduce ideas of Western economic and ecological modernization while also striving to coproduce alternative ways of thinking and being, thus attempting to undo entrenched relationships of violent colonial appropriation of land and life? The narratives I encountered expressed a real sense of exclusion from the promised modernity, perceived as an ideal to attain, and an appeal to membership in global society (see also Ferguson 2006). They represented material, emotional, and bodily experiences and needs with consequences on resource creation and use, even if they contradicted postcolonial critiques of (eco)modernization and alternative worlds. In encountering these views, "engagement" meant needing to embrace the "disconcertment" (Law and Lin 2010) I felt as my critical assumptions were shaken and to adopt a "hyper self-reflexive practice" regarding my personal and institutional positioning, that is, the social construction of my own critical discourses (Kapoor 2004). My encounters and collaborations in DRC have pushed me to reflect on how socioeconomic privilege might work to permit and enable a "strong" critique of (eco)modernization, as well as on my own complicity within a system I critique (see also Kapoor 2004). Engaging through praxis with divergent or even contradictory understandings, rather than sticking to the tidiness of one theoretical frame, requires what Mignolo (2009) refers to as "epistemic disobedience." In my research, this has entailed engaging with thick description in order to avoid essentializing the struggles of our action-research participants and colleagues and providing a better account of how such struggles are articulated in terms of distribution and inclusion in the system, not only in terms of alternatives or resistance to it. Being disconcerted through this process of reflexive engagement is, I believe, a first step to avoid treating the places where we conduct (action-)research as repositories of critical thinking, moving instead toward (co)producing knowledge and transforming unequal relationships.

Playing games for engagement in Nicaragua—reflections by Gert Van Hecken

I am currently an assistant professor in Development Studies at the University of Antwerp, Belgium. I have lived and worked almost half of my adult life in Nicaragua as a researcher and later as a representative for a Belgian rural development nongovernmental organization (NGO). In this long-term engagement with processes of social and environmental change in rural Nicaragua, I have worked closely with *campesino* and environmental movements as well as with research and governmental organizations. These experiences, especially those in which I was involved as an action-researcher, have taught me that reflecting critically, while crucial for transforming structural injustices, is a necessary, but not sufficient, approach to bring about material changes in people's everyday lives. This realization has struck me most clearly when sharing my critiques regarding PES as neoliberal conservation tools with partner organizations in the field. Although they most often welcome these discussions, they also ask what practical changes these critical framings and insights suggest for actors on the ground or in generating workable alternatives.

I have collaborated for many years with a small-scale environmental NGO in southeastern Nicaragua, close to the Indio Maíz Biological Reserve. In 2013, I engaged in a collaborative action-research process with several researchers and practitioners in which we jointly reflected on the implementation of the different strategies that the NGO had applied over two decades. The reflections this process generated revealed the difficulties of creating consensus-based, practical ways to collaborate with *campesino* men and women to arrest the deforestation and social disruption associated with an expanding agricultural frontier. The group concluded that the more "conventional" participatory methods initially employed (e.g., workshops, interviews, focus groups) could not sufficiently capture the many social-economic as well as livelihood struggles that *campesinos* faced in their daily lives that greatly influenced their decision-making processes around land use change and deforestation. Nor could they adequately reveal how decision-making and practices are embedded in and shaped by local, power-laden institutional arrangements. We also felt that these methods were insufficient to produce the types of "knowledge encounters" necessary for stimulating open debate in which all involved actors (not only *campesinos*, but also researchers and NGO staff) engage in a questioning and deconstruction of their own worldviews and (implicit) assumptions, while recognizing alternative ways of knowing and doing, with the intent of offering a platform to collaboratively construct and discuss alternative social-environmental perceptions and practices.

In response, the NGO staff and I jointly experimented with the development of new tools for meaningful engagement through the cocreation of shared "actionable knowledge." One such tool was a simulation board game intended to enable users/players to co-construct deeper understandings of how local historical, sociocultural, and economic dynamics shape farmers' decision-making, fostering cooperation in the creation, testing, and discussing of new (or hitherto marginalized) alternative practices. The game, in which participants take up the roles of *campesino/campesina* households, mimics historical processes of agrarian change and social differentiation, simulates a range of potential alternative practices, and creates space for participants to collectively reflect on the often hidden motivational and sociopolitical dynamics triggered by policy tools such as PES. Multiple iterations and applications of the simulation game demonstrated its potential. We found that the game created a collective learning platform where different perspectives from different actors (including ourselves) could be compared, where links to real-life situations could be made, and where alternative views could be openly discussed and jointly interpreted (Merlet, Van Hecken, and Rodriguez-Fabilena 2018). First, when NGO practitioners and researchers played the game, they could observe firsthand and in real time how *campesinos'* production decisions are constrained by broader structural-historical processes in which they are embedded but which are often overlooked or disregarded from an "outsider" perspective. Playing the game with local groups not only encouraged the NGO and researchers to be more humble in comparing their (theoretical/policy-informed) knowledge to *campesinos'* and *campesinas'* deeply ingrained knowledge of human-nature relations but also encouraged all of us to pay more attention to the importance of mutual relations based on empathy, which in the postgame feedback sessions often emerged as a crucial condition for creating meaningful and respectful collaborations with *campesinos* and *campesinas*. It also offered new entry points for discussion of sensitive issues related to power differences in local communities, such as land grabbing by richer *campesinos*. Ultimately, the game not only provided a platform and an impetus for discussions amongst the NGO and *campesinos* about why unequal power relations are so persistent and difficult to challenge but also stimulated reflections on possible alternative strategies to transform them. (For more details, see Merlet, Van Hecken, and Rodriguez-Fabilena 2018.)

Reflecting on this case highlights a number of tensions in engaged critical scholarship. Meaningful engagement with local actors requires recognition and transparency about uneven power relations in efforts toward co-constructing a common "humanity" that is not differentiated into expert academics and "research subjects" but rather generates epistemic plurality by ensuring that everyone is responsible for the production of knowledge and for transitioning toward alternative lifeways. In the same way, a situated approach, which allows for greater nuance in the interpretation of human-nature relations and takes into account differentiated conditions and contexts, can safeguard against theoretical interpretations of PES schemes divorced from the lived experience of those directly involved (Van Hecken et al. 2018). But there are additional questions that these experiences evoke. To what extent are we, as "engaged" researchers, willing and able to commit to supporting transformations for justice with "Others" (Dussel 2011)?

What would that commitment entail in relation to the collective critical agreements reached for transforming existing hegemonic structures or relations that commodify nature and devalue life? Critical scholars must grapple with how we can invest energy and (emotional) labor for the long run in critical processes and practices that give priority to building consensus—going beyond what is conventionally expected from an academic and relinquishing the all-too-common institutional and colonial research practice of simply entering and exiting into people's lives at will to "get the academic job done." Based on this and other collaborative experiences, I strongly believe that engaged critical scholarship in resource geography requires taking responsibility for collectively fomenting and strengthening ways that push beyond traditional academic practice to build relationships of empathy, care, and commitment to a common struggle, working away from reproducing exploitative and extractivist forms of knowledge production.

Relational accountabilities in a disciplined capitalist landscape of Canada—reflections by Vijay Kolinjivadi

As part of my postdoctoral research within an academic department at a Canadian university, I gathered stories from indebted farmers on Prince Edward Island (PEI) who receive payments from the Canadian government to plant trees, widen buffer zones between their fields and waterways, and retire production from sloping land in order to prevent soil erosion. The program is described as PES, in which services of "nature" to "humans" (seen as outside of it) in the form of conservation of soils to maintain intensive agricultural production are increasingly being valued over more "multifunctional" landscapes (Kolinjivadi, Zaga-Mendez, and Dupras 2019). The research project's explicit goal was to evaluate and improve the PES program on PEI.

PEI is a small island province with a major agricultural industry—potato production and processing—to which the provincial government is closely aligned. Many farmers on PEI have transitioned from small-scale diverse production systems to increasingly specialized agriculture, especially in potatoes. The majority belong to families that have farmed the same piece of land for several generations. The farmers I spoke to on PEI were largely ambivalent about the government payment scheme that would require them to implement certain land-use practices (e.g., grassed waterways, retiring production on sloped land, installing livestock fences), mostly because their contracts with the monopolistic potato-processing industry on PEI do not recognize or promote soil rehabilitation and biodiversity considerations. From the industry's perspective, until consumers (e.g., food retailers and fast-food chains) begin placing value on "environmental sustainability," no premium can be made to ensure that soil protection is prioritized in contracts with farmers. However, it is unlikely that consumers of PEI French fries in Latin America and Asia will be willing to pay more for a better protected environment in an island already broadcasted to the world as being idyllic and in a country (mistakenly) acclaimed for being peaceful, tolerant, and orderly (Howell 2005). By framing alternative agricultural practices as "delivering ecosystem services," market-based environmental governance strategies often sustain and enhance processes of translocal capitalist accumulation to respond to new challenges. Identifying novel and potentially lucrative strategies to internalize environmental externalities to maintain production is one such example. The consequence is that other, nonproduction-oriented relationships that intimately characterize farmers' connections to the land become neutralized and instantly rebranded as new resources that might ideally fetch a premium along the supply chain.

For the farming families of PEI, surviving in the induced scarcity of capital relations has necessarily meant collective action and cooperation among neighbors, including sharing equipment to defray costs, comparing contracts, collectively generating knowledge around best practices of soil retention, and reasserting agency in solidarity to establish their own agro-environmental groups that do not involve the provincial government's intervention. Thus, what actually gets articulated is the autonomous emergence of socio-ecological relations required to survive in an otherwise cutthroat market environment. Farmers do not frame these relations as "ecosystem services" that will further enhance value on their property but as a continuous desire, perhaps unconsciously, to reestablish the conditions that give meaning to their vocation, both symbolically and biophysically.

Assuming responsibility as a researcher for the relations in which I am entangled in this project has been and continues to be a challenge. Two directions seem possible: one reflects collective accountabilities with research partners, as Van Hecken noted from his experiences in Nicaragua, and another the more structural contradictions of an academic "producing" publications by examining how the concealed labor of farmers and the extra-human natures, framed as ecosystem services, are perceived as "resources" for continued production. On an island as small as PEI, the fact that everyone knows each other provides a significant deterrent for farmers to voice their opinions in public. Despite being an outsider, the interest I showed in understanding their difficulties in farming offered space to "vent," as one producer confided to me during an interview that went on longer than planned.

While no easy shortcuts exist to understanding my complicity as an academic researcher in further patterning resource logics and state strategies, I have attempted to translate farmers' experiences, perceptions, and activities to illustrate how relations of care and attachment to fellow neighbors and nonhuman others do not necessarily fall within the ambit of "delivering ecosystem services." In doing so, I have to draw attention to the resistance of both people (in this case farmers who refuse to be disciplined by the state and the market as "machine operators") and non-people (who can never be straightjacketed into ecosystem services for humans), even in the most disciplined capitalist landscapes.

The translation of farmers' experiences and their everyday forms of resistance has been a priority in the published research papers and presentations that have resulted from my research. However, fostering self-reflection is becoming a greater priority for me beyond making resistance "visible" through academic products, as Harney and Moten (2013) have advised against. In solidarity with resistance against turning people and non-people into resources for production, this intention has meant directly interrogating how knowledge is produced and packaged within the neoliberal academy. It has entailed initiating interdepartmental discussions with colleagues about the ways in which we (academics) seek to identify the uneven social and ecological impacts of resource production "out there" while simultaneously replicating such uneven impacts in the logics of worker precarity, adherence to the pressure to produce faster, in greater quantity, and through ever-tighter evaluation metrics to compete with others and justify our positions within the university itself. With no pretense of falsely equating farmer struggles on PEI (or worse still, threats of land dispossession of the rural peasantry by neoliberal logics more broadly) with the precarity of academia, my aim is to foster self-reflexivity on the part of academics over a common root to collective struggles. In doing so, greater potential emerges to brainstorm ways to structurally alter how knowledge coproduction takes place in order to build upon connected, though highly differentiated, struggles.

Engaging with carbon offsetting in Mexico—reflections by Elizabeth Shapiro-Garza

I am a critical human geographer and have served on the faculty of a multidisciplinary school of the environment at a private university in the United States for the last ten years. I spent the first eight years of my career in Latin America working in and with rural communities on natural resources issues, and my scholarship has been greatly informed by these early, highly engaged experiences.

I have collaborated with an NGO, the Integrator of Campesino and Indigenous Communities of Oaxaca (ICICO), since 2005, when I began fieldwork for my dissertation on PES in Mexico. Governed by an elected committee of representatives from 15 indigenous communities, ICICO's mission is to attain recognition for and benefit from the "ecosystem services" provided through traditional stewardship of their territories. These communities, while holding relatively strong communal land tenure and maintaining the vitality of their traditional governance systems and cultural practices, have also been hard hit by neoliberal reforms in Mexico that undermined rural economies and spurred significant out-migration.

My dissertation research on the then newly formed federal PES programs in Mexico built upon a body of critique in resource geography that posits that the commodification of ES, folding previously unrecognized resources into capitalistic markets, will result in processes of "accumulation by dispossession" (Harvey 2004). My early experiences with ICICO and many other communities enrolled in the national programs seemed to support this critique. In meetings to discuss the concept of PES, community members openly worried that selling something that was continuously produced and tied to their land (e.g., carbon sequestration, biodiversity conservation, etc.) would give the buyers of these "services" rights to their territory. The lack in these early years of any connection to actual markets for ES also supported the critique that those marginalized from capitalist economies will never be able to negotiate on equitable terms.

And yet... as time went by and other communities with whom I had partnered had long given up on the possibility of "selling" their ES, ICICO persisted and, in the process, did much to "decolonize" both the concept and the practice of PES. Framing the concept of PES as a means to "revalue the rural," ICICO represented potential payments as a recognition of the value of the stewardship of these communities for their land by urban areas and the global North. In 2008, in partnership with a Mexican environmental NGO, ICICO created a national-level, voluntary market for carbon offsets. In doing so, ICICO was able to establish the "rules of the game" through which carbon sequestration would be produced (through management practices, such as agroforestry systems, that would produce additional local economic and environmental benefits), measured (through protocols that best suited their forests and their capacities), and sold (by setting a high fixed price per ton). ICICO insisted on replicating this approach when they partnered with the California Air Resources Board (CARB) to develop the Mexican protocol for monitoring and verification of forest-based carbon offsets for that state's newly created voluntary market. The ICICO communities have since largely invested the significant funds from carbon offsetting into community infrastructure and services, cultural activities, and the generation of educational and employment opportunities to counter the flow of out-migration.

In 2018 and again in 2019, I worked to link my university's Office of Sustainability to ICICO to purchase carbon offsets as part of a broader plan to become "carbon neutral" by 2024. Was my "engagement" as a broker within the very system I had critiqued a betrayal of my principles as a critical scholar? This question, and this dilemma, represents for me a complex set of responsibilities that often come into contradiction, if not conflict, with each other, generating questions with which I continue to struggle. While I still view the underlying rationale of PES as inevitably linked to an overtly neoliberal political project and the expansion of capitalist rationalities and systems of accumulation, I also feel a strong responsibility to report on and support the ways in which these indigenous communities, operating within their particular context and with incredible persistence and acumen, have been able to *aprovechar* (take advantage of), and even to some extent decolonize, the rationalities and practices of PES to support the reproduction of their own cultural practices and to strengthen territorial claims.

However, in attempting to honor my relationship with ICICO and the communities it serves by linking them to carbon funding, was I serving as a "handmaiden" to pragmatic, immediate needs instead of taking on a more daunting, but profound, praxis by working with them to explicate and contest the problematic structural issues at the root of this approach (Mohan 2006; Brenner 2009) as demonstrated by Van Hecken's engagement with collective knowledge production in Nicaragua? Similar to Kolinjivadi, I did make some attempts to challenge the reproduction of capitalistic logics and potentially dispossessing practices of market-based approaches to climate-change mitigation within my own academic institution: teaching a course and organizing an international workshop at my university that brought together scholars and practitioners, including ICICO members, to explore critical theorizations, contestations, and alternative practices in PES. In an attempt to ensure that ICICO's "success" was understood within a broader structural context, I coauthored a peer-reviewed article that explored the factors that allowed these communities in Oaxaca to benefit from forest-based carbon offsetting while so many others did not (Osborne and Shapiro-Garza 2018). Were these the morally defensible responses to the dilemmas facing me? Was my "epistemic disobedience" (Mignolo 2009) within my own institution sufficient to counter my direct participation in systems and structures I critique? My answer, perhaps unsatisfactory, is that I understand my responsibility as an engaged critical scholar to be willing to live in, as noted by Windey, a state of "disconcertment" (Law and Lin 2010), continuously and iteratively grappling with these and other tensions between engagement and critique.

Conclusions

In this chapter, we explored how the relationship between critique and engagement can be fraught. However, the dynamics and tensions we describe also provide insight into the ways in which both critical theory and grounded practice can inform and transform, and in turn be informed and be transformed by, engaged scholarship in resource geography.

Our reflections present the tensions of engaged critical scholarship in resource geography as ever-evolving, complex, and above all requiring continuous reflexivity related to roles and relational accountabilities of multiple types and scales: individual, collective, and institutional (see also Delgado, Chapter 25 this volume). In Windey's narrative, this struggle is manifest in her attempts to understand and represent the complexity of local conceptualizations and reactions to PES projects in the DRC, including those embraced for the perceived benefits of capitalist eco-modernization that defied her critical assumptions of grounded resistance. She argues that one of the first steps to critical engagement should be hyper self-reflexivity related to our own positionality in "the world order" and refraining from projecting oneself and our critical assumptions onto the "Others" with whom we engage (Dussel 2011). Van Hecken's reflections on the inability of critical framings alone to generate concrete alternatives for the community-based organization implementing PES in Nicaragua led him to collaboratively design games intended to enact "knowledge encounters." In doing so, he and his collaborators cocreated conditions through which all actors could better access each other's perspectives and challenge their own assumptions. Further, it advanced a move from individual self-reflexivity to collective reflexivity with the possibility of enabling alternatives in resource management. Kolinjivadi delves into the role of the academy in bringing the concept of "ecosystem services" as a novel resource into being in the context of a PES program with potato farmers in Canada. He demonstrates not only how the meaning and value of the work of both the potato farmer and the academic is valued, structured, and measured in terms of the market but also the ways in which this commonality can foster unexpected affinities of resistance and solidarity, generating the conditions for coproduced knowledge. Finally, Shapiro-Garza explores the moral and epistemological dilemmas generated through her role as a broker in a PES exchange between her academic institution and indigenous communities in Mexico with whom she collaborates, namely, when her understanding of structural inequities is in conflict with her sense of responsibility to walk with and support those with whom she engages on the pathways they have chosen.

All four reflections also reveal that the praxis embodied in engaging with (multiple) meanings and experiences with marginalized communities requires moving beyond the hegemonic subjectivity of the critical but normative expert and its authority. Moreover, they shatter the myth of scholars being fully consistent in their work and unable to allow empirical context and engaged relationships to influence the evolution of how learning takes place (Law and Lin 2010). We find that, instead of trying to resolve and rationalize discrepancies according to one's own paradigms and subjectivities, critical resource scholarship must embrace conflicting ideas, countercurrents, tensions, and contradictions in order to ensure that insights that emerge from the relationships between scholars and their partners are attentive to the messy ways in which resources are created, defined, and valued (Harney and Moten 2013).

As we continually (re)learn, being accountable to these uneven processes of resource formation requires constant, careful attention and a willingness to embrace, or at least accept, some level of ambiguity and dissonance. At times, the messiness itself is disguised or unclear, pushing us to disassemble those assumptions at the foundation of recognized intellectual traditions, promoting deeper critique (Windey). It may mean recognizing and understanding that the relations potato producers have with the land and/or academics with the process of doing research can never quite be translated into the production of an ES resource or a published knowledge product, respectively, but rather can have multiple coexisting and/or disputed definitions and values (Kolinjivadi). Even in the pursuit of approaches and processes of action research that might uncover the undisciplined nature of resource creation, we can be confronted with the unevenness of knowledge valuation, which in turn pushes us to even deeper accountability (Van Hecken). This reflexive attentiveness to the messy ways of engaged scholarship also may mean throwing our critical lot in with the collective decisions and processes of the communities with whom we collaborate, such as the sale of carbon offsets from the indigenous and *campesino* communities in Oaxaca, Mexico, to simultaneously support and make legible the accomplishments of this one group while continuing to critique the system as a whole (Shapiro-Garza).

The brightest, strongest thread woven through all four narratives is that engaged critical scholarship in resource geography demands reflexivity and a commitment to co-constructing life-enhancing materialities with those with whom we engage while continuously and iteratively struggling with the inherent contradictions and messiness that those processes generate. The responsibility of engagement involves reconstituting new relations of coproduced action and knowledge between researchers and their interlocutors as well as within their own institutions, similar to what Derrida (1994, 31) termed "originary performative" acts, contesting state and market attempts at classification and enrollment into the logics of dispossession and accumulation (Bastian 2012). Based on the concept of praxis, such engagement is rooted in a commitment to relationship building, above and beyond academically professionalized discursive reproduction, whether as critique or support (Freire 1970). Relational accountability in resource geography therefore serves as the foundation undergirding engaged scholars' attempts to walk the line between serving as "handmaidens" to pragmatic needs or "socially weightless" critique in the professional academy, all the while remaining skeptical of the modern conceit of human-nature relations as fixed and unchangeable "resources" to be sustained or enhanced (Brenner 2009; McNay 2014).

References

Bastian, Michelle. 2012. "Fatally Confused: Telling the Time in the Midst of Ecological Crises." *Environmental Philosophy* 9 (1): 23–48.

Brenner, Neil. 2009. "What Is Critical Urban Theory?" *City* 2 (3): 198–207.

Derrida, Jacques. 1994. *Specters of Marx: The State of the Debt, the Work of Mourning and the New International.* New York, NY: Routledge.

Dussel, Enrique. 2011. "From Critical Theory to the Philosophy of Liberation: Some Themes for Dialogue." Translated by George Cicciariello-Maher." *Transmodernity* 1 (2): 16–43.

Ferguson, James. 2006. *Global Shadows: Africa in the Neoliberal World Order.* Durham, NC: Duke University Press.

Freire, Paolo. 1970. *Pedagogy of the Oppressed.* New York, NY: Continuum International Publishing Group.

Gibson-Graham, Julie Katherine. 2008. "Diverse Economies: Performative Practices for 'Other Worlds.'" *Progress in Human Geography* 32 (5): 613–632.

Gómez-Baggethun, Erik, and Manuel Ruiz-Pérez. 2011. "Economic Valuation and the Commodification of Ecosystem Services." *Progress in Physical Geography* 35 (5): 613–628.

Harney, Stefano, and Fred Moten. 2013. *The Undercommons: Fugitive Planning and Black Study.* New York, NY: Minor Compositions.

Harvey, David. 2004. "The 'New' Imperialism: Accumulation by Dispossession." *Socialist Register* 40: 63–87.

Howell, Alison. 2005. "Peaceful, Tolerant, and Orderly? A Feminist Analysis of Discourses of 'Canadian Values' in Canadian Foreign Policy." *Canadian Foreign Policy Journal* 12 (1): 49–69.

Huber, Matt. 2018. "Resource Geographies I: Valuing Nature (or Not)." *Progress in Human Geography* 42 (1): 148–159.

Kapoor, Ilan. 2004. "Hyper-Self-Reflexive Development? Spivak on Representing the Third World 'Other'." *Third World Quarterly* 25 (4): 627–647.

Kolinjivadi, Vijay, Alexandra Zaga-Mendez, and Jérôme Dupras. 2019. "Putting Nature 'to Work' Through Payments for Ecosystem Services (PES): Tensions between Autonomy, Voluntary Action and the Political Economy of Agri-Environmental Practice." *Land Use Policy* 81: 324–336.

Law, John, and Wen-yuan Lin. 2010. "Cultivating Disconcertment." *The Sociological Review* 58 (s2): 135–153.

McNay, Lois. 2014. *The Misguided Search for the Political. Social Weightlessness in Radical Democratic Theory.* Cambridge, UK: Polity Press.

Merlet, Pierre, Gert Van Hecken, and René Rodriguez-Fabilena. 2018. "Playing before Paying? A PES Simulation Game for Assessing Power Inequalities and Motivations in the Governance of Ecosystem Services." *Ecosystem Services* 34: 218–227.

Mignolo, Walter D. 2009. "Epistemic Disobedience, Independent Thought and Decolonial Freedom." *Theory, Culture & Society* 26 (7–8): 159–181.

Mohan, Giles. 2006. "Beyond Participation: Strategies for Deeper Empowerment." In *Participation: The New Tyranny*, edited by Bill Cooke, and Uma Kothari, 134–167. London: Zed Books.

Osborne, Tracey, and Elizabeth Shapiro-Garza. 2018. "Embedding Carbon Markets: Complicating Commodification of Ecosystem Services in Mexico's Forests." *Annals of the Association of American Geographers* 108 (1): 88–105.

Refstie, Hilde. 2018. "Action Research in Critical Scholarship — Negotiating Multiple Imperatives." *ACME* 17 (1): 201–227.

Salzman, James, Genevieve Bennett, Nathaniel Carroll, Alle Goldstein, and Michael Jenkins. 2018. "The Global Status and Trends of Payments for Ecosystem Services." *Nature Sustainability* 1 (3): 136–144.

Shapiro-Garza, Elizabeth, Pamela McElwee, Gert Van Hecken, and Esteve Corbera. 2020. "Beyond Market Logics: Payments for Ecosystem Services as Alternative Development Practices in the Global South." *Development and Change* 51 (1): 3–25.

Straubhaar, Rolf. 2015. "The Stark Reality of the 'White Saviour' Complex and the Need for Critical Consciousness: A Document Analysis of the Early Journals of a Freirean Educator." *Compare: A Journal of Comparative and International Education* 45: 381–400.

Van Hecken, Gert, Vijay Kolinjivadi, Catherine Windey, Pamela McElwee, Elizabeth Shapiro-Garza, Frédéric Huybrechs, and Johan Bastiaensen. 2018. "Silencing Agency in Payments for Ecosystem

Services (PES) by Essentializing a Neoliberal 'Monster' Into Being: A Response to Fletcher & Büscher's 'PES Conceit'." *Ecological Economics* 144: 314–318.

Windey, Catherine. 2020. "Abstracting Congolese Forests: Mappings, Representational Narratives, and the Production of the Plantation Space under REDD+." *IOB Discussion Paper, 2020-01.* Antwerp: Institute of Development Policy, University of Antwerp.

Windey, Catherine, and Gert Van Hecken. 2019. "Contested Mappings in a Dynamic Space: Emerging Socio-Spatial Relationships in the Context of REDD+. A Case from the Democratic Republic of Congo." *Landscape Research.* doi:10.1080/01426397.2019.1691983.

21

Negotiating the mine

Commitments, engagements, contradictions

Anthony Bebbington, Ana Estefanía Carballo,
Gillian Gregory, and Tim Werner

Introduction

The idea of "engagement" can invoke varied meanings: of involvement, of commitment (including emotional commitment), of attracting attention, of being locked in.[1] While these meanings have much in common, they are not the same. This range of meanings suggests that what is meant by doing "engaged" work is less obvious than it may first seem and that when scholars talk of their work as being "engaged," they are not necessarily invoking the same meanings associated with the word (compare with Kirsch 2018; see also Courtheyn and Kamal, Chapter 23 this volume). Indeed, no one subdiscipline, critical resource geography included, has a special claim on the idea of being engaged. Commitments to "justice," "equity," or "sustainability" characterize the work of a wide range of scholars; what varies is how different scholars define these concepts or understand the pathways that might lead to these goals. Meanwhile, emotion can be as present in commitments to scientific rigor or to purist nature conservation as it is in commitments to particular distributions of rights and resources. If all, or at least most, resource geography seeks to be engaged and engaging, is there any specific dimension to engagement that defines critical resource geography? What is it about something called "critical" resource geography that defines how it might engage? Or, conversely, what aspects of engagement would define a resource geography as being critical?

The answers to these questions are far from obvious. Furthermore, engagement in critical resource geography, however defined, can characterize teaching and service as much as research and knowledge generation (see Fabricant, Chapter 26 this volume). And, complicating the issue further, engagement in critical resource geography can be as intramural as it is extramural. To be engaged might just as easily imply profound commitment to student learning and self-realization, or to giving of oneself, one's time, and one's ideas for the collective good of the subdiscipline, as it indicates a commitment to particular social change goals or societal transformation.

In this chapter, we make no attempt to define engagement for others, instead focusing our attention to discussion of experiences that we consider "engaged" in various forms. Underlying this breadth is the idea that (for us) engagement has something to do with work that seeks to broaden the public sphere and to open up discussion of resource issues in the face of multiple efforts by a wide range of actors to close them down and silence certain knowledges. Our focus is on engagements beyond the campus, and insofar as we touch on engaged service, we address service outside the subdiscipline. These choices are purely related to word-length, and we wish to be explicit in stating our view that resource geographers' work can be engaged (in our understanding of the term) and expand the range and inclusiveness of public debate while being entirely intramural and/or focused on teaching and service (compare with Batterbury 2015).

To pursue these ideas, we focus on mineral extraction. While each of us has worked on a variety of resource issues, we suggest that mining—both large- and small-scale—raises particularly acute challenges for engagement. Mines often become sites of symbolic and physical violence, painfully illustrating the tensions around resource governance. At their most extreme, these tensions are reflected in the increasing numbers of environmental defenders killed in relation to mining projects that Global Witness' (2016, 2017, 2018, 2019) reports have documented in recent years. Such violence brings a particular urgency to reflect on the nature of our engagement as critical resource geographers. Mining is a form of resource governance in which the very nature of the problem frequently means that scholars cannot help but become involved: the topic enrolls them and does so in a way in which emotions figure prominently. Emotions affect how different parties act and position themselves. The legitimacy of the knowledge claims that anyone makes is frequently determined by who you—as a researcher—are perceived to be and to support, more than the credibility of the procedures through which those knowledge claims have been arrived at.

As such, we reflect on some of the risks, dilemmas, and opportunities that surround scholarly involvement with mining. Our observations are based on experiences in which we ourselves have sought to engage with one or more of the following: rights-based activist groups, governments with whose aspirations we identified, indigenous land councils and indigenous communities in areas of small-scale mining, nongovernmental organizations (NGOs), and agenda setting international study groups. We must also note that our observations and perspectives are based in different positions. We are two geographers, an environmental engineer, and a political theorist—though all now based in geography departments. We have engaged as NGOs, consultants, and academics. We have mostly worked in Latin America, though also Australia and Southeast Asia. Our collaborations have been diverse. Furthermore, three of us are early career, meaning that more of our collective engagements have been shorter term, with fewer of them stretching over periods of many years. Our experiences do not lead to singular conclusions nor are we necessarily of one mind on the issues we discuss. These are, instead, reflections on the challenges of engagement of which we have become aware.

In the following section, we elaborate a little more on the issue of engagement, with a focus on the work of geographers. We then explore the complexity of issues surrounding scholars' engagements with mining. The fourth section discusses specific experiences that throw further light on the problematic of engagement. The final section suggests that the task of expanding public debate on mining remains a critical issue for engagement, particularly given the levels of violence and silencing that surround the activity.

Forms of engagement around mineral extraction

While the number of geographers working on mineral extraction has increased substantially over the last 15 years,[2] it remains modest (certainly in comparison to those working on agriculture or forests). The work of these geographers illustrates a range of approaches to working beyond the academy. Without forcing any formal typology, certain orientations help map these different approaches.

First is the distinction between commitments to ideas/principles and commitments to particular actors. While the distinction is not clear-cut, there is a difference between: work which commits to certain principles (e.g., transparency) and will part ways with actors when their actions deviate from these principles (albeit sometimes in pursuit of other principles) versus work which commits to particular actors and will remain committed to them when their actions deviate from certain principles, on the grounds that it is more important to sustain the relationship than to take the moral high-ground. Neither of these is an easy position to take, and presumably many scholars try to muddle through some compromise position. But the tension is real, and certainly the four of us constantly grapple with this divide and the choices it can force onto us.

A second distinction is that between commitments with what one might call boundary organizations, and those with actors on the so-called front line. Boundary organizations run a wide range and include high-level panels whose role is to organize knowledge such that it is more likely to influence policy and action (for instance, the International Resource Panel; the Mbeki Panel on Illicit Financial Flows; or the International Lead-Zinc, Nickel, and Copper Study Groups); philanthropic bodies that combine the funding of initiatives with the convening of debates on mineral policies, directly or through their grant making (e.g., the Ford Foundation); university centers that are financed and organized more like think tanks and policy brokers than orthodox departments and seek above all to engage business and government actors (for instance, the Sustainable Minerals Institute in Queensland or the Columbia Center on Sustainable Investment); and national bodies such as parliamentary commissions, committees, or ombudspersons' offices. Frontline organizations and actors are perhaps more obvious: community-based organizations, government ministries, NGOs, companies (especially company social responsibility and community development offices), journalists, and social movements at different scales.

Third, and continuing with the same idea, the sorts of actors with whom engagement occurs can vary. In some cases, engagement may be with single organizations, while in others, scholars may seek to engage with several organizations of different types at the same time (though this can be a difficult balancing act). Sometimes scholars may view these engagements as being primarily with individuals rather than the organizations of which they are a part, and this can also influence how they negotiate the tensions referred to previously. In practice, our sense is that critical resource geographers tend toward engagements with NGOs and some social movement organizations more frequently than with governments and companies. Indeed, these latter engagements may be more complicated because of factors such as political changes and clientelisms in government or the desire of companies to manage reputations, control information flows, and govern legal risks. Critical resource geographers also tend to start out from more critical positions of companies and states, dissuading them from such engagements. That said, geographers' host universities might be more interested in sustaining relationships with business or governments with a view to generating revenue, placing students in jobs, and managing the universities' own perceived legal risks. Such contexts can put pressure on scholars who are critical of industry, again requiring them to balance their own motivations and their institution's pressures—a negotiation that can be especially complex for early career scholars and, all the more so, for those employed on soft money. Indeed, early career researchers and graduate students may conceal their engaged work while seeking jobs and building a research career for fear that public knowledge of their commitments might compromise their prospects for employment or research funding.

Fourth, the activities on which engagement is based can also vary. Most obviously, perhaps, are those relationships in which research is the vehicle for engagement. This research might be wholly conceived and implemented by the scholar, conceived by the organization with whom the scholar is engaging but implemented by the scholar, or jointly conceived and implemented (though "jointly" does not mean that the relationships of power involved are symmetric).[3] In other instances, the scholar might engage as a sharer of the knowledge or insights they have, advising on an organization's tactics or strategy, or simply sharing data and information. Scholars in some cases become formal parts of organizations or even (albeit rarely) establish nongovernmental bodies through which to act on their ideas and commitments.[4] Scholars have advised on litigation or served as expert witnesses for one or other party (Kirsch 2014). In yet other contexts, scholars might play the role of public intellectual, placing their ideas in the public sphere with the explicit intent of attracting attention and fostering debate (Acosta 2010; Svampa 2017).

Fifth, the length of engagement may also vary. At one extreme it can be punctual and short-term, an engagement around a particular problem, report, or policy debate, for instance. At another extreme, engagements can be multi-decadal, and in some instances, the critical resource geographer becomes part of another organization in some formal or semiformal way—as advisor, board member, research associate, etc. Over these extended periods, the very nature of engagement is likely to evolve and may go through tense and difficult moments as well as collaborative ones. Of course, earlier career resource geographers will often, by definition, have had less scope for such extended relationships, though they may be building shorter term engagements with a view to building long-term commitments.

All engagements map somewhere onto these axes. They can also be resourced in different ways: wholly by the scholar, wholly by the actor with whom the scholar is engaging, by a third party, or some mix of these. In laying out this diversity, we again remain agnostic on what options are preferable—the choice is that of the scholar, of the organizations, and individuals with whom the scholar is engaging (though the absence of funding and a range of political and personal concerns can limit the range of such choices). Instead, we want to make the point that just as there are different meanings of engagement, how engagement happens is also diverse. Of course, forms of engagement that critical resource geographers follow are also likely to vary over time and may well be multiple at any one point in time.

All engagements also involve asymmetrical relations of power, though the degree and direction of these asymmetries may vary across space and time. The researcher tends to have particular power over forms of representation, groups with which they engage have more power over relationships of access. The researcher may often have more power over financial resources (though certainly not always), and perhaps more importantly the scholar tends to have more powers of mobility, and the power to leave particularly dangerous sites, than might be the case for actors with whom they are collaborating. The nature of power within engagements is not straightforward, but it is always present and demands reflexivity.

Overall, if one looks at the work of scholars doing resource geographies that are critical of normal extractive industry practices, it does not take long to encounter a wide range of engagements with different actors in the sector. A parallel here is with the nongovernmental world in which some NGOs assume more adversarial and advocacy-oriented forms of engagement ("hostile engagement") and others assume more conciliatory or, at least, less confrontational forms. Some engage off-camera, others in public view. This range may or may not be discussed beforehand in any strategic way, but it has been recognized that the combination of forms and postures can be productive, opening up more possibility for change than the use of just one strategy or channel of engagement (Fox 2007). Similarly, it may be that a variety of forms of scholarly engagement is more effective than all scholars engaging in the same way.

The mine ... or is it?

Work on mining raises particularly challenging issues around engagement because there is often so much at stake. Potential profits can be very high. The prospect of capturing and controlling some of these "rents" can elicit self-interested behaviors and a proclivity to use violence to secure and protect these rents (as the Global Witness [2016, 2017, 2018, 2019] reports document). Meanwhile, levels of landscape transformation can be enormous, and the physical extent of these transformations coupled with the economic power of a mining operation—or of an extended network of artisanal and small-scale mining (ASM) operations—can easily lead to the displacement of populations and local governance institutions (Elmes et al. 2014; Peluso 2018). These landscape and social transformations disrupt whole systems of livelihood and of signification, the most extreme cases of the latter being those where the implantation of a mine destroys worlds and sacred sites (compare with Blaser 2010; de la Cadena 2015).

Indeed, one of the challenges in working on mining is that very frequently what looks to some like a mine site, or potential site, is something quite different to other actors. By the sheer magnitude of their endeavors, extractive projects often become sites where different visions of life or worldviews collide (Escobar 2008). Mines become terrains of ontological disagreement, where the capacity to articulate—let alone to negotiate—the distribution of benefits and costs is challenged by a deep dissonance in the understandings of the implications of extractive projects. For some people, especially indigenous populations, what a company or government calls a mine site may also be a living being or entity with which they engage in relationships of reciprocity (Nash 1979; de la Cadena 2015). Actors within the state may look at a mine site—or a map of mining concessions—and see an instrument for financing a local political campaign, a means for resolving a balance of payments problem, or a fiscal instrument for generating the resources needed to finance cash transfer programs—which they may, in turn, view as being instruments of social development or of political clientelism (or both). The state's view of mining may also change over short periods of time as a result of changing political ambitions and balances of power. To some local elites, a mining operation can look like the chance to accumulate wealth, while to some community members, the prospect of a mine looks like the loss of territory and of personhood. To the miner and the geologist, the mine is the result of many years of research, collaboration, prospecting, and exploration to discover an area with a unique geological history that is also economically and technically feasible to mine: something that is not easily found and, hence, worth fighting to secure. And to the firms and individuals behind finance capital, the mine is simply rate of return.

These bigger issues animate actors to protect or promote what they value in relation to the mine. Furthermore, they often pursue these objectives within the context of hugely asymmetric relations. Even in the case of ASM, the relationships between the often many miners and smaller, indigenous communities (for instance) are asymmetric in terms of numbers and of economic power. This is sometimes further complicated because community members may also be miners (Gregory and Vaccaro 2015). One potential consequence of such asymmetries can be the greater propensity to use force, or the threat of force, to impose a position (because powerful actors have the capacity to do so) or to defend a position (because those at a disadvantage may have few other alternatives for making their concerns visible). And even if physical force is not used, the temptation to use slander, character assassination, counter-insurgency techniques, and criminalization as a means of weakening opponents can be real (Dunlap 2019).

It is therefore not surprising that mining has become so conflictive (Martinez-Alier et al. 2016). Socio-environmental conflicts have become an increasingly significant proportion of all conflicts, and extractive industry conflicts are an important share of socio-environmental conflicts. This is perhaps especially the case in Latin America (De Echave et al. 2009; Svampa 2017). As many have noted, in such contexts, actors take sides very quickly, often well before a mine site comes into being, and in such contexts, it becomes extremely difficult for all actors, scholars included, to resist becoming partisan and to keep open lines of communication with a wide range of parties. The scope for friendly critique can become very limited. Sometimes it is also extremely difficult to know where one stands. For instance, this helps explain, perhaps, why different "critical" scholars of resource extraction in Bolivia had quite distinct views on the policies of the Evo Morales government. For some, the fact that increased natural gas extraction can finance social protection on a national scale makes extraction an important path to pursue. For others, the fact that this policy leads to increased silencing of dissent and drilling in indigenous territories and protected areas makes it a policy to be questioned.

Engaging in such contexts carries many risks. Beyond the obvious ones of safety are the risks that engaging can do harm. Harm can derive from misperception. For instance, scholars and scientists might engage with mining in an effort to provide "expert" knowledge with a view to helping bridge controversies and disagreements—for example, in the construction of environmental impact assessments—but actually end up alienating communities involved and undervaluing their knowledges (Li 2015). Harm can also be the result of "epistemic extraction," witting or unwitting, as when the resource geographer assumes ownership of the knowledge gained through engagement without recognition of its sources—sometimes also depoliticizing this knowledge in the process (Rivera Cusicanqui 2010; Grosfoguel 2016). Harm can come from work that focuses on damage rather than regeneration (Tuck 2009). Trying to follow the precautionary principle of "do no harm" may not help either, as what a researcher views as "no harm" may, in the eyes of others, still look like harm made possible by imbalances of power between the scholar and others.

Mines and minefields

Over the course of our research on mining, each of us has engaged with actors beyond the academy and have done so in the hope of contributing to explicit normative concerns (explicit, at least, in our own minds) (see also Shapiro-Garza et al., Chapter 20 this volume). These concerns have included: to reduce the environmental impact of resource extraction; to contribute to indigenous community claims for territorial autonomy and control; to stop mining in particularly fragile social and ecological contexts; or simply to broaden the scope of debate on some aspect of mining governance. In this section, we present three vignettes that throw light on a series of challenges and dilemmas that surround engagement and discuss several themes that arise from these experiences. It is, of course, important to note that these vignettes are told from our perspectives and not those with whom we were engaged.

Collaborations with indigenous land councils and communities: Canada and Guyana

In addition to having to navigate relationships among various actors, researchers conducting engaged work are often trying to juggle and think through the relationships between their personal models of research integrity and their commitments to community-based organizations. For example, as part of a rights claim commission, Gillian was asked by an indigenous land council in northern Canada to conduct research on historical land use in the area surrounding a large mining project. The results of this research were to be used to determine whether there was sufficient evidence documenting indigenous presence in the area prior to the commencement of mining activity—information that would be used along with other materials to determine compensation owed to the land council for loss of ability to practice treaty rights. Despite her history of work and affiliation with the land council on similar themes, the council distanced themselves from Gillian when it became clear that her research findings were not as definitively in their favor as expected. First, those alive in the community at the time of the mine's development had since died, and their kin had limited knowledge of events that transpired before they were born or when they were young children. Moreover, the oral history that did exist carried much less weight with the justice system than did written records, though often these also did not exist or were difficult to secure because of the state's refusal to provide timely access to key archival information about territorial development or treaty-making. The absence of a definitive conclusion in the terms rendered authoritative by the judicial system was interpreted as reflecting her "obviously pro-mining" stance. Nonetheless, following a long process of negotiations, the state judicial system recently awarded compensation to the land council, demonstrating that the notion of what constitutes evidence, as well as the authority given to engaged researchers, is a dynamic part of resource (and indigenous-state) governance processes.

In work in Guyana, a different dilemma emerged as a consequence of the differences between the narratives of a national indigenous organization and those of communities. In collaboration with three indigenous communities as well as a national organization representing indigenous interests across Guyana, Gillian conducted research on historical changes to the boundaries of indigenous territories in relation to resource extraction. The national organization was lobbying for greater recognition of land rights in the context of growing incursions on land titles by informal (or unregulated) gold mining in the northern Amazon. These land use dynamics were framed by the organization as a clear-cut conflict between indigenous community members ("good") and nonindigenous outsiders ("bad"). However, many community members were themselves actively involved in mining on their ancestral land titles and frequently explained that their ability to participate in mining reinforced a sense of cultural identity and continuity, as artisanal and small-scale gold mining was considered a traditional livelihood activity. This attempt to engage with multiple groups, even though all identified as indigenous, placed the research in a precarious position, as inevitably the study's conclusions regarding territorial boundaries would clash with one or other indigenous perspective. Moreover, the process of mapping territorial boundaries according to a colonial view of lands and borders was a frequent point of discussion and of frustration for all involved given that it required a singular way of "seeing."

Engaging an Australian government agency

Geoscience Australia is a government agency whose mining work involves the provision of geographical and geoscientific information that will inform future mineral activity in the country. One of its projects assessed availability of "critical minerals" in Australia, defined as those which are strategically important to the country, but (for a number of possible reasons) difficult to supply. Tim engaged with this study with a view toward demonstrating that these minerals are already extracted during the normal mining and processing of base metals like copper and zinc but are routinely wasted during these processes. The goal was to highlight this wastage and to encourage the development of domestic refining capacity for these metals as well as for reprocessing tailings and slags in which the metals are already present. This would reduce the need for new mining and reduce waste volumes at sites that have operated for a long time.

While this message was made clear in a report produced for the study, it was apparent that a primary interest of Geoscience Australia in critical minerals was to attract new investment into the country's mining sector and also to reduce Australia's vulnerability in the event of a supply disruption from other countries. The report was completed over seven months prior to its release; the delay in publication was due to a desire for it to be made public at a time that was deemed politically beneficial. Australia's Minister for Industry, Innovation, and Science subsequently emphasized one figure in particular: the estimated economic value of critical metal resources in Australia and sidelined environmental considerations. While Tim was attracted to this project because it offered a direct line to decision makers (and gave Tim, as an early career researcher, coauthorship of a government report), he observed that the socioenvironmental aspects of the research would not be given much public attention compared to what might have been emphasized in a scientific publication outside this engagement. While the authors could influence the content of the report (though it was edited and checked by multiple government employees), the public narrative and presentation of the report was out of their hands. Ultimately, the authors agreed that it was better to have decision makers read a watered-down report than to publish a more pointed journal article where their argument might not be read by anyone other than scholars. In this case, this attitude paid dividends, as following the release of the report (Mudd et al. 2019), Tim was directly contacted by mining companies interested in aspects of the work he initially believed were buried. This positively shifted his perspectives of the report and of the collaboration, though it is unclear how far chance had played a role in this or whether this was to be an expected outcome all along.

Engaging multiple audiences in a context of contentious politics: Peru and El Salvador

In 2006, Tony was approached by a UK-based human rights group to lead a fact-finding mission into a mine site in Peru operated by a British-based company accused of human rights violations as well as of illegal occupation of community lands. The mission was to include three researchers, a British Member of Parliament, and a journalist, and would conduct its work in coordination with the mining company (which was necessary in order to access the mine site and company workers) along with the social justice arm of the local Roman Catholic Church. As the project was potentially one of the largest in the government of Peru's investment pipeline, there was also coordination with government actors—both those promoting mining and above all the human rights ombudsman (La Defensoría del Pueblo). In the course of doing the work, the company insinuated a thinly veiled threat of legal action against the UK human rights organization, while different actors themselves combined support to the mission with attempts to lobby its findings. Once written, Tony was involved in two public debates of the report in Peru with the company's national director and lawyer, one closed-door debate, one discussion in the UK Houses of Parliament committee rooms, and multiple briefings to journalists. The report (Bebbington et al. 2007) was denigrated by the company as foreign meddling, while some activist groups had hoped it would be more critical. Some of its findings subsequently became one of various inputs into litigation in the UK representing community members claiming human rights violations by the company. For the researchers, the experience presented the challenge of negotiating a range of pressures, preexisting relationships with organizations, and their own models of research integrity and principled research, while also producing a representation that the team believed reflected the diverse types of evidence encountered.

In a similar process in 2011–2012, Tony was asked by the Ministry of Environment of El Salvador to be part of a team overseeing the quality of a Strategic Environmental Assessment of the mining sector and to provide input into subsequent drafting of a legislative proposal for new mining policy.[5] These processes once again involved engagement with activist communities as well as with other arms of government concerned with managing relationships with the private sector and addressing acute levels of social conflict in the country. It also required engagement with Washington-based lawyers defending El Salvador in an investment dispute with an international mining company. The final products, especially the draft law, were subject to intense negotiation. Parts of the Executive as well as DC-based lawyers insisted that the law could not propose a ban on mining, otherwise it would undermine the government's case in the investment dispute and potentially alienate the private sector. The draft law instead called for indefinite suspension of all mining activity, but this was rejected by activist groups who felt their views had not been adequately attended to and that an indefinite suspension was too weak a policy. Those groups presented their own draft law for a ban on mining. Ultimately both proposals got stuck in the committee stage at the national assembly. It was only when the legal dispute in Washington, DC, was resolved in the Government of El Salvador's favor several years later that political space opened again for legislation banning mining. This law was ultimately passed in 2017 (Spalding 2018; Bebbington, Fash, and Rogan 2019).

The minefields of engagement

In each of these experiences, engagement has not always culminated in the outcomes we first imagined and has involved difficult moments and relationships. If there is one thing that can unequivocally be said about engagement, it is that it is humbling. Consistently we have fallen short, either in general, or at least with one or more of the constituencies with whom we have engaged. The reasons for this are many: political positioning by members of these constituencies as they defend their constituents' interests, or their own; our own lack of capacity, time, and resources to do everything that people had hoped of us; our own decisions (power) to prioritize one or other dimension of what we perceived to be research rigor, rather than the claims of one of these constituencies; our tendency to commit to an idea and principle rather than to any particular actor; the impossibility of any serious engagement with some of these constituencies because our starting points were so different; and perhaps most importantly, the impossibility of sustaining engagement with a range of constituencies with contending positions. Ultimately, one of these constituencies will part ways when they view the costs of engagement as being greater than the benefits—and also because they can be every bit as stubbornly committed to their views as is the stereotypical hard-core academic. Engagement in scholarship is, like most relationships in life, at moments satisfying and at moments painful; and also like many relationships, engagement runs its course and comes to an end when the different parties cease to see its value (relative to the costs involved) and maybe also begin to see things differently from how they saw them at the beginning of the engagement.

Sustaining engaged scholarship may also become more challenging as academic life becomes more precarious. There are particular challenges for early career geographers who have to move between jobs without security, have to change roles, and have to alter their research foci and the geographic regions in which they are working—all as consequences of the next temporary, fixed-term position to which they have to move. Similarly, the demands of the tenure process (or other evaluation processes for conversion to permanent academic positions) still militate against engagement, given the continuing, and arguably deepening, emphasis on formal publication in high impact journals. These pressures constrain the possibilities of engagement—regardless of levels of personal commitment. Indeed, to expect "engagement and knowledge transfer" of precariously employed resource geographers on top of everything else can end up being one more demand of the contemporary (corporatized, neoliberalized, uncaring, …) university, and its increasing tendency to see scholars as vehicles of performance metrics more than of sustained social and intellectual creativity.

Regardless of constraints and failures, we have learned a great deal from our engagements, and we would do them all again, albeit perhaps with more humility and fewer expectations of the other. (We presume that the same would apply for those with whom we engaged in their expectations of us.) In each instance the engagement left us with the hope of having contributed just a little bit more than would have been the case had we limited our work to scholarly writing (not that we ever gave up on the scholarly components of these experiences). The experience also left us with lessons about engaged work which are perhaps too infrequently discussed or written about among us, especially in a way that is reflexive and self-critical. Even when engagements ultimately unravel, as is so often the case, their mere existence often represents progress made and a willingness among different parties to understand one another better. In some instances, they may also generate productive ideas about how to better survive/contest/govern the expansion of the extractive frontier.

Of course, this is the selfish side of engagement, and we have each gained enormously from these and other experiences. Above all is the lesson that "thinking with" is always more challenging and rewarding than "thinking alone." Of course, this "thinking with" can happen intramurally too—that is the core of interdisciplinary enquiry, team research, and student-faculty collaborative research. But thinking with extramural actors has opened up to us ways of seeing that we would not otherwise have understood, or felt, and this has been enormously helpful for us in our other work and our teaching. Engaging has broadened the debates in which we are involved, widening our small corners of the public sphere. In that sense, it has convinced us again of the importance of a vibrant public sphere and public debate for a democratic society.

Countering silencing: engagement for the public sphere

Regardless of one's position as to whether mining and natural resource extraction are generally beneficial or harmful for society, we should *all* be appalled by the figures that Global Witness publishes each year documenting the worldwide killings of environmental defenders (Global Witness 2016, 2017, 2018, 2019). We note this while also recognizing that Global Witness's methodology for classifying killings of environmental defenders is strict, and so the numbers that they report are quite conservative, especially when compared with country-level data on killings related to socio-environmental disputes.

Homicide is the most extreme form of silencing debate and one to which community members, activist lawyers, and environmental defenders are much more subject than are most engaged researchers (while recognizing the many academics who have been killed for their commitments). These murders lie at one extreme of a spectrum of tactics that are repeatedly used to rein in public debate on issues associated with mining. The issues run the gamut: some are related to the overall human-environmental impacts of mines; some are related to struggles for power over territory; and some are related to intra-community and intra-movement arguments over local preferences, corruption, and rent-capture. In all instances, they constitute attempts to discipline thinking, to make it more homogenous, and to align it with the thinking of those with the power and predisposition to kill, litigate, threaten, imprison, and slander. These silencings diminish life, make democracy and civil society "thinner" (compare to Fox 1996), and impoverish the scope for exploration of alternatives. If development is freedom (Sen 1999), then these are systematic and sustained exercises in creating unfreedoms.

We would suggest that this is the terrain for engagement in critical resource geography, and it is a terrain that is under pressure everywhere—in liberal democracies, in post-neoliberal regimes, in authoritarian regimes, in religious regimes (McCarthy 2019). In such contexts, as important as engagement in defense of particular ideas and actors is, engagement to promote ideas, and the right of everybody to express them, matters enormously: for human dignity, for diverse ways of being, for the quality of democracy, and for the expansion of freedoms. Advocating for the importance of ideas, the right to express them, and the value of public debate is *ipso facto* an advocacy for the rights of those who bear these ideas. Commitment to, and enactment of, such a principle would appear to be a least common (yet vitally important) denominator of an engaged critical resource geography.

Notes

1   This chapter is dedicated to the memory of Ricardo Morel who was committed to protecting inclusive debate about mining against all the odds. It was made possible by an Australian Laureate Fellowship to Bebbington (FL160100072).
2   Prior to 2000, there were only a handful of Anglophone geographers working on mining in any sustained way (e.g., Emel, Bridge, and Krueger 1995; Emel, Angel, and Bridge 1995; Bridge 2000; Auty 2001).
3   See for instance Laura Sauls' diverse engagements with the Central American organization PRISMA, as well as with community-based organizations such as the Mesoamerican Alliance for Peoples and Forests (AMPB in Spanish) and Moskitia Asla Takanka (MASTA) in Honduras (Sauls 2019).
4   For instance, Liza Grandia and Paige West, who have been involved in establishing organizations, albeit with foci that go well beyond natural resource extraction.
5   This is discussed in more detail in Bebbington (2015).

References

Acosta, Alberto. 2010. "Andean Voices: Alberto Acosta," interview by John Vidal, *The Guardian*, December 2. https://www.theguardian.com/environment/2010/dec/02/andean-voices-alberto-acosta.

Auty, Richard M. 2001. *Resource Abundance and Economic Development*. Oxford: Oxford University Press.

Batterbury, Simon. 2015. "Who are the Radical Academics Today?" *The Winnower*, March 17. https://thewinnower.com/papers/327-who-are-the-radical-academics-today.

Bebbington, Anthony. 2015. "At the Boundaries of *La Política*: Political Ecology, Policy Networks and Moments of Government." In *Handbook of Political Ecology*, edited by Tom Perreault, Gavin Bridge, and James McCarthy, 198–208. London: Routledge.

Bebbington, Anthony, Michael Connarty, Wendy Coxshall, Hugh, O'Shaughnessy, and Mark Williams. 2007. *Mining and Development in Peru, with Special Reference to the Rio Blanco Project*. Piura. London: Peru Support Group.

Bebbington, Anthony, Benjamin Fash, and John Rogan. 2019. "Socio-Environmental Conflict, Political Settlements, and Mining Governance: A Cross-Border Comparison, El Salvador and Honduras." *Latin American Perspectives* 46 (2): 84–106.

Blaser, Mario. 2010. *Storytelling Globalization from the Chaco and Beyond*. Durham, NC: Duke University Press.

Bridge, Gavin. 2000. "The Social Regulation of Resource Access and Environmental Impact: Production, Nature and Contradiction in the US Copper Industry." *Geoforum* 31 (2): 237–256.

De Echave, Jose, Alejandro Diez, Ludwig Huber, Bruno Revesz, Xavier Ricard, and Martin Tanaka. 2009. *Minería y Conflicto Social*. Lima: Instituto de Estudios Peruanos.

De la Cadena, Marisol. 2015. *Earth Beings: Ecologies of Practice Across Andean Worlds*. Durham, NC: Duke University Press.

Dunlap, Alexander. 2019. "Agro sí, mina NO!' the Tía Maria Copper Mine, State Terrorism and Social War by Every Means in the Tambo Valley, Peru." *Political Geography* 71: 10–25.

Elmes, Arthur, Josué Gabriel Yarlequé Ipanaqué, John Rogan, Nicholas Cuba, and Anthony Bebbington. 2014. "Mapping Licit and Illicit Mining Activity in the Madre de Dios Region of Peru." *Remote Sensing Letters* 5 (10): 882–891.

Emel, Jody, Gavin Bridge, and Rob Krueger. 1995. "The Earth as Input." In *Geographies of Global Change: Remapping the World in the Twentieth Century*, edited by Ron J. Johnston, Peter Taylor, and Michael Watts, 318–332. Oxford, UK: Basil Blackwell Press.

Emel, Jody, David Angel, and Gavin Bridge. 1995. "New Models of Exhaustible Resource Development." *Business Strategies and Greening* 4 (4): 200–207.

Escobar, Arturo. 2008. *Territories of Difference: Place, Movements, Life, Redes*. Durham, NC: Duke University Press.

Fox, Jonathan. 1996. "How Does Civil Society Thicken? The Political Construction of Social Capital in Rural Mexico." *World Development* 24 (6): 1089–1103.

Fox, Jonathan. 2007. *Accountability Politics: Power and Voice in Rural Mexico*. Oxford: Oxford University Press.

Global Witness. 2016. *On Dangerous Ground*. June 20. London: Global Witness. https://www.globalwitness.org/en/campaigns/environmental-activists/dangerous-ground/.

Global Witness. 2017. *Defenders of the Earth: Global Killings of Land and Environmental Defenders in 2016*. July 13. London: Global Witness. https://www.globalwitness.org/en/campaigns/environmental-activists/defenders-earth/.

Global Witness. 2018. *At What Cost? Irresponsible Business and the Murder of Land and Environmental Defenders in 2017*. July 24. London: Global Witness. https://www.globalwitness.org/en/campaigns/environmental-activists/at-what-cost/.

Global Witness. 2019. *Enemies of the State: How Governments and Businesses Silence Environmental Defenders*. July 30. London: Global Witness. https://www.globalwitness.org/en/campaigns/environmental-activists/enemies-state/.

Gregory, Gillan, and Ismael Vaccaro. 2015. "Islands of Governmentality: Rainforest Conservation, Indigenous Rights, and the Territorial Reconfiguration of Guyanese Sovereignty." *Territory, Politics, Governance* 3 (3): 344–363.

Grosfoguel, Ramón. 2016. "Del 'extractivismo económico' al 'extractivismo epistémico' y al 'extractivismo ontológico': una forma destructiva de conocer, ser y estar en el mundo." *Tabula Rasa* 24: 123–143.

Kirsch, Stuart. 2014. *Mining Capitalism: The Relationship Between Corporations and Their Critics*. Berkeley, CA: University of California Press.

Kirsch, Stuart. 2018. *Engaged Anthropology: Politics Beyond the Text*. Berkeley, CA: University of California Press.

Li, Fabiana. 2015. *Unearthing Conflict: Corporate Mining, Activism, and Expertise in Peru*. Durham, NC: Duke University Press.

Martinez-Alier, Joan, Leah Temper, Daniela Del Bene, and Arnim Scheidel. 2016. "Is There a Global Environmental Justice Movement?." *Journal of Peasant Studies* 43 (3): 731–755.

McCarthy, James. 2019. "Authoritarianism, Populism, and the Environment: Comparative Experiences, Insights, and Perspectives." *Annals of the American Association of Geographers* 109 (2): 301–313.

Mudd, Gavin, Tim Werner, Zhehan Weng, Mohan Yellishetty, Yue Yuan, Sarlae McAlpine, Roger Skirrow, and Karol Czarnota. 2019. *Critical Minerals in Australia: A Review of Opportunities and Research Needs*. Record 2018/051. Canberra: Geoscience Australia. doi:10.11636/Record.2018.051.

Nash, June. 1979. *We Eat the Mines and the Mines Eat Us: Dependency and Exploitation in Bolivian Tin Mines*. New York, NY: Columbia University Press.

Peluso, Nancy. 2018. "Entangled Territories in Small-Scale Gold Mining Frontiers: Labor Practices, Property, and Secrets in Indonesian Gold Country." *World Development* 101: 400–416.

Rivera Cusicanqui, Silvia. 2010. *Ch'ixinakak utxiwa: una reflexión sobre prácticas y discursos descolonizadores*. Buenos Aires: Tinta Limón Ediciones.

Sauls, Laura Aileen. 2019. "Becoming Fundable? Converting Climate Justice Claims into Climate Finance in Mesoamerica's Forests." *Climatic Change*. 161 307–325.

Sen, Amartya. 1999. *Development as Freedom*. Oxford: Oxford University Press.

Spalding, Rose. 2018. "From the streets to the chamber: social movements and the mining ban in El Salvador." *European Review of Latin American and Caribbean Studies*. 106: 49–76.

Svampa, Maristella. 2017. *Del Cambio de época al fin de ciclo. Gobiernos progresistas, extractivismo, movimientos sociales en América Latina*. Buenos Aires: Edhasa.

Tuck, Eve. 2009. "Suspending Damage: A Letter to Communities." *Harvard Educational Review* 79 (3): 409–427.

22

Intergenerational equity and the geographical ebb and flow of resources

The time and space of natural capital accounting

Patrick Bond and Rahul Basu

Introduction

The *uncompensated extraction of nonrenewable minerals, oil, and gas* is siphoning wealth from future generations to the current as well as from poorer, resource-dependent regions to the core of the world economy. As Andreas Mayer and Willi Haas (2016, 351) put it: "The disproportionately high levels of resource consumption in some parts of the world have created a debt towards the environment, towards other parts of the world, and also towards future generations through the excessive consumption of non-renewable resources."

This chapter reveals ways of conceptualizing resource injustices that follow from this particular form of "unequal ecological exchange" across both time and space. Measurement of this unequal exchange immediately raises controversies, such as those associated with "natural capital accounting" methodologies and, specifically, their utilization by major corporations and allied multilateral institutions to "neoliberalize nature." Hence, we take very seriously the importance of a dialectical application of these ideas. In our applied work with organizations committed to eliminating or minimizing damage associated with the resource-extractive industries, we find critical resource geography important for both empirical and theoretical critique of both the *status quo* and its ideology. But as discussed in the conclusion, the merits of a radical reappropriation of resource accounting for intergenerational and spatial justice are enormous.

To begin, the most unjust abuse of resources is obvious: fossil fuels. Oil and gas are already responsible for catastrophic greenhouse gas emissions and what is termed "loss and damage" caused by climate change. The benefits of fossil fuel consumption accrue to those alive *today* versus costs paid by those who will suffer droughts and drying soils, forest fires, floods, sea-level increases, and ocean acidification amidst fast-rising temperatures *tomorrow*. Not only temporal but also geographical implications are obvious too: those who benefit most have a higher carbon footprint and come from the "Global North," versus those in the Global South, who not only did not cause the crisis but are least able to finance adaptation and resilience as well as cover loss and damage (Bond 2012). In this chapter, we will focus on not only such obvious climate injustices of this sort but also two other features associated with resource dependencies: the intergenerational and the geographical transfers of nonrenewable resource wealth. These transfers are extremely important. The Pacific island nation of Nauru, where a century of phosphate mining left a legacy of barren land and environmental catastrophe along with near bankruptcy, is a classic example of consuming intergenerational wealth (Gowdy and McDaniel 1999). In banal quantitative terms, according to a 2018 World Bank report, unequal ecological exchange through resource depletion collectively costs sub-Saharan African countries more than $100 billion annually in minerals and oil extraction as well as trade via transnational corporations (Bond 2018; Lange, Wodon, and Carey 2018).

There are, in short, massive "ecological debts" owed by current generations and by the Global North (see also Ciccantell, Chapter 15 this volume). These include debts owed by elites within the South—e.g., what might be considered "sub-imperial" Brazilian, Russian, Indian, Chinese, and South African companies operating within Africa—to future generations and the poorest peoples. These are not currently on the agenda to be repaid, even if resistance to firms from such countries is rising (Bond and Garcia 2015; van der Merwe, Bond, and Dodd 2019).

A just society's version of a natural resource legacy should, in contrast, be provided to future generations and to the poorest regions from which the resources are extracted. One example of ecological debt paid to compensate pollution-related damage to natural resources is the US Superfund program, but it is limited to local processes. Globally, as another unsatisfactory example, the Green Climate Fund supports projects in poor countries but with no recognition of the funders' liability in the event they are major greenhouse gas emitters (ignoring such "climate debt" liabilities was a condition of signing the 2015 Paris Climate Agreement; Bond 2016).

Natural resources are typically depleted without appropriate compensation mainly because of adverse power relations. The depletion of wealth is most extreme in the Global South and especially in cases in which dictatorships or authoritarian regimes are in league with global and local mining corporations. Such sites' reinvestments of resource rents compare unfavorably to democratic countries, some of which (e.g., Norway) partially plough back the proceeds of resource extraction into national wealth funds. That process occurs because of the state-owned and managed character of resource ownership and exploitation. For example, Equinor, Norway's state-owned oil company, provides funds to a sovereign wealth fund that invests the proceeds and distributes its income. In turn, this makes it possible to engage in more rational use of such income for future generations via educational and social-infrastructural investments and to curb uneven geographical development with compensatory payments.

There is a long-standing difference between the "weak-sustainability" conception of the substitutability of capital—i.e., reduced natural capital being compensated for by higher productive or human capital (Hartwick 1977; Solow 1974)—and the "strong-sustainability" position that aims to protect the overall stock of natural resources, as argued, for instance, by Herman Daly (1996). Regardless of this debate, the recirculation of resource-sourced wealth is extremely unusual and is feasible in a few sites, like Norway, because of a long tradition of social democracy based on the society's commitment to fairness. That commitment followed the struggles of trade unions and rural people's organizations to unite in red-green coalitions that won power during the twentieth century, eventually constructing a welfare state (Esping-Andersen 1990).

In contrast, resource-endowed areas that have long witnessed repressive, super-exploitative conditions of accumulation are exemplars of appropriating the commonwealth. In these sites, current wealthy owners of the extractive industries facilitate both the temporal and spatial shifts in ways that impoverish descendants and the "resource-cursed" regions. The resource curse meme has been abused, particularly because it blames the minerals, oil, and gas when it is in reality powerful extractive and rentier forces that require attention. The ecological determinism associated with the resource curse literature also undermines relational ways of understanding nonrenewable resources within changing historical and sociocultural contexts. As one example, Cyril Obi (2010, 483) suggests Africans replace the resource curse premise with a "radical political economy which lays bare the class relations, contradictions, and conflicts rooted in the subordination of the continent and its resources to transnational processes and elites embedded in globalized capitalist relations."

Again, as in the weak-versus-strong sustainability debate, our setting aside semantic disputes over the two words "resource curse" leaves intact our critique of helter-skelter extraction. This critique combines concerns about unequal exchange in both the temporal and spatial ebbs and flows of mineral wealth. These two levels of analysis—temporal and spatial—have as their core political strategy the "valuation" of mineral wealth so as to empower both younger generations and current inhabitants.

The point we want to establish, in the process, is that a radical reappropriation of these concepts is feasible so as to develop new tools to better resist exploitative resource extraction. These critiques of unequal ecological exchange rely upon natural capital accounting as the main conceptual tool, one which we believe adds to anti-extractivists' traditional concerns about the excessive political-socio-economic-environmental costs of nonrenewable resource depletion.

This is not Don Quixote at the windmill, for there is at least partial support growing in some official quarters—for example, the Indian judiciary (Supreme Court of India 2014) and Africa's 2012 Gaborone Declaration (Bond 2018). There are also activist initiatives that use ecological economics as a basis to account for nature at least in the struggle against extractivism (including in Environmental Impact Assessments), even if it does not yet influence policy decision-making. In all these settings, the imperatives of capital accumulation and class formation have together, thus far, overwhelmed both the grassroots activists who resist extraction and those technocrats from the state and civil society who are engaged in policy critique using natural capital accounting. One opportunity to raise awareness and address extraction is during engagements with both resource curse analysis and NGO "transparency" advocacy. Both could improve dramatically if the concepts of intergenerational equity (IE) and spatial equity were applied vigorously.

At the same time, it is vital not to promote "neoliberal nature," and so we agree with critics of natural capital accounting—e.g., Sian Sullivan (2014)—that there is a danger in ecological modernizationists attempting to resolve the extractive industries' environmental externalities with market mechanisms (such as offsets or emissions trading). That battle must also be joined, as the revitalization of global climate policy by the Biden Administration re-empowers neoliberal strategists and financiers, especially those who believe that having 22 percent of current world emissions under some form of carbon pricing is a positive sign—no matter that the 2020 market prices for emitting a ton of carbon are nearly universally under $35 and that there is huge variance (e.g., South Africa's at just $0.43 per ton compared to Sweden's at $132 per ton). The CO_2 emissions market still swings wildly, following closely the wild fluctuations of stock markets, including during the 2020 Covid-19 roller-coaster.

Mainstream discourses on resource justice need revitalization

Paul Burkett (2006) confirms how one of the major challenges to neoclassical economics starting in the early 1970s was the notion of "limits to growth": scarcity associated with excessively rapid depletion of nonrenewable resources. A new discipline soon emerged as a result, based on the reconceptualization of the environment as natural capital. According to Burkett (2006, 94), "although the origins of natural capital are unambiguously neoclassical, ecological economists have been at the forefront in developing and popularizing its usage... as a core paradigmatic concept." In short, argued neoclassical economists, "an economy can indefinitely maintain a positive level of consumption by investing its savings in capital, so long as capital can be substituted for the natural resource," according to Burkett. Within this logic, the concept of "Intergenerational Equity" (IE) gradually dawned. The next step, following Robert Solow's

(1974) suggestion that various forms of capital are substitutable, was his student John Hartwick's (1977, 972) call to:

> Invest all profits or rents from exhaustible resources in reproducible capital such as machines. This injunction seems to solve the ethical problems of the current generation short-changing future generations by "overconsuming" the current product, partly ascribable to current use of exhaustible resources. Under such a program, the current generation converts exhaustible resources into machines and "lives off" current flows from machines and labor.

Taking the "Hartwick Rule" argument from natural to productive capital and then to human capital investment (the "knowledge stock") was not a major leap. One ready objection is that the *substitutability* of these different kinds of capital is not at all straightforward (Berkes and Folke 1992; Burkett 2006). Also important, conceptually, is the division of natural resources into two types: renewable and nonrenewable (exhaustible). For accounting purposes, Robert Costanza and Herman Daly (1992, 38) suggested, "Renewable natural capital is analogous to machines and is subject to … depreciation; nonrenewable natural capital is analogous to inventories and is subject to liquidation."

Notwithstanding these complications, as well as the ethical problem of monetizing life forms, the framing of natural resources as capital has taken hold. In terms of temporality, Gro Harlem Brundtland's World Commission on Environment and Development (1987) defined "sustainable development" as meeting "the needs of the present without compromising the ability of future generations to meet their own needs." In 1993, summing up two decades of mainstream environmental-economics thinking, Solow (1993, 170) asked, "What should each generation give back in exchange for depleted resources if it wishes to abide by the ethic of sustainability? … [W]e owe to the future a volume of investment that will compensate for this year's withdrawal from the inherited stock."

The calculation of nonrenewable natural capital as a negative "withdrawal" became one of Herman Daly's (1996) objectives inside the World Bank, even though his agenda was much more radical: reversing the neoliberal policy agenda using natural capital accounting as a lever. His 1996 farewell speech beseeched colleagues to "stop counting natural capital as income" without a corresponding debit to account for depletion, and to instead, "maximize the productivity of natural capital in the short run, and invest in increasing its supply in the long run." That would also necessarily entail a "move away from the ideology of global economic integration by free trade, free capital mobility, and export-led growth." That application of natural capital accounting hit a brick wall inside the Bank, Daly admitted, because it "just confirmed the orthodox economists' worst fears about the subversive nature of the idea and reinforced their resolve to keep it vague" (Daly 1996, 88–93).

It took more than a decade for Daly's words to resonate sufficiently that the counting of natural capital began in earnest. Impressive organizing efforts by the World Bank and Conservation International led to a Wealth Accounting and the Valuation of Ecosystem Services (Waves) project. A parallel pro-mining interstate institution is the Committee for Mineral Reserves International Reporting Standards (2020), self-described as an "international initiative to standardize market-related reporting definitions for mineral resources and mineral reserves."

As one example of Waves, the 2012 Gaborone Declaration for Sustainability in Africa was signed a month before the Rio+10 Earth Summit, at a time when world environmental policy leaders anticipated a major shift toward the "green economy" and associated techniques of ecological modernization. According to the Declaration, due to "limitations that GDP has as a measure of well-being and sustainable growth," the ten African signatories (including Botswana, South Africa, and Zambia) would begin "integrating the value of natural capital into national accounting and corporate planning." Although supposedly an "African-led initiative for sustainable development," the Bank and Conservation International took up the primary duties of implementation. Indeed, because of the often-corrupting ties between multinational mining corporations and states, few if any African elites exhibited interest in the project—quite logically, for this sort of information certainly *doesn't serve their interests*.

At this point, it is again important to distinguish between the two divergent conceptions of sustainability—strong and weak. Strong sustainability considers as critical those types of capital that cannot be substituted by other forms of capital, such as the wealth of Earth's biodiversity and species diversity. Strong sustainability requires that the stock of natural capital does not decrease and motivates the precautionary principle—do not risk a catastrophe—that environmentalists apply to climate change, nuclear power, and the like. Sociocultural artifacts, burial sites, spiritually important places, and sacred mountains would also fall in this category. (For example, in the context of Goa, India, the Western Ghats are both a biodiversity hotspot as well as the water tower of peninsular India.)

Although, following Hartwick (1977), weak sustainability advocates do acknowledge that natural capital is depleting, typically without adequate compensation to those losing wealth, this perspective assumes that different types of capital (natural, produced, cultural, and so on) can be substituted for each other. Their standpoint is merely that the total stock of capital should not decline. Thus, while extracting minerals reduces or depletes the available quantity of mineral resources for use by future generations, the Hartwick Rule holds that consumption can be maintained "sustainably" if the values of nonrenewable resources are continuously invested—e.g., in human capital through higher education subsidies—rather than used for consumption.

The Hartwick Rule is quite intuitive: to keep inherited capital at least constant when a mineral (a nonrenewable resource) is extracted thereby reducing a country's mineral wealth that country should create or invest in another asset of at least the same value as the mineral that is no longer there. Taking the economic value of mineral resources to be the difference between the price paid in the market versus the total cost of producing it, including a proper return on capital ("resource rent" or "economic rent"), the Hartwick Rule thus requires that this value of extracted mineral resources be captured and continually reinvested.

After Hartwick, many others expressed concerns with depletion, including Christian Azur and John Holmberg (1995, 11) who grappled with appropriate costings over time, "Since we are dealing with intergenerational pollution and resource depletion, the choice of discount rate is crucial to the cost analysis." This value-laden choice continues to aggravate environmental economics, such as when the Nobel economics prize was given to William Nordhaus in 2018 in spite of his controversial discounting of future life (Hickel 2018).

The Hartwick Rule is one way to implement the IE principle. But the concept has deep roots in human civilization, which unfortunately have been obscured over time. Consider inheritance law: inheritors of property are simply custodians for future generations, especially if the inheritance involves entailment, which constrains the present heir from consuming the inheritance by recognizing the rights of subsequent heirs. (To illustrate, in most cultures, there is the rich good-for-nothing heir who lives by selling off the family silver, unfairly impoverishing his future generations.)

Or, consider endowment funds, where the capital is conserved and only the income used. The deepest rationale in this case is the idea of stewardship, the idea that capital must first be conserved. Indeed, the accounting and economics professions generally define income not as revenue but as the residual after we ensure that the capital is held constant.

Further, environmental economics provides us with the "sustainable yield" principle: we can only consume that amount that does not endanger the capital. And, in most countries, natural resources—including forests, streams, beaches, oceans, atmosphere, and minerals—are owned by the state as a trustee on behalf of the people and especially future generations. This is the public trust doctrine and represents the implementation of the IE principle in the public domain. It is often derived from natural law and considered more fundamental than the Constitution. For the trustees, the foremost obligation is to ensure the corpus of the trust is kept whole. Moreover, there is a duty to treat all the beneficiaries equally. To illustrate, in 2017, the Pennsylvania Environmental Defense Foundation won a judgment in the Supreme Court of Pennsylvania (2017) that the state must consider natural resources in the role as trustee, not proprietor, and therefore must use the proceeds from extracting oil and gas for restoring the environment, not general government expenditure.

Given the powerful demands by youth activists that IE be central to future climate politics, this is a vital time to revisit IE and apply it to all society-nature relations.

Evidence from natural capital accounts

There is increasing empirical work on both spatial and temporal inequity associated with extractivism. In Goa, India, for example, intergenerational inequity was identified based upon audited financials of the largest mining company, which revealed a system of mining leases that resulted in the loss of over 95 percent of the value of local minerals (after deducting extraction expenses and a reasonable profit for the extractor). Over eight years, the state Government of Goa received less than 5 percent of the mineral value as royalty, a pittance in comparison to the windfall profits raked in by the mining companies. The amount lost is a redistributive per-head tax, contributing to spatial inequity. Worse still, even the royalty was treated as windfall revenue and frittered away, cheating future generations of their entire inheritance (Basu 2015, 44–48).

In such contexts of extreme resource depletion, there are at least two steps in achieving, at minimum, weak sustainability of mineral resources: ensuring no loss in the value of the natural resources and investing those amounts received in exchange for selling the mineral wealth in productive assets so that overall wealth does not decline (Basu and Pegg 2020, 1). To measure whether these two steps have been taken, an "adjusted net savings" (ANS) variable was adopted by the World Bank, and *The Changing Wealth of Nations* series reported on how "increasing standards of living lies in building national wealth, which requires investment and national savings to finance this investment" (World Bank 2011, 37). The level of national savings reflects whether there has been an increase in wealth, net of natural capital extraction. By definition, income minus consumption = savings = increased wealth. Increases in national wealth usually enable higher levels of subsequent income; rising wealth is a stronger requirement than Hartwick's Rule, which simply requires the maintenance of wealth.

The Bank modified the orthodox measure of savings to account for nature and education, terming this ANS or "genuine savings": gross national savings adjusted for the annual changes in volumes of all forms of capital, including natural capital and human capital. ANS is measured as net national savings minus the value of physical capital's depreciation ("fixed capital consumption," due to wear and tear), environmental degradation (including pollution), depletion of subsoil assets and deforestation, and credited for education expenditures (World Bank 2011, 37).

If ANS is negative, then the country is running down its capital stocks and reducing future well-being, social welfare, and future capacity to maintain extant standards of living; if ANS is positive, then the country is adding to wealth and future well-being (World Bank 2011, 37). Evidence indicates that countries that are more dependent on mineral extraction have underinvested—their ANS tends to be lower. In the Bank's calculations, all countries where mineral rents account for 15 percent or more of their gross domestic product (GDP) have underinvested in other forms of capital. In other words, these countries are simply using up their natural resources to finance consumption rather than investing in productive assets, thereby making themselves poorer in aggregate (World Bank 2011, 11) and cheating future generations of their inheritance. (Though not typically conceded by the Bank, it is vital to define "consumption" as profit expatriation plus Illicit Financial Flows associated with

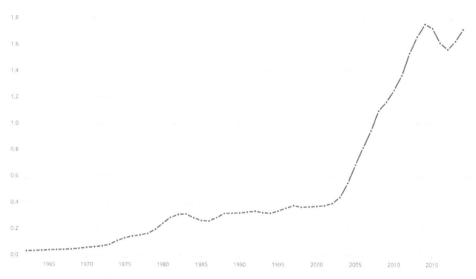

Figure 22.1   Sub-Saharan Africa's gross national income, US$ (trillions), 1962–2019.
Source: World Bank (2020).

multinational corporates' extraction in poor countries, thus allowing inordinately high consumption by the firms' executives and shareholders, and not within the country of origin.)

Most of sub-Saharan Africa, for instance, has in recent decades suffered negative ANS, especially during the rapid commodity-related rise in gross national income (GNI) beginning in the early 2000s (Figure 22.1). By adding North African countries where mineral wealth is poorly managed—especially Libya since 2011—the overall African depletion of wealth is formidable given that natural capital ranges between 40 and 55 percent of Africa's overall wealth (Figure 22.2, but noting that World Bank researchers were confounded by platinum and diamond markets so simply left them out, hence dramatically underestimating the wealth and depletion process in

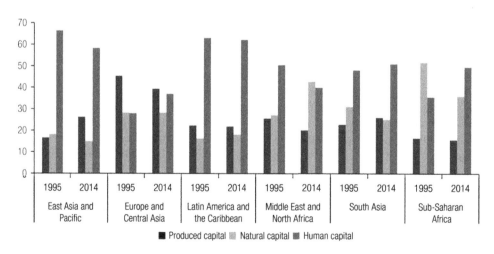

Figure 22.2   Regional composition of wealth by types of capital, 1995 and 2014 (percent).
Source: Lange, Wodon, and Carey (2018, 51).

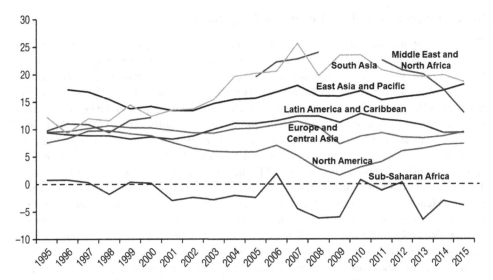

Figure 22.3 Adjusted net saving (as share of gross national income) by region, 1995–2015. Middle East/North Africa have data gaps, thus broken lines.

Source: Lange, Wodon, and Carey (2018, 63).

some countries, such as South Africa, Zimbabwe, Botswana, Namibia, Angola, the Democratic Republic of the Congo, Sierra Leone, and Liberia). Even with neglected platinum and diamond extraction caveats, since 2000 between 2 and 3 percent of the region's reported income has been depleted without offsetting compensation, an amount typically exceeding $100 billion (Figure 22.3). In 2008, for example, the subcontinent's GNI hit $1 trillion, but the natural resource depletion account (the second bar category below zero) was negative 15 percent, i.e., $150 billion. Likewise at the point of least net natural resource depletion (6 percent), in 2015, as commodity prices crashed, GNI was $1.5 trillion so the total net outflow on resource accounts was at least $90 billion (Figure 22.4).

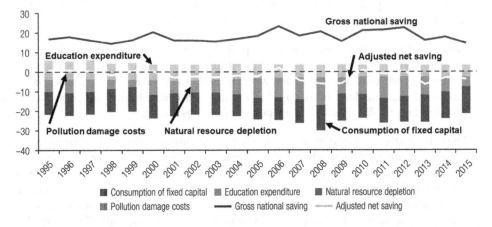

Figure 22.4 Adjusted net saving for sub-Saharan Africa, 1995–2015 (percent of GNI).

Source: Lange, Wodon, and Carey (2018, 66).

Uses, nonuses, and abuses of natural capital

There are temporal and spatial aspects of this natural capital accounting methodology, especially in allocating ecological debt owed to future generations and those currently in the sites of extraction. But what is most important to point out here is that in subsequent years, a related concept—the resource curse—took two distinct turns without referring to this work: on the one hand, a blame-the-victim concern with Third World elite inadequacies during negotiations with transnational corporations that is evident both in the academic literature (e.g., Humphreys, Sachs, and Stiglitz 2007; Robinson, Torvik, and Verdier 2006) and applied projects (e.g., the Extractive Industries Transparency Initiative and Publish What You Pay); and on the other, a grassroots anti-extractivist agenda witnessed in thousands of anti-mining and anti-oil protests. The moderate NGOs rely on a simple philosophy: *Transparency* can sanitize the process of corruption. Hence, they pay most attention to payment flows, project damage mitigation, Free Prior and Informed Consent (a United Nations mandate that genuine community consultation and permission should be acquired before extraction), and other assimilationist reforms, typically ignoring both grassroots resistance and natural capital depletion.

What the two strands of this argument now suggest should be obvious: both IE and North-South equity are at stake. Hence we adopt the strategy of "counting nature" in order to identify withdrawals from poor areas and countries—not as an endorsement of ecological modernization, and certainly not so as to price and therefore market the environment, but to *value* it and halt the unjust extractivism still ravaging resource-rich countries. In contrast, Pavan Sukhdev's strategy, termed "The Economics of Ecosystems and Biodiversity" (TEEB 2020, 1), is to "make nature's values visible" and thus "help decision-makers recognize the wide range of benefits provided by ecosystems and biodiversity, demonstrate their values in economic terms and, where appropriate, capture those values in decision-making." The mainstream thinking is that such "capture" can be accomplished by market mechanisms (such as carbon trading), which in many cases are utterly inappropriate, for instance, because natural assets have characteristics unsuitable to speculative financial markets (Bond 2012).

Indeed, there is an extraordinary tension built into this debate, with vital strategic implications about whether anti-extractivist grassroots movements as well as transparency-based NGOs and solidarity advocacy groups should take up this methodology in coming years. Those from a critical resource geography perspective who advocate rejecting natural capital accounting on methodological and political grounds, such as Sullivan (2014), are not necessarily opposed to demands for reparations against those engaged in environmental damage:

> In a move with which I am in broad sympathy, Sharife and Bond (2013) also argue that natural capital accounting and associated calculations might be mobilized in the course of reckoning ecological debt reparations, whereby retributive payments for "ecological debt" are based on both "loss and damage" accounting and environmental justice, and made through fines for damages and prohibitions on further pollution.

The latter strategies—i.e., fines and prohibitions—are entirely different than the objective of ecological modernizationists who insist on fees (e.g., the "payment for ecosystem services" concept) so that pollution can continue. And as for the methodology associated with ecological debt quantification, Joan Martinez-Alier (2002, 228) concedes, "mea culpa. My excuse is that the language of chrematistics is well understood in the north." In other words, by taking the neoliberal conception of market valuation *on its own terms,* the case can still be made that extractivism should be halted in most situations, given that the declining wealth cannot justify the (maldistributed) rise in short-term income.

But aside from a conceptual agreement about the validity of ecological debt, Sullivan and other critics of monetizing nature do not do justice to our concerns about temporal (intergenerational) and spatial (unequal ecological exchange) injustices associated with depleted natural wealth. Nor does she/they envisage a route by which halting such injustices through activist critique may entail recourse to full-cost accounting—e.g., in Environmental Impact Assessment critiques of resource-extractive projects (see, e.g., Basu 2017; South Durban Community Environmental Alliance 2019).

However, we do agree with profound aspects of the critique of natural capital accounting. For example, Sullivan (2017, 408) is absolutely correct to remark on other deficiencies of ecological modernist conceptions. She observes, "diversities are lost in the world-making mission to fashion and fabricate the entire planet as an abstracted plane of (ac)countable, monetizable and potentially substitutable natural capital." Adds Patrick Bresnihan (2017, 44), "The enthusiasm for economic valuation methodologies should not obscure the continued significance of questions of ownership, uneven distributions of power, and the distribution of environmental goods and bads."

Resistance narratives mature

It should now be evident that exceptionally high outflows of resource-related wealth with which we are concerned can be understood as both temporal and spatial modes of systemic impoverishment, or in another vocabulary that harks back to Rosa Luxemburg's 1913 critique of capitalist/noncapitalist relations in resource-rich colonies, "accumulation by dispossession" (Harvey 2003). The most important reflection of the heightened intensity of this process since the early 2000s is that citizen groups, journalists, and researchers are beginning to recognize how important it is to halt the illegitimate offshoring and intergenerational transfers of wealth from these resources.

Perhaps the most urgent and profound intergenerational message for future advocacy narratives is that climate breakdown requires fossil fuels to be left underground—and with a compensation arrangement for governments and peoples in poor countries, as was attempted in the Yasuní National Park of Ecuador from 2007 to 2013, albeit unsuccessfully (Leave Fossil Fuels Underground 2018). We do not need to dwell upon this obvious conclusion, aside from remarking that the "Blockadia" concept—i.e., disrupting fossil fuel extraction, refining, and transport—is well underway (EJAtlas 2019; Klein 2014;). Of course, the extraction of certain classes of hydrocarbons can certainly be justified even in a climate-conscious society, assuming this could be accomplished without contributing greenhouse gases—i.e., the fossil fuels would not be combusted for wasteful energy or transport purposes—for the sake of the myriad ways they are used in the production of necessary goods (pharmaceutical products, synthetic materials, and even some vital forms of plastic). Indeed, we anticipate the phrase "minimally necessary extraction" to become increasingly important, especially as the "degrowth" philosophy becomes more popular in coming years (Bond 2019; Kallis and March 2015).

As for the value transfer problem, the rise of citizen awareness is occurring in part through campaigns against Illicit Financial Flows associated with multinational corporate tax dodges, especially in the resource-extractive industries. The international Publish What You Pay NGO network feeds into the Extractive Industries Transparency Initiative—innocuous enough to be endorsed by many states and extractive-sector corporations—and "Stop the Bleeding" campaign in Africa. However, such transparency and anti-corruption narratives alone have not proven to be a solution, since they explicitly ignore the depletion dilemma.

The same problem exists in most community-based struggles against the local pollution and social damage that is done through resource extraction. In some such cases, to grapple with resource depletion as a concept would drastically interfere with an NGO reform agenda that simply polishes the rough edges of mining so that it can continue with more legitimacy. This latter process is evident in the annual Cape Town Alternative Mining Indaba, which became so stultifying (Maguwu and Terreblanche 2016) that an alternative Thematic Social Forum on Resisting Mining and Extractivism—stressing "the right to say no!"—was founded in 2018, in Johannesburg. This was not merely theoretical: in 2015, a year in which $80 billion in new mining was undertaken, those advocating for leaving minerals underground managed to block a vast share of mining. Anglo American Corporation Chief Executive Mark Cutifani estimated, "There's something like $25 billion worth of projects tied up or stopped" (Kayakiran and Janse van Vuuren 2015).

In other cases, there is also a growing awareness about the dangers of *licit*—albeit immoral—resource value transfers, in which formal contracts legalize the (uncompensated) extraction of wealth across time and space. This was the conclusion that even a Financial Times Writer, Tom Burgis, arrived at: "keep resources in the country and implement high tariffs to protect domestic industries" (quoted in Monks 2018). Others have called for sovereign wealth funds to continue extraction but slow the geographical flight of wealth by ensuring community trusts or national funds are replenished en route.

We argue for a much more depletion-conscious narrative that pays tribute to future generations and to local citizenries who are not well served by the current mode of extractivism. We have learned of creative resistance to extractivism from both activists and intellectuals. The former have driven public policy in all spheres of life, and anti-extractivism is one of the most powerful recent trends. The Goa experience includes advocacy that resulted in a 2019 National Mineral Policy statement: "natural resources, including minerals, are a shared inheritance where the state is the trustee on behalf of the people to ensure that future generations receive the benefit of inheritance. State Governments will endeavour to ensure that the full value of the extracted minerals is received by the State" (Government of India 2019, 11–12).

While few activists have publicly grappled with the enormity of the continental-scale resource depletion problem—i.e., the conservatively estimated $100 billion drawn down from sub-Saharan Africa's natural wealth (not including the platinum and diamond sectors)—at least at the local level, their anti-extractivism is enhanced by showing how economic degeneration results from the net decline in natural capital.

To illustrate, in eastern Zimbabwe's notorious Marange diamond fields, bottom-up resistance has generated not only courageous protests but also arguments against extraction that deploy natural capital accounting. According to Marange's main civil society watchdog, Farai Maguwu (2016, 5) of the Center for Natural Resource Governance, "mining is a disaster unfolding across Zimbabwe. Mining is creating an enclave economy full of white-collar criminals, who make virtually no positive linkages to the broader Zimbabwean economy. They simply deplete our natural capital and provide an inconsequential return." Likewise, the Zimbabwe Environmental Law Association finds that "Diamond revenue represents natural capital depletion and, therefore, its expenditure should be judicious" (Sibanda and Makore 2013, 29). In 2016, then President Robert Mugabe admitted an exceptional level of injudicious extraction: "We have not received much from the diamond industry at all. I don't think we have exceeded $2 billion, yet we think more than $15 billion has been earned" (Magaisa 2016).

Conclusion

A *combination* of the geographical and generational justice agendas will be required to make headway against excessive resource depletion. For those engaged in critical geographical scholarship who have been skeptical of natural capital accounting (for very good reasons reviewed previously), our appeal is to consider whether that opposition risks becoming politically disempowering: first, it allows ecological-debt denialists to avoid facing up to the scale of damage done by an extractivism that favors the Global North and current generations and, second, this technique as we advocate it can usefully add pressure against mindless extractivism, whether as national economic policy or to augment campaigning by grassroots activists (such as we believe occurs fruitfully in Goa, South Durban, and eastern Zimbabwe as just three example sites).

To move the agenda back to a general ideological level, away from the environmental economists and other ecological modernizationists, the market techniques of which will continue to distract attention from justice, it may be that an *ecofeminist* approach will be required to compel civil society to take up these matters properly. In Africa, the leading network that advocates further praxis-oriented research along these lines is the Johannesburg-based African Women Unite Against Destructive Resource Extraction, or WoMin. In coming years, WoMin proposes a full-cost accounting sensitive to ecofeminist principles, one that extends from a "social reproduction" (Mitchell, Marston, and Katz 2012) analysis of how male labor gets to the point of production within the extractive circuits (see also Fent, Chapter 9 this volume). The analysis incorporates (gendered) small-scale agricultural production systems on the land that are increasingly threatened by extractivism (e.g., land grabs and air-water-land pollution), to women's burdens during climate breakdowns due to fossil-fuel extractivism, to the *de facto* responsibility that women are given to steward life itself (including ecological inheritances) into future generations:

> WoMin will deepen its efforts to foreground a feminist analysis of costs, showing that this places particular burdens on the cheap and unpaid labor of impacted women. We will grapple further with the problematic of costing damage and impacts, immediately and on a cumulative basis, to show that an extractivist model of development does not advance people and their economies, but rather destroys and immiserates them. We will show the inter-generational costs of extractivism and we will work to argue that Africa and African nations are losing sovereign wealth through extractivism and only becoming poorer. These efforts lay the basis for advocacy and campaigns to build wider popular and public consciousness, build the grounds for advocacy on development alternatives, as well as advocate and campaign to force the internalization of real costs, which would render the majority of projects unsustainable.
>
> (WoMin 2019, 9)

If many of the world's mining and fossil-fuel projects are indeed determined to be unsustainable—once a full accounting of costs is accomplished, with consciousness about who wins and loses, across time and space—then ultimately this is the basis for moving from the field of ideas to a more confident rejection of *status quo* extractivism by activists and perhaps even conscientized policy makers. Only then can an end be brought to what is now the uncompensated and often unnecessary extraction of nonrenewable minerals, oil, and gas. Only then can ecological reparations for the debilitating injustices of resource extraction be justly demanded. In sum, the times and the spaces in which we engage in *radical* forms of natural capital accounting are, potentially, just scratching the surface of a more general strategy for environmental, sociocultural, political-economic, geographical, and intergenerational justice.

References

Azur, Christian and John Holmberg. 1995. "Defining the Generational Environmental Debt." *Ecological Economics*, 14: 7–19.

Basu, Rahul. 2015. "Catastrophic Failure of Public Trust in Mining: Case Study of Goa." *Economic & Political Weekly*, L (38): 44–51.

Basu, Rahul. 2017. "What is the Future We Need?", *Medium*, 14 July. https://medium.com/@thefutureweneed/what-is-the-future-we-need-8ae3de8d55a3.

Basu, Rahul, and Scott Pegg. 2020. "Minerals Are a Shared Inheritance." *The Extractive Industries and Society*. doi: 10.1016/j.exis.2020.08.001.

Berkes, Fikret, and Carl Folke. 1992. "A Systems Perspective on the Interrelations Between Natural, Human-Made and Cultural Capital." *Ecological Economics* 5 (1): 1–8. doi: 10.1016/0921-8009(92)90017-M.

Bond, Patrick. 2012. *Politics of Climate Justice: Paralysis Above, Movement Below*. Pietermaritzburg: University of KwaZulu-Natal Press.

Bond, Patrick. 2016. "Who Wins From 'Climate Apartheid'?: African Climate Justice Narratives about the Paris COP21." *New Politics* 15 (4): 122–129.

Bond, Patrick. 2018. "Ecological-Economic Narratives for Resisting Extractive Industries in Africa." *Research in Political Economy* 33: 73–110.

Bond, Patrick. 2019. "Degrowth, Devaluation, and Uneven Development from North to South." In *Towards a Political Economy of Degrowth*, edited by Ekaterina Chertkovskaya, Alexander Paulsson, and Stefania Barca, 137–156. London: Rowman and Littlefield.

Bond, Patrick, and Ana Garcia, eds. 2015. *BRICS*. London: Pluto Press.

Bresnihan, Patrick. 2017. "Valuing Nature—Perspectives and Issues" Research Series, Paper No. 11. Dublin: National Economic and Social Council Research Series. https://www.eesc.europa.eu/ceslink/sites/default/files/document-file-uploads/research_series_paper_11_pbresnihan_valuingnature.pdf.

Burkett, Paul. 2006. *Marxism and Ecological Economics*. Amsterdam: Brill.

Committee for Mineral Reserves International Reporting Standards. 2020. Home. Accessed September 3, 2020. http://crirsco.com/welcome.asp.

Costanza, Robert, and Herman Daly. 1992. "Natural Capital and Sustainable Development." *Conservation Biology* 6 (1): 37–46. doi: 10.1046/j.1523-1739.1992.610037.x.

Daly, Herman. 1996. *Beyond Growth: The Economics of Sustainable Development*. Boston, MA: Beacon Press.

EJAtlas. 2019. "Mapping Environmental Justice." Accessed September 3, 2020. https://ejatlas.org/.

Esping-Andersen, Gosta. 1990. *The Three Worlds of Welfare Capitalism*. Princeton, NJ: Princeton University Press.

Government of India. 2019. "National Mineral Policy." Delhi: Ministry of Mines. https://mines.gov.in/writereaddata/Content/NMP12032019.pdf.

Gowdy, John, and Carl McDaniel. 1999. "The Physical Destruction of Nauru: An Example of Weak Sustainability." *Land Economics* 75 (2): 333–338.

Hartwick, John. 1977. "Intergenerational Equity and the Investing of Rents from Exhaustible Resources." *American Economic Review* 67 (5): 972–974.

Harvey, David. 2003. *The New Imperialism*. Oxford: Oxford University Press.

Hickel, Jason. 2018. "The Nobel Prize for Climate Catastrophe." *Foreign Policy*, 6 December. https://foreignpolicy.com/2018/12/06/the-nobel-prize-for-climate-catastrophe/

Humphreys, Macartan, Jeffrey Sachs, and Joseph Stiglitz, eds. 2007. *Escaping the Resource Curse*. New York, NY: Columbia University Press.

Kayakiran, Firat, and Andre Janse van Vuuren. 2015. "Miners Offer Bull Rings, Clinics as Protests Ice $25 Billion." *Bloomberg News*, 17 March. Accessed September 3, 2020. https://www.bloomberg.com/news/articles/2015-03-17/miners-offer-clinics-bull-rings-as-protests-tie-up-25-billion.

Kallis, Giorgos, and Hull March. 2015. "Imaginaries of Hope: The Utopianism of Degrowth." *Annals of the Association of American Geographers* 105 (2): 360–368. doi: 10.1080/00045608.2014.973803.

Klein, Naomi. 2014. *This Changes Everything: Capitalism vs the Climate*. New York, NY: Simon & Schuster.

Lange, Glenn-Marie, Quentin Wodon, and Kevin Carey, eds. 2018. *The Changing Wealth of Nations 2018: Building a Sustainable Future*. Washington, DC: World Bank. doi: 10.1596/978-1-4648-1046-6.

Leave Fossil Fuels Underground. 2018. *Yasuni Forum*. Quito, Ecuador, 28 August. https://vimeo.com/287927514.

Magaisa, Alex T. 2016. "Mugabe and the $15 Billion Question." *The Standard*. 14 March. Re https://www.thestandard.co.zw/2016/03/14/mugabe-and-the-15-billion-question/.

Maguwu, Farai. 2016. "An Open Letter to President Robert Mugabe." Harare: Centre for Natural Resource Governance. https://impacttransform.org/wp-content/uploads/2017/09/Annex_3_-_An_Open_Letter_to_President_Robert_Mugabe.pdf.

Maguwu, Farai, and Christelle Terreblanche. 2016. "We Need a Real 'Alternatives to Mining' Indaba." *Pambazuka News*, 12 February. https://www.pambazuka.org/global-south/we-need-real-%E2%80%9Calternatives-mining%E2%80%9D-indaba.

Martinez-Alier, Joan. 2002. *The Environmentalism of the Poor: A Study of Ecological Conflicts and Valuation*. Cheltenham: Edward Elgar.

Mayer, Andreas and Willie Haas. 2016. "Cumulative Material Flows Provide Indicators to Quantify the Ecological Debt." *Journal of Political Ecology*, 13:350-363.

Mitchell, Katharyne, Sallie Marston, and Cindi Katz, eds. 2012. *Life's Work: Geographies of Social Reproduction*. London: John Wiley.

Monks, Kieron. 2018. "Why the Wealth of Africa Does Not Make Africans Wealthy." *CNN*, 2 January. https://edition.cnn.com/2016/04/18/africa/looting-machine-tom-burgis-africa/index.html.

Obi, Cyril. 2010. "Oil as the 'Curse' of Conflict in Africa." *Review of African Political Economy* 37 (126): 483–495.

Robinson, James, Ragnar Torvik, and Thierry Verdier. 2006. "Political Foundations of the Resource Curse." *Journal of Development Economics* 79 (2): 447–468.

Sharife, Khadija, and Patrick Bond. 2013. "Payment for Ecosystem Services versus Ecological Reparations." *South African Journal on Human Rights* 29 (1): 144–169. doi: 10.1080/19962126.2013.11865069.

Sibanda, Mukasiri, and Gilbert Makore. 2013. *Tracking the Trends: An Assessment of Diamond Mining Sector Tax Contributions to Treasury with Particular Reference to Marange Diamond Fields*. Harare: Zimbabwe Environmental Lawyers Association.

Solow, Robert. 1974. "The Economics of Resources or the Resources of Economics." *The American Economic Review* 64 (2): 1–14.

Solow, Robert. 1993. "Sustainability: An Economist's Perspective." In *Economics of the Environment: Selected Readings*, edited by Robert N. Stavins, 179–187. New York, NY: Norton.

South Durban Community Environmental Alliance. 2019. "Appeal in Terms of the Section 43 (2) of the National Environmental Management Act, 106 Of 1998 Against the Environmental Authorisation Granted to Eni South Africa Bv & Sasol Africa Limited." Durban, 9 October.

Sullivan, Sian. 2014. "The Natural Capital Myth; or Will Accounting Save the World." Leverhulme Centre for the Study of Value Working Paper Series No. 3. Manchester: University of Manchester Leverhulme Centre for the Study of Value.

Sullivan, Sian. 2017. "On 'Natural Capital', 'Fairy Tales,' and Ideology." *Development and Change* 48 (2): 397–423.

Supreme Court of India. 2014. *Goa Foundation vs UOI & Ors*. WP(civil) 435 of 2012. 6 SCC 590.

Supreme Court of Pennsylvania. 2017. "Pennsylvania Environmental Defense Foundation v. Commonwealth." 20 June 2017. http://www.pedf.org/uploads/1/9/0/7/19078501/supreme_court_opinion_062017.pdf.

The Economics of Ecosystems and Biodiversity (TEEB). 2020. Home. Accessed on September 3, 2020. http://www.teebweb.org/.

van der Merwe, Justin, Patrick Bond, and Nicole Dodd, eds. 2019. *BRICS and Resistance in Africa*. London: Zed Books.

WoMin. 2019. "WoMin Five Year Strategy (2020–2024)." Johannesburg.

World Bank. 2011. *The Changing Wealth of Nations*. Washington, DC: World Bank.

World Bank. 2020. "GNI, Atlas Method (Current US$)—Sub-Saharan Africa." World Bank National Accounts Data, and OECD National Accounts Data File. https://data.worldbank.org/indicator/NY.GNP.ATLS.CD?end=2019&locations=ZG&start=1962&view=chart.

World Commission on Environment and Development. 1987. *Our Common Future*. New York. https://sustainabledevelopment.un.org/content/documents/5987our-common-future.pdf.

23

Research as action and performance

Learning with activists in resource conflicts

Christopher Courtheyn and Ahsan Kamal

Introduction

How do conflicts around nature and resources influence our attempts to learn about them?[1] We wrestle with this question by engaging in a dialogue guided by our ethical commitments and experiences working with activists and communities who are fighting to defend their socio-natural worlds in two countries of the Global South. We view social movements as rich sites of knowledge production where diverse epistemological approaches unveil different underlying realities that constitute "resource geographies" through embodied spatial practices. Instead of considering activists as objects of investigation, we look at the tensions between academic and activist spaces, attend to the performance of research, and consider how navigating resource conflicts potentiate our attempts to learn about them (on shared vulnerabilities and ethnographic sensibilities, see also Berman-Arévalo, Chapter 17 in this volume).

This chapter is a methodological intervention in crossing the border between the Academy and activist spaces. Part of a broader project on academic and activist exchanges across the Global South and the souths in the Global North, our conversations are based on collaborative reflections as we attempt to cross such borders. The chapter is presented as a conversation, like Alexander Craft et al. (2007), to performatively interrupt usual forms of scholarly exchanges.

We begin the performance by stating our intentions and commitments to certain struggles and describing our research methodologies via two characters, Ahsan and Chris, who loosely represent the authors. The characters reflect on a set of questions in a dialogic fashion: What is the benefit of viewing research as performances and conflicts as performative spaces? What are the design implications of our approach, particularly in relation to trust, access, and reciprocity in violence-filled spaces? What are the challenges of navigating movement space and time, especially the tensions between academia and activism with respect to the myth of the individual researcher, spatial and temporal frictions, and the messiness of ground realities?

The performance of research

Chris

How about we begin by introducing readers to our respective engagements researching resource conflicts with social movements?

Ahsan

I had been working on issues of development, disasters, and conflicts since 2005 but shifted to social movements in 2010 when I started teaching at a public university in Islamabad. This was a period when Pakistan saw an alarming rise in violence, distressing even for a postcolonial state with a long history of deploying violence against its people. Despite the increase in state and non-state militancy targeting activists and marginalized communities, a number of them continue to engage in nonviolent movement building against evictions, dispossession, discrimination, and the ontocide[2] of marginalized groups. Most of my work with activists resides on the borders of academia and activism: In a global south university, in activist-academic collectives, and amidst social movements and communities in resistance.

My dissertation focused on the death and defense of the Sindhu *darya*, which is what Adivasi and riverine activists call the Indus River in Pakistan. I studied three social movement organizations in the northern, central, and southern Indus valley that emerged at the turn of the twenty-first century. This is a time when Southern movements against dams and dispossession were on the rise. I was interested in learning how activists take Southern ideas from other places and incorporate these into existing struggles. I am fascinated by how activists receive, transform, and travel with these ideas to develop new praxes in conversation with local cultures, histories, and spiritual traditions. I classify my methodology as *action research* that draws from defiant mystical traditions, radical politics, and decolonial praxis. The goal is to study schemes of resistance and dreams of re-existence that are inaccessible through the utopian drives of capital and coloniality.

Chris

Since 2008, I have worked in solidarity with and researched the Peace Community of San José de Apartadó. Located in the war-torn Urabá region in northwestern Colombia, this is a peasant community who refuses to support any armed group. The latter include state, paramilitary, and guerrilla forces competing to control Urabá's population and resources, including fertile land, water, and mineral reserves. The region's farmers have suffered hundreds of assassinations and repeated evictions at gunpoint from their lands, including forced displacement to pave the way for the Urrá hydroelectric project. The Peace Community rejects extractivist development and opposes state-corporate plans to enlarge the Urrá dam and build another. Meanwhile, in contrast to an extractivist "peace" advanced by the capitalist state, they forge a grassroots form of peace as ecological dignity through communal farming toward food sovereignty and solidarity networks that support communities' autonomy.

My first engagement in San José was as an international protective accompanier with the Fellowship of Reconciliation (FOR) Peace Presence, a US-based peace organization dedicated to nonviolent resistance to war and documenting human rights violations associated with US military aid in Latin America. Threatened activists in certain parts of the Global South invite such accompaniers, who are primarily from the Global North, to walk alongside them as "unarmed bodyguards" and report human rights abuses (Koopman 2011). This experience inspired me to write a book manuscript on San José's peace praxis amidst resource conflict and war. Entering the Academy from this activist experience, I was interested in exploring how local processes interact with structural conditions enacted by global capitalist and colonial systems, and how researcher and activist positionalities influence these spaces of conflict. I came to articulate my research methodology as *radical performance geography*, which combines critical spatial theory with performance ethnography to attend to the performative aspects of geographic research and activism.

Ahsan

There's a convergence in our approaches with the emphasis on self-reflexivity and positionality. Both action and performance research may be recognizable in other guises, such as engaged scholarship and militant research. For me, this approach grew out of the renunciations and the rebellious lyrics of Sufi mystics against powerful actors—what we now may call decolonial praxis. A modern genealogical strand stretches from Marx's dictum of knowing the world to change it, to postcolonial reflections on subalternity, participation, and representation. This sets the agenda for critical research that requires, as Orlando Fals Borda (2006, 353) noted, a "connection with grand theory and the pursuit of alternative paradigms." In my view, these alternative paradigms require introspection, engagement, and possibly a co-laboring between the active researcher and their interlocuters. Learning about resource conflicts is a reflexive activity laced with our ethical commitments.

Our understanding as scholars is heightened not only simply by observing, interpreting, and reporting, but also through embodied performances of research and action. In this, our dual responsibility is to know-act and act-know: Knowing through and for action as well as acting with and for knowledge (see also Bryan, Chapter 37 in this volume).

Chris

Radical performance geography is also rooted in collaboration and deep dialogue between researchers and interlocutors. This requires a profound reflexivity about one's academic and political positionality, including the effects of one's presence in the field and publications. It draws from the repertoire of performance ethnography, which Soyini Madison (2012, 16) deems "critical theory in action." Challenging the academic monopoly on knowledge, it foregrounds the voices of social movement interlocutors through evocative and poetic transcriptions of ethnographic data. It uses staged theater and role-play performances as embodied complements to text, which are designed for participants and witnesses to *feel* the stakes of toxicity and displacement toward reflections about political strategy and theory. Further, when I join a social movement mobilization as a researcher, I perform solidarity with the cause, because armed groups view me as an international witness who reports human rights violations. Therefore, armed actors can be less likely to attack. Fusing analysis and action is all the more important in the study of conflicts around nature.

These conflicts are ontological and epistemological, and social movements seek to protect and change their worlds. The emphasis on performativity highlights how geographers do not simply analyze and write about space, especially in conflicts around nature. Working in or with social movements, we also perform and produce space.

Ahsan

Your discussion of the production of space through the performance of research really drives home the point that resource conflicts are "active" zones where different actors—activists, communities, state, corporations, and scholars—all analyze the situation based on their political commitments and ontological assumptions. So, while all research has performative elements—we are situated in fields of knowledge production that gain validity through institutionally sanctioned performances in classrooms, seminars, publications, and so on—in research on resource conflicts, two types of representations overlap. Claims of representation of realities or analyses become inseparable from the attempts to represent human and nonhuman entities, i.e., what we call politics and some refer to as political ontologies (De la Cadena 2015). An analysis is then also a performance of a political position, which seeks to conceptualize certain beings and relations as "resources," "nature," "environment," "ecology," or so on, thus fixing fluidities of reality into concrete socio-material objects of contestation. In describing the different realities, we are inevitably and simultaneously *speaking for* someone and from certain positions.

Chris

Yes, and there is a tension between *speaking for* versus *serving as a conduit* through which a community can spread their knowledge, because knowledge always transforms as it moves. We work with this tension because there is no guarantee of resolution. This is precisely why attentiveness to the performativity of our writings is crucial, i.e., whose voices are highlighted and how the power of prose and poetry can reveal violence and resistances in embodied and politically engaging ways. Active attention to and engagement with the performed realities of movements allow us to illuminate ideas springing from beings and relations of nature that some methodologies are unable to decipher.

Ahsan

Performative elements are evident in all aspects of our work, from walking amidst activists to writing up our results. I would say that many academics *perform* action research even if they eschew the label and its attendant responsibility, i.e., a shying away from acknowledging the double bind of representation. Perhaps this is a symptom of a too-narrow definition of activism as direct action, even when we know activism includes care work and other labors—even encompassing theorizing—as long as this aligns with our ethical commitments to justice. Such commitments are visible in classrooms where teachers provide students the tools to learn the facts and the discourses of resource conflicts. Nevertheless, by calling for an expanded definition, we are also pointing to the need for certain additional considerations by our academic colleagues.

Trust, access, and reciprocity

Chris

Let us talk about some of these additional concerns, which stem from our emphasis on action and performance as our research unfolds among activists and communities in resistance. Our embedded methodologies raise certain pragmatic questions about establishing trust and gaining access in violence-filled spaces, which are indicative of conflicts around nature. Activists and communities are often pitted against those with more power, such as the state, militaries, corporations, and dominant or hegemonic elites. Gaining access and establishing trust is a significant consideration in situations where people are fighting for their lives and their lifeworlds. And this is not only an issue for those researching a community different from their own.

The question of trust and access also applies when we are part of a community or a movement, where we have to negotiate access amidst internal power asymmetries (Smith 1999). For instance, researchers may be asked or expected to exercise discretion regarding what information to share.

Ahsan

In a way, what you are pointing out goes beyond standard research design practices. As academics, we often need activists and community members to help us make sense of their worlds. Institutionalized research provides access through the credibility established by institutional affiliations. Good research practices entail rigorous design that wrestles with issues of potential harm, consent, confidentiality, and anonymity of research subjects. But activists and communities often do not trust the Academy because of the troubled and ongoing history of imperial researchers who mine data, build careers, and support colonial orders of rule. So, there is a need for building deeper trust with activists to gain access to things that are not, say, put out in press releases that give a limited view of the movements' aspirations and dynamics.

I have experienced this working with non-violent Baloch activists from conflict zones in Pakistan, where certain Baloch groups have taken up arms against the state, in defense of their land against state occupation and rampant exploitation of natural resources. Stuck between the military, courts, politicians, and separatists, the non-violent Baloch activists have to navigate a dangerous line in their public performance. We once organized a hunger-strike camp in the federal capital where dam affected communities and activists traveled for days through conflict zones to give their testimonies in a people's tribunal. A student, feeling distraught after hearing their testimonies, stood up to say that given all their suffering, it would made sense for them to take up arms against the state. This incident stimulated a conversation that evening, around navigating public performances and drafting press releases. There were real risks of the non-violent activists being labeled as traitors or terrorists and of being "disappeared," just because of their ethnicity.

Our access to such conversations tells us about important dimensions of the conflicts: the complexities of reality, the strategic articulation of public discourses, and the anxieties these situations generate. But to access the inner workings, these in-between moments that help us learn, we have to establish deep trust and recognize our responsibilities. Sometimes these are obvious, sometimes we fumble and make mistakes, but at no time can we afford to ignore this responsibility. Negotiating access requires frank conversations beyond reading approved interview scripts. It may even require setting aside certain aspects of our research agenda and refusing to disclose certain information.

Chris

That is to say, activists in resource conflicts are vulnerable and may trust us to understand the complexities and respond accordingly. In a way, we are always responding to how social movement actors see us fitting into their world. In the same vein, activists may make certain asks of the researcher, utilizing the latter's skills to help with translations, media outreach, legal petitions, fundraising, or, in my case, to accompany them as an "unarmed bodyguard" in their travels along a river between villages. An activist asking a researcher to do something is a relational and performative act, to ask at the same time, "What is your commitment to what we are doing?" I have also experienced a similar dynamic when community leaders ask my opinion about a particular issue, which tests my political stance as well as commitment to the movement. Certain Colombian activists oftentimes measure intellectual sophistication by gauging outsiders' ability to critique government-led "peace processes" deemed to be more akin to war by other means. In highly politicized ecological contexts, researchers inevitably situate themselves or are situated alongside or against certain actors, and activists are keen to ascertain one's position.

Ahsan

Indeed, these lines are drawn between powerful actors and activists as well as within communities. It is common in resource conflicts to have local "beneficiaries" of development interventions, such as local landowners befitting from irrigation canals, along with with people who are dispossessed, like the landless or traditional irrigators. The different groups may have conflict of interests, and might not get along. While activists often work hard to build solidarity among diverse actors, researchers may nevertheless be confronted with many gatekeepers of different factions. Activists are cognizant of who we as researchers interact with, and we need to be honest with them and ourselves about our political positions and commitments.

Let us also clarify that we've been speaking about positionality in the very sense of our political commitments. Other dimensions of our performance—our race, class, gender, ethnicity, and nationality—influence our ability to gain access and establish trust in all research situations. For instance, in Pakistan, my ability to access these spaces is tied to my gender performance that gives me privileges of mobility but limits a deeper understanding of rural women activists, for instance. Our research is impacted by all aspects of our positionality.

Chris

You are rightly pointing out that our emphasis is on ethical commitments and political positionality, and that research is indeed always gendered. The same goes for citizenship and race. In Colombia, government officials, including army officers, happily share their perspective with researchers from the Global North like me as part of their international campaign to posit themselves as human rights victims rather than victimizers. Therefore, my position as a white US citizen facilitates access to military brigades. However, when conducting research with social movements, close proximity with dominant actors can arouse suspicion among activists, potentially compromising one's engagement with them. While the Peace Community is keen to use me as a witness due to my racial and national positionalities, my ability to continue to walk alongside and critically understand their peace praxis of ecological dignity is made possible by my political and intellectual solidarity with their project. Our awareness of such performative dimensions is important if we are to negotiate access, which is an ongoing process where activists evaluate our trustworthiness and utility.

Ahsan

To the point about our usefulness, most research on social movements does little for the movements directly, so how do we participate? I found helpful Gillan and Pickerill's (2012) distinction between *general reciprocity*—the cumulative contribution of individual research for our knowledge of things—and *immediate reciprocity* where researchers offer something directly to the movements or communities that they study. Immediate reciprocity is not always possible, but researchers may participate by responding to the asks and needs of activists, as you observed. For instance, researchers may play the role of a translator of sorts when meeting a team of development experts from the World Bank or helping to counter silencing attempts by translating technocratic language. Action researchers in Pakistan often comb through planning and design documents and find ways to convey complex ideas of hydrology, mining, engineering, and legalese to community members and local activists, so that they are better prepared to fight back against deliberate obfuscation in "expert" discourse.

Chris

In my case, "participatory research" takes a particular form. I am simultaneously an "insider" within the US anti-militarism movement in FOR and an "outsider" in San José de Apartadó. In FOR accompanier trainings, my academic theoretical grounding allowed me to lead workshops on Colombian political history and the complicated ethics of international solidarity; my work in the university enriches what I contribute internally as part of the movement and vice versa. Meanwhile, in San José, rather than a participatory action research approach (Fals Borda 2006), the Peace Community has never been interested in co-designing my academic project. My interviews and publications address concepts of peace, territory, and race; these questions arise through my time in the field but were formally developed in the university. Rather than a collectively designed study, Peace Community leaders are most interested in the protective value of my international presence. Joining them on a pilgrimage to commemorate a past massacre or in a caravan across villages to confront army/paramilitary soldiers is used to increase visibility and safety for their community struggle.

If we recognize ourselves to be in a global struggle, we also have to acknowledge how people are differentially situated amid the current socioecological crisis and can thus play strategic roles. For instance, as a radical performance geographer, when I conducted interviews with army and police officers, I was interested in their perspectives on peace and extractivism. But the Peace Community and I also understood the interviews of these officers—many of whom have received training from US military aid packages—as a performance of international solidarity, as communicating to the commanders that I was present in the region and spreading my analysis to international academic and human rights audiences, as a deterrence to armed attack in San José. Social movements understand and critique the differential privilege of particular bodies tied to colonialism, but they also mobilize such asymmetries in their favor.

Navigating movement space and time

Ahsan

Indeed, spatially mapped asymmetries are visible if we think through questions of positionality. In my case, navigating conflicts was a bit different because my ethnicity made me a party to the conflict and local activists were keen on keeping things on the down-low. Our bodies send different signals in different situations and how we relate to these influences what we can learn. I am reminded of two incidents regarding activist-state interactions. The first one, an all too serious matter, was when I accompanied community activists to meet a top-level bureaucrat in his office, where he performed state authority by alternating between rehearsed benevolence and outright aggression. Knowing that I was a researcher, he was also eager to recruit me as an ally, ignoring community activists and addressing me directly to talk about project financing decisions. When I deferred to the community activists, he shifted gears and started to question my credentials, asked me about ethnicity, and threatened to get me fired from my job at a public university. The second encounter was on the opposite end of the spectrum. We were at a people's tribunal where state officials were completely out of their element, sitting on the ground, uncomfortably shifting about, and being mocked in theater performances. I was there as a witness, one among many academics and journalists. In these two interactions, we see an emotional unraveling of the state officials, humanized momentarily as they performed benevolence, discomfort, aggression, and embarrassment. My understanding of the different roles that I was asked or assumed to play influenced what I was able to learn about the performative enactments of state and peoples' power. I can think of several other examples from Pakistan where solidarity work resulted in violence and threats. The lines become gray and things get messy. These are difficult terrains for both activism and scholarship to navigate—a burden that an isolated researcher may not be able to carry.

Chris

Indeed, navigating the complexities and asymmetries of the field is not an easy task and points to the fact that we never act in isolation. This speaks to the myth of the "independent" academic researcher on a journey of discovering truth. There is a certain egocentric construction of this heroic figure with all its corresponding downsides. I recall periodic feelings of inadequacy, feeling that I could/should go beyond the role of physical presence and documentation by contributing as a mapmaker or producing environmental impact studies. However, for the most part, social movements already have a variety of people involved in doing these tasks. For instance, when the Peace Community makes legal claims or denounces human rights violations, they work with Colombian lawyers, not ethnographers. The idea of personally "doing more" is a research-driven idea, because we are trained to be in the field and do "our" project. It is important to acknowledge that we are trained to work in isolation, although we rarely do so. Action research works to build a collective that brings people with diverse skills together.

Ahsan

Yes, and there is another tension built into academic research: its temporality. I remember a conversation with a community organizer where they expressed dismay at the inability of academics to align with movement time. Social movement temporalities range from the slow-moving long durations of activity and inactivity to a sudden surge on short notice. Academic researchers work with time frames established by grant, publication, and semester cycles. Movement time does not fit neatly with this academic time. Inevitably, academics end up privileging academic time. This is not a problem in and of itself, but once again, we can be more open about our capacities and discuss issues of reciprocity with activists. I have also been in situations of wanting to do more, even promising to do more and not delivering. Viewing these as temporal conflicts and limits rather than individual inadequacy is a good starting point.

Chris

We can certainly tap into the creative potential that emerges out of our unique experiences and skills. But there is something to be said about sustained engagements following movement time and to be in an ongoing dialogue with other researchers and activists to determine what is to be done and when. Even what we do in the short term is driven by ongoing dialogues. Ultimately, given the Peace Community's lack of faith in any state justice given Colombia's structural impunity, they utilize witnesses like me to document their peace praxis and thus contribute to what they deem will be the ultimate justice of history (Lanchero 2002). This is due to their emphasis on creating an autonomous space, rather than a politics of reforming the state. What participation/action means is dependent upon what the particular movement's goals are.

Ahsan

In many areas in Pakistan, activists strive for autonomy and building alternative institutions, like the Peace Community. But they also work with state-centric forums to bring communities together. We know that resource conflicts often alter social-spatial formations and destroy existing forums of solidarity. However, things on the ground do not map onto the dichotomies, often grounded in the literature, of movements "who engage with the state" and those "who seek autonomy." When we engage with the complexity of movements—and their different temporalities that represent contrasting aspirations—we are setting aside dualist modes of thinking. Rather than merely studying the everyday, investigating the big strategic goals of campaigns, or measuring the "success" or "failure" of building alternatives or reforming institutions, we can engage more deeply with these different strategies.

Chris

I think you are highlighting the supposed tensions between "theory" and "reality" in all fieldwork-oriented research but more so on social movements. Sometimes we expect our fieldwork to neatly confirm or refute macro-level theories. Yet, realities on the ground are typically messy. Nevertheless, encountering contradictions—even within the movements themselves—should not lead us to discount either theory or activists' attempts at ecological justice. Our grand theories ideally provide critical analyses of reality while also empowering political thought of what could be.

Ahsan

In a way to confront "theory" with "reality," we try to account for the messiness of asymmetries and the tensions of temporalities in our general approach to research, even if our particular research questions are narrower in scope.

Chris

Precisely. And while it is important to critique how we/social movements reproduce violent systems of power, our focus should be to discern, analyze, and then communicate what is "new" or "radical" in our/their geographical imaginaries and practices. And rather than simply limiting ourselves to local particularities, engagement with movements' critiques of extractivism is an opportunity to complexify and enhance global theories about the political economy of resources. Attending to spatial performances, temporal misalignments, and the messiness of ground realities that we encounter is necessary for us to navigate the field and to come out with a better understanding of resource geographies.

Conclusion

In an earlier version of this chapter, we tried to conclude the conversation with a joke: "An activist, a scholar, and an editor walk into a bar. The activist starts drafting a pamphlet for the protest that weekend, while the scholar observes them and takes notes in her field journal. She asks the editor to review what she wrote, and the editor takes a quick look and responds: 'But you are over the world limit?'" Upon reading this, our wonderful and kind editors of this handbook asked: "Don't you mean *word* limit?"

We are grateful to the editors who helped in this performance, guiding and nudging us, helping us bring clarity and coherence, ignoring our bland humor, and gently pointing out our mistakes. But what wonderful mistakes we make.

Word limits are world limits. Critical approaches to knowledge appear out of mistakes. Or they attain refinement through many missteps, misstatements, and errors that are ideally identified and removed. The ideas of this chapter first emerged as we reflected on the performative presentation of our research in the "conflict zones" of elite US universities, fumbling into the intersections of our anxieties—anxieties of bodies white and brown, male, and male and queer, imperial citizens and imperialized subjects, protected and tortured, intellectual and emotional, voicing and silencing, and acting and performing to the tunes of institutionalized dances. Three hundred years ago, Bullah the mystic wanted to accost his beloved, a representation of divine truth, not through words but by dancing and chanting *aleph*. He sang *aleph*, the first letter of the alphabet that represents the oneness of being. But the Mullah—the religious scholar—confronted Bullah and insisted that the path to knowledge was by learning the entire alphabet: The grammars and concepts and the distinctions. As activist academics, we carry the Bullah-Mullah tension: In our quest for clarity, we are reminded that our conversations are laced with discursive ambiguity, repetitions, and multiple meanings with respect to the key terms that we use: Performance, resources, activists, nature, trust, we, us, love.

Our words limit our worlds and our worlds limit our words, but they also set us into motion. What have we set into motion? We have noted that all research may be deemed as performative. However, there are certain additional considerations in resource conflicts that emerge from the dual-bind of representation, where the tension between technical resource language and alternative cosmovisions of nature intersect in analyses and politics. If we wish to gain access to the messiness of the realities that we seek to represent, we need the help of activists to make sense of their choices and understandings of their worlds. This form of access relies on building deep trust through negotiated reciprocity, where we respond to the asks of various actors (see also Johnson, Chapter 24 in this volume). Shunning both an idealized and romanticized notion of the activist on the one hand, and the insistence on highlighting the contradictions and conflicts within communities in resistance on the other, we advise that scholars develop the skills to respond to the demands and needs of social movement actors. Equal reciprocity may not be possible, but scholars will do well to be willing to work with the temporal, spatial, and ethical demands of movements as much as they do with those of the university. We do not offer a single formula for doing critical resource geography, because working through and amidst tensions without seeking ultimate resolution is central to researching with social movements. The praxis of action and performance research invites us to act with the ability to reflect and think with the intention to act.

At the beginning of this chapter we asked: How do conflicts over nature and resources influence our attempts to learn about them? Through our dialogic approach, we have unveiled some of our anxieties as we navigate the borders between academia and activism. In conclusion, we invite you into motion with a final question: What conflicts and anxieties do you face when learning about nature and resources?

Notes

1 We acknowledge the debt of numerous colleagues and mentors in shaping these ideas. They are too many to mention, but to name a few: Mushtaq Gadi, Mahvish Ahmad, Arturo Escobar, Saydia Gulrukh, Stevie Larson, Yousuf Al-Bulushi, Jonathan Hogstad, Pavithra Vasudevan, Mayra Sofía Moreno, and those who attended the summer 2016 Action Research Convergence in Carrboro, North Carolina.

2 Our appreciation to Arturo Escobar for this term to label the world-killing tendency of colonialist/capitalist modernity.

References

Alexander Craft, Renée, Meida McNeal, Mshaï S. Mwangola, Queen Meccasia, and E. Zabriskie. 2007. "The Quilt: Towards a Twenty-First-Century Black Feminist Ethnography." *Performance Research* 12 (3): 54–83.

De la Cadena, Marisol. 2015. *Earth Beings: Ecologies of Practice Across Andean Worlds*. Durham, NC: Duke University Press.

Fals Borda, Orlando. 2006. "The North-South Convergence: A 30-Year First-Person Assessment of PAR." *Action Research* 4 (3): 351–358.

Gillan, Kevin, and Jenny Pickerill. 2012. "The Difficult and Hopeful Ethics of Research on, and With, Social Movements." *Social Movement Studies* 11 (2): 133–143.

Koopman, Sara. 2011. "Alter-Geopolitics: Other Securities Are Happening." *Geoforum* 42 (3): 274–284.

Lanchero, Eduar. 2002. *El amanecer de las resistencias*. Bogotá: Editorial Códice.

Madison, D. Soyini. 2012. *Critical Ethnography: Method, Ethics, and Performance*. 2nd ed. Los Angeles, CA: Sage Publications.

Smith, Linda Tuhiwai. 1999. *Decolonizing Methodologies: Research and Indigenous Peoples*. New York, NY: Zed Books.

24

Engaged research with smallholders and palm oil firms

Relational and feminist insights from the field

Adrienne Johnson

Introduction

This chapter examines possibilities for an "engaged research" agenda in industrial agriculture scholarship, drawing on my research on the politics of extraction and smallholder-company relations in Ecuador's palm oil industry. For scholars in critical resource geography and cognate fields, engaged research seeks to understand the social and political-economic drivers of pressing environmental issues *and* devise community-based, praxis-oriented interventions and solutions rooted in social and ecological justice (Osborne 2017). I offer two intersecting approaches that can guide scholarly engagement in the agro-industry, and assist critical resource geographers in thinking about, and through, the instabilities of firm-community relations. First, an engaged research agenda that starts from a feminist political economy perspective emphasizes relational connections and interdependencies between actors as shaping economic outcomes (McDowell 2001; Nagar et al. 2002; Marchand and Runyan 2011). This kind of network analysis of indus-trial agriculture (compare to Busch and Juska 1997) destabilizes fixed understandings of agricul-tural firms and smallholders: the firm is not a monolithic, capital-seeking entity but is dynamic and structured by community relations and environmental conditions of production (Campling, Chapter 16 this volume), and smallholders are active agents who strategically work to forge livelihoods amid tense economic relations and ecological uncertainty. Second, feminist insights on researcher positionality and power reject the idea of scientific objectivity and emphasize how researchers' social identity influences the collection of data and the orientation of research conclusions in sites of resource extraction (Billo and Hiemstra 2013; Johnson et al. 2020). These contribute to an engaged research approach by examining researcher bias and researchers' ten-dency to inadvertently simplify complex relationships while denying actor agency, despite best intentions. Together, these insights can assist researchers as they navigate relationships with firms and smallholders to pursue research with transformational potential and meaningful impact.

The chapter proceeds as follows. First, I discuss engaged research and the importance it holds for critical resource geographers. Second, drawing on my own research, I lay out the relations of Ecuador's palm oil industry and the specific concerns between companies and smallholders, examining these in relation to my own commitments to research engagement. Third, I illus-trate an engaged research plan for critical resource geographers focused on the interdependent relations between firms and smallholders by explaining and elaborating relational and feminist perspectives through the lens of my work on Ecuador's palm oil sector. I conclude the chapter with several insights on how to build an engaged research program in critical resource geography going forward.

Engaged research in critical resource geography

Engaged research speaks to critical geography's concern with "the analysis of power, domination and injustice, [which] involve[s] a commitment to progressive social and environmental change through an embrace of critical reflexivity and an engagement with politics" (McGuirk and O'Neill 2012, 1376). Engaged research can emerge from the "bottom-up" where strategies and empowerment are focused on grassroots actors, or from the "top-down," where change is pursued from the policy level (Brewer 2013). In both cases, nonacademic actors are seen as partners throughout the research process where the identification of problems and offered solutions are developed according to varying degrees of collaboration. Overall, the objective of engaged research is to establish public utility for scholarly work (Brewer 2013) by forging meaningful working relationships and collaborations with actors outside of the academy (Whitmer et al. 2010). Scholars doing engaged research seek to make constructive interventions for the purposes of empowering marginalized groups and transforming our world according to ideals of justice and equality.

Critical social scientists working in environmental fields such as political ecology have made explicit efforts to work with marginalized communities in order to produce knowledge and outputs that are useful to their environmental struggles. For example, engaged researchers often participate in radical protest or calls for the end of industry activities in solidarity with the communities they are studying (e.g., Sawyer 2004; Li 2009; Kirsch 2018). Other political ecologists have been key in setting up initiatives such as the Public Political Ecology Lab to collaborate with Indigenous organizations (among other actors) to create tools such as maps that are useful to anti-fossil fuel campaigns (Osborne 2017, 2020).

However, critical resource geographers have not been as quick to foreground engaged research as a feature in their scholarly programs (for exceptions, see Valdivia and Lu 2014; Havice 2020). Perhaps, this is because of the unique fieldwork challenges faced by critical resource geographers. First, academics working in this vein often study resource sectors that involve large firms, where information and personnel can be hard to access (Ghomeshi and Zalik 2013; Billo 2015; Zalik 2020). Second, resource sectors often involve actors whose interests might converge and/or also clash with others. Studying these relationships is difficult as is sustaining long-term working relationships with participants (Oglesby 2010; Johnson 2020). Third, resources are dynamic combinations of material and social dimensions that transform over time and space (Bakker 2004; Bakker and Bridge 2006; Bridge 2009; Sultana 2009). By extension, the locations and identities of those who use and depend on resources also change and this can impact the types of engagements that those both directly involved in a sector and researchers pursue. Fourth, researchers often wield much power and privilege in relation to those in communities with which they work. In contentious sites of resource extraction, researchers witness the life-threatening situations experienced by their interlocutors, but they can always leave (Johnson et al. 2020). In the following section, I address these hard-to-navigate concerns that shape engaged research by referencing research I have conducted in Ecuador's palm oil industry. I use this example to explore and reflect on how critical resource research engagement, grounded in relational and feminist insights, can take root and emerge.

Researching the dialectical relations between firms and smallholders in Ecuador's palm oil industry

Ecuador is a major player in the global palm oil industry and is currently ranked seventh among top-producing countries, with over 240,000 ha dedicated to palm cultivation (Cardona 2019). Ecuador's palm oil companies largely play a contracting and purchasing role in the industry. Within the country, palm fruit is mostly grown by smallholders who make up over 87 percent of the country's total producers and who control a significant portion of the country's 257,120 ha of land that is currently under cultivation (Peñaherrera 2015). Smallholders have plantations that range from 1 to roughly 50 ha in size.[1] Palm oil companies buy unprocessed palm fruit from smallholders at a set price and then process the fruit on-site at one of their refineries and sell the resulting oil either to domestic food companies or export the resource to other countries in Latin America (e.g., Venezuela, Chile) or Europe (Naranjo 2013).

Relations between companies and smallholders can be power-laden and conflictual; they also demonstrate mutual dependence. On the one hand, since companies are the main buyers of the fruit, they ultimately have the "final say" as to which fruit and from whom they will buy. This places immense pressure on farmers to adhere to the different preferred growing practices of specific companies. On the other hand, companies are almost entirely reliant on contracting small producers for production. Smallholders in Ecuador who grow palm have a history of being well organized via the country's smallholder industry association and are not contractually obligated to sell their fruit to just one company, even though that company may have provided them with equipment through loans. As I learned from an interview in 2014, these features can lead to conflict between smallholders and companies especially if farmers choose to sell their fruit to a competing firm. To some company employees, smallholders have "too much" selling autonomy demonstrated by them "playing" companies off one another in order to get higher prices. Further, the lack of loyalty smallholders have to companies sometimes prevents a steady stream of fruit from being collected by the company to satisfy the demands of international buyers. A respondent shared with me in 2013 that, in some cases, palm oil companies have had to raise their prices to attract smallholders despite some firms believing that many of the smallholders are not deserving of the "high prices" due to their "backward" or "archaic" cultivation techniques (Johnson 2017).

Research encounters

I came to identify and study the dialectics between firms and smallholders in Ecuador's palm oil industry in 2011 as part of a PhD project that aimed to examine how actors, particularly smallholders, participated (or not) in industry decision-making processes. This concern was driven by previous activist-oriented research conducted alongside smallholders in Indonesia where I examined their exclusion and marginalization from an emerging sustainability certification initiative. Given my critiques of smallholder exclusion and the industry more broadly, I wanted to forefront these themes in my Ecuador research. In the early stages of the project, I contacted the country's National Palm Oil Cultivators' Association (ANCUPA), which represents the financial, technical, and environmental interests of smallholders, for research leads. An Ecuadorian biologist friend also connected me to several smallholders he was working with in the northwest province of Esmeraldas and eventually helped me establish living arrangements in a local smallholder community. Initially, I was interested in the national status of small producers and the relations they have with companies buying their products. After interviewing ANCUPA members and other smallholders, I received invitations to attend industry-sponsored events.

Through these events, I met sustainability directors and corporate social responsibility managers of Ecuadorian palm oil companies. Through conversations with firm personnel, I became fascinated by the logics and actions of companies. Additionally, I became interested in the ways firms are connected to violent palm oil industry expansion and the tendency of companies to minimize resulting wildlife and forest loss, and land dispossession (Oslender 2007; Mingorría 2018; de Vos, Köhne, and Roth 2018).

At one particular industry-sponsored event, I was introduced to a director of a large palm oil firm and I explained to him my growing interest in palm oil companies. He immediately suggested that I would benefit from working *within* a company and offered to set me up with a desk in his company's main office outside of Quito. The offer was tempting—I missed the structure and routine of working in an office environment after living in a smallholder community for several months. However, this temptation was overpowered by concerns. For example, would working *within* a company prevent me from exerting autonomy over my research project's objectives and direction? Over time, would the smallholders I was originally working with begin to see me as someone who "sided" with corporate interests? More importantly, by making this move, would I be legitimizing an industry that is often accused of facilitating social and ecological violence?

To navigate these concerns, I nervously asked some of the smallholders I knew what they thought of me studying the logics of palm oil firms instead of solely focusing on the smallholders' experience. I explained that I did not want to be based at the company's office (to maintain independence and reduce the possibility of my research being co-opted); however, I emphasized that I saw value in visiting company operations and accompanying various employees to learn about company practices. Many of the smallholders responded that they supported my new research direction and felt that my connection to the inner workings of the company could assist their efforts, for example, to advocate for higher and more stable prices for their fresh fruit in addition to helping them obtain financial assistance from the company to pay for sustainability certifications.

Their approval surprised me because I assumed the smallholders would take issue with me closely interacting with local palm oil firms. In retrospect, this thought was informed by my simple reading of, and assumptions about, farmer-firm dynamics where I saw smallholders as only "victims" of the capitalist palm oil economy and companies as the perpetrators of environmental and social harm. As I explain next, this simple reading is indicative of the two categories of engagement, informed by feminist reflexive work, that I outline in what follows. First it obscured—and simplified—the dynamics between smallholders and firms as each strategically navigates challenges to position themselves in economically and politically advantageous ways. Second, it signified the imposition of my personal beliefs onto a situation that was more complex than I understood. By assuming that all smallholders would be against my research relationship with firms, I inadvertently denied their agency and efforts to forge better livelihood conditions given resources available to them. Reflecting on this moment and several others of fieldwork dilemmas, in the following section, I discuss two insights to crafting an engaged research agenda in critical resource geography, focusing on firm-smallholder relations. These are (1) the relationality of firm and non-firm actors and (2) reflections on researcher positionality in the field.

A relational approach to the firm and non-firm actors

Firms, in political ecology and related fields, are often depicted as monolithic entities driven simply by the logics of capital accumulation (Busch and Juska 1997; Campling, Chapter 16 this volume). This is evident in political-ecological studies that portray companies as pursuing profits at the expense of ecosystem health and human well-being (e.g., Horowitz 2009; Zalik 2010; Kirsch 2014). While the environmental change and social vulnerabilities that companies inflict upon communities cannot be denied, such flat renderings of companies gloss over their multiple governing logics, visions, and desires (O'Neill 2001; Welker 2014). Furthermore, they do not acknowledge the dynamic interplay among firms, non-firm actors, and environmental conditions of production (Havice and Campling 2017). A feminist political economy interpretation of the firm understands it as being constituted by a relational, uneven, and interdependent network of social processes and actors whose complex interactions shape its economic outcomes (Mullings 1999; O'Neill and Gibson-Graham 1999; McDowell 2001). These insights support more general economic geography findings that argue that social relations among economic actors drive and structure economic performance (Boggs and Rantisi 2003; Yeung 2005). While economic geographers understand that firm strategies are strongly influenced by locational conventions, regulatory regimes, business contexts, and ecological conditions (Gibbon and Ponte 2005; Havice and Campling 2017), feminist political economists assert that social difference and uneven relations of power are also integral to the shaping and functioning of political-economic network actors and their strategies (Werner et al. 2017).

Using actor-network theory, critical environmental scholars also emphasize how *non-firm actors*—such as consumers, small-scale producers, laborers, and even nature—influence the uneven spatialities of firms and, in return, how firm interests structure the actions of other network actors (Busch and Juska 1997). In the example of agriculture, strict cultivation practices (as demanded by northern consumers) have profoundly affected firm-producer relations in Africa, as sourcing has gradually shifted from smallholders to predominantly large firms better equipped to handle such demands (Freidberg 2003). In the case of grape production in Brazil, the international demand for producers to meet a range of high-quality production standards has provided fruit production workers better labor conditions and permanent employment and laborers have strategically organized for better working conditions going on strike precisely at the time that the fruit is ripe for picking; as a result, laborers are now empowered to unionize as firms seek out a knowledgeable and stable workforce to invest in (Selwyn 2007). Nonhuman influences, including ecological conditions and the materialities of nature, also contribute to shaping economic objectives (Havice and Campling 2017). Firm and non-firm actors act strategically in relation to one another to pass off risk and position themselves politically and commercially (e.g., Havice and Campling 2017). Overall, a relational lens informed by feminist political economy and actor-network insights highlights the dynamic and codependent ways firm and non-firm actors function in relation to one another in unpredictable environments. Additionally, it underscores the flexible agency of such actors and argues that relations are an outcome of changing environmental conditions, economic objectives, and livelihood strategies all shaped by changing power relations.

Palm oil plant disease and relational actors

The value of a relational reading of the agro-industry for engaged research is evident in the examination of a recent plant disease outbreak in northern Ecuador (for more on disease and relations of production, see Bustos-Gallardo, Chapter 31 this volume). In 2015, a state of emergency was called in Esmeraldas due to the massive effects of *pudrición del cogollo* (bud rot) on palm plantations owned by companies and small-scale producers, respectively. For over 30 years, the disease has baffled scientists who remain unsure if it is a fungus, bacteria, or a spore (LaHora. com 2014). Overall, the disease has killed 23,000–38,000 ha of planted palms and is the main driver behind the loss of approximately US$150 million in investments (ElProductor.com 2014;

Naranjo 2014). As I have previously documented (Johnson 2017), although the actual cause of the disease is unknown, palm oil firms often blame farmers for being the main vectors of the disease due to their "archaic" farming practices and lack of knowledge surrounding methods of transmission or treatments. According to an interview from 2014, smallholders often respond to these claims by arguing that companies are to blame since they own the largest contiguous plantations in the country comprising homogenous landscapes while smallholders are more likely to diversify their landscapes with bananas or yucca thereby reducing chances of widespread disease transmission.

I entered into this debate with preexisting beliefs about the firm-smallholder relationship. Having witnessed the intercropping practices of smallholders firsthand, I initially viewed firms as largely responsible for the spread of the disease. However, after spending time with company technicians and learning about the science underlying disease transmission, I learned that smallholders too play a major role in the acceleration of bud rot, mainly via cross-contamination that occurs through the sharing of small plantation equipment. In interviews, smallholders explained they commonly share boots, fertilizer spray tanks, or *malayas* (fruit cutters) with neighboring farms as a cost-saving measure when faced with decreasing incomes due to lowered fruit output. In an attempt to slow down the transmission of disease among smallholders and address their concerns, firms devised trainings to teach smallholders about proper sanitation practices as well as ways to identify early signs of infection. Firms, in conjunction with the government, also dispatched teams of environmental technicians to visit smallholder plots to assess their plantations and provide solutions. Many of these solutions required smallholders to invest in new equipment and agrochemicals to prevent transmission and treat infections. Although some smallholder interviewees quickly abided by the suggestions, many others remained ambivalent about the solutions as they felt that by following firm instructions, they were conceding to company blame. Some smallholders could not afford the solutions (and this explained their rejection of them), while others willfully chose not to invest in any new inputs because they felt it was a waste of money and that the company was trying to sell them something they did not truly need. In these cases, smallholders continued to engage in risky cultivation practices.

An examination of the interplay between palm oil firms and smallholders demonstrates how both actors negotiated their positions during a time of disease outbreak, revealing several insights about the relational character of community-company dynamics. First, smallholders, and not only companies as I originally thought, were responsible for ecological devastation as linked to the spread of plant disease. Second, smallholders were active agents who chose to strategically reject firm assistance based on their livelihood needs and political viewpoints—viewpoints that were conditioned by historical firm-producer interactions and environmental conditions. Third, firms provided farmers with technical assistance and educational opportunities to combat the disease. Although these measures were offered to ensure the continuity of capitalist production and sourcing of palm fruit, a relational perspective reveals firm action as also being driven by farmers' loss of livelihood and, perhaps, a notion of care. This was indicated by a statement made by a firm technician during a 2014 interview who justified company assistance provided to smallholders by stating, "many of us have known these smallholders for years, it's difficult to see them lose everything."

Understanding the relationality of actors is a crucial aspect of engaged scholarship as it drives more researchers' interventions that are informed by and attuned to the interdependent complexities of actors and their strategic actions. These actions are always shaped by connections to other network actors, power relations, and the geo-social histories and biophysical realities of changing environments. Studying the relationality of actors with the ultimate goal of constructing an engaged project requires researchers to implement methodological approaches attuned to research dilemmas and personal bias. I turn to this topic in the following section.

Positionality and reflexivity in research with firms and smallholders

Feminist insights have provided me with tools to assess my personal power, privilege, and assumptions in order to position myself critically and activate an engaged research program. My approach is informed by the work of feminist geographers who have long studied the dynamics between "researcher" and "the researched" in order to construct critically engaged projects that facilitate the production of knowledge in nonexploitative ways (Katz 1994; Moss 2002; Sultana 2007; de Leeuw, Cameron, and Greenwood 2012; England 2015). To do this, feminist scholars challenge ideas of researchers as being disembodied, rational subjects who are unaffected by interactions with other actors (McDowell 1997). Rather, feminist geography encourages researchers to develop an introspective awareness of themselves as beings constituted by multiple and intersectional identities of race, gender, class, and sexual orientation (Kobayashi 1994; Mollett and Faria 2013; Di Feliciantonio, Gadelha, and DasGupta 2017). Feminist geographers understand that all researchers enter the field with a set of social beliefs and assumptions (McDowell 1997; Mullings 1999) and it is the layering of complex identities and perspectives, mediated by power and privilege, which influences how researchers "see" the social and environmental aspects of the world. Reflexivity informs engaged research by revealing how researchers' social identity influences the collection and analysis of data. It opens possibilities in critical resource geography to discuss the challenges of "doing" fieldwork in politically and environmentally contentious sites of extraction as part of efforts to devise socially just interventions. Furthermore, it impels researchers to confront their preconceived ideals brought to the research experience, despite best intentions. Focusing on reflexivity and positionality as integral components of engaged research can assist researchers in coming to terms with unresolved tensions in the field (Billo and Hiemstra 2013; Faria and Mollett 2016); this can contribute to forging a transformational and engaged research agenda in critical resource geography.

Employing a reflexive lens revealed the biases that shaped the initial stages of my research. I held preconceived notions about palm oil firms and assumed they exercised complete control over smallholder livelihoods. These beliefs were informed by my previous work with smallholders in Indonesia whose relationships with large companies are often characterized by abuse-and-debt peonage (Friends of the Earth, Life Mosaic, and Sawit Watch 2008; Li 2015). However, after spending time with several Ecuadorian firms during the plant disease outbreak, I learned about the complex relations between smallholders and firms that simultaneously comprise mutual dependence, tension, and opposition. This observation prompted a need to examine my positionality and biases while in the field.

My support for smallholders was unexpectedly challenged by the plant disease outbreak when I saw farmers question and even reject solutions offered by companies—solutions that could potentially protect or restore their sick plantations. These moments of "uncomfortable witnessings" (Valdivia and Lu 2020) were challenging because the smallholders were making decisions I would not personally make. To me, accepting company assistance would provide an immediate "fix" to plant disease problems. Upon closer examination, however, these field moments highlighted how I failed to see rejection of company assistance as a smallholder livelihood strategy to exist alongside, or in opposition to, palm oil companies. My privileged position—as someone who could leave the plantations at any time—propelled me to make quick judgments that glossed over how smallholders negotiate their daily livelihoods, which are constantly jeopardized by the entanglements of plant disease, climate change, volatile fruit prices, and fluctuating market demand. I failed to see how their rejection of company solutions were bold attempts to assert their autonomy and shape their collective reputation among firms as a well-organized force to contend with. Even in the infrequent instances where smallholders accepted firm support, this was an expression of autonomy since it emerged upon, and according to, terms formulated by smallholders themselves.

Relational research engagements

Forefronting the needs of smallholders and firms and their codependent dynamics alongside my own assumptions shaped how I enacted (and continue to enact) an engaged research program. My critical approach is driven by my commitment to understand the standpoints and agendas of smallholders and firms and requires that I attune myself to their changing needs while I assess my own privileges in efforts to devise responsible engagements. This requires an ongoing, iterative, and oftentimes "bumpy" process of personal reflection and position realignment. My engaged research is informed by, and more sensitive to, the needs of smallholders and their desire to forge better working relations *with* local palm oil companies in order to sustain livelihood opportunities. It employs a much more flexible approach than what I initially envisioned and does not aim to "make sense" of smallholder goals but rather seeks to communicate to firms the daily social and ecological struggles small producers face in the region, and vice versa. Drawing on my knowledge of both smallholders and firms, I have worked to identify areas where smallholder interests and company goals converge and points where companies are likely to loosen their stance (to the benefit of smallholder demands and interests) going forward.

My research engagements have evolved into a policy-oriented path and reflect a reformist approach rather than a radical "take-down" of the palm oil industry. As I have noted elsewhere (e.g., Johnson 2019), one possibility to facilitate better working relations between smallholders and firms currently resides in the growing area of certification for "sustainably produced" palm oil. Many of the smallholders I work with view certification as financially promising as it may facilitate the establishment of new partnerships with international buyers (especially from Europe) thereby increasing smallholder revenue. Their interest in becoming certified growers motivates my policy-oriented engaged work that seeks to make smallholder certification more attainable. I am currently working on two independent reports that will be shared with a leading sustainability palm oil certification initiative as well as with Ecuadorian palm oil firms. The first is a technical report that targets palm oil companies and suggests ways for them to assist smallholders in their preparations for certification. With global demand for sustainable palm oil rising, there is major incentive for farmers and firms to work together to produce this highly desired resource. The second is an evaluation report to be sent to the sustainability certification's governing body. It documents barriers to certification entry encountered by Ecuadorian smallholders, such as high costs, lack of certification knowledge, and disease crises. It suggests how sustainability certification decision-making—which is largely dominated by firms—can better involve smallholders. Overall, these outcomes of my engaged research program do not reflect what I initially had in mind. I originally envisioned my work contributing to more radical, activist-oriented critiques of the palm oil industry. However, my policy-based interventions are important because they are informed by local smallholder realities and emerge out of the two "principles" that I outlined earlier, namely, that smallholders and firms are network actors whose complex interdependencies shape their economic outcomes, and reflections on fieldwork positionality are integral to crafting responsible and impactful critical research engagements.

Toward engaged research in critical resource geography

Relational and feminist insights are core components of an engaged research agenda on firm-smallholder relations in critical resource geography, and they may be useful for the study of other resource sectors as well. These components lend support to conceptual and practical thinking around relations among companies, communities, and individuals, and the challenges associated with research on industrial agriculture. Like many agro-industries, the palm oil sector comprises close and contingent relations between small producers and firms who rely on each other in multiple ways. In Ecuador's palm industry, smallholders periodically rely on the technical assistance provided by companies to protect their palm oil plants and to assist in their cultivation. They also rely on firms to ultimately purchase their fruit for income. Further, companies rely on smallholders to supply them with consistent palm fruit in order to satisfy global demand. This relationality is evident in the ways that all actors respond to disruptions, such as disease crises, which require actors to act strategically in order to minimize ecological damage and financial loss.

Engaged research that is supported by feminist insights is sensitive to both the dialectical relations between actors such as smallholders and firms during times of environmental crisis (and otherwise) and researchers' power and privilege. The examination of researchers' assumptions and understandings when conducting fieldwork is key, as they can obscure certain research insights and ultimately lead to interventions that reflect the desires of the researcher rather than the community and people involved in the sector. By interrogating researchers' power and privilege, interventions can better reflect the livelihood choices and desires of communities. Centering relationality and feminist methodological approaches in my research has opened up opportunities for engagement with multiple stakeholders and has created more opportunities for my work to make a difference on the terms of those involved in the sector, rather than on my own terms. I have chosen to build working relations with firms and communities in order to maintain communication and long-term relationships with both. Because of this, complex power inequalities are an inherent part of my take on engaged research. I maintain a critical perspective by paying close attention to how these dynamics may be exacerbated by new initiatives (such as emerging sustainability certification arrangements). Overall, my research approach has enabled me to develop a sense of the needs of both smallholders and firms while identifying areas where interests converge and where new reconfigured relations between both actors can emerge.

Note

1 Smallholders in Ecuador own less than 50 ha of cultivated palm. Medium producers own between 51 and 99 ha, and large producers (usually companies) own 100+ ha of palm.

References

Bakker, Karen. 2004. *An Uncooperative Commodity*. Oxford: Oxford University Press.

Bakker, Karen, and Gavin Bridge. 2006. "Material Worlds? Resource Geographies and the `Matter of Nature." *Progress in Human Geography* 30 (1): 5–27. doi: 10.1191/0309132506ph588oa.

Billo, Emily. 2015. "Sovereignty and Subterranean Resources: An Institutional Ethnography of Repsol's Corporate Social Responsibility Programs in Ecuador." *Geoforum* 59: 268–277.

Billo, Emily, and Nancy Hiemstra. 2013. "Mediating Messiness: Expanding Ideas of Flexibility, Reflexivity, and Embodiment in Fieldwork." *Gender, Place & Culture* 20 (3): 313–328. doi: 10.1080/0966369X. 2012.674929.

Boggs, Jeffrey S., and Norma M. Rantisi. 2003. "The 'Relational Turn' in Economic Geography." *Journal of Economic Geography* 3 (2): 109–116.

Brewer, Jennifer F. 2013. "Toward a Publicly Engaged Geography Polycentric and Iterated Research." *Southeastern Geographer* 53 (3): 328–347.

Bridge, Gavin. 2009. "Material Worlds: Natural Resources, Resource Geography and the Material Economy." *Geography Compass* 3 (3): 1217–1244. doi: 10.1111/j.1749-8198.2009.00233.x.

Busch, Lawrence, and Arunas Juska. 1997. "Beyond Political Economy: Actor Networks and the Globalization of Agriculture." *Review of International Political Economy* 4 (4): 688–708.

Cardona, Antonio José Paz. 2019. "Researchers Urge Sustainability as Palm Oil Tightens Its Grip on Latin America." *Mongabay.* https://news.mongabay.com/2019/11/researchers-urge-sustainability-as-palm-oil-tightens-its-grip-on-latin-america/.

de Leeuw, Sarah, Emilie S. Cameron, and Margo L. Greenwood. 2012. "Participatory and Community-Based Research, Indigenous Geographies, and the Spaces of Friendship: A Critical Engagement." *The Canadian Geographer/Le Géographe Canadien* 56 (2): 180–194. doi: 10.1111/j.1541-0064.2012.00434.x.

de Vos, Rosanne, Michiel Köhne, and Dik Roth. 2018. "'We'll Turn Your Water into Coca-Cola': The Atomizing Practices of Oil Palm Plantation Development in Indonesia." *Journal of Agrarian Change* 18 (2): 385–405. doi: 10.1111/joac.12246.

Di Feliciantonio, Cesare, Kaciano B. Gadelha, and Debanuj DasGupta. 2017. "'Queer(y)ing Methodologies: Doing Fieldwork and Becoming Queer'—Guest Editorial." *Gender, Place & Culture* 24 (3): 403–412. doi: 10.1080/0966369X.2017.1314950.

ElProductor.com. 2014. "Ecuador: Sector Palmicultor Está en Emergencia." http://elproductor.com/2014/10/28/ecuador-sector-palmicultor-esta-en-emergencia/.

England, Kim. 2015. "Producing Feminist Geographies: Theory, Methodologies and Research Strategies." In *Approaches to Human Geography*, edited by Stuart Aitkin, and Gill Valentine, 361–373. London: SAGE Publications.

Faria, Caroline, and Sharlene Mollett. 2016. "Critical Feminist Reflexivity and the Politics of Whiteness in the 'Field.'" *Gender, Place & Culture* 23 (1): 79–93. doi: 10.1080/0966369X.2014.958065.

Freidberg, Susanne. 2003. "Cleaning Up Down South: Supermarkets, Ethical Trade and African Horticulture." *Social & Cultural Geography* 4 (1): 27–43. doi: 10.1080/1464936032000049298.

Friends of the Earth, Life Mosaic, and Sawit Watch. 2008. *Losing Ground: The Human Rights Impacts of Oil Palm Plantation Expansion in Indonesia.* Amsterdam: Friends of the Earth. https://www.foei.org/wp-content/uploads/2014/08/losingground.pdf.

Ghomeshi, Kimia, and Anna Zalik. 2013. "Corporate Privacy and Environmental Review at Export Development Canada—Intervention." *Antipode.* https://antipodeonline.org/2013/10/23/corporate-privacy-and-environmental-review-at-export-development-canada/.

Gibbon, Peter, and Stefano Ponte. 2005. *Trading Down: Africa, Value Chains, and the Global Economy.* Philadelphia, PA: Temple University Press.

Havice, Elizabeth. 2020. "Methods and Mobility in Extractive Tuna Fisheries." *Environment and Planning E:* 1–46. doi: 10.1177/2514848620907470.

Havice, Elizabeth, and Liam Campling. 2017. "Where Chain Governance and Environmental Governance Meet: Interfirm Strategies in the Canned Tuna Global Value Chain." *Economic Geography* 93 (3): 292–313. doi: 10.1080/00130095.2017.1292848.

Horowitz, Leah S. 2009. "Environmental Violence and Crises of Legitimacy in New Caledonia." *Political Geography* 28 (4): 248–258. doi: https://doi.org/10.1016/j.polgeo.2009.07.001.

Johnson, Adrienne. 2017. "Pudrición del Cogollo and the (Post-)Neoliberal Ecological Fix in Ecuador's Palm Oil Industry." *Geoforum* 80: 13–23. doi: http://dx.doi.org/10.1016/j.geoforum.2016.12.016.

Johnson, Adrienne. 2019. "The Roundtable on Sustainable Palm Oil's National Interpretation Process in Ecuador: 'Fitting' Global Standards into Local Contexts." *Journal of Rural Studies* 71: 125–133. doi: https://doi.org/10.1016/j.jrurstud.2019.02.013.

Johnson, Adrienne. 2020. "Engaged Research with 'Oppositional' Corporate Actors." *Environment and Planning E: Nature and Space* 1–46. doi: 10.1177/2514848620907470.

Johnson, Adrienne, Anna Zalik, Sharlene Mollett, Farhana Sultana, Elizabeth Havice, Tracey Osborne, Gabriela Valdivia, Flora Lu, and Emily Billo. 2020. "Extraction, Entanglements, and (Im)Materialities: Reflections on the Methods and Methodologies of Natural Resource Industries Fieldwork." *Environment and Planning E: Nature and Space* 1–46. doi: 10.1177/2514848620907470.

Katz, Cindi. 1994. "Playing the Field: Questions of Fieldwork in Geography." *The Professional Geographer* 46 (1): 67–72. doi: 10.1111/j.0033-0124.1994.00067.x.

Kirsch, Stuart. 2014. *Mining Capitalism: The Relationship Between Corporations and Their Critics*. Berkeley, CA: University of California Press.

Kirsch, Stuart. 2018. *Engaged Anthropology: Politics beyond the Text*. Berkley, CA: University of California.

Kobayashi, Audrey. 1994. "Coloring the Field: Gender, 'Race,' and the Politics of Fieldwork." *The Professional Geographer* 46 (1): 73–80. doi: 10.1111/j.0033-0124.1994.00073.x.

LaHora.com. 2014. "Agrocalidad contra el PC." *La Hora*. Accessed July 5, 2015. http://www.lahora.com.ec/index.php/noticias/show/1101731026/-1/Agrocalidad_contra_el_PC.html#.Vgf6OvlVikp.

Li, Fabiana. 2009. "Documenting Accountability: Environmental Impact Assessment in a Peruvian Mining Project." *PoLAR: Political and Legal Anthropology Review* 32 (2): 218–236. doi: 10.1111/j.1555-2934.2009.01042.x.

Li, Tania Murray. 2015. *Social Impacts of Oil Palm in Indonesia: A Gendered Perspective from West Kalimantan*. Bogor Barat, Indonesia: Center for International Forestry Research (CIFOR).

Marchand, Marianne H., and Anne S. Runyan. 2011. *Gender and Global Restructuring: Sightings, Sites and Resistances*. London: Routledge.

McDowell, Linda. 1997. "Women/Gender/Feminisms: Doing Feminist Geography." *Journal of Geography in Higher Education* 21 (3): 381–400. doi: 10.1080/03098269708725444.

McDowell, Linda. 2001. "Linking Scales: Or How Research about Gender and Organizations Raises New Issues for Economic Geography." *Journal of Economic Geography* 1 (2): 227–250. doi: 10.1093/jeg/1.2.227.

McGuirk, Pauline, and Phillip O'Neill. 2012. "Critical Geographies with the State: The Problem of Social Vulnerability and the Politics of Engaged Research." *Antipode* 44 (4): 1374–1394. doi: 10.1111/j.1467-8330.2011.00976.x.

Mingorría, Sara. 2018. "Violence and Visibility in Oil Palm and Sugarcane Conflicts: The Case of Polochic Valley, Guatemala." *The Journal of Peasant Studies* 45 (7): 1314–1340. doi: 10.1080/03066150.2017.1293046.

Mollett, Sharlene, and Caroline Faria. 2013. "Messing with Gender in Feminist Political Ecology." *Geoforum* 45: 116–125. doi: 10.1016/j.geoforum.2012.10.009.

Moss, Pamela. 2002. "Taking on, Thinking about, and Doing Feminist Research." In *Feminist Geography in Practice: Research and Methods*, edited by Pamela Moss, 1–17. Oxford: Balckwell Publishers.

Mullings, Beverly. 1999. "Insider or Outsider, Both or Neither: Some Dilemmas of Interviewing in a Cross-Cultural Setting." *Geoforum* 30: 337–350.

Nagar, Richa, Victoria Lawson, Linda McDowell, and Susan Hanson. 2002. "Locating Globalization: Feminist (Re)readings of the Subjects and Spaces of Globalization." *Economic Geography* 78 (3): 257–284. doi: 10.1111/j.1944-8287.2002.tb00187.x.

Naranjo, F. 2013. "Palma Aceitera en Ecuador y su experiencia frente a RSPO." Presented at IIII Latin American RSPO Meeting. Honduras. Aug 6–8.

Naranjo, Francisco. 2014. "Plan de Acción Contra la PC." In *Palma Ecuador*, edited by Francisco Naranjo, Gustavo Bernal, Rommel Gargas, Susana Naranjo, and Roberto Burgos, 8–13. Quito: ANCUPA.

O'Neill, Phillip. 2001. "Financial Narratives of the Modern Corporation." *Journal of Economic Geography* 1 (2): 181–199. doi: 10.1093/jeg/1.2.181.

O'Neill, Phillip, and JK Gibson-Graham. 1999. "Enterprise Discourse and Executive Talk: Stories that Destabilize the Company." *Transactions of the Institute of British Geographers* 24 (1): 11–22. doi: 10.1111/j.0020-2754.1999.00011.x.

Oglesby, Elizabeth. 2010. "Interviewing Landed Elites in Post-War Guatemala." *Geoforum* 41 (1): 23–25. doi: 10.1016/j.geoforum.2009.07.008.

Osborne, Tracey. 2017. "Public Political Ecology: A Community of Praxis for Earth Stewardship." *Journal of Political Ecology* 24 (1): 843–860.

Osborne, Tracey. 2020. "Decolonizing Methodologies for Climate Justice Research." *Environment and Planning E: Nature and Space* 1–46. doi: 10.1177/2514848620907470.

Oslender, Ulrich. 2007. "Violence in Development: The Logic of Forced Displacement on Colombia's Pacific Coast." *Development in Practice* 17 (6): 752–764. doi: 10.1080/09614520701628147.

Peñaherrera, Silvana. 2015. "La Cadena Productiva de la Palma Aceitera en el Ecuador." *ANCUPA*. http://www.sharp-partnership.org/RSS/ANCUPA_Palma_aceitera_Ecuador.pdf.

Sawyer, Suzana. 2004. *Crude Chronicles: Indigenous Politics, Multinational Oil, and Neoliberalism in Ecuador.* Durham, NC: Duke University Press.

Selwyn, Ben. 2007. "Labour Process and Workers' Bargaining Power in Export Grape Production, North East Brazil." *Journal of Agrarian Change* 7 (4): 526–553. doi: 10.1111/j.1471-0366.2007.00155.x.

Sultana, Farhana. 2007. "Reflexivity, Positionality and Participatory Ethics: Negotiating Fieldwork Dilemmas in International Research." *ACME: An International Journal for Critical Geographies* 6 (3): 374–385.

Sultana, Farhana. 2009. "Fluid Lives: Subjectivities, Gender and Water in Rural Bangladesh." *Gender, Place & Culture* 16 (4): 427–444. doi: 10.1080/09663690903003942.

Valdivia, Gabriela, and Flora Lu. 2014. "Crude Entanglements." *Anthropology News.* http://crudeoil.web.unc.edu/files/2015/03/In-Focus-2014-Anthropology-News-crude-entanglements.pdf.

Valdivia, Gabriela, and Flora Lu. 2020. "Extractive Entanglements: Environmental Justice and the Realpolitik of Life-with-Oil." *Environment and Planning E: Nature and Space* 1–46. doi: 10.1177/2514848620907470.

Welker, Marina. 2014. *Enacting the Corporation: An American Mining Firm in Post-Authoritative Indonesia.* Berkeley, CA: University of California Press.

Werner, Marion, Kendra Strauss, Brenda Parker, Reecia Orzeck, Kate Derickson, and Anne Bonds. 2017. "Feminist Political Economy in Geography: Why Now, What is Different, and What for?" *Geoforum* 79: 1–4. doi: https://doi.org/10.1016/j.geoforum.2016.11.013.

Whitmer, Ali, Laura Ogden, John Lawton, Pam Sturner, Peter M. Groffman, Laura Schneider, David Hart, Benjamin Halpern, William Schlesinger, and Steve Raciti. 2010. "The Engaged University: Providing a Platform for Research that Transforms Society." *Frontiers in Ecology and the Environment* 8 (6): 314–321.

Yeung, Henry Wai-chung. 2005. "Rethinking Relational Economic Geography." *Transactions of the Institute of British Geographers* 30 (1): 37–51. doi: 10.1111/j.1475-5661.2005.00150.x.

Zalik, Anna. 2010. "Oil 'Futures': Shell's Scenarios and the Social Constitution of the Global Oil Market." *Geoforum* 41 (4): 553–564. doi: https://doi.org/10.1016/j.geoforum.2009.11.008.

Zalik, Anna. 2020. "A Politics of Representation for Extractive Industry Research?" *Environment and Planning E: Nature and Space* 1–46. doi: 10.1177/2514848620907470.

25

Renewable energy landscapes and community engagements

The role of critical resource geographers beyond academia

Elvin Delgado

Introduction

On July 11, 2018, *The New York Times* published an article discussing the political implications of the collision between large-scale solar development and traditional farming lifestyles in Kittitas County, Washington (Johnson 2018).[1] This issue, which had been developing for about three years before it reached national coverage, is the result of energy policies developed at the state level to incentivize the adoption of renewable energy to reduce greenhouse gas emissions (GHGe). It is now quite common for problems associated with the location, scale, and potential environmental and economic impacts associated with large-scale energy projects around the world to appear in the news media outlets. This time, however, I was not reading about contentious relations between residents, government officials, and private firms in another country. The story portrayed in *The New York Times* was happening in the place where I live and work.

Kittitas County, a rural area located approximately 107 mi southeast of Seattle on the eastern slope of the Cascade Range, has ideal geographical conditions to develop solar projects. Between 2015 and 2016, the county received several proposals from private firms in Seattle to develop large-scale solar projects that would cover hundreds of acres of productive irrigated agricultural land. County residents have had mixed reactions to these proposals. Some farmers believe that allowing these projects on their land will provide additional income. Others argue that these projects will negatively affect the rural character of the region. Local government officials have been concerned with the geographical extension of the projects and their compliance with county laws and zoning codes. Meanwhile, a local environmental group, Our Environment of Kittitas County (Our Environment), submitted a resolution to the mayor of Ellensburg (county seat of Kittitas County) to increase the city's commitment to solar power.

What is the role of critical resource geographers in debates like these? What contributions can critical scholars provide to discussions about the governance of renewable energy projects in the places where we live? This chapter offers a reflection on local engagement in critical resource geography through an analysis of my own engagement in debates involving government officials, community members, private firms, and environmental groups regarding the adoption of solar projects in Kittitas County. I refer to local engagement as the process by which a scholar interacts and gets involved with the community *where she/he lives and works*.

My engagement in Kittitas County solar controversies is twofold. First, I am a county resident who will experience the potential socio-environmental and political-economic implications of large-scale solar projects on a daily basis. Second, I am an Associate Professor in the Department of Geography at Central Washington University (CWU, located in Ellensburg) and have long-standing research and teaching interests in energy issues. When I started my tenure-track job in 2012, I led the creation of CWU's Institute for Integrated Energy Studies (IIES), which I directed for four years. I also led the creation of a new B.S. in Integrated Energy Management (IEM) and am still the director of this program. In the context of these institutional positions, I have been drawn into not only debates about solar development in Kittitas Country but also local energy governance, as in 2017, I was appointed by the mayor of Ellensburg to the city's Utility Advisory Committee (UAC).

I use my own history of involvement with debates surrounding solar development in Kittitas County to offer my perspective on the opportunities, challenges, and lessons learned from engaging in local issues in and around where one lives and works. I focus on two separate moments of this debate that occurred around the same time: public hearings and community conversations regarding the Columbia Solar Project (CSP) and the Our Environment resolution to increase the use of solar power in Ellensburg. These scenarios provide valuable opportunities to understand how resource geographers engaging in local issues can mediate opposing views regarding the socio-environmental implications and political-economic contexts of renewable energy projects. My experiences suggest that resource geographers can engage in academic praxis outside the confines of academia and apply critical social theory to shed light on the social issues affecting communities where they live and work. However, doing so requires institutional support that promotes communication between that faculty and her/his department to articulate the importance of local engagement for the department, the university, and the community. I further contend that the nature of community engagement shifts when it happens at home rather than in a distant place where it may be contained by the time-space of fieldwork. My experiences of local engagement have prompted me to reflect critically about these questions in the research I conduct in both near- and far-field sites.

The chapter proceeds as follows. The next section provides a brief overview of Washington state's policies to incentivize renewable energy projects to reduce GHGe, the county's moratorium and final decision over the CSP, and the challenges solar projects face when their siting does not comply with state and county zoning codes. Focusing on my engagement, the third section further explores the solar debate in Kittitas Country and discusses the resolution submitted by Our Environment and the final outcome. The fourth section examines the opportunities and challenges of engaging in community issues and the lessons learned from critical resource geographers. The conclusion reflects on our roles and responsibilities as critical resource geographers engaging in local issues to make a difference.

Incentivizing renewable energy in Washington and the protection of critical environmental areas in Kittitas County

Washington State has been committed to reducing GHGe since the approval of the Energy Independence Act of 2006 that set Renewable Portfolio Standards (RPS)[2] requiring utilities to use renewable energy and energy conservation initiatives (WA Department of Commerce 2019a). Unfortunately, the state had not reached its target by 2014. Governor Jay Inslee furthered the state's goal of reducing GHGe by signing the Subnational Global Climate Leadership Memorandum of Understanding at the United Nations Conference of the Parties in Paris in 2015. This forced the state to reduce its GHGe from more than 90 million tons per year to less than 20 million tons by 2050 (WA Department of Commerce 2019b, 7). With the approval of Senate Bill 5116 in 2019, the state required that all electric utilities transition to 100-percent carbon neutral by 2030 and 100-percent carbon free by 2045. As the state continues to incentivize the adoption of green technology to achieve their RPS and GHGe targets, more large-scale solar projects have been proposed in areas with the highest solar potential, such as Kittitas County.

Kittitas is a rural county located in central Washington with a total surface area of 2333 square miles, which makes it one of the largest counties in the state. Agriculture is the major land use and thus one of the most important on privately owned land. In 2017, the county produced approximately $83 million in market value from agricultural products (USDA 2017). The county's large extensions of flatland, strong and consistent winds, and an average of 300 days of sunshine create the right geographical conditions to take advantage of renewable energy.

However, the siting of these projects may conflict with state and local land use codes. The state protects critical areas that perform key functions that enhance the environment, protect its citizens, and preserve sensitive ecosystems under the Growth Management Act (GMA). The GMA protects the rural character of counties and assists local governments to manage growth and development. It requires counties and cities to designate and protect critical environmental areas, including natural resource lands of commercial significance such as agriculture. The conservation of productive forest and agricultural lands is at the core of the GMA and all incompatible uses in these areas are discouraged. Adopted by the county in compliance with the state's GMA, the Kittitas County Comprehensive Plan (KCCP) is the centerpiece of local planning in Kittitas County. Land use is a key component of the KCCP, which protects the rural character of the county. Its policies focus on future growth and economic development as a strategy to protect lands from conflictive land uses. The location of large-scale solar projects may conflict with the state's GMA and the KCCP depending on how different decision-makers interpret the state and local land use codes and planning policies.

In this policy context, Kittitas County received two applications for conditional use permits (CUPs)[3] from OneEnergy Renewables (OneEnergy) in 2015 and TUUSSO Energy (TUUSSO) in 2016 to develop large-scale solar projects. OneEnergy proposed the development of the approximately 10,379 MWh Iron Horse Solar Project on 47.5 acres zoned as Rural Working-Agricultural. However, the application was denied by the Kittitas County Board of Commissioners (Commissioners) on January 10, 2017, in a two-to-one vote. As noted in a *Daily Record* article on January 11, 2017, Commissioners Osiadacz and O'Brien voted to deny the permit, because the project did not fit in with land uses as stipulated by the KCCP and, as a result, would affect the rural character of the county, which is protected under the state's GMA (Buhr 2017). Conversely, Commissioner Jewell voted in favor, stating that the county code allows utility-scale solar projects as a CUP in land zoned as rural agriculture and that the county could be in violation of the KCCP if it denied the permit (*Daily Record* Editorial Board, January 12, 2017). Commissioner O'Brien contended that the county did not anticipate the demand and size of the project when language allowing solar farms in agricultural lands was included in the KCCP. Then, on January 10, 2017, the Commissioners voted unanimously to place a 60-day moratorium on all large renewable projects in the county. This would give them time to develop new rules to evaluate how these projects affect the rural character of the county. It also allowed for public hearings and conversations with county residents where TUUSSO's representatives could discuss the CSP and answer questions from the Commissioners and the public. The moratorium was extended two more times on July 18, 2017, and January 3, 2018, until July 20, 2018.

TUUSSO proposed the development of the $50-million Columbia Solar Project (CSP). It would include five separate projects (Camas, Fumaria, Penstemon, Typha, and Urtica) with an

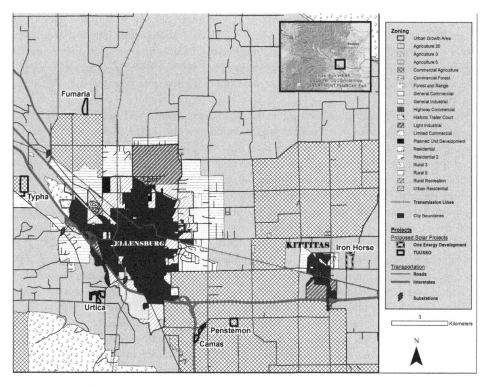

Figure 25.1   Columbia Solar Project and Iron Horse Solar Project. Map by Dusty Pilkington.

annual production capacity of 56.5 GWh (see Figure 25.1). The projects would be located on five separate leased sites (totaling 232 ac of prime irrigated agricultural land) near existing Puget Sound Energy (PSE) substation and electrical distribution lines. The company signed a 15-year power purchase agreement (PPA) with PSE beginning in December 2016 for electricity produced by the CSP (EFSEC 2018). This would allow PSE to include the electricity produced by this project as part of their RPS, which would also help the state achieve its GHGe targets.

As stated earlier, the proposals submitted by OneEnergy and TUUSSO received mixed reviews from county officials. While the application submitted by OneEnergy was denied by the Commissioners, the application submitted by TUUSSO was delayed by the moratorium. This affected TUUSSO's bottom line because the CSP was not going to generate the electricity required by the PPA signed with PSE. The main concern associated with TUUSSO's application centered around zoning issues in critical environmental areas as stipulated by the GMA and the KCCP, because the geographical extension and location of the CSP would jeopardize prime irrigated agricultural land and the rural character of the county.

The moratorium forced TUUSSO to apply in October 2017 to the state's Energy Facility Site Evaluation Council (EFSEC)—which can bypass the county's permitting process—for an expedited review of the CSP. EFSEC was created in 1970 to streamline the evaluation, siting, and permitting process for large energy projects in one state agency that makes a final recommendation to the governor. EFSEC conducted a land use hearing at the Kittitas Valley Event Center Armory in Ellensburg on December 12, 2017, to hear testimony regarding whether the location of the CSP followed the county's land use provisions. They also received written comments from county residents expressing their concerns about the CSP. After evaluating all the information, on April 17, 2018, EFSEC recommended approval for an expedited review and certification of the CSP. Two months later, EFSEC approved the development of the CSP in a five-to-one vote. Finally, on October 17, 2018, Governor Inslee approved the project on 232 ac of productive irrigated agricultural land in Kittitas County.

Academic praxis and local engagement: debating solar power in Kittitas County

There are differences in how a resource geographer can "engage" with a community when this engagement happens in the place they live and work as opposed to a nonlocal field site. I do not think this difference lies in the methods of data collection, as the systematic process of collecting, analyzing, and triangulating data should be the same regardless of where the process happens. There may, however, be an advantage if the data collection happens near home, if a scholar is more familiar with the place, its history, and its people and knows where to get the data, which reduces the time it takes to collect it.

From its inception, I looked at the solar debate in Kittitas County from a political ecology and critical resource geography perspective. However, I did not initially consider it a potential research opportunity. This was not because I thought local issues were unimportant or that the solar debate did not have academic merit. Rather, it was because I had an active research agenda studying hydraulic fracturing operations in northern Patagonia, Argentina and wanted to focus my energy on that. Nonetheless, as I became more involved in the debate over time, I decided to include it within my research agenda. Systematic data collection had not been part of my plan when I started engaging in this issue. Nevertheless, I took copious notes on people's comments and demeanor during my participation in public hearings and community meetings. And, before writing this chapter, I collected additional archival data from newspaper articles, government reports, and UAC's official minutes. The data was used to corroborate my notes, identify official statements from community members and government officials, and provide a chronology of events. As a result, this chapter is based on archival work; document analysis; participant observation of the day-to-day life of Ellensburg residents as well as informal conversations at grocery stores, the local farmers' market, and public events; and my official discussions as a member of the UAC. I now draw on this information to illustrate how a political ecology and critical resource geography approach allowed me to understand controversies over the adoption of large-scale solar projects.

Public hearings and community conversations regarding the CSP

Public hearings and community meetings took place in the Commissioners auditorium, Hal Holmes Community Center, and Kittitas Valley Event Center between March 17, 2017, and March 1, 2018. I attended one public hearing and several community meetings where I took notes and was able to identify major themes from public comments. Opinions about the CSP were mixed and extensive. Some of the issues discussed revolved around the footprint of solar farms, the potential transformation of the rural character of the county, and visual impacts resulting from the solar panels. Others focused on the economic incentives for the county and farmers leasing their land. People debated whether to produce energy instead of crops and the potential loss of agricultural land. Other issues highlighted the concentration of solar projects near residents who are against them. Some were concerned with the time these infrastructures will stay at the site (approximately 30 years) and questioned what would happen with the land after the project is decommissioned. Others debated the potential of marginal lands as an alternate location for this project. Those favoring the CSP saw it as an opportunity to develop clean energy and move away from fossil fuels. TUUSSO representatives answered all questions. However, some local government officials and county residents were skeptical.

Political ecology and critical resource geography provided a useful framework to study the complexity of solar power development and shaped my understanding of the debate over the CSP in particular ways. For instance, a critical resource geography approach allows me to study the complex ways capitalism transforms natural resources and society. In particular, it helps me understand the political-economic forces that shape the development of solar power in Kittitas County; how the biophysical characteristics of solar energy define where it can be developed; how it can be commodified in the form of electricity; and how its incorporation into local, regional, and national circuits of capital production and accumulation create social, political, and environmental problems at multiple scales in Washington State.

Core concepts of political ecology and critical resource geography also helped me analyze the socio-environmental problems resulting from resource development in rural communities and the varied positions that people take on controversial resource issues. First is the idea that resource users are embedded in social relations that may impose excessive pressure of production on the environment (Watts 1983). Social relations at various scales define where solar projects are placed in Kittitas County, who has access to and control over the electricity produced, and who receives the economic benefits or socio-environmental burden associated with such projects. Second is the recognition of multiple positions and rationalities in relation to the environment (Blaikie 1985; see also Bebbington et al., Chapter 21 in this volume). The CSP could be economically beneficial for TUUSSO and the farmers leasing their land or detrimental to those who are concerned about losing agricultural land and the rural character of the county. There is no singular "community opinion" in regard to the adoption of large-scale solar projects in Kittitas County. Instead, the solar debate is shaped by disparate actors who pursue different kinds of individual and collective objectives depending on their particular needs and desires. What differentiates one perception from the other is the plurality of positions, rationalities, and interests of each actor. The following section further highlights how opposing views toward the adoption of solar power have shaped policy discussions by focusing on the resolution submitted by Our Environment to the city council.

Our Environment's resolution to the city of Ellensburg

In February 2017, a member of Our Environment who is an administrator at CWU asked me to give a presentation to the group about energy issues and my goals for IIES and IEM as the Director of both. I did so in May 2017. Meanwhile, the Chair of the city of Ellensburg's Utility Advisory Committee[4] (UAC) asked if I, or one of my students, would consider becoming a member of the UAC. I accepted the request, at the time not anticipating that it would again lead me to be in discussions with Our Environment, but this time as a member of the UAC.

The prompt for this new interaction was the resolution Our Environment submitted to the city council requesting the city increase its commitment to solar power. The resolution was delivered in June 2017, while the city council was debating TUUSSO's application. Our Environment asked the city to establish a code to produce 25 percent of the electricity from solar; that all public buildings moving forward be solar ready; that 25 percent of all commercial and residential buildings adopt solar systems; that all new construction in Ellensburg be solar ready; and that the Environmental Commission (EC) submit a recommendation to the city council by the second quarter of 2018 to achieve these goals. City council requested the UAC and the EC to provide a recommendation on the proposed resolution in July 2017. The UAC (with me as a member), the EC, and Our Environment met on August 16, 2017, to discuss the resolution.

During the joint meeting,[5] the city's Energy Services Director, Larry Dunbar, discussed the details of the 20-year contract between the city and Bonneville Power Administration (BPA), which is divided into two Tiers of electricity prices (Tier 1 equals 3.5 cent per kWh and Tier 2 equates to about 7 cents per kWh). In order to avoid Tier 2 electricity, the city incentivized its customers to use natural gas as a heating source during the winter. Mr. Dunbar explained that 97 percent of the power purchased from BPA is renewable or carbon free. The remaining power came from solar either from the Ellensburg Solar Park or rooftop solar systems from city customers. He stated that, in order to achieve 25-percent solar, the city would need a solar farm covering about 175 ac of land or about 3600 residential solar systems.

A representative from Our Environment stated that the 25 percent coming from solar was not flexible. He then explained the negative impacts associated with the extraction and use of natural gas—a point I had, in fact, discussed during my presentation to Our Environment—and how their resolution would help reduce that impact. A commissioner mentioned the advances in technology that made renewable energy more efficient and economic.

I expressed my views regarding the environmental impacts of fossil fuels, global warming, and the potential implications of relying only on renewable energy. I agreed with Our Environment's arguments about climate change but said that we could not focus exclusively on renewable energy to solve the problem. We should *transition* to renewable energy. While I agreed that green technology was becoming cheaper and more efficient, unfortunately, the energy return on investment on fossil fuels is still higher and the price cheaper when eliminating the hidden costs of extraction, transportation, and petrochemical transformations. I further suggested that if we wanted to include the hidden costs of fossil fuels, we should do the same for green technology by tracing the socio-environmental impacts throughout their commodity chain, especially at production sites. (This is especially true if we consider the rare-earth minerals needed for manufacturing solar panels and batteries, which have negative geopolitical ramifications too.) I concluded by stating that the problem was complex, that I was not in favor or against the resolution, and that Our Environment should consider revising their proposal.

A member of the EC expressed concern that the resolution could add an economic burden to low-income members of the community who could not afford an increase in the price of electricity or buy a home under the proposed codes. I seconded his point and explained that this would raise energy security issues in the city, which could marginalize vulnerable residents. The meeting was ending, and it was time to vote. After a long deliberation, the UAC and the EC voted and forwarded an unfavorable recommendation to the city council.

Opportunities, challenges, and lessons learned

Opportunities

Political ecologists and critical resource geographers have research agendas that explore the social and environmental justice issues produced by resource extraction in communities located around the world. I believe that there is further need for these scholars to bring academic practices and theoretical conceptualizations of community engagement to resource issues and controversies where they live and work. Staeheli and Mitchell's (2008) discussion of the politics of relevance in geography helps us understand the dichotomy that exists between the discipline's turn to social theory and what they see as a lack of direct community engagement. They argue, "while rapid theoretical development has perhaps drawn geography closer to the heart of main debates in the social sciences… it may have also drawn it further from the social movements, political formations, policy makers, and lay people many of us hope to reach" (Staeheli and

Mitchell. 2008, 357). Moving away from the confines of academia to engage in local issues presents many opportunities. Professionally and intellectually, engaging in local issues allowed me to use my theoretical training to provide nuance opinions about solar development: for instance, when I applied my expertise in public hearings and community conversations to underscore how the biophysical characteristics of solar energy defined its commodification in the form of electricity or how social relations embedded in the development of large-scale solar projects shaped the local debate.

My engagement in local discussions about solar power allowed me to learn diverse views and perspectives. Engaging in local issues as a scholar who advises and provides recommendations to local government officials allowed me to connect with different actors in the community in new ways. I moved from being a county resident passively observing how the debate developed to one with a more active role in which I engaged in sometimes uncomfortable discussions about different issues in my community. It also provided a space where I could share my opinion, listen to others, and see politics playing from the front seat. In these moments, I learned about the complexities of local politics, the perspectives of different actors, and the implications of my position about an issue with great nuance. I observed how policies and laws are interpreted differently and gained a greater understanding of the local implications of energy policies developed at the state level. I also was able to bring my academic training to bear. In the case of Our Environment's resolution, for example, my knowledge of climate change and energy justice allowed me to speak to the complexity of the difficult decision that the UAC had to make in terms of the desire to be carbon free in a way that would not affect vulnerable county residents. I came to consider public hearings, official community conversations, and UAC discussions as educational spaces where actors learned from each other's views and experiences.

These experiences also influenced how I teach my courses, mentor my graduate students, and conduct research. My training as a critical resource geographer and political ecologist helps me identify the multi-scalar interconnections associated with resource development. These interconnections became evident in the solar debate in Kittitas County, and I decided to incorporate some aspects of the debate in courses I teach in the form of class discussion, final research projects, and field trips. I ask students to think about how a global issue such as anthropogenic climate change shapes global energy policies to reduce GHGe and how these connect to and have implications over national, state, and local energy issues. The fact that the solar debate in Kittitas was something concrete that students could grasp and relate to made these in-class exercises more meaningful. My experiences also shaped the way I mentor graduate students. For example, it further underscored the need to ask them to think carefully about their role and responsibilities as a researcher during fieldwork and the implications of their interactions for the community when they come back. My engagement in this debate has had implications in the way I think and go about my research too. As I describe later, I am now more concerned than ever about the implications of my actions for the people I interview or interact with during fieldwork.

Challenges

The first challenge is time management: how to balance my institutional responsibilities while engaging in local issues. I was a tenure-track Assistant Professor directing IIES and IEM when the debate was developing. In this context, I received several requests to participate in community events and round-table discussions to provide my academic opinion about energy issues, especially solar development in Kittitas County. My tenure and promotion application was also looming. Juggling all these responsibilities was difficult and stressful: how much time could I dedicate to these commitments while maintaining a research agenda in northern Patagonia, Argentina?

The second, related, challenge has been making sure that my department and university consider my local engagement part of my service requirements. I included documentation of my participation and the time required in these commitments in my workload plan and activity report every year. I also explained the potential benefits of my engagement in my reflection on service included in my file for tenure and promotion. I will do the same for my participation in the UAC in my post-tenure review and application for promotion to full professor when the time comes. Undergirding these reporting activities is my belief that engaging in local issues is not an external endeavor outside a faculty member's responsibilities. Instead, as Butin (2007, 35) contends, "we should reconceptualize and refocus it as inherent to faculty members' work as teachers and scholars." Perhaps one of the reasons why there is a lack of engagement in local issues among critical resource geographers is the institutional constraints over what counts as service and scholarship for tenure and promotion purposes. Klein et al. (2011, 433) state that universities should support research agendas that benefit the community "through financial means, relief from other professional obligations and recognition of professional and learning achievements." However, as Butin (2007, 37) contends, faculty are often not "rewarded for link-ing their courses to their communities; grounding their research in real-life community dilem-mas; or disseminating their research to non-academic audiences."

The third challenge relates to the views I represent when engaging in local issues. How can I maintain consistency in the message I communicate to different audiences when providing my perspective as the Director of IIES and IEM (representing the views of CWU), or as a member of the UAC (representing the views of the city), or my personal opinion? These interactions are messy and sometimes uncomfortable, especially if my positions are seen as contradictory. For instance, my presentation to Our Environment showed the socio-environmental impacts of oil and gas extraction in vulnerable communities in Venezuela and Argentina and ended with a call to transition from fossil fuels to alternative energy. Yet, three months later, I found myself, as a member of the UAC, voting to forward an unfavorable recommendation to the city council regarding Our Environment's resolution, not because I had reversed my general position on renewable energy, but because of energy justice concerns. This is by far the most difficult part of my job, because the issues are about politics and integrity, which highlights the connections and gaps between policy and academia. This example illustrates the difficulty of translating academic understandings of the complexity of energy politics into a policy setting. I tried to overcome this challenge by being clear about what hat I was wearing when answering a question or express-ing my opinion; by crafting consistent answers; and by staying true to my ethics and basing my opinions on facts.

Lessons learned

Many critical resource geographers engage with communities where their research takes place. In my case, this was the fishing community of Ancón de Iturre in Lake Maracaibo, Venezuela (Delgado 2017), and more recently the community of Añelo in northern Patagonia, Argentina (Delgado 2018). In both cases, I sought to develop relationships based on mutual trust and respect. I worked to accommodate the need of the interviewees by creating opportunities for them to set the terms of our interactions by deciding how, where, when, and for how long the interviews would take place. However, I learned that there are differences between going to live with a community for an extended period to conduct research and engaging in local issues where I live. One difference is the fact that, in the latter case, my views are scrutinized not only by community residents but also by my colleagues and administrators of the university where I work. In other words, with local engagement, there may be no leaving "the field."

This experience also has led me to rethink how I interact with the nonlocal communities with which I work. More than ever, I think carefully about the questions I include in interviews and aim to be sensitive about how I ask them. And, I aim to provide the space to discuss with interviewees the implications of their participation after I leave, with the goal of pursuing more nonhierarchical power relations between me and the participants. Kitchin (1999; quoted in Fuller and Kitchin. 2004, 4) reminded us that our role "is not simply as expert but primarily as enabler or facilitator, and the role of the participants is one of co-researcher or co-activist. This arrangement allows the research to become more reflexive, reciprocal and representative." Such partnership can be more emancipatory and politically progressive, resulting in "knowledge co-generation, and at minimum knowledge exchange" (Klein et al. 2011, 427).

I also learned that engaging in local issues does not fit neatly within a fieldwork period and that the interactions can be difficult. The demands associated with local engagement can seep into different aspects of the daily routine. For example, I interact (within and outside of working hours) with other county residents, colleagues, and administrators who might have disagreed with my views about the CSP or the city's need to rely more on solar power. The same can be said about my research in Argentina. However, there, I only do fieldwork for a short amount of time. That is not the case when I am engaging with community debates about resource issues where I live and work. In this context, the dynamics of academic praxis and community engagement has shifted because I am not limited by the space and timeframe constraints of conducting research in a distant place.

As a critical resource geographer, while my academic experience can be useful to conceptualize resource development issues, sometimes it does not fit perfectly with local cultural norms, historical trajectories, and local politics (see also Shapiro-Garza et al., Chapter 20 in this volume). This is true regardless of where I conduct my research. Nevertheless, my understanding of these mismatches may be clearer (or different) when I do work where I live. This allows me to be more open to confront my own preconceived notions and rethink my role as a scholar and member of the community.

Conclusion

Moving beyond the confines of academia to engage in local resource issues is physically, intellectually, and emotionally demanding. Indeed, engaging in community issues "at home" is not easy, in part, because it "forces faculty members to confront the limits of their identity as productive and effective scholars" (Butin 2007, 35). This is especially true if a faculty member's engagement does not result in scholarly publications that count toward tenure and/or promotion. In this context, local engagement requires a keen understanding of the balance between one's responsibilities as a faculty member and the costs, benefits, and other implications of their interactions as a citizen and community member.

The siting of energy projects is the source of social controversy among different actors when their views about landscape characteristics or policy interpretation do not align. Along these lines, Cowell (2010, 223) states, "achieving reflexivity between energy policy goals and contextual conditions raises exacting questions about how landscape characteristics become pulled into the state's strategic planning processes, what is omitted, and the consequences of the compromises that are struck." This can be seen in Kittitas County where the desires of some residents to protect the rural character of the county collide with the state's goals to reduce GHGe and the capital imperatives of private firms wanting to develop solar projects in irrigated agricultural land. Critical resource geographers can use their expertise to engage with the community and advise local governments about resource development issues, which may also provide them a better understanding of how policy is interpreted and decisions are made, especially when outcomes conflict with what they believe is right. In this sense, McCusker (2015, 195) argues, "taking the time to engage with and understand some of the internal dynamics helps explain what appear to be contradictory or inexplicable policies." Regardless of how stressful, contradictory, or messy academic praxis might be, I believe it is important to move beyond the confines of academia, because scholars have a social responsibility to use our expertise to engage in controversial local issues and make a difference on the ground.

Notes

1 I want to thank Dr. Craig Revels for his insightful comments and edits on an early version of the manuscript. Thanks to Dusty Pilkington for the cartography. Special thanks to the editors for their helpful comments and suggestions. All mistakes and omissions are my own.

2 Policies designed to incentivize the generation of electricity from renewable sources.

3 A CUP is needed for all projects proposed in a zone classification that requires a review and hearing under the provisions of the Kittitas County Code. Conditional uses would not be allowed in specific zones unless the proponent of the project can demonstrate that there is a compliance with each of the CUP criteria at that particular site.

4 The UAC is composed of seven representatives and provides recommendations regarding energy policy and operations to the city council. My participation fills the position of customer of city utility systems.

5 Information presented here is a summary of my recollections and notes I took during the joint meeting between Our Environment, the UAC, and the EC. I used the EC/UAC official minutes to corroborate the information.

References

Blaikie, Piers. 1985. *Political Economy of Soil Erosion in Developing Countries*. London: Routledge.

Buhr, Tony. 2017. "County Commissioners Deny Solar Farm Permit Near Kittitas." Daily Record, January 11. Accessed 20 February 2019. https://www.dailyrecordnews.com/news/county-commissioners-deny-solar-farm-permit-near-kittitas/article_408ef8f5-6339-5a1d-9ee2-2a33485a00dc.html

Butin, Dan. 2007. "Focusing Our Aim: Strengthening Faculty Commitment to Community Engagement." *Change: The Magazine of Higher Learning* 39 (6): 34–39.

Cowell, Richard. 2010. "Wind Power, Landscape and Strategic, Spatial Planning — The Construction of 'Acceptable Locations' in Wales." *Land Use Policy* 27: 222–232. doi: 10.1016/j.landusepol.2009.01.006.

Daily Record Editorial Board. 2017. "Editorial: Complex Solar Farm Decision." Daily Record, January 12. Accessed 20 February 2019. https://www.dailyrecordnews.com/opinion/editorial/editorial-complex-solar-farm-decision/article_fb6c1a86-3ad1-584d-ac44-1a30cfa43664.html

Delgado, Elvin. 2017. "Conflictive Energy Landscapes: Petrocasas and the Petrochemical Revolution in Venezuela." In *The Routledge Companion to Energy Geographies*, edited by Stefan Bouzarovski, Martin Pasqualetti, and Vanesa Castán Broto, 330–346. New York, NY: Routledge.

Delgado, Elvin. 2018. "Fracking Vaca Muerta: Socio-Economic Implication of Sale Gas Extraction in Northern Patagonia, Argentina." *Journal of Latin American Geography* 17 (3): 102–131. doi: 10.1353/lag.2018.0043.

EFSEC. 2018. "Report to the Governor on Application No. 2017-01." Accessed 19 December 2018. https://www.efsec.wa.gov/Tuusso_Solar/Recommendation/ReportGovInslee.pdf.

Fuller, Duncan, and Rob Kitchin. 2004. "Radical Theory/Critical Praxis: Academic Geography Beyond the Academy?" In *Radical Theory/Critical Praxis: Making a Difference Beyond the Academy?*, edited by Duncan Fuller, and Rob Kitchin, 2–20. New York, NY: Praxis Publishing.

Johnson, Kirk. 2018. "Solar Plan Collides with Farm Tradition in Pacific Northwest." *The New York Times*, July 11. Accessed 18 October 2018. https://www.nytimes.com/2018/07/11/us/washington-state-rural-solar-economy.html.

Kitchin, Rob. 1999. "Morals and Ethics in Geographical Studies of Disability." In *Geography and Ethics: Journeys in a Moral Terrain*, edited by James Proctor and David Smith. London: Routledge. 223–236

Klein, Phil, Munazza Fatima, Lindsey McEwen, Susanne Moser, Deanna Schmidt, and Sandra Zupan. 2011. "Dismantling the Ivory Tower: Engaging Geographers in University-Community Partnerships." *Journal of Geography in Higher Education* 35 (3): 425–444. doi: 10.1080/03098265.2011.576337.

McCusker, Brent. 2015. "Political Ecology and Policy: A Case Study in Engagement." In *The Routledge Handbook of Political Ecology*, edited by Tom Perreault, Gavin Bridge, and James McCarthy, 188–197. London: Routledge.

Staeheli, Lynn, and Don Mitchell. 2008. "The Complex Politics of Relevance in Geography." *Annals of the American Association of Geographers* 95 (2): 357–372. doi: 10.1111/j.1467-8306.2005.00464.x.

USDA. 2017. "Census of Agriculture: County Profile — Kittitas County Washington." United States Department of Agriculture. Accessed 9 December 2019. https://www.nass.usda.gov/Publications/AgCensus/2017/Online_Resources/County_Profiles/Washington/cp53037.pdf.

WA Department of Commerce. 2019a. "Energy Independence Act (EIA or I-937) — Washington State Department of Commerce." Accessed 5 January 2019. https://www.commerce.wa.gov/growing-the-economy/energy/energy-independence-act/.

WA Department of Commerce. 2019b. "2019 Biennial Energy Report: Issues, Analysis, and Updates." Washington State Department of Commerce. Accessed 5 January 2019. https://www.commerce.wa.gov/wp-content/uploads/2013/01/COMMERCE-Biennial-Energy.pdf.

Watts, Michael. 1983. *Silent Violence: Food, Famine, and Peasantry in Northern Nigeria*. Berkeley, CA: University of California Press.

26

Learning about coal frontiers

From the mountains of Appalachia to the streets of South Baltimore

Nicole Fabricant

Introduction

Resource Wars is an upper level anthropology elective that I designed and regularly teach at Towson University. The course covers theories of global political economy, critical race theory, environmental inequity, and natural resources, incorporating both a macroeconomic analysis of resource economies with a microlevel analysis of the experience of living in a resource-rich zone. Since 2011, I have taken approximately 150 students on multiple trips to visit an active site of mountaintop removal (MTR) coal mining on Kayford Mountain in West Virginia. After three to four days at the site, we then trace the transport and processing of nonrenewable fossil fuels from Kayford Mountain back to Baltimore, where coal is burned for energy. At both sites, students learn about Just Transitions, which involve place-based education principles, as well as processes and practices that build economic and political power to shift from an extractive to a regenerative economy.

This "living classroom," where ideas and theories from the classroom are brought to life through experiential learning guided and led by environmental activists (see also Courtheyn and Kamal, Chapter 23 this volume), provides students with the tools to understand: (a) the long history of coal and low-wage labor forces; (b) labor militancy and its demise in the 1980s; and (c) the rise of MTR as a cheap, quick fix for our energy needs. The critical pedagogy approach of "witnessing" and "experiencing" these processes alongside activists raises students' political consciousness and moves them toward critical self-reflection and political action.

My pedagogical and methodological praxis comes from my work as a scholar-activist with movements like the Landless Peasant Movement in Latin America (known by its Portuguese acronym, MST), which use reading and critical reflection as a praxis for envisioning alternative worlds. MST borrows from a Freirean educational model, (Freire 1994) whereby peasant farmers use their own experiences to think about and critique the broader capitalist agrarian system. Organizers use witnessing, or as American Studies scholar Fouts (2020) puts it, "critical accompaniment," to teach allies to listen to and accompany individuals to better respond to injustices and help break down barriers through movement-building strategies. The educational strategy behind the MST curriculum is to use the power of narrative and storytelling to transform dispossessed farmers into active and critical thinkers. My experience with MST challenged me to think about the question of dispossession in urban environments like Baltimore and to incorporate similar pedagogical strategies into my university teaching.

This chapter is organized as follows. The first three sections outline the journey that I and students in my Resource Wars course take from Baltimore toward West Virginia and back to Baltimore. I rely on moments from our 2015 trip to illustrate how this course connects students with the historic context of mining towns in the twentieth century. We start at the Whipple Company Store, a historic mining company store turned into a museum in Beckley, West Virginia, where students learn through a demonstration of archives and storytelling that connects them to a longer history of exploitation and environmental degradation. Then we move to Kayford Mountain. The last section discusses the export zones where coal is transported to, and the consequences of living amidst it in Curtis Bay, Maryland, a low-income neighborhood in South Baltimore. The concluding section reflects on how this course has shaped students' lives.

Whipple Company Store: "The building is a living breathing history book"

On our 2015 trip, we left Towson University at 8:45 a.m. on a warm April morning. A total of 30 students piled into 12-seat passenger vans with their sleeping bags, tents, backpacks, and snacks for a seven-hour journey to Beckley, West Virginia. The radical geographical, economic, and cultural transformations from the streets of Baltimore to West Virginia often catch students by surprise: paved roads lead to unpaved ones as the landscape rapidly turns from semi-urban and suburban to rural, while the vans mark the complex terrain from consumption to production zones. Before making the trip, students rarely think about charging their iPhones, iPads, or computers. They have a kind of innocence and naiveté surrounding the world of electronic consumption and how it relates to commodity-chain analysis.

As we approached Beckley, with a population of about 16,000, students noticed as we scanned the local radio stations that each program ended with sponsorship announcements like: "This program was made possible by friends of coal." Others pointed to the large and flashy billboards, with messages like "Coal Supports our Schools, Do You?" and "Coal is America's Energy. It's Good for America and It's Good for West Virginia." This publicity prompted students in the van to make connections with works we'd read in class about how the hegemony of the big coal industry in Appalachia mystifies the uneven landscapes of class, race, and environmental inequities that characterize the fossil fuel industry.

The students made connections between the history and hegemony of coal and the history and hegemony of whiteness as national identity, tying back to Lipsitz (2006). Others thought of Pulido's (2016) work on uneven "landscapes of whiteness." The intensity of white poverty in West Virginia shocks students, and some immigrant students from Central America even remarked that the roads and the housing structures reminded them of neighborhoods back in Guatemala and El Salvador.

The coal industry has transformed the region from a place of globally noteworthy natural diversity to an increasingly polluted, degraded landscape with few future economic possibilities. As anthropologist McNeil (2011, 141) has written:

> Economic development in West Virginia, especially southern West Virginia, is still guided by a coal first philosophy. Beyond actual mining, coal's extractive industry model profoundly influences economic development in general. West Virginia's economic development policies follow a pattern in which the state systematically favors industries over citizens regardless of the consequences.

McNeil compares West Virginia to the banana industry in the Caribbean, where local production, tied to the fate of one industry, is dependent on unpredictable distant consumers. In both
contexts, this dynamic has led to intense poverty and inequality. As West Virginia environmental
activist Junior Walk noted in 2015, "We have always been a resource colony. We do not reap the
benefit of the rich natural resources in our backyard" (personal communication, April 2015).

Visiting the Whipple Company Store—founded in 1890 by Justus Collins—allowed students
to learn about the complex history of coal and the low-wage labor regimes of miners.[1] Company stores were retail stores in the early 1900s that sold a limited range of food, clothing, and
daily necessities to employees of a company. Typically, a company town was in a remote area
where everyone was employed by one firm, such as a coal mine. Miners lived in housing owned
by the company, and the company store only accepted scrip or non-cash vouchers issued by
the company in advance of weekly paychecks. Miners were thus both producers and consumers
for the economic gain of coal barons. Our trip to the Whipple Company Store brought to life
the uneven formations of coal extraction throughout the late-nineteenth and early-twentieth
centuries—the devaluation of the lives of white impoverished West Virginians inside the mines
and the total control of their bodies and homes outside the mines—and provided insight into
the contemporary impoverishment of Appalachia.

As students learned about early regimes of capital accumulation through coal mining at
the Whipple Company Store, students also connected the history of underground mining to
contemporary MTR extraction. The miners who relied on exploitative systems exemplified by
the Whipple Company Store suffered physical abuse and mistreatment and lived at the complete mercy of the coal baron. MTR, on the other hand, destroys bodies of water and physical
landscapes while wreaking a slower violence in the bodies of Appalachians as they drink water,
breathe air, and use polluted soils for agriculture (on water and environmental racism, see Purifoy, Chapter 10 this volume). Exposure to danger and harm is less visible and explicit today but
deeply embedded in the history of coal extraction (McNeil 2011).

Students learned about other aspects of the area's labor history, including the Baldwin–Felts
Detective Agency, a private detective agency active from the 1890s until 1937, which violently
confronted labor unions that tried to organize during West Virginia's coal wars.[2] Students learned
that guards surveilled miners' bodies, movements, and even personal relationships. Women's
bodies, too, served to repay miners' debt, through sexual favors to the company.

Students were mesmerized by Joy Lynn—a native of Fayette County and the museum cura-
tor—who told colorful stories to create a virtual reality of early miners, their everyday lives, and
their organizing efforts. Walking into the store, the steps fan out like a triangle, which we learned
was so that the guards could have a close view of all who entered it. Some of my students remarked
that it looked like a kind of "panopticon." In parts of the store, guards could overhear miners'
conversations because room design amplified voices. "This was early 1900s and they already had
early forms of state-based surveillance, or should we say coal-based surveillance," Joy explained. She
shared patches of a quilt that women made in the early 1890s to secretly communicate where and
at what time union meetings would be held (personal communication, April 2015).

I have found that while students engage with the history of coal in the classroom, they do
not always fully empathize with Appalachian miners until they experience the Whipple Company Store, demonstrating what is repeatedly shown in studies of experiential education (hooks
1994; Shor and Freire 1987). Joy's presentations drew from testimonials from older miners about
growing up in a coal camp and the hardships they faced. Her tour underscored the physical
abuse and hunger the miners faced and their dependency upon coal barons like Collins. As she
shared, "Coal to a coal miner is personal," a perspective reflected in isolated mining communities
worldwide. The tour felt intimate and personal to the students too. I could see many reflecting

in new ways about this history, putting theories learned in classes into motion, and developing empathy. One student whispered to me as we moved through the store, "I have never felt so connected to Appalachian people as I do now. Their stories are so powerful." The power of narrative and storytelling brought students at times to anger and to tears, but always into intense conversation. It activated their minds and challenged them to search for interlocutors to share thoughts and ideas in ways that the static space of a classroom fails to provoke.

The mechanisms of control that coal barons exercised over miners in the twentieth century have parallels today in the practice of MTR coal mining. On this trip, the body of the miner and the body of the mountain came together in powerful ways.

MTR coal mining: "Sostalgia"[3]

After leaving the Whipple Company Store, we drove to Kayford Mountain, an active site of MTR blasting about a half hour away from the museum. Students got settled at the Stanley Heirs camp site, 50 acres of land that environmental activist Larry Gibson has preserved despite coal companies offering him millions to sell his family heirloom and open the land to mining. The land has been in his family since the 1700s and is now established as a land trust where student groups come to learn about MTR coal mining. I met Gibson in 2011 on one of my first trips to the region. As an environmental activist, he has lived constantly fearing attacks by pro-coal groups, to such an extent that he has cameras tied to the rearview mirrors of his four-wheel-drive pickup. When I asked about the cameras, he said, "When these coal barons kill me… at least I'll have proof that they did it" (personal communication, 2011).

Gibson has preserved his family's land in a context in which much of the region's history has been destroyed due to the detonation of mountains, which leaves large craters and rocks that disrupt the grave sites of generations. As blasting has left a landscape of pockmarks and holes, the region no longer looks or feels like home for many who live there. As organizer Dustin White, from Ohio Valley Environmental Coalition, told us, "this is called 'sostalgia:' The idea is that the landscape due to extraction has so radically changed that one no longer recognizes the place of home" (personal communication, 2015).

Within the last 20 years, rapid transformations have occurred, from expanded underground mining to new forms of extracting coal through MTR, which removes up to 800 feet of a mountain's top to gain quick access to the coal beneath. It is the cheapest and easiest way to extract bituminous coal in Appalachia and provides the perfect recipe for businesses like Massey Energy—which was the fourth-largest coal producer in the United States in the early 2000s and the largest in central Appalachia—and their successor Alpha Natural Resources, Inc.—currently the world's third-largest metallurgical coal producer—to accumulate capital quickly.[4] Several advances in technology facilitated the rise of MTR, among them computer engineering, large machinery such as draglines—$100 million machines that can move 100 t of material with each scoop—and powerful explosives of ammonium nitrate and fuel oil (McNeil 2011). Explosives remove up to 400 vertical feet of mountain to expose the underlying coal seams with excess rock and soil laden with toxic mining byproducts often dumped into nearby valleys. Although valley fills are supposed to be carefully terraced and engineered with water diversion ditches, they frequently bury streams. More than 1200 mi of streams have been buried in the Appalachian region, possibly as many as 700 mi in West Virginia alone (Appalachian Voices 2020).

During our 2015 trip, students listened attentively as residents and environmentalists narrated the ways in which MTR has affected their water and air quality, exposing them to ever greater risks. Residents have emphasized that burying the headwaters of streams causes irreversible damage to regional ecosystems, and blasting away layers of mountains removes layers of the aquifer.

Strip mining also increases the possibility of flash floods, and there have been instances when rocks from blasts fall into people's yards and damage property. Fly ash, a common by-product of MTR coal mining, has caused all sorts of health problems in the area. Exposure to heightened levels of lead from the ash has been linked to kidney diseases, swelling of the brain, and reproductive problems (McNeil 2011).

The second and third days of our 2015 trip focused on the geographic, environmental, and local health consequences of MTR. We hiked to an active blasting site with Elise Keaton, Acting Director of Keepers of the Mountain, and Junior Walk of Coal River Mountain Watch. Junior, a 20-something activist with a full beard and thick West Virginia accent, became our teacher and guide. He was born in Whitesville, West Virginia, a short 20-min drive from Kayford Mountain. Gathering the group in a large semi-circle, he explained:

> When I was a little kid, you could walk through Whitesville, West Virginia, and there were two movie theaters, a bowling alley, and a bar on every corner [...] There has been a mass die-off in West Virginia, and Whitesville has been one of the hardest areas hit by the opioid crisis. When coal companies began employing this method of mining, many [people] lost their jobs, moved away, or fell ill"
>
> (personal communication, April 2015).

Students sat on the edge of Kayford Mountain with Elise and other youth leaders from Keepers of the Mountain looking at a crater in the middle of the Appalachian Mountain range, a casualty of MTR. Some described it as a "moonscape," while others simply sat and stared in disbelief. Elise explained that the Appalachian region is the water source for the entire eastern seaboard, "from Maine to Georgia." She said over 40 percent of these rivers have been compromised. (personal communication, April 2015).

Elise discussed the Elk River chemical spill that had occurred in January of that year, when a chemical company spilled crude methylcyclohexanemethanol—a toxic chemical used to clean coal, also called crude MCHM—into the Kanawha River. She didn't find out about the spill until after she'd drunk a whole glass of tap water. Elise told us that she had been planning to have a baby over the next year but was reconsidering trying to get pregnant, because the medical community had warned families that methylcyclohexanemethanol in the water supply could result in birth defects and deformities. From Elise's experience, students made connections between the earlier history of control that the Baldwin Felts Guards had exerted over coal miners in the twentieth century and how coal companies continued to control the bodies of low-wage miners who lived in this region through environmental degradation and associated hazards.

Next, we took part in a water workshop in which community residents shared different water samples. First, students heard from Dustin White, an environmental organizer for Ohio Valley Environmental Coalition.[5] He showed students samples of orange and green water contaminated by coal slurry. Analysis of this water revealed iron, mercury, arsenic, and other heavy metals. As students passed around jars of water, they heard Dustin's story of growing up "pro-coal" until he went on a helicopter fly-over of Boone County and observed the cancerous sores on the landscape, where machines were "eating away at the lush green forests that I used to play in as a kid. I saw the graves of my ancestors in an island of trees in the middle of a vast, barren, lifeless moonscape" (personal communication, April 2015).

These workshops, surrounded by the active sights and sounds of blasting, challenged students to suspend regional, ethnic, and class barriers. Often in the classroom, middle-class white suburban Towson students have a hard time connecting with the lived experience of urban Baltimore youth of color. Often, students of color are the ones who have to explain how and in what way environmental injustice is occurring in their communities. However, this engaged learning experience challenges students across race and ethnicity to see inequalities in new ways. As one student shared, "I thought that low-wage labor and exploitation only occurred within the restaurant industry in Baltimore. I had no idea that we could be connected to West Virginia."

Students translated what they learned in the classroom into active conversations about the human consequences of MTR and about how low-income Appalachians have become collateral damage for coal companies. Yet the youth organizers from Keepers of the Mountain and Coal River Mountain Watch also had an alternative vision of what a Just Transition from coal would look like. Coal River Mountain Watch organizers have been working toward a 440-MW wind farm consisting of 220 wind turbines to be constructed on land slated to be blown away for coal mining, and the coalition has done extensive studies to show just how viable an alternative this wind farm would be in creating more than 200 local jobs and powering more than 150,000 homes. This is a source of clean, renewable energy that would also produce sustained tax income that could be used for the construction of new schools, Junior explained. These plans have faced much resistance from the coal community.

Around a campfire that evening, students discussed the connections they felt with the activists around a shared sense of struggle—many of Towson's working-class students have certainly struggled, particularly in an era in which education is more and more expensive, and they are working longer hours. One student spoke about how she too grew up without running water and plumbing and said that many of the activists reminded her of her own father. But most of our Towson students do not live in "toxic zones" or "contaminated environments," so witnessing the activists' experience gave them immediate insight into the socio-geographically segmented system of capitalism. Most concluded that despite their struggles, they were privileged compared to their West Virginian counterparts.

We closed the West Virginia portion of the course by drawing inspiration from bell hooks and other Black radical feminists who discuss "critical pedagogy," one that is never neutral but rather transgressive. In employing this concept within our experiential learning, students move from relative spaces of privilege to feeling, experiencing, and tasting the effects of MTR coal mining on local communities (see also Harrison and Snediker, Chapter 37 this volume).

Over the years, I have taken more than 150 students to Kayford Mountain. Many students from my 2011, 2013, 2015, and 2019 trips have translated their education into political action as they have organized anti-coal forums back in Baltimore. Some 30–40 of those students have used photography and film to communicate what they saw and heard on the mountain. In 2015, around ten students, upon their return from West Virginia, connected with other student groups at Johns Hopkins University and other large universities to interrogate where their electricity and power comes from and encouraging administrators to divest from dirty coal. These campaigns continue at Hopkins and other private institutions in the area.

Tracing coal from West Virginia to South Baltimore

On our way back to Baltimore from West Virginia in April 2015, students turned their cell phones back on and realized that Baltimore had blown up in the news over the weekend. Freddie Gray, a 25-year-old Baltimore resident, had been found dead in a police vehicle after being violently arrested by the Baltimore Police Department. Protests had erupted and students' Twitter feeds alerted them that Maryland Governor Larry Hogan had called the national guard into the city while cars burned, houses and stores were looted, and the entire city was put on lockdown. Students were shocked—and began to connect what they had seen on the mountain to what was going on in the streets of West and South Baltimore.

Many students recognized in the experiences of people in West Virginia the same dynamics as those living in places like Sandtown-Winchester, the community where Freddie Gray died. After his death, high levels of lead were found in his bloodstream, which can negatively affect a person's ability to learn, focus, and achieve (McCoy 2015). Junior Walk from Coal River Mountain Watch had told us that drinking water filled with heavy metals had meant a lifetime of taking medication to control stomach ailments. The residual consequences of our industrial system, whether through mining or other heavy industries, has left bodies contaminated with harmful chemicals—from low-wage laborers of West Virginia to dispossessed Black, brown, and white working-class residents of Baltimore (on industrial toxicity and bodies, see also Purifoy, Chapter 10 this volume).

As an organizer from United Workers (2019), a grassroots human rights organization in Baltimore, who accompanied my class on the 2019 trip summed it up: "The abandoned towns and burnt down buildings in West Virginia remind me of our communities in South Baltimore. I thought often about the sense of abandonment here and how that connects me back to Baltimore. We too have boarded up and abandoned areas" (personal communication, May 2019). Students on my West Virginia trips have made connections between the "devaluing" of poor white bodies in West Virginia with the dynamics of racial capitalism on poor brown and Black bodies in urban Baltimore.

Toxic zones of coal processing and export

Through the classroom portion of the Resource Wars course, students traced coal from its points of extraction in Appalachia to transport via CSX trains to Curtis Bay, Maryland, which lies about 30 mi south of Towson University. Seven miles south of downtown Baltimore, Curtis Bay is a world unto itself: a working class and mixed-race community (45 percent white, 40 percent Black, 5 percent Latinx, and 1 percent Asian) living alongside an open-air coal pit that has generated significant health problems. According to the Environmental Integrity Project (2012), Curtis Bay ranks in the ninety-first percentile for risk of developing cancer from toxic air pollution, and in the ninety-second percentile for respiratory risk, one of the highest across the entire United States. Curtis Bay has several coal-fired power plants, which students in my course visit and where they hear directly from environmental activists about the health consequences tied to coal processing.

Curtis Bay has been a hub of resource movement, transport, and waste disposal in the northeastern United States since the mid-twentieth century and was a central site for shipbuilding during the First and Second World Wars. Today, Curtis Bay is an amalgamation of polluting industries, including but not limited to ship demolition as well as chemical, pesticide, and fertilizer production. Curtis Bay is also a portal for shipping coal from Appalachia to other parts of the United States, making Baltimore the second-largest coal-export pier in the country after Newport News, Virginia. The last few years have seen an increase in coal arriving via CSX trains and piling up in front of Curtis Bay's recreation center, local homes on Hazel Street, and the playground. Curtis Bay also hosts two coal-fired power plants, a medical waste incinerator, a wastewater treatment plant, and three chemical manufacturing companies.

Destiny Watford, an English and Mass Communications major who graduated from Towson University in 2017, is now a full-time organizer for United Workers, where she focuses on environmental justice issues in her own community and training the next generation of leaders in Curtis Bay. Her presentation to students in my Resource Wars class and other environmental organizers begins by asking participants to "take a deep breath. Breathe it in … Breathe in and now breathe out. What do you feel? Anything different?" In a 2019 presentation, she continued, "Every breath I take in … I feel it. We have more than double the rate of asthma in this community of Curtis Bay" (personal communication, 2019). Indeed, Destiny's community has some of the highest rates of asthma and respiratory illnesses in Maryland due to the cumulative effects of diesel emissions from heavy truck traffic, an enormous open-air coal pit, and the combination of incinerators and other coal-fired power plants in the area. Curtis Bay's zip code was among the top 10 in the country for the highest quantity of toxic air pollutants released by stationary (nonmobile) facilities. In 2007 and 2008, Curtis Bay ranked first in the country for the quantity of toxic air pollutants, with 20.6 and 21.6 million pounds released respectively each year.

In order to better understand the health issues in the area and advocate for change, a team of Towson University anthropology students and I are collaborating with United Workers' Free Your Voice campaign and teenagers from the Curtis Bay community to measure heavy metals in the local soils as part of a participatory action research project. Additionally, Morgan State University scientists have helped students measure fine particulate matter and toxins in the air. These relationships came out of our organizing around a proposed trash-to-energy incinerator in Curtis Bay in 2012–2015. Informally, a group of United Workers youth and I were researching together and utilizing the research to build a broader campaign. In 2015, this informal relationship turned into a class at the local Ben Franklin High School, where youth are trained in participatory action research. Towson University undergraduates assist with "collective" research on environmental injustice, and youth translate their findings into action-based projects in the community. Students have recorded and videotaped their conversations, which have then been transcribed and incorporated into popular and academic texts we've written.

Curtis Bay struck many of the participating Towson undergraduates as a twenty-first century company town, similar to that created by Justus Collins to control every aspect of his coal miner's lives. Students pointed out the billboards in the community with CSX logos, which reminded them of West Virginia's coal propaganda. While CSX runs coal trains through Curtis Bay, the company also provides employment and summer internships for teenagers at the local high school. As Dr. Lawrence Brown, Community Health Scholar from Morgan State University, noted in our popular education class with students, "There is a difference between justice and charity. These [CSX donations] are acts of charity by private corporations" (personal communication, March 2019). Grace Chemicals, a large chemical company housed in the Curtis Bay community, also provides scholarships for youths to attend summer camp in the South Baltimore area. As with Collins' Whipple Company Store, this clientelism is another way corporations seek to quell resistance to corporate expansionism and toxicity.

Towson University students participated in a "toxic tour" in 2019 through the neighborhood, led by Destiny, which provided a history of industrial development, displacement, and environmental injustices in the community. Although many Towson students grew up nearby, many have not traveled through Baltimore city, let alone South Baltimore. The last stop on the tour was the Filbert Street Garden run by Ms. Rodette, a lively community member who has invested years working to turn the plot of land into a community-owned and community-run enterprise. Planters are on raised beds to avoid some of the toxic runoff; solar panels produce energy; honey is extracted from a bee sanctuary; and goats and ducks, whose manure is used as fertilizer, roam. While the coal pile alongside the train tracks is still visible from Filbert Street, many students noted that the garden felt like "an alternative landscape." One student in 2019 said, "This is like a utopia surrounded by environmental injustice or dystopias."

At the back of the garden, students visited the Baltimore Compost Collective, run by Mr. Marvin Hayes, who grew up in Sandtown-Winchester. He regaled students with stories of growing up in intense poverty and in a food desert. He told us, "We learn so we do *not* burn," referencing the garden's focus on recycling rather than sending food waste to an incinerator (personal communication, April 2019). Students worked alongside Mr. Marvin to learn how to compost, sorting and separating food scraps, and learning about the three-bin system. One student from my class exclaimed during a 2018 visit, "I didn't even know that the majority of our food waste is being burned by an incinerator in Baltimore."

In addition, students' interaction with the grassroots group Just Transitions led to vibrant and important conversations about alternatives to our fossil fuel–dominated energy empires (Sullivan 2018). As with our 2015 trip, students had heard about Just Transitions in West Virginia from Junior Walk, in terms of his work convincing coal miners that wind and solar energy is a feasible alternative to coal. In Baltimore, students witnessed grassroots and community-based groups fully embracing Just Alternatives (Shen 2018), as with through the Baltimore Compost Collective's efforts to turn food scraps into "black gold" (Marvin Hayes, April 10, 2019) or mulch for gardens, in the process serving as an alternative "economy" of scale. Indeed, Filbert Street Garden has become a political hub for broader zero-waste efforts which activists have been scaling up to the city-level.

Afterlives of Resource Wars

Coming home to Baltimore to see coal piled up and ready for export allows students to understand the ways that the fossil fuel regime takes advantage of low-income communities of color as well as impoverished white communities. During the tour she led, Destiny Watford narrated her journey from innocence to political awareness and action, all through understanding the history and political economy of a coal pile in her community, which she used to feel numb toward, but then moved toward an awakening as she read and studied.

This kind of teaching and learning is also about raising a Freirean consciousness, since it challenges students to step outside of their comfort zones and thrust themselves into someone else's world. As critical resource scholars, our teaching must be emancipatory, pushing students into spaces of deep learning and discovery. It must connect spaces of extraction with those of transport and processing. In the case of my Resource Wars class, this meant connecting communities in Appalachia, where residents cannot breathe due to coal dust, to places where urban Black Baltimoreans cannot breathe. The circulation, transport, and processing of nonrenewable fossil fuels have real consequences: They are matters of life or death for local communities and students who experience this personally are profoundly affected.

Experiential learning can transform students' lives through building critical social relationships, networks, and even opportunities for future employment in social, racial, and environmental justice. Many of my former Towson students have folded themselves into campaign and grassroots organizing work across Baltimore. Some have gone on to work at United Workers as organizers of the state-wide Poor People Campaign, while others have become more deeply involved in building out Just Transitions initiatives, alongside zero-waste experts. For students, their personal emotional and intellectual journey through engaged learning with teenagers in Curtis Bay is about learning how to support frontline communities through collective inquiry and new forms of solidarity. A collective envisioning of alternatives allows students to imagine with and alongside community members what a fossil-free future might look like and how to build Just Transitions from a grassroots perspective.

Notes

1 Collins was a coal baron and an entrepreneur who opened his first coal mine in southern West Virginia. He headed a coal sales agency; speculated in coal and timber lands; headed a cement company; and invested in rubber, oil, and gas companies. He built and operated the Whipple Colliery Company, which became known as the New River Company Store #4 Whipple in 1907.

2   The West Virginia coal wars (1912–1921), also known as the mine wars, arose out of a dispute between coal companies and miners. The first workers' strike in West Virginia was the Cabin Creek and Paint Creek strike of 1912–1913.

3   "Sostalgia" (missing home without leaving home), introduced to the author by a West Virginia resident, expresses a sentiment close to "solastalgia," introduced by Albrecht (2007), which describes the existential melancholia experienced with the desolation of a home environment.

4   Bituminous coal is a black soft coal containing a tarlike substance called bitumen or asphalt. It is of higher quality than lignite coal.

5   The Ohio Valley Environmental Coalition was formed in 1987 to stop a toxic waste dump and incinerator project proposed for a low-income community near Huntington, West Virginia—a campaign they won. By supporting organized voices and empowered communities, they've been winning environmental and social justice campaigns ever since.

References

Albrecht, Glenn. 2007. "Solastalgia: the Distress Caused by Environmental Change." *Australasian Psychiatry* 15: 95–98.

Appalachian Voices. 2020. "Ecological Impacts of Mountaintop Removal." Accessed February 25. http://appvoices.org/end-mountaintop-removal/ecology/.

Environmental Integrity Project. 2012. "Air Quality Profile of Curtis Bay, Brooklyn and Hawkins Point, Maryland." Unpublished paper. https://www.environmentalintegrity.org/wp-content/uploads/2016/11/2012-06_Final_Curtis_Bay.pdf.

Fouts, Sarah. 2020. "When "Doing With" Can be Without: Employing Critical Service-Learning Strategies in Creating the 'New Orleans Black Worker Organizing History' Digital Timeline." *Journal of Community Engagement and Higher Education* 12 (1), 29–38.

Freire, Paulo.. 1994. *Pedagogy of the Oppressed*. New York, NY: Continuing Publishing Company.

hooks, bell. 1994. *Teaching to Transgress: Education as the Practice of Freedom*. New York, NY: Routledge.

Lipsitz, George. 2006. *The Possessive Investment in Whiteness: How White People Profit from Identity Politics*. Philadelphia, PA: Temple University Press.

McCoy, Terrance. 2015. "Freddie Gray's life a study on the effects of lead paint on poor blacks." *The Washington Post*, April 25. https://www.washingtonpost.com/local/freddie-grays-life-a-study-in-the-sad-effects-of-lead-paint-on-poor-blacks/2015/04/29/0be898e6-eea8-11e4-8abc-d6aa3bad79dd_story.html?utm_term=.0481b153ea0a.

McNeil, Bryan. 2011. *Combating Mountaintop Removal: New Directions in the Fight Against Big Coal*. Champaign, IL: University of Illinois Press.

Pulido, Laura. 2016. "Flint, Environmental Racism, and Racial Capitalism." *Capitalism, Nature, Socialism* 27 (3): 1–16. doi: 10.1080/10455752.2016.1213013.

Shen, Fern. 2018. "Update from the Compost Collective." *Baltimore Brew*, November 5. https://www.baltimorebrew.com/2018/11/05/update-from-the-compost-collective-new-shed-new-goats-new-customers/.

Shor, Ira, and Paulo Freire. 1987. *A Pedagogy for Liberation*. Westport, CT: Bergin & Garvey.

Sullivan, Zoe. 2018. "The Hidden Powers of Composting in Baltimore City." *Next City*, September 21. https://nextcity.org/daily/entry/the-hidden-powers-of-composting-in-baltimore-city.

United Workers. 2019. "Community+Land+Trust: Tools for Development without Displacement." Accessed January 22, 2019. http://www.unitedworkers.org/community_land_trust_tools_for_development_without_displacement.

27

Teaching critical resource geography

Integrating research into the classroom

Conor Harrison and Kathryn Snediker

Introduction

While "getting out into the field" has long been viewed as the singular formative learning experience in academic geography, the vast majority of teaching in geography takes place in the classroom setting. Given this reality, how can students study the geographically dispersed and socially and culturally disparate worlds of resources from the confines of the classroom? In this chapter, we describe an in-class research project that students complete by drawing on data sources available through the university library and the Internet. The project design is the outcome of a five-year collaboration between the authors, a professor of geography and environmental science, and an academic librarian. We came to this collaboration when the faculty instructor sought out the expertise of his department's liaison librarian. The library liaison model is prevalent in the academic environment, usually assigning librarians with an interest or subject specialization to support the specific research needs of the various disciplines or departments at the institution. Through this collaboration, we have been able to combine concepts and approaches used by economic and critical resource geographers with those put forward under the umbrella of critical librarianship that we believe encourages students to develop skills in both the critical analysis of resources and the critical consumption and use of data and information. The outcome is an active learning approach to the study of resources that is easily translatable to a diverse range of classroom settings.

Our approach to this project is rooted in the need for classroom-based research projects, in part due to challenges posed by field experiences (on field experiences in teaching critical resource geography, see Fabricant, Chapter 26 in this volume). Fieldwork programs are saddled with difficulties posed by funding, organizing, and staffing while serving as a barrier to a more inclusive program of study. Many students juggle a variety of obligations, including conflicting time pressures from other coursework, family, and/or employment, and fieldwork can at times exacerbate existing inequalities between racially and generationally diverse students (Hughes 2016). In addition, while field experiences are often touted as the premier active learning experience in geography, poorly planned and executed field experiences can reinforce existing ideologies and prejudices (Golubchikov 2015).

While the project we describe here is typically taught in an economic geography course, our approach is also grounded in recognition of the challenges of studying resources. Because undergraduate field experiences are often limited to areas close to the college or university, examining the networked geographies that allow resources to "become" (Zimmermann 1933) is difficult, particularly because resource networks increasingly span the globe and draw together a vast array of actors, institutions, and materials. Further, as Bridge (2009, 1219) makes clear, while the location and availability of resources is often assumed to be "fixed and given," what counts as resources shifts with "cultural appraisals about utility and value" that result from technological and cultural change. Resources, then, must be examined as things—both material and social— that have a historical and geographical context that can be obscured during a single field visit to a mine, power plant, or water treatment facility. The project we describe here is therefore an excellent complement to field experiences, as it allows students to trace commodities across scales and to draw a broader set of connections between various economic sites, places, and processes.

In this chapter, we describe an assignment that asks students to trace commodity chains from start to finish, an approach that helps elucidate the human and more-than-human relationship that make local and global economies. Chain tracing prompts students to expand their understanding of where the products they use come from, and to consider what resources are and how they are made. This chapter begins by reviewing the pedagogical underpinnings of our approach: critical commodity chain analysis and critical information literacy (IL). We then outline the research assignment and describe some of the key data sources we use before turning to a discussion of how the assignment has evolved to address some of the common challenges we have faced. In the conclusion, we point to several ways in which the assignment can be adjusted to address current research by critical resource geographers.

Critical analysis of resources *and* information

The research project we describe in this chapter draws from the geographic research method of commodity chain tracing that seeks to defetishize commodities by unveiling "the often-exploitative relationships of production hidden behind capitalist commodities" (Goodman 2008, 368). Commodity chain tracing has a range of advocates. Coe and Yeung (2006, 400), for example, argue for "using the lives of particular commodities as a thread to connect different topics and themes." Walker (2006, 430) urges that the best approach to getting students to understand the economy is to start with the "system of production and circulation of goods (commodities) that undergirds everything else." Robbins and Moore (2015) echo these sentiments, proposing teaching through case objects as a way to begin thinking about human-environment relationships.

However, as Goodman (2008, 369) makes clear, expecting students to "become magically politicized and caring about … human and ecological exploitation" after yet another lecture from the "sage on the stage" is likely fruitless. Indeed, as Heyman (2001) argues, many students may have little in common with left-wing instructors who pay more attention to theory, politics, and practice in their research than in their teaching, although, as Cook et al. (2007, 1118) points out, "they may have a go for the sake of an essay or exam answer." As a result, pedagogical best practices indicate that rather than lecturing to students about the commodity chain of this or that resource, an active learning approach can enable a more embodied study of the networks of resource production in which students are already entangled (Goodman 2008).

Given the above, Cook et al. argue that commodity chain research and pedagogy need not be didactic, but instead "mobilize rather than dictate meaning, to give audiences/participants material to think with, to provoke questions about what this work is about, what the 'moral of the story' might be, how they 'ought' to respond" (2007, 1118). While we broadly agree with the active learning approach suggested by Cook et al. and others, one particular challenge often emerges. Outside of enabling students to engage in sustained field-based research analyzing resource networks, the challenge falls on specifically *what material* to give students to think with. Attempting to address this question has formed the basis of our collaborative approach to this assignment, an approach that we believe brings together two complementary skillsets: critical geographical analysis with critical librarianship focused on IL.

Active learning assignments based around critical inquiry create the opportunity for academic librarians to incorporate IL instruction into the disciplinary classroom (Fosmire and Macklin 2002; Spence 2004; Weaver and Tuten 2014; Maybee, Doan, and Flierl 2016). Most traditional IL instruction involves a librarian facilitating a single class session (often referred to as a "one-shot") on search strategies and information-seeking behaviors that are largely decontextualized from the course content. However, numerous examples within the library and information science literature describe case studies of librarian-faculty collaborations in a variety of subject disciplines (Ford and Williams 2002; Garcia 2014; Cote and Juskiewicz 2014; Hulseberg and Versluis 2017). Advocates of this disciplinary-based approach to IL argue convincingly that IL "cannot be effectively taught as a value-added addition to the regular course-based curriculum" and that librarians and classroom faculty share the responsibility for teaching the skills that are required for developing subject-specific knowledge through research and inquiry (Grafstein 2002, 202). Disciplinary faculty looking for an introduction to the core concepts of IL instruction should start with the *Framework for Information Literacy for Higher Education* (Association of College & Research Libraries 2016). The *Framework* delineates six frames, each consisting of a concept central to IL (Box 27.1).

Box 27.1   Overview of information literacy

Definition of information literacy

"Information literacy is the set of integrated abilities encompassing the reflective discovery of information, the understanding of how information is produced and valued, and the use of information in creating new knowledge and participating ethically in communities of learning."
  ACRL Framework for Information Literacy for Higher Education

Framework for information literacy for higher education

Six frames, or core concepts:

- **Authority is constructed and contextual**
  Information resources reflect their creators' expertise and credibility and are evaluated based on the information need and the context in which the information will be used. Authority is constructed in that various communities may recognize different types of authority. It is contextual in that the information need may help to determine the level of authority required.
- **Information creation as a process**
  Information in any format is produced to convey a message and is shared via a selected delivery method. The iterative processes of researching, creating, revising, and disseminating information vary, and the resulting product reflects these differences.

- **Information has value**

  Information possesses several dimensions of value, including as a commodity, as a means of education, as a means to influence, and as a means of negotiating and understanding the world. Legal and socioeconomic interests influence information production and dissemination.
- **Research as inquiry**

  Research is iterative and depends upon asking increasingly complex or new questions whose answers in turn develop additional questions or lines of inquiry in any field.
- **Scholarship as conversation**

  Communities of scholars, researchers, or professionals engage in sustained discourse with new insights and discoveries occurring over time as a result of varied perspectives and interpretations.
- **Searching as strategic exploration**

  Searching for information is often nonlinear and iterative, requiring the evaluation of a range of information sources and the mental flexibility to pursue alternate avenues as new understanding develops.

Frames are supported by both knowledge practices and dispositions

Knowledge practices describe some of the activities that learners would engage in as they develop information literate abilities related to that frame.

Dispositions describe the preferences, attitudes, and intentions of learners who are developing their information literate abilities related to that frame.

Source: Association of College & Research Libraries. 2016. *Framework for Information Literacy for Higher Education*. Chicago, IL: American Library Association. http://www.ala.org/acrl/standards/ilframework.

We have also been influenced by research on critical theories of IL. Traditional IL instruction has been criticized for focusing on "skills-based technical aspects of information seeking [while] it does not adequately address critical thinking skills which enable people to critically assess the information they encounter and the structures in which the information and knowledge is held" (Smith 2013, 16). Rather than demonstrating search processes and encouraging unquestioning acceptance of "authoritative" sources as such, a critical IL approach "embrace[s] a collective questioning of how information is constructed, disseminated, and understood" (Cope 2010, 25). Smith proposes that incorporating a critical approach into IL "could improve young people's political knowledge, critical awareness and ability to challenge structures of power and control" (2013, 23). It is following these guidelines that we have over the course of the last five years developed the research project we describe in what follows.

The research assignment: tracing commodity chains

This assignment is part of an undergraduate economic geography course that commonly enrolls around 25 students at a large public university in the American South. As such, the course is not explicitly focused on resources, but in the fifth section, we discuss how a resource-focused course might adapt to this assignment. Typically, around half of the students are geography majors, with the remaining using the course to fulfill elective or cognate requirements from fields as diverse as real estate, supply chain management, public health, and global studies. In general, most of the students come from suburban or urban areas, and while few have lived in or been directly involved in resource extraction, many students have some connection (often via family members) to the large manufacturing sector still present in the southeastern United States. As is the case in many geography and environmental courses and programs (Pulido 2002; Taylor 2018), the vast majority of the students are white and have interacted with commodities primarily as consumers. Thus, for many students, the predominant relation with the commodities they choose to trace in this assignment has been through prices and advertising and, in some cases, social and/or sustainability concerns.

The overall course is organized around geographical political economy and focuses on understanding the geographical underpinnings of the capitalist system. Following the text of Coe, Kelly, and Yeung (2013), we first spend several weeks providing conceptual foundations for thinking geographically about economic activity and then consider some of the ways in which economic space is organized (e.g., clusters and global production networks). We then turn to introducing some of the key inputs and actors in the economy—namely, the state, nature, labor, finance, the environment, and technology. The final weeks of the course examine the intersections of people, identities, and economic life.

The final project instructs small groups of students (two to three students in a group) to select a product that they use and to analyze the process and impacts of producing and consuming the product. This includes tracing the production, processing and packaging, distribution, consumption, and waste streams, including the extraction of the raw materials used to produce the product. Students ultimately deliver a final presentation and develop a fact sheet that summarizes their findings. The final products are largely based on a series of detailed memos that ask the students to think through how their particular commodity chain interacts with concepts they have learned about various facets, processes, and actors in global economy. The assignment is introduced relatively early in the semester, typically around the 6th week of a 15-week semester. By this point, students have been introduced to some key vocabulary (e.g., scale, location, territory) and concepts (e.g., commodity chains, transnational corporations, clustering). The assignment is introduced when the course content shifts to discuss some of the key inputs and actors in the capitalist space economy (e.g., labor, the state, and environment).

The class meeting that follows the assignment's introduction occurs in one of the multimedia classrooms in the library that are equipped with desktop computers to allow hands-on practice during and after the presentation. Groups are asked to come to this meeting with an idea of the product that they would like to analyze. We begin by briefly introducing some of the available library resources, and we then work individually with students to home in on their chosen product and begin finding some key information sources. These sources include company and product websites and the company's annual report if available, industry market research reports, trade publications, news articles along with other media reports, academic research studies, and company profiles. (Box 27.2 provides additional details on research resources our library provided for this project.)

It is in this classroom session that we begin introducing core IL concepts from the *Framework* (recall Box 27.1). Students are instructed to start by gathering a variety of information sources to gain a broad understanding of the company and issues involved, rather than attempting to search immediately for the specific answers required for their subsequent memo assignments (described next). Our aim is to avoid a checklist approach to searching, wherein students stop their inquiry as soon as they feel they can meet the minimum source requirements and instead to develop the skills and dispositions associated with the frame "Searching as Strategic Exploration," including

---

Box 27.2   Recommended library resources

Company information

These databases provide information on public, private, American, and foreign companies that can help students understand the corporate structure of their parent company and its worldwide divisions and subsidiaries. They also generally provide at least brief company descriptions, financial data, and other details to supplement information from websites and annual reports.

- Corporate Affiliations
- Hoover's Online
- Mergent Intellect
- Nexis Uni
- PrivCo

Industry market research

Students should look at reports for the main manufacturing and/or retailing industries for their product as well as the supply and demand industries. Industry research reports from these sources are especially helpful for understanding the structure of the industry and the market as well as the operating conditions, including the regulatory environment.

- IBISWorld
- Passport (from Euromonitor)
- BMO (from Fitch Solutions)

Trade publications and news media

Excellent sources for recent trends and investigative reporting, trade publications, and news articles can be found using any number of library databases. These have particularly good coverage of business and international news.

- Business Source Complete
- Nexis Uni
- Factiva

Scholarly journals

Students can benefit from academic research in a variety of disciplines depending on their selected product, including business, geography, environmental studies, and applied science. However, providing too many options can be overwhelming, so this list is limited to the most broadly useful databases our library has for this specific project.

- Academic Search Complete
- Business Source Complete

- GreenFILE
- Agricultural & Environmental Sciences

Government information

Easily found online, government information provides vital environmental, trade, and regulatory information for students to incorporate into their projects. Some *primary* examples include:

- USA Trade Online (provided by the US Census Bureau—free account required for access)
- Environmental Protection Agency
- Department of Agriculture
- Food and Drug Administration

These resources were chosen based on the subscriptions available from the university library at the time this project was developed. There are many other commercial database options that could replace or supplement those we listed here. We recommend consulting with your institution's librarians.

going beyond initial search attempts and evaluating the relative value of different sources based on their information needs. We also encourage them to reflect on why and how each of their data and information sources was produced and how those motives shape their evaluation of the information, addressing the frames of "Information Creation as a Process" and "Authority Is Constructed and Contextual."

Two weeks after the library meeting, each group turns in an annotated bibliography that includes some of the key scholarly articles, trade publications, and data sources they will use to analyze their products. We review these and often make suggestions on additional resources they should consider. Each group of students then produce a series of four 500-word memos that are the building blocks for the analysis of their chosen product. As shown in Box 27.3, the memos ask students to consider their product in relation to the economic inputs and actors we are discussing during class. For instance, in the case of the state, students analyze state engagement in a particular production process, both in the particular governance functions the state may play (resource owner and regulator) and the scale of state intervention (federal ownership, and local government control). It is at this point that students realize the usefulness of the information they are collecting for the project. Many groups return to the databases to find other data sources, which at times requires that they make appointments to meet with the librarian who provides additional assistance.

The assignment does not dictate that students select products exclusively defined as resources, for example, potato chips, peanuts, baseball gloves, almonds, bottled water, jeans, and tomato soup. However, most products include a raw material or resource dimension and therefore encourage engagement with questions of interest to resource geographers. Peanut production provides an apt example. As one student astutely analyzed, the federal government plays an important role in providing subsidies for peanut production by acting as a production "guarantor" via a price loss coverage program. While the federal role in agriculture is relatively well known in the United States, the same student pointed out the slightly obscured role that state governments play in peanut production by funding agricultural extension offices that support peanut farmers with information and advocacy.

Box 27.3   Memo content overview and structure

Memo 1—the state

1. At which scale is the state involved (local, state, federal, etc.)?
2. Through which governance functions (ownership, guarantor, regulation, etc.) is the state involved? Is your product governed by state and/or national laws?
3. What are the current trends in state involvement? Is the level of intervention and/or regulation increasing or decreasing?
4. Are their non-state entities that are heavily involved in the governance of this product? What functions are they performing?

Memo 2—environment and labor

Environment:

1. How is the environment employed, manipulated, and/or harmed in the production of this product?
2. What is the ownership status of the raw materials and land used to produce this product (e.g., is it leased, and a common pool resource)?
3. Do the resources gain value through the use of standards, labeling, etc.?

Labor:

1. Describe the general working conditions at each stage in the production of this product.
2. Is the workforce unionized? Was it historically?
3. What sort of mobility does labor have in this industry? How mobile is the industry itself compared to labor?

Memo 3—finance and technology

Finance:

1. Track the stock value of key firms over time, if possible. How has it changed in the long and short run?
2. According to financial analysts, what is the outlook for this industry?
3. Who are the key investors and/or shareholders in the firm?
4. To what extent is the market for these products sold on commodity markets? Are there futures prices? Is there concern about price fluctuation because of futures trading?
5. Who is making lots of money in this production chain? Who is not?

Technology:

1. How has technology been used to change the commodity chain? What are the space-shrinking technologies that have been used? Where are the centers of product innovation for this product?

2. Is lean distribution an important part of the production process? How so?
3. How have innovations reshaped the product that is ultimately produced and delivered? Is this product shaped by a product life cycle?
4. Can you link changes in production process to the shift from Fordism to neoliberalism?

Memo 4—transnational corporations and retailing

Transnational corporations:

1. Is this product produced by transnational corporations? Why (or why not) has this firm transnationalized?
2. How is this firm and/or production process part of the production network of other transnational corporations? What is the relationship between those firms? Do they use sub-contracting relationships?
3. Are they part of any strategic alliances or joint ventures?

Retailing:

1. Where is this product sold? Describe the sales/retailing environment through which this product is purchased. Has this changed over time?
2. How does this product fit within the broader retail environment?

An additional memo asks students to investigate labor and the environment. One recent student chose to investigate a pair of jeans produced by Everlane, an "ethical fashion" company. The student quickly identified an interesting conundrum posed by the jeans. While Everlane publicized their efforts to source products from environmentally sound production facilities, the company does not hold any of the international certifications (e.g., Fairtrade, B-Corp, Goodwell). In addition, while the company aims to provide transparency into its production process, Everlane has, in the words of the student, "been criticized due to its lack of transparency in the raw materials portion of the production chain and in fact they have not publicly released their code of conduct." As such, in this memo the student was able to explore how the impacts of cotton production—on the environment, on labor, and otherwise—have been ignored by a company explicitly marketing itself as embracing transparency and ethical production.

In sum, the purpose of the memos is threefold. First, completing the memos ensures that students are making steady progress on the assignment and continuing to engage with their data sources. Second, the memos require that students immediately engage with material discussed in class by asking them to apply it to their particular product. Finally, the memos provide ample opportunities for us to give feedback and enable students to correct course before going too far down an unfruitful path.

Once the memos have been completed and feedback has been provided by the instructor, the groups will begin preparing two final products. The first is a two-page fact sheet that covers each of the memo topics and forces students to distill their key findings. The fact sheet is distributed to the audience at the final presentation, and we have used both traditional Power-Point presentations as well as ESRI Story Maps for the presentations. While ESRI Story Maps provides a dynamic platform to demonstrate the movement and spatiality of commodities and resources across space and makes for compelling interactive webpages, students have found them clunky as a presentation tool compared to the more familiar PowerPoint. While some students are excited about the chance to try new technologies such as Story Maps, successfully introducing and training students on new technologies takes away classroom instructional time. Successfully using new technologies also depends on the instructor having existing knowledge of the software or having outside technological assistance. However, in our experience, this sort of outside pedagogical assistance is often among the first positions cut by universities during times of budget austerity.

Encouraging *critical* commodity chain analysis and information literacy

Two common problems have emerged during our five years running this project, and we have sought to adjust the assignment accordingly. The first challenge typically occurs when students are picking a product. A common problem is selecting something with too many components. For some products, having too many components is quite obvious—a car or mobile phone contains hundreds or thousands of parts. This is less obvious for a product like eye shadow or other make-up products, which seem homogeneous but contain numerous compounds ranging from pigments and oils to shark liver and fish scales. While a fascinating mix, tracing the product network of each of those products is beyond the scope of what can be accomplished in a single semester.

In addition, students often select products that are produced by privately held companies. In the United States, these companies are required to release far less information than those that are publicly traded, and as such, students struggle to find adequate information. In one example, a student selected Bacardi rum, which is produced by Bacardi Limited, a privately held spirits company. Despite rum having few ingredients—only sugarcane—Bacardi Limited and other spirits companies offer very little transparency into their financial accounting, labor, and environmental practices. While this does not absolutely preclude students from studying these companies, it has proven to be a barrier to gathering some of the key initial information.

The second challenge we have faced is how to encourage students to make a *critical* examination of the product and not simply regurgitate information and claims from company websites, annual reports, and 10-K filings. As Goodman (2008) and Cook et al. (2007) have argued, students frequently do not come to these projects steeped in critical approaches to research, and nor should we be didactic in telling them what to think. While this will always be a challenge for students who have come of age in an era of neoliberal capitalism, we have sought to address this challenge in two primary ways.

The first is by tasking students with collecting data from a variety of sources and emphasizing the need to critically evaluate what they find. For each source, students should consider not just the relevance of the content to their project, but also *who* is responsible for creating and disseminating the information and what perspectives, motivations, and goals they may have had when producing it. The case of Everlane jeans mentioned previously is an example of a student confronting the dissonance between, in this case, information produced by the company for marketing purposes and research on cotton production from academic and industry sources. This is the environment we want to create: one in which students can actively engage with sources and question familiar types of "authorities" rather than simply acting as passive consumers of information.

The second method to address this problem is to find ways to tie the assignment more closely into the course material. It is here that the memos have proven critical, as they aim to weave into and reinforce the course material as students are being exposed to it. Simply put, we are asking students to apply the concepts and ideas that they are introduced to during the course. However, rather than forcing them to do this at the culmination of the semester after having been provided with little feedback along the way, completing the memos requires them to use the concepts as they are learning the material. In our experiences so far, the memos have been successful in moving students away from "binge and purge" methods of assessment and toward developing deeper levels of content knowledge through more learner-centered forms of assessment (Biggs and Tang 2011).

Toward a critical geographic analysis of resources

In this chapter, we describe an assignment that has been developed by combining the concepts and approaches of critical librarianship with critical geographic analysis. The goals of this project are for students to critically engage with primary materials in order to investigate and analyze how commodities and resources are made, and to do so using data sources available through the university library and the Internet. The assignment has adapted and improved in large part because it is a collaboration that draws on different areas of faculty expertise and is a working relationship between the authors that is, put simply, fun.

In concluding this chapter, we offer several thoughts regarding how this assignment, which has been taught in an economic geography course, might be adapted to a course that specifically examines critical resource geographies. In doing so, we point to some of the ways such an assignment articulates with several of the key debates and lines of research within critical resource geography.

First, in a course organized around critical resource geography, this assignment should provide more sophisticated engagement with the state and state theory. While in our assignment we typically urge students to think through how the state is involved in governing, owning, or managing the resource, Lu, Valdivia, and Silva (2017) and others have made clear that resources are also a crucial part of the emergence of new forms of citizenship and the processes of state formation (see also, Bustos-Gallardo, Chapter 31 in this volume; Perreault, Chapter 11 in this volume). How can we get students to engage with this idea in their work? One method might be a critical investigation of state media, advertising, and discourses through an examination of social media and other popular outlets (see, for example, the recent rebranding of natural gas as "freedom gas" [US Department of Energy 2019]). This would, of course, require the introduction of another set of sources and analysis methods to teach students but could be possible in the structure of our course.

Second, adapting this assignment to a course on critical resource geography should include direct engagement with resource conflicts (Gergan 2017). The assignment could be shifted so that rather than starting with a particular product and working backwards, students could use a particular resource conflict as an entry point to analyze the commodity chain of that product. For example, conflict stemming from diamond mining could be the starting point to investigate the broader diamond trade and the cultural constructions of scarcity and value commonly attributed to diamonds. In another example, starting with more local controversies over resource transportation systems, like natural gas pipeline expansion, can help students make connections with potentially distant extractive conflicts such as fracking.

Third is a question of methodology. By restricting the assignment to an analysis of secondary data, students do not engage in many of the methods currently employed to study resource geographies. However, this type of project can serve as a building block for longer term and more in-depth projects or field experiences, such as undergraduate theses or a capstone class, that would draw on other methodologies being used by resource geographers. This includes, for example, recent ethnographic analyses of water (Barnes 2014) and copper (Kneas 2018) that help us to understand the ambivalence behind official statistics and to acknowledge the uncertainty behind resource assessment and governance. Other methodological approaches include expert interviews with key figures in firms and institutions of resource governance, like those that have yielded valuable insights into fisheries management (Havice and Campling 2017). Finally, given longer time frames—and better grounding in secondary sources—students can partake in more extensive archival excavations that can illuminate the provisional and impermanent ways that commodity chains form and evolve (Hecht 2012).

Fourth and finally, our assignment typically involves students investigating products that they experience directly, and thus those biological or geological entities that we are most familiar with classifying as resources. Yet, a critical resource geography course should incorporate a focus on resources in the making, such as the emergence of "ecosystem services" (e.g., carbon sequestration and flood management) as resources that can be, and in many cases already have been, produced as commodities in and of themselves (Dempsey and Robertson 2012; see also Carton and Edstedt, Chapter 34 in this volume). Better including resources in the making can enable a greater focus on the processes of privatization and commodification. It can also shift our attention to questions of waste in resource value chains, because, after all, many ecosystem services function as a sink for the externalities of production (Moore et al. 2018). Bringing resources in the making into the conversation can also unsettle what is sometimes viewed as a clear distinction between resource geographies and geographies of waste and help to raise questions about how we value some landscapes over others.

In sum, this chapter provides an overview of the in-class research assignment we run, outlines some of the data sources we use, and considers some of the challenges and adjustments we have made to the assignment over the years. Our goal is for students to analyze and engage with information critically and, we hope, to further develop as critical producers and consumers of information more generally. It is our hope that this chapter proves informative for others seeking not only to build research projects into their courses, but also to help build the subfield of critical resource geography through undergraduate teaching.

References

Association of College & Research Libraries. 2016. *Framework for Information Literacy for Higher Education*. Chicago, IL: American Library Association. http://www.ala.org/acrl/standards/ilframework.

Barnes, Jessica. 2014. *Cultivating the Nile: The Everyday Politics of Water in Egypt. New Ecologies for the Twenty-First Century*. Durham, NC: Duke University Press.

Biggs, John B., and Catherine So-kum Tang. 2011. *Teaching for Quality Learning at University: What the Student Does*. 4th ed. Maidenhead, UK: Open University Press.

Bridge, Gavin. 2009. "Material Worlds: Natural Resources, Resource Geography and the Material Economy." *Geography Compass* 3 (3): 1217–1244. doi: 10.1111/j.1749-8198.2009.00233.x.

Coe, Neil M., Philip F. Kelly, and Henry Wai-Chung Yeung. 2013. *Economic Geography: A Contemporary Introduction*. 2nd ed. Hoboken, NJ: Wiley.

Coe, Neil M., and Henry Wai-Chung Yeung. 2006. "Revitalizing Economic Geography Through Teaching Excellence: Some Pedagogic Reflections." *Journal of Geography in Higher Education* 30 (3): 389–404. doi: 10.1080/03098260600927161.

Cook, Ian, James Evans, Helen Griffiths, Rebecca Morris, and Sarah Wrathmell. 2007. "'It's More than Just What It Is': Defetishising Commodities, Expanding Fields, Mobilising Change…." *Geoforum* 38 (6): 1113–1126. doi: 10.1016/j.geoforum.2006.08.015.

Cope, Jonathan. 2010. "Information Literacy and Social Power." In *Critical Library Instruction: Theories and Methods*, edited by Maria T. Accardi, Emily Drabinski, and Alana Kumbier. Duluth, MN: Library Juice Press.

Cote, Conor, and Scott Juskiewicz. 2014. "Two Approaches to Collaborative Information Literacy Instruction at a Small Engineering School." *Collaborative Librarianship* 6 (2): 73–81. https://digitalcommons.du.edu/collaborativelibrarianship/vol6/iss2/3.

Dempsey, Jessica, and Morgan M. Robertson. 2012. "Ecosystem Services: Tensions, Impurities, and Points of Engagement within Neoliberalism." *Progress in Human Geography* 36 (6): 758–779. doi: 10.1177/0309132512437076.

Ford, Madeline, and Clay Williams. 2002. "Research and Writing in Sociology: A Collaboration Between Classroom Instructor and Librarian." *Public Services Quarterly* 1 (3): 37–49. doi: 10.1300/J295v01n03_05.

Fosmire, Michael, and Alexius Macklin. 2002. "Riding the Active Learning Wave: Problem-Based Learning as a Catalyst for Creating Faculty-Librarian Instructional Partnerships." *Issues in Science and Technology Librarianship* 34. doi: 10.5062/f4c53htb.

Garcia, Larissa. 2014. "Applying the Framework for Information Literacy to the Developmental Education Classroom." *Community & Junior College Libraries* 20 (1–2): 39–47. doi: 10.1080/02763915.2014.1013399.

Gergan, Mabel. 2017. "Living with Earthquakes and Angry Deities at the Himalayan Borderlands." *Annals of the American Association of Geographers* 107 (2): 490–498. doi: 10.1080/24694452.2016.1209103.

Golubchikov, Oleg. 2015. "Negotiating Critical Geographies Through a 'Feel-Trip': Experiential, Affective and Critical Learning in Engaged Fieldwork." *Journal of Geography in Higher Education* 39 (1): 143–157. doi: 10.1080/03098265.2014.1003800.

Goodman, Michael K. 2008. "'Did Ronald Mcdonald Also Tend to Scare You as a Child?': Working to Emplace Consumption, Commodities and Citizen-Students in a Large Classroom Setting." *Journal of Geography in Higher Education* 32 (3): 365–386. doi: 10.1080/03098260802221157.

Grafstein, Ann. 2002. "A Discipline-Based Approach to Information Literacy." *The Journal of Academic Librarianship* 28 (4): 197–204. doi: 10.1016/S0099-1333(02)00283-5.

Havice, Elizabeth, and Liam Campling. 2017. "Where Chain Governance and Environmental Governance Meet: Interfirm Strategies in the Canned Tuna Global Value Chain." *Economic Geography* 93 (3): 292–313. doi: 10.1080/00130095.2017.1292848.

Hecht, Gabrielle. 2012. *Being Nuclear: Africans and the Global Uranium Trade.* Cambridge, MA: MIT Press.

Heyman, Richard. 2001. "Pedagogy and the 'Cultural Turn' in Geography." *Environment and Planning D: Society and Space* 19 (1): 1–6. doi: 10.1068/d1901ed.

Hughes, Annie. 2016. "Exploring Normative Whiteness: Ensuring Inclusive Pedagogic Practice in Undergraduate Fieldwork Teaching and Learning." *Journal of Geography in Higher Education* 40 (3): 460–477. doi: 10.1080/03098265.2016.1155206.

Hulseberg, Anna, and Anna Versluis. 2017. "Integrating Information Literacy into an Undergraduate Geography Research Methods Course." *College & Undergraduate Libraries* 24 (1): 14–28. doi: 10.1080/10691316.2017.1251371.

Kneas, David. 2018. "Emergence and Aftermath: The (Un)Becoming of Resources and Identities in Northwestern Ecuador: Emergence and Aftermath." *American Anthropologist* 120 (4): 752–764. doi: 10.1111/aman.13150.

Lu, Flora, Gabriela Valdivia, and Néstor L. Silva, eds. 2017. *Oil, Revolution, and Indigenous Citizenship in Ecuadorian Amazonia.* Latin American Political Economy. New York, NY: Palgrave Macmillan.

Maybee, Clarence, Tomalee Doan, and Michael Flierl. 2016. "Information Literacy in the Active Learning Classroom." *The Journal of Academic Librarianship* 42 (6): 705–711. doi: 10.1016/j.acalib.2016.07.005.

Moore, Sarah A., Heather Rosenfeld, Eric Nost, Kristen Vincent, and Robert E. Roth. 2018. "Undermining Methodological Nationalism: Cosmopolitan Analysis and Visualization of the North American Hazardous Waste Trade." *Environment and Planning A* 50 (8): 1558–1579.

Pulido, Laura. 2002. "Reflections on a White Discipline." *The Professional Geographer* 54 (1): 42–49.

Robbins, Paul, and Sarah A. Moore. 2015. "Teaching Through Objects: Grounding Environmental Studies in Things." *Journal of Environmental Studies and Sciences* 5 (2): 231–236.

Smith, Lauren. 2013. "Critical Information Literacy Instruction for the Development of Political Agency." *Journal of Information Literacy* 7 (2): 15–32. doi: 10.11645/7.2.1809.

Spence, Larry. 2004. "The Usual Doesn't Work: Why We Need Problem-Based Learning." *Portal: Libraries and the Academy* 4 (4): 485–493. doi: 10.1353/pla.2004.0072.

Taylor, Dorceta. 2018. "Enhancing Racial Diversity in the Association of Environmental Studies and Sciences." *Journal of Environmental Studies and Sciences* 8 (4): 379–384.

US Department of Energy. 2019. "Department of Energy Authorizes Additional LNG Exports from Freeport LNG" [Press Release]. May 28. Accessed 4 June 2019. https://www.energy.gov/articles/department-energy-authorizes-additional-lng-exports-freeport-lng.

Walker, Richard. 2006. "Teaching (Political) Economic Geography: Some Personal Reflections." *Journal of Geography in Higher Education* 30 (3): 427–437. doi: 10.1080/03098260600927310.

Weaver, Kari D., and Jane H. Tuten. 2014. "The Critical Inquiry Imperative: Information Literacy and Critical Inquiry as Complementary Concepts in Higher Education." *College & Undergraduate Libraries* 21 (2): 136–144. doi: 10.1080/10691316.2014.906779.

Zimmermann, Erich W. 1933. *World Resources and Industries: A Functional Appraisal of the Availability of Agricultural and Industrial Materials*. New York, NY: Harper.

Section IV
Resource-making/world-making

28

Soy, domestication, and colonialism

Gustavo de L. T. Oliveira

Soy as resource, and domestication as resource-making

Over the last century, soy became one of the world's most important crops. It now covers an area equal to France, Spain, Portugal, and Ireland combined (about 1.3 million km²), yielding over 350 million tons each year for a total value of about 130 billion US dollars (according to average international prices).[1] Its significance for international markets is illustrated by its key role in the US-China trade war, but its strategic importance roots down into the soil where this leguminous plant doubles as "green fertilizer" that replaces nitrogen taken up by maize, cotton, and other crops with which it is usually rotated. Above all, soy has become the key nexus between human food production and livestock feed, as well as biodiesel and myriad industrial products (Oliveira and Hecht 2016; Oliveira and Schneider 2016). The rapidly multiplying uses of soy provide an excellent touchstone for analysis of how resources are not only created, but also constantly recreated and redesigned. In turn, these transformations reflect the shifting political-cal ecological foundations that undergird resource creation—above all the territorial expansion of an extractive mode of production intimately associated with colonialism—and illustrate how processes of resource-making generate new socio-ecological relations of their own (see also Curley, Chapter 7 in this volume).

My theoretical framework draws upon political ecology and critical agrarian studies, as I frame the *domestication* of plants and animals as the quintessential process of resource-making. While geographers and other social scientists (re)turn to the examination of resources and materiality, Richardson and Weszkalnys (2014) remark that little attention has been given to insights from political ecology and critical agrarian studies, likely because many believe the bulk of this literature "applies to resources—rather than derives from them—theoretical approaches" (Bakker and Bridge 2006, 7). Yet critical agrarian studies and political ecology have contributed significant theoretical advancements that can inform research on resources and materiality, including approaches that transcend facile distinctions between production and consumption (Anderson 1997; Freidberg 2004; Mandelblatt 2012) and native/invasive species, orienting multispecies ethnography (Rocheleau, Thomas-Slayter, and Wangari 1997; Robbins 2007; Ogden, Hall, and Tanita 2013). I build upon this literature to frame *domestication* as key for understanding resource-making. I define it as the material *dialectic of coevolution* between social organizations and ecological relations, extending this dialectic from a perceived "moment" of domestication through a continuum of socio-ecological interventions—otherwise identified as agriculture and livestock husbandry, industrialization, etc.—through a historical analysis of the processes that make and remake "resources," their "users," and the world in which they coexist.

Adopting domestication as theoretical framework for critical resource geography must not be confused with an uncritical assumption of shallow binaries between humans/nature or domination/subjection as though describing a process in which humans gain control over an external environment and overlooking the parallel autonomy and agency of the nonhuman (Ogden, Hall, and Tanita 2013; Richardson and Weszkalnys 2014). Rather, my analysis makes more explicit how geographers are contributing to the vibrant field of transdisciplinary studies about domestication, which already resists any dualism "intrinsic" to the topic. After all, while there is no consensus on the definition of domestication and there is lively debate about why/when/where/how it occurs (and does not occur), there is unanimous agreement that it is *not* simply a practice undertaken "by humans on nature" as attested by well-known cases of leaf-cutting ants and the fungi they cultivate as well as "dairying" ants and their associated aphids (Stadler and Dixon 2005). Moreover, as domestication results from a coevolutionary process of mutation-induced behavioral, physiological, and morphological changes in both species, it is a fundamentally dialectical process that requires recognition of agency and adaptation on behalf of all beings who become engaged in this increasingly mutualistic relationship (Larson et al. 2014). Domestication involving humans may occur faster than mutation-induced transformations among other species, but contemporary humans also display coevolutionary traits dialectically related to their nonhuman domesticates, such as the continued production of lactase into adulthood among humans who coevolved with dairying livestock, and the absence of this phenomenon in other groups. As it rests fundamentally on the generative capacity of the nonhuman to participate in the coproduction of biophysical and socioeconomic processes, domestication as the dialectic of coevolution can revitalize frameworks of the "production of nature" in ways that surpass their supposed "diminishing returns" (*pace* Bakker and Bridge 2006).

This is most blatantly evident when the domestication of plants and animals develops further into agriculture and livestock husbandry. This ongoing process of domestication undergirds the transformation of both humans and our planet into a dialectical whole characterized on the one hand by a novel ecosystem and historical era usually called (albeit undialectically) the Anthropocene (Smith and Zeder 2013) and on the other by the origins of private property, class society, the state, and colonialism (Scott 2017). The fact that agriculture is intrinsically associated with colonialism is worth highlighting. Once agricultural cultivation of floodplains and state conscription of non-settled peoples increased populations of early sedentary societies beyond their capacity for self-sufficiency, humans began to clear hillsides and other marginal land for agricultural production. Overtime, pests coevolved with crops, soil fertility became depleted, and deforestation for incorporating new areas into agricultural production accelerated soil erosion and the sedimentation of rivers, causing crop failures, increasingly catastrophic flooding, and a further drive of agricultural societies to expand their territory. This cycle repeats itself world-historically, as the coevolutionary process of domestication becomes the dialectic of making-and-depleting agricultural soils, creating-and-destroying resources for food, fiber, and shelter, and conquering territory through colonialism (Moore 2017; Scott 2017).

In this chapter, I trace the historical geography of soy from its origins in Chinese tributary colonialism, through Japanese imperialism, to contemporary US-led neocolonial globalization and the expansion of "Western" diets worldwide. I identify each moment's key driving processes—domestication, industrialization, flexing, and genetic modification—and the most prominent resources created, including vegetable protein, nitrogen fertilizer, edible oil, livestock feed, biofuel and myriad industrial inputs, and a veritable agroindustrial neo-nature in its own right.

Domestication, Chinese tributary colonialism, and vegetable protein

Wild precursors of soybeans began to be used in present-day northeast China at least 9000 years ago and were domesticated between 6000 and 3500 years ago (Hu 1963; Hymowitz 1970; Ho 1975). This region was inhabited by groups of nomadic proto-Tungusic tribes, who transformed a scrambling leguminous plant with small black seeds encased in easily shattered pods (which facilitated long-distance seed dispersal) into an essential resource for humans, namely, a source of edible protein that could be first gathered from wetlands and then cultivated on the mountainous areas in which they lived. The transformation of a wild legume into an important supplementary source of food took place through a dialectical process of coevolution over millennia, just like all processes of domestication. At first, the small black seeds were gathered by nomadic tribes from the fragile pods of soy's wild ancestor and then competing plants were removed from their surroundings (becoming "weeds" in the dialectical process of resource-making) as these sites became frequented more often by increasingly seminomadic descendants of the tribes that first identified uses for soy's precursor. Gradually, plants with larger seeds were selected for replanting, encased in stronger pods that prevented dispersion by wind and rain, progressively developing the originators of modern varieties of soybeans. In turn, this process of resource-making enabled the gradual transformation of seminomadic tribes into increasingly more settled communities of part-time farmers (Hu 1963; Hymowitz 1970; Ho 1975).

Meanwhile, the ancestors of the Han people who compose the majority of modern-day Chinese had already advanced much further the cultivation of millet and sorghum and established the earliest East Asian networks of city-states and dynastic kingdoms on the Yellow River valley and loess plateau of northern China.[2] Written evidence of domesticated soybeans only appears during the early Zhou dynasty (c. 3000 years ago), and etymological and historical studies indicate early varieties of soy were still identified as a rambling crop sticking close to the ground but valued as well for the nitrogen-fixing nodules in its root. Zhou elites first came into contact with (early varieties of) soybeans from tributes provided to them by the Tungusic tribes to their northeast. They never subjected those tribes to their direct rule or conquered their territory but integrated soy into their own agricultural production systems, because it provided not only a high-protein vegetable food source, but did so in a way that also addressed the crisis of soil fertility that contributed to the collapse of China's first dynasty (Shang) during the previous century. After all, millet and sorghum cultivation were already driving the exhaustion of the loess plateau soils upon which the Neolithic ancestors of the Shang originally domesticated them. The introduction of soybeans into millet/sorghum rotation systems enabled the nitrogen-fixing bacteria that coexist symbiotically in the nodules of soy roots to provide increasing food yields while postponing further depletion of soil fertility (Hu 1963; Ho 1975).

This was a unique moment in the historical geography and political ecology of soybeans. The crop did not expand through settler colonialism, simplifying the agroecosystems into which it was integrated for the purpose of advancing the territory of its producers and their extractivist mode of production. In fact, the coevolution of soybeans with the Tungusic tribes assured their relatively independence in face of Zhou rulers, who were satisfied with receiving this newly created resource as tribute. Meanwhile, Zhou elites began to mandate the integration of soybeans among their Han ethnic subjects. As a result, soybean domestication advanced at relatively fast pace, spreading to the entire territory controlled by the Zhou (north and central China). Farmers in Zhou territories began to select for reproduction not only plants with larger (and yellower) seeds, but also plants with stronger and taller stems, which prevented loss of seeds, facilitated harvest, and provided postharvest kindling for the winter (Ho 1975).

During the Zhou dynasty, this expansion, intensification, and new techniques for soybean domestication transformed this plant from an exotic form of edible protein into a crucial source of nitrogen for the depleted soils of millet/sorghum fields, a supplementary source of fuel, and an essential resource for human diet. By the end of the Zhou dynasty, techniques to ferment soy had been developed, producing soy sauce, *jiang* (Jp.: *miso*, fermented soybean paste), and soy-based wines, in addition to steamed, boiled, or fried whole soybeans, which caused soy to match millet as the staples of Chinese society (Chang 1977).[3] In turn, soy transformed the Zhou from an adventuring clan who seized power over a decaying kingdom, into the longest-lasting (c. 1046–256 BCE) and most formative dynasty in China. During their rule, the Chinese script emerged, as well as China's leading religions/philosophies of Confucianism and Daoism, and the notion that the dynasty's rule is maintained by a "mandate of heaven," evidenced through their ability to sustain their subjects and their agricultural production—fed, warmed, and fertilized by the domestication of soybeans.

Industrialization, Japanese imperialism, and fertilizer/vegetable oil

There is no sharp distinction between domestication and continued agricultural development through seed selection: these are better understood as a continuum in the process of resource-making (compare to Larson et al. 2014). As agricultural practice rewarded cultivators with more useful resources—soy plants with larger edible seeds, encased in stronger pods, held by taller and stronger stems, harboring more nitrogen-fixing root nodules—seed selection extended to improvements in taste, nutritional value, resistance to droughts/floods, and multiple pests that began to thrive in this newly made world of soy fields. By the early 1900s, on the eve of industrialization in China, there were over 6000 varieties of soybeans throughout the country, and an equally rich diversity of food products and artisanal methods for producing them, including various types of soy sauce, paste, milk, flour, as well as multiple dishes and snacks made of pickled, boiled, steamed, or fried soybeans, and above all, various types of tofu (Chang 1977). Animal products accounted for as little as 1 percent of most people's diet, so soybeans had become their most vital source of protein. While soy production and consumption as human food, fuel, and fertilizer expanded along with Chinese (i.e., ethnic Han) imperial expansion and settler colonialism (and Buddhist proselytization across east Asia, promoting vegetarian diets), this rich diversity collapsed with the process of industrialization at the hand of Japanese and US imperialists during the twentieth century (Oliveira and Schneider 2016).

Soybeans first reached Japan during China's Tang dynasty (618–917 CE), one of the periods of furthest Han ethnic imperial expansion, and strongest influence of Buddhism (and its associated promotion of vegetarianism). Soybeans and fermented soy foods became integral to Japanese farming practices and diets over the next centuries. But to understand how the Japanese transformed soy into an industrial resource, which in turn assisted in the transformation of Japan into a regional empire, it is necessary to first grasp how new resources were made through preindustrial crushing of soybeans in China itself.

As soy production increased and varieties with large oily seeds were developed, Chinese people began to crush soybeans (c. 980 CE) in order to supplement vegetable oil production from sesame and rapeseed. These were utilized in cooking, as fuel for lanterns, as sealant for caulking boats, and as lubricant for cartwheels and other purposes. The soy meal by-product was considered of little importance, given to livestock as feed or discarded as waste. However, increased domestic trade along the north-south canal of central China during the Qing dynasty (1616–1912 CE) led to increased attention to this by-product. Cotton, tea, sugarcane, and other cash-crop production intensified in southern China to supply Beijing and other cities in the north (and for export) but depleted soil fertility in the process. Traders began to ship soy meal back south in otherwise empty barges, transforming this waste by-product into a cheap source of agricultural fertilizer that also eliminated the wasted cost of navigating empty ships (Hiraga and Hisano 2017).

Meanwhile, as Japan industrialized during the late 1800s, the combination of rural exodus and depletion of soil fertility created a dire need for fertilizers there as well. Domestic fishmeal supplies were insufficient and imported guano from the United States and European colonies were unaffordable, so Japanese traders began importing Chinese and Korean soy meal instead. A poor soy harvest in 1889 led to a partial restriction of Korean soy meal exports to Japan, precipitating the Japanese invasion of Korea and the First Sino-Japanese War (1895–1896) and their subsequent war against Russia (1904–1905) for influence over the Korean peninsula and northeast China, then known as "Manchuria." The procurement of coal to supply Japan's growing industrial economy remains widely recognized as a major factor in these confrontations, but almost forgotten was the equally important pursuit of soybeans as a food and agricultural resource. After these wars, Japanese trading companies and fertilizer manufacturers created the first industrial soybean crushing operations in the world in Manchuria. Eighty percent of exports (by value) from Manchuria consisted of soybeans and soy meal for Japan, and soy oil for Europe and the United States, where this versatile product was becoming a new resource in the manufacturing of soap, margarine, and several industrial products (Seth 2011; Hiraga and Hisano 2017).

The Japanese structured colonization of Korea and Manchuria around an extractivist mode of production that hinged on control of ports and railroads. Integrating Chinese (and Korean) soybeans into a global market through industrial processing, Japanese imperialism transformed soy into an essential resource for its further industrialization, intensifying agricultural production domestically and increasing food supplies even while its own peasantry was dispossessed in the process. The South Manchuria Railway Co. (SMR), "Japan's *de facto* colonial institution" in northeastern China during the early 1900s (Hiraga and Hisano 2017, 12), derived as much as 57 percent of its profits from soy trade. Moreover, the SMR established a network of soybean crushing factories, warehouses, hotels for soybean traders, and deep-water port facilities to facilitate extraction of this newly created industrial resource. Significantly as well, Japanese industrialization promoted the *standardization* of soybeans, rewarding production of varieties with high oil and protein contents and marginalizing other varieties that may have been more suitable for food production, more nutritious, more resistant to pests or environmental hazards, or richer in other socio-ecologically desirable features.

Demand for vegetable oil and initial experimentation with soy meal for concentrated livestock feed increased along with industrialization in Europe and the United States during the first decades of the 1900s. Meanwhile, after the collapse of the Qing dynasty in China, warlords in Manchuria began to assert greater control over the region's soy economy from the SMR. In response, the Japanese imperial army occupied China during the 1930s, and with the onset of war against the United States during the 1940s, it restricted soybean, oil, and meal exports. This triggered US government, farmer, and agribusiness interest in the establishment of a domestic soybean agroindustry, which would come to dominate global markets after the United States defeated Japan in 1945 (Hiraga and Hisano 2017).

Flexing, neocolonial globalization, and agroindustrial integration

The multiple uses of soy that emerged as it became an industrial resource, and the flexibility of agroindustrial processors to incorporate it into existing formulations of vegetable oil products and meal for livestock feed, advanced vertiginously once the US government adopted soy as part of its total war effort. An increasingly integrated agroindustrial economy consolidated the process we call "flexing" (Oliveira and Schneider 2016; compare with Goodman, Sorj, and Wilkinson 1987), that is, (re)creating numerous uses and markets for a resource, malleably interchanged with other agroindustrial inputs according to variations in prices, timing of harvests, and other conditions. This enabled not only US soybean agribusiness to surpass its historical center of production in northeastern China and the pioneering trading and processing companies from Japan but also to control the technologies and markets developed through neocolonial globalization along the twentieth century.

Many imagine the use of soybeans for the production of biofuels and industrial products is a recent invention, and contemporary agribusinesses often portray themselves as entrepreneurial innovators of a "green economy." Gustavo Grobocopatel, for example, the patriarch of an Argentinian soy conglomerate, proudly proclaimed in 2014 that:

> What is to come in ten years is a sort of Green Industrial Revolution, plants begin to be transformed into factories. That is, a plant that until now produced grain begins to produce energy, bioplastics, molecules, and enzymes for industrial use.
>
> (Quoted in Oliveira and Schneider 2016, 168)

Yet this entrepreneurial triumphalism is utterly revisionist. Albeit eccentrically, Henry Ford already boosted the use of soy as a dietary supplement and industrial input and imagined his car's nonmetallic structures composed mostly of soy derivatives (Oliveira and Hecht 2016). In fact, his famous Model T was a flex-fuel vehicle that could be adjusted to run on ethanol, gasoline, or a "gasohol" blend. But corn-based ethanol and soy-based biodiesel were displaced after the 1920s when the antiknock properties of tetraethyl lead enabled the powerful petroleum industry in the United States to monopolize automotive fuel markets (Oliveira, McKay, and Plank 2017). Moreover, decommissioned bomb factories in the United States enabled the production of abundant synthetic fertilizers. So during the twentieth century, it was no longer the fertilizer industry that drove demand for soy (meal) as in Japan but the vegetable oil industry in the United States, Europe, and newly industrializing countries like Brazil and Argentina where soy production also began to take root (Oliveira and Schneider 2016). As in the United States, wheat and corn farmers in those South American countries welcomed soy production because this nitrogen-fixing legume assisted their efforts to reduce fertilizer use and delay the depletion of soil fertility (Oliveira 2016).

The main driving force of soybean flexing shifted with the rapid integration of soy meal as key ingredient for livestock feed in newly created concentrated animal feeding operations (CAFOs). Prior to the twentieth century, livestock feed was a minor supplement mainly utilized during winter months, and efforts to concentrate livestock failed because animals became sick too quickly. This was due to the absence of sunlight (which induces absorption of vitamins A and D), and the increased vulnerability and exponential proliferation of infectious diseases that resulted from their agglomeration in confined spaces. When industrialization rendered antibiotics, vitamins, and other nutritional additives cheap enough for use in livestock feed, year-round confinement of large numbers of livestock in CAFOs became possible. The soy meal by-product of the burgeoning vegetable oil industry in the United States, therefore, was quickly identified as a cheap and abundant source of protein for livestock feed. These technological advancements soon spread from the United States to other countries, especially Brazil (Oliveira 2016).

The United States did not colonize Brazil and Argentina directly, but it did support the integration of agribusiness experts—agronomists, biologists, chemists, food engineers, economists, etc.—into leading land-grant universities in the United States, where they fostered technical expertise in the adaptation of soy for industrial farming practices and the flexing of this resource for multiple markets and industries. Moreover, the United States also trained, armed, and supported the Brazilian and Argentinian militaries to put down peasant uprisings and communist movements advocating for the redistribution of land and agrarian reforms aimed at reducing the power of the landed oligarchy and the export-dependence of these South American postcolonial economies (Oliveira 2016). Through brutal military dictatorships that ruled these countries for decades during the twentieth century, an increasingly technified and capitalized agroindustrial production system consolidated the neocolonial patterns of land distribution, labor relations, extractivist production, and export-dependent global integration of this region, which the Swiss-based agrochemical and biotechnology company Syngenta advertised as the "United Soybean Republic" (Oliveira and Hecht 2016).

Neocolonial globalization drove the expansion of soy production and the advancement of Brazilian and Argentinian agroindustrial elites over their own hinterlands and thereby transformed them into sub-imperial powers in their own right. Brazilian farmers took advantage of close ties between military dictatorships to effectively colonize Paraguayan and Bolivian lowlands with soybeans, and Argentinian firms partnered with US agrochemical and biotechnology giants like Monsanto to gain control over seed and other agroindustrial input markets throughout the continent (Oliveira 2016; Craviotti 2016). Brazilian and Argentinian soybean conglomerates are now major vehicles for United States, European, and Japanese agribusiness investments, collectively controlling over a million hectares across South America (Oliveira and Hecht 2016). Their efforts are expanding to Uruguay, Colombia, Venezuela, Cuba, and various African countries, where Japanese financiers—who promoted Brazilian soy since the 1970s to reduce dependence upon US exports, and now seek to reclaim international agroindustrial markets—are coordinating a new round of neocolonial globalization of soybean agribusiness (Oliveira 2016).

Genetic modification, dietary colonization, and neo-natures

The latest intervention remaking soybeans into new resources delves into the very genetic materials of this plant itself and spans the entire landscape over which soy production dominates, generating a veritable neo-nature. While agroindustrial integration and neocolonial globalization expanded soy production astonishingly, it simultaneously eradicated the biodiversity of the ecosystems it engulfed in South America (as wheat and corn had done in the US Midwest before) and reduced genetic diversity of soybeans to a meager handful of varieties. After China opened up its markets for soybean imports, effectively outsourcing the production of resources for its rapidly growing network of CAFOs and vegetable oil industry, the invaluable wealth of thousands of soybean landraces domesticated over millennia has virtually disappeared (Oliveira and Schneider 2016). Now, agricultural research companies and institutes value nonindustrial varieties of soybeans for the genetic material they may provide for biotechnological developments, particularly necessary to address increasingly devastating pest outbreaks, progressively troublesome herbicide-resistant "super weeds," and the unpredictability of climate change—vulnerabilities largely created by the expansiveness of monocultures and their agroindustrial production practices in the first place. The appropriation of soybean landraces and transgenic seeds through patents and other intellectual property rights has itself created a new resource in genetic material (Kloppenburg 2010).

Despite these challenges, the present-day abundance of soy has transformed it into one of the cheapest and most versatile agroindustrial inputs. Among its uses are adhesives, analytical reagents, antibiotics, asphalt emulsions, anticorrosive agents, antistatic agents, binders for wood/resin, caulking compounds, cosmetics, core oils, disinfectants, dispersing agents, dust control agents, electrical insulation, epoxies, films for packaging, foams and anti-foaming agents, fungicides, herbicides, inks and crayons, insecticides, linoleum backing, leather substitutes, metal casting, oiled fabrics, paints, plastics and plasticizers, plywood, protective coatings, polyesters, pharmaceuticals, putty, rubber manufacture, soaps/shampoo/detergents, textiles, vinyl plastics, waterproof cement, and wallboards. Similarly, there are innumerable processed foods that contain soy products in small amounts, such as soy flour, stock, lecithin, glycerol, fatty acids, and sterols. These are used in the production (or low-cost "extension") of baby food, beer and ales, breads, cookies, pancakes and other bakery products, candies, chocolates, and confections, cereals and grits, creamers, dietary products, frozen desserts, instant milk drinks, juices, liquid shortening and yeast, noodles, salad dressing/oils, sandwich spreads, sausage casings, toppings, vegetable shortening, and, of course, mayonnaise and margarine. Soy products are not only used as processed food ingredients, but also as emulsifying or stabilizing agents, shortening, and coatings (Oliveira and Schneider 2016).

The amount of industrial products and processed foods that include soy inputs is astounding, yet the soy and food processing industries continue unsatisfied with the "slow uptake" of this new resource, and actively lobby TV stars, health advocates, medical doctors, and nutritionists to recommend soy foods and soybean-infused products. Ironically, however, these new processed foods—or age-old East Asian dishes associated with (mostly) vegetarian diets—do not account for the largest share of human-consumed soy. That is the indirect consumption of soy meal converted into meat through livestock feed. The "meatification" of diets—particularly in China—has become the single largest driver of global soybean demand (Oliveira and Schneider 2016). There, this dietary transformation is intimately associated with a broader neocolonial "Westernization" of diets, including rising amounts of meat consumption, dairy replacing soy milk, and "Western-style" breads, which combine above all in "Western" fast foods like burgers and pizzas.

These industrial and processed food markets are extensive due to their variety, but the second-largest market for soybeans (after livestock feed and vegetable oil) is biofuel. This return to soy as a fuel resource reemerged with the energy crisis of the 1970s and accelerated with more recent concerns about fossil fuel–driven climate change and the need of agroindustrial processors to find new and ever-expanding markets to sustain demand and high prices for their key resource (Oliveira and Schneider 2016; Oliveira, McKay, and Plank 2017). Agroindustrial processors justify the production of soy-based biofuel mainly through the idea that agricultural intensification and biofuels can limit deforestation and reduce carbon emissions, but this argument is scientifically discredited or at least highly disputed (Oliveira and Hecht 2016; Oliveira and Schneider 2016; Oliveira, McKay, and Plank 2017). Still, just as agroindustrial resource-making has contributed to an anthropogenic planet, so too does "green entrepreneurialism" around climate change now contribute to the dialectical remaking of soy into "new" resources.

Most encompassing is the valuation of soybean-transformed landscapes as sites of further agroindustrial investment and capital accumulation, whereby agribusiness companies do not profit from soy production itself but utilize this resource in the making of "developed" farmland for resale. This occurs because soy prices tend to fall with agroindustrial expansion and intensification, while production costs rise with treadmill effects that require new "improved" transgenic seeds, increasingly more toxic and expensive agrochemical inputs (to manage progressively more agrochemical-resistant pests and weeds), and ever-growing amounts of fertilizer to sustain production on soils that lose their organic matter and biological capacity for nutrient recycling as constant agrochemical use effectively reduces it to a barren substrate. Moreover, environmental regulations in Brazil during the 2000s and 2010s made the conversion of standing forests into agricultural fields increasingly prohibitive, so "developing" deforested landscapes like degraded pastures for agroindustrial crop production became a profitable business practice. Multiple investors from the United States, Europe, and Japan—including pension funds like TIAA and Harvard University's endowment—are partnering with South American agribusinesses to use soybeans as a key instrument for making "developed" farmland (Oliveira and Hecht 2016). They may lose money on soy production itself, but profit from the speculative valorization of farmland "development" and resale, as neo-Malthusian narratives of future food and farmland scarcity sustain conjecture that farmland prices will continue to rise as population grows and production conditions collapse. Thus, soy contributes to the making of the Anthropocene, which dialectically remakes soybeans into an agroindustrial resource for the production of a neo-nature in its own right.

Conclusion

The coproduction and coevolution of soy farmers and soybeans is an under-acknowledged force in the making and remaking of a world-historically important set of resources—food, feed, fuel, and fertilizer above all, as well as myriad industrial products, genetic resources, and even a neo-nature that extends to the commodification of land and whole environments themselves. I have argued that it provides a strong basis for theorizing domestication as a process of resource-making and fertile ground for examining the dialectic between resource-making and world-making through colonialism, imperialism, and neocolonialism.

Not all resource-making is rooted in agriculture, nor are all forms of agriculture (e.g., agroecology) embroiled in some form of colonialism and imperialism. Yet, when domestication is theoretically understood as resource-making in a continuum that extends through agricultural development and livestock husbandry, it brings attention to the interrelated practices of class formation, state-making, and colonialism (Scott 2017) as crucial for the analysis of the process of resource-making as well (see also Bustos-Gallardo, Chapter 31 in this volume). This brief but sweeping history of soy domestication and resources reveals the dialectic of coevolution between resource-making and world-making and provides fruitful fodder for the further theoretical development of global political ecology, critical agrarian studies, and, of course, critical resource geographies. Above all, this approach lays to rest the critique that political ecology and critical agrarian studies "appl[y] to resources—rather than derives from them—theoretical approaches" (Bakker and Bridge 2006, 7), as my analysis derives a distinctive and theoretically robust account of "domestication" from the historical geography of soybean resources, and thus places the human-environment dialectics of domestication as a core theme for research in critical resource geographies.

Notes

1 Estimated volume of global production multiplied by average international price of whole soybeans, based on US Department of Agriculture 2019 data.
2 Although the name "China" would not become common until the 1800s, as the kingdoms and empires that waxed and waned over this region usually took the name of ruling dynasties, I employ this term anachronistically for the sake of brevity.
3 Rice, domesticated in southern China, was restricted to elite and ceremonial consumption in northern China until the recent development of high-yield varieties, but soy was consumed across northern and southern regions equally (Chang 1977).

References

Anderson, Kay. 1997. "A Walk on the Wild Side: A Critical Geography of Domestication." *Progress in Human Geography* 21 (4): 463–485.

Bakker, Karen, and Gavin Bridge. 2006. "Material Worlds? Resource Geographies and the 'Matter of Nature'." *Progress in Human Geography* 30 (1): 5–27.

Chang, Kwang-chih. 1977. *Food in Chinese Culture: Anthropological and Historical Perspectives*. New Haven, CT: Yale University Press.

Craviotti, Clara. 2016. "Which Territorial Embeddedness? Territorial Relationships of Recently Internationalized Firms of the Soybean Chain." *Journal of Peasant Studies* 43 (2): 331–347.

Freidberg, Susanne. 2004. *French Beans and Food Scares: Culture and Commerce in an Anxious Age*. Oxford, UK: Oxford University Press.

Goodman, David, Bernardo Sorj, and John Wilkinson. 1987. *From Farming to Biotechnology: A Theory of Agro-Industrial Development*. London: Basil Blackwell.

Hiraga, Midori, and Shuji Hisano. 2017. "The First Food Regime in Asian Context? Japan's Capitalist Development and the Making of Soybean as a Global Commodity in the 1890s–1930s." AGST Working Paper Series No. 03. November 2017. Kyoto: Kyoto University. https://agst.jgp.kyoto-u.ac.jp/workingpaper/1240.

Ho, Ping-ti. 1975. *The Cradle of the East: An Inquiry into the Indigenous Origins of Techniques and Ideas of Neolithic and Early Historic China, 5000–1000 B.C.* Chicago, IL: University of Chicago Press.

Hu, Daojing. 1963. "Interpretation: On the Understanding of Soybean Root Nodules by Ancient Chinese Peasants." [In Chinese.] *Essays on Chinese Literature and History* 3: 111–115.

Hymowitz, T. 1970. "On the Domestication of the Soybean." *Economic Botany* 24 (4): 408–421.

Kloppenburg, Jack. 2010. "Impeding Dispossession, Enabling Repossession: Biological Open Source and the Recovery of Seed Sovereignty." *Journal of Agrarian Change* 10 (3): 367–388.

Larson, Greger, Dolores R. Piperno, Robin G. Allaby, Michael D. Purugganan, Leif Andersson, Manuel Arroyo-Kalin, Loukas Barton, et al. 2014. "Current Perspectives and the Future of Domestication Studies." *Proceedings of the National Academy of Sciences* 111 (17): 6139–6146.

Mandelblatt, Bertie. 2012. "Geography of Food." In *The Oxford Handbook of Food History*, edited by Jeffrey M. Pilcher, 154–171. Oxford, UK: Oxford University Press.

Moore, Jason. 2017. "The Capitalocene, Part I: On the Nature and Origins of Our Ecological Crisis." *Journal of Peasant Studies* 44 (3): 594–630.

Ogden, Laura, Billy Hall, and Kimiko Tanita. 2013. "Animals, Plants, People, and Things: A Review of Multispecies Ethnography." *Environment and Society* 4 (1): 5–24.

Oliveira, Gustavo de L. T. 2016. "The Geopolitics of Brazilian Soybeans." *Journal of Peasant Studies* 43 (2): 348–372.

Oliveira, Gustavo de L. T., and Susanna Hecht. 2016. "Sacred Groves, Sacrifice Zones, and Soy Production: Globalization, Intensification and Neo-Nature in South America." *Journal of Peasant Studies* 43 (2): 251–285.

Oliveira, Gustavo de L. T., Ben McKay, and Christina Plank. 2017. "How Biofuel Policies Backfire: Misguided Goals, Inefficient Mechanisms, and Political-Ecological Blind Spots." *Energy Policy* 108: 765–775.

Oliveira, Gustavo de L. T., and Mindi Schneider. 2016. "The Politics of Flexing Soybeans: China, Brazil, and Global Agroindustrial Restructuring." *Journal of Peasant Studies* 43 (1): 167–194.

Richardson, Tanya, and Gisa Weszkalnys. 2014. "Introduction: Resource Materialities." *Anthropological Quarterly* 87 (1): 5–30.

Robbins, Paul. 2007. *Lawn People: How Grasses, Weeds, and Chemicals Make Us Who We Are*. Philadelphia, PA: Temple University Press.

Rocheleau, Dianne, Barbara Thomas-Slayter, and Esther Wangari. 1997. *Feminist Political Ecology: Global Perspectives and Local Experience*. London: Routledge.

Scott, James. 2017. *Against the Grain: A Deep History of the Earliest States*. New Haven, CT: Yale University Press.

Seth, Michael. 2011. *A History of Korea: From Antiquity to the Present*. Lanham, UK: Rowman & Littlefield.

Smith, Bruce D., and Melinda A. Zeder. 2013. "The Onset of the Anthropocene." *Anthropocene* 4: 8–13.

Stadler, Bernhard, and Anthony Dixon. 2005. "Ecology and Evolution of Aphid-Ant Interactions." *Annual Review of Ecology, Evolution, and Systematics* 36: 345–372.

29

From gold to rosewood

Agrarian change, high-value resources, and the flexible frontier-makers of the twenty-first century

Annah Zhu and Nancy Lee Peluso

Introduction

The promise of gold brought tens of thousands of migrant laborers from China to the forested and marshy shores of western Borneo in the mid-eighteenth and early-nineteenth centuries. Settling in for the short or long term, along and inland from the sparsely populated coasts, those early miners created patchworks of small mines and farms—and farms on top of spent mines— as they produced one of the first smallholder-driven gold frontiers in the world. In the province of West Kalimantan (Indonesian Borneo) today, a new kind of gold frontier is being carved, again by small-scale miners, digging out new pits and shafts or working the planted-over sites of those centuries-old mines. Since the early 1990s, men and women struck with gold fever do not focus solely on mining or farming livelihoods; they are operating within a more financialized, and consequently more volatile, global economy than that of centuries prior. They now engage with a broader range of mining, farming, entrepreneurial, and wage labor activities that ebb and flow along with the massive transformations in the landscape as well as in the province's political economy. If returns from the farming or agroforestry commodities flag or fall through the sensitive global price floor, some household members turn to mining gold. If the gold deposits they are working become exhausted, or their bodies are exhausted from the stresses of mining gold, they can turn back to agroforestry, dig elsewhere for gold, engage in wage labor in the city or countryside, travel to other villages by bicycle or motorbike to sell food or clothes, or seek work across Indonesian Borneo's land border with Malaysia. If all else fails, they remain in place, at home, succumbing to what they call "*kuli* labor on their own lands"—wage labor on the oil palm plantations arriving in their neighborhoods in the last decade or two—and wait for commodity prices to rebound. In all cases, miners, farmers, and workers alike have become flexible laboring subjects in order to survive.

On the other side of the Indian Ocean, some 7000 km away, gold miners and other small producers in northeastern Madagascar are creating new resource frontiers of their own. Abandoning their gold panning and cash cropping operations, many have turned to the booming rosewood economy. As with West Kalimantan's gold fever, this new economic opportunity arose suddenly, when demand for rosewood was on the rise in China and a 2009 *coup d'état* left

Madagascar with no formal government to regulate the trade. Rather than mining or farming, villagers entered the region's national parks by the thousands in search of rosewood to log and carry to the coast. The price of the timber rose more than a hundredfold since the early 2000s. By 2013, rosewood logs from Madagascar were worth over USD 60,000 per metric ton in Chinese timber markets, and around USD 26 per metric ton in the forest where they were first cut.[1] By 2014, a market downturn in China had weakened demand and the price for rosewood, while political restructuring in Madagascar and pressure from international conservation organizations incited stronger enforcement of existing logging and trade bans. Newly minted loggers returned to their abandoned mines, fields, and villages. There, like the smallholder miners of West Kalimantan, they struggled to make ends meet before seeking greater fortunes in subsequent opportunities afforded by the vagaries of the global economy.

Indonesian small-scale miner-farmer-workers and Malagasy miners-turned-loggers complicate the categories of "peasant," "smallholder," "petty commodity producer," "industrial laborer," or "working class" that feature so prominently in classic accounts of agrarian change (Thompson 1963; Scott 1976, 1985; Bernstein and Byres 2001). Although many maintain or adopt a profile of wage-labor work when their own entrepreneurial, cash cropping, or subsistence activities fall short, they challenge forecasts of the peasantry's proletarianization while also failing to do justice to the idea of a persistent peasantry (Roseberry 1976; Johnson 2004).[2] Most are not seeking to "re-peasantize" their families' livelihoods by a return to subsistence agriculture. Rather, they or other family members maintain an agrarian identity as rural smallholders through a broader portfolio of work as laborers and entrepreneurs, supplementing farming income with off-farm labor outside of agriculture. Developing small-scale mining technologies and exchanging or supplementing farming with other forms of commodity production and labor, the composite worker-farmers of Indonesia, Madagascar, and many rural spaces across the globe are no longer the "peasants," "proletarians," or "shifting cultivators" of yore.

Some agrarian change scholars have proposed new hybrid terms. Rather than peasant or proletarian, we (and they) now speak of "smallholder miners" (Peluso 2017), "extractive peasants" (Lahiri-Dutt 2018), and "worker-smallholders" (Barral 2018). As has been true for centuries, these actors assemble their daily livelihoods through a complex bricolage based on the opportunities of the moment (Scoones 2009). What have changed, however, are the diversity and locations of opportunities, the rates at which work may emerge or vanish, the sources of work—in terms of regional or local investors—and the increasingly speculative nature of the local and global political economies within which these flexible agrarian subjects operate. The "reach" of China's speculative demand for rare timbers and rare earths, other forest products, and gold, for example, has penetrated Madagascar's forests halfway across the world, inspiring a new wave of trade and extraction and transforming Malagasy rural lives in ways resonant of China's enduring interactions in Southeast Asia. These sites illustrate how worker-smallholders of the Global South navigate long-standing and emergent global connections, moving back and forth among extraction, subsistence, cash cropping, and wage labor with dexterity.

Beyond peasantry and proletariat, resource territories have become fragmented in new ways as well. Global commodity markets have splintered, with pioneers and wealthy investors from China, India, Malaysia, Saudi Arabia, and other emerging economies vying for resources and territory alongside well-established networks leading to the Global North (North America, EU, Australia, Japan). Along with these new global actors, new resources, new ways of accessing them, and new ideas about how they should be valued have emerged as well. Operating amidst these tensions, rural subjects find themselves at the edges of multiple, sometimes recursive commodity frontiers, cyclically recurring in postcolonial settings

(Middleton 2019)—or, we suggest, reappearing in new spaces more sensitive to rapidly changing global dynamics. Smallholders also create new resource territories, as laborers, claimants, and challengers of their larger competitors.

This chapter builds on the hybrid concepts of "extractive peasants" and "smallholder miner-farmers" in order to rethink agrarian environments and rural producers in this highly speculative, late capitalist global economy. We use the examples of gold and rosewood booms to argue that the global geographies of late capitalism are generating flexible agrarian subjects who increasingly navigate, often out of necessity, volatile price dynamics, emergent export opportunities, and multiple, overlapping, and recursive resource frontiers. These flexible subjects provide an agrarian parallel to the flexible late capitalist subjects of the Global North; they comprise a new class of flexible frontier-makers that will be a defining feature of the rural twenty-first century moving forward.

The fragmentation of rural agrarian livelihoods and the flexibility of rural subjects in the Global South bear strong parallels with the flexible knowledge and service workers that define our current era of late capitalism (Mandel 1975; Jameson 1991) and post-Fordist "flexible accumulation" (Harvey 1989). Yet, late capitalism and the subjective flexibility it requires is typically theorized within the urban and suburban contexts of the postindustrial Global North—not their rural counterparts in the Global South. As much as late capitalism depends on labor regimes in the Global South and new forms of global production connecting the North and South, theories of the late capitalist subject are still predominantly associated with the context of the Global North.

Yet, as our cases show, rural subjects in the Global South experience the speculative dynamics and ephemeral consumer desires of late capitalism just as keenly as those in the North, if not more. Materially, they may live in industrial or even preindustrial landscapes, but in terms of economic volatility and the flexible subjectivities required to respond to such volatility, the hybrid worker-farmer-entrepreneurs of the Global South are quite advanced. The increasingly rapid oscillations in prices of gold, rosewood, rubber, palm oil, and vanilla, and the increasing need to navigate between these volatile global economies, all provide prime examples coming from Indonesia and Madagascar. Southern agrarian subjects, such as those described here, experience these unpredictable and capricious trajectories in the form of land, jobs, and other resources won or lost; exploitative wage arrangements; clever credit schemes (speculation on failures); and influxes of capital that might come in rapid succession, pull away quickly, or never arrive. Their everyday work and the rhythms of their lives have been altered dramatically. Rural laboring subjects *must* be flexible, or they founder.

The rest of this chapter unfolds in four parts. The next section discusses gold as a "made" resource in order to demonstrate the concept of resource-making/world-making. The third section refocuses on gold at its points of extraction, using the contemporary example of small-scale gold mining in West Kalimantan, Indonesia, to demonstrate a new "flexible agrarian subject" or "flexible frontier-maker" that can be understood as the product of late capitalist world-making processes. The fourth section then turns to rosewood, the trade name for a group of timber species that are both globally endangered and highly valued as a luxury timber commodity. In China in particular, some species of rosewood have become more valuable than gold due to speculative demand. This section uses rosewood extraction in Madagascar and its crafting for consumption in China to demonstrate how emerging global dynamics add new high-valued resources to the global milieu, molding ever more flexible agrarian subjects increasingly adept at responding to the demands of the late capitalist moment. The final section concludes with a reflection on what this means for rural life and livelihoods in the twenty-first century.[3]

Gold made the world and the world made gold

The value attached to all resources is neither inherent nor universal but is a product of social imagination. Gold provides an iconic example. Although gold has been widely valued, not all societies have valued it in the same way. In ancient Egypt and ancient Java, gold was used for beads, jewelry, and elaborate rituals, serving as one of the primary signifiers of social status—only certain people could wear or use it (Miksic 1990, 2011; Schoenberger 2011; Ross 2016). In ancient China, in contrast, gold was commonly known and used but was considered inferior to other substances, such as jade and bronze (Froncek 1969, 275). Similarly, the Aztecs valued a single lump of jade as being worth two loads of gold (Clark 1986, 10). When European ships sailed to the New World in search of gold and other treasure, the Aztecs they encountered were confused by the explorers' acute desire for this particular metal, not quite sure what purpose it might serve (Clark 1986). Even the Incas, famous for their affinity for gold, did not find this metal to be *the most* valuable substance. Rather, it was cloth (though, better yet, gold-lined cloth) that afforded the greatest prestige in the Incan Empire (Murra 1962).

When gold became a global currency, things changed. Gold began to assume a more "universal" value—that is, it was recognized across cultures and places as embodying a similar unit of worth. Beginning in the early-eighteenth century, gold rapidly outpaced silver as the preeminent currency of the British Empire. By the 1870s, an international gold standard was established (Schoenberger 2011, 16). For the first time, gold as a global standard of value connected countries across six continents in a unified international monetary system with few exceptions (Schoenberger 2011). It is no exaggeration to say "the world" as we know it today—networked across all hemispheres—was in fact *made* by the establishment and circulation of gold currency.[4]

Even before the gold standard was adopted, as gold coins flowed freely through global cosmopolitan centers, the search for new deposits across the world was intensified. A growing network of explorers, colonial institutions, small-scale miners, traders, and financers was required to supply this new global monetary system and keep it running smoothly. Thus, just as gold made the world, miners and minters across the world made gold.

Today, gold is still in high demand, but it no longer serves as *the* universal standard of value. By the end of the twentieth century, no country could keep a stock of gold equivalent to the value of its national economy. Instead, the world has transitioned to a more flexible and dematerialized system of storing, exchanging, and creating value. Speculators the world over invest in gold when currencies are weak, monetary crisis is rampant, or other commodities experience a downturn. The same is true for rosewood. In China, speculators turn to rosewood and other traditional Chinese art forms to diversify their portfolios when returns from housing or the stock market wane (Zhu 2020). When conditions change—over the course of months, days, moments—they abandon these investment avenues just as quickly as they embrace them.[5]

Rural sites of gold and rosewood production in the Global South feel these volatile dynamics. Rural and migrant laborers experience the ups and downs of the virtual economy while still providing the resources and embodying productive forces that anchor the global economy in material worlds. The remainder of this chapter discusses this type of global connection—where the speculative dynamics of a late capitalist, dematerialized economy meet the material productions and flexible producers of the countryside. First, we examine gold rushes in West Kalimantan, where discoveries of gold deposits make new smallholder mining frontiers and flexible territories across the region (Peluso 2017). Then we examine a rosewood boom in Madagascar, where another flexible gold frontier has given way to rosewood's new world-making powers (Zhu 2017, 2018).

Indonesia's small-scale mining and the flexible frontier

Ironically, the gold stores of subterranean Borneo pulled the region's residents (migrants or indigenous) into global capitalist markets earlier than did the island's renowned rain forests. As long ago as 1750, large and small pit mines became classic frontiers of gold production by tens of thousands of Chinese men coming to Borneo and the local people they allied with (Jackson 1970).[6] These miners and the traders arguably brought early capitalist relations to local social relations in Borneo's forest expanses (Heidhues 2003; Peluso 2018). Worlds changed as well, after their arrival. Many of these miners made their lives in Borneo rather than returning to China with their savings and gold finds. They married local women and continued transforming the forests and peat swamps into working landscapes of gold mines and rice paddies, many traveling eastward into Borneo's interior from the western coast. This early transoceanic gold rush fueled by Chinese laborers bringing water control technologies initiated a pattern of both long-distance labor mobilities and smallholder gold production that would be repeated in California, Australia, and the Yukon in the nineteenth and twentieth centuries as well as Africa's "gold coast" and other parts of the Global South in the twenty-first.

One hundred years after these migrants arrived in Borneo, colonial Dutch territorial aspirations put a violent stop to this older form of gold world-making. Because they had successfully resisted Dutch imperialist aspirations, mining zones that had grown into small but self-governing "Chinese Republics" were thorns in the sides of a different set of world-makers expanding the Netherlands East Indies (de Groot, 1885; Yuan 2000; Heidhues 2003). Waging a four-year war to take control of the territory, the Dutch put an end to the resource world that had constituted northwestern Borneo's small-scale gold-mining and farming territories, forcing about half the resident Chinese population to abandon their mining enterprises (Yuan 2000). Those who remained moved into the commodification of forest products and tree crops, expanding and commodifying other classic rural smallholder productions of Borneo: pepper and later, rubber (Ozinga 1940; Heidhues 2003).

Since the late 1980s, many of Indonesia's Borneo provinces have witnessed the resurgence of small-scale gold mining, albeit within an entirely different global political economic and geopolitical milieu. Small-scale gold mining exploded in the 1990s and 2000s, woken by the efforts of two entrepreneurial industrial gold-mining ventures in the eastern and western reaches of Indonesian Borneo. In East Kalimantan, the Bre-X mine project spectacularly captured the world's imagination, seducing stock marketeers from Wall Street to Toronto and Hong Kong (Tsing 2004) as well as mom-and-pop investors in California before scandalously crashing. Meanwhile in West Kalimantan, with the blessing of President Suharto, a small but internationally funded company began exploring some of the old Chinese mining sites for their industrial mining potential. Their mapping of sites and boring did not go unnoticed by locals living in the region's villages and towns, some of whose families had remained engaged in artisanal mining without fanfare over the years. A pogrom against rural Chinese saw them evicted and replaced by Indonesians of other ethnic heritages in the 1960s and '70s (Peluso 2009). In the late 1980s, gold did not dominate their resource world but served as an occasional, artisanal livelihood option.

Some 300 km upriver from the west coast, villagers continued to engage in small-scale mining in forests and on the banks of the river's tributaries while industrial-scale state and corporate projects for logging, plantation establishment, and resettlement projects were clearing forest and unwittingly revealing new sources of gold. Observing some of the dredging operations and equipment used by foreign firms, one local mechanic worked with his miner friends to perfect a portable dredge that could efficiently suck up the heavy peat soils in

Figure 29.1 "Digging" a mine hydraulically. West Kalimantan, Indonesia. Photo by Nancy Peluso.

which the region's gold was often hidden (Peluso 2018). They field-tested prototypes of the dredging technology under the rivers and in swamps, fields, and forests, eventually birthing a homegrown, flexible mining technology (Peluso 2018).[7] Small- and medium-sized dredges imported from China followed later. By then, these upriver miners had already brought their dredges to the many sites between their interior home and work spaces and the Borneo coast, partnering with local people willing to mine (Figure 29.1).

For a decade before the world gold price began to rise more steeply than ever before, informal gold-mining ventures provided capital and work for thousands of miners, diggers, buyers, and financiers. Pioneers from these historic gold-mining districts in northwest Borneo, in turn, created new mining territories in the still vast forests of the southern part of the province. Gold production once again dominated this Bornean resource-world. Tens of thousands of mining laborers, crews, traders, investors, shopkeepers, bar maids, water and oil delivery men, itinerant medicine salesmen, and every kind of frontier market trader imaginable endured the grueling 24-hour trip by land and sea from the northwest corner of the island to these new southern sites, attracted by friends' and competitors' bragging of pulling up clumps of grass with shallow roots adorned with golden nuggets. Crews mined gold under rivers or moved the rivers to extract gold from beneath. Gold prices rose during the 2008 global economic downturn; they peaked in the summer of 2012. Crew bosses had unprecedented access to credit- or cash-driven techno-organizational options, ranging from single-operator 1.5-horsepower dredges to much larger dredges powered by truck engines and worked by crews of 12–15 men. One mining boss described his excitement at watching used truck engines traversing the South China Sea between Singapore and West Kalimantan on huge rafts. Landing on Borneo's southwest coast, they helped construct flexible, adaptive, and opportunistic resource worlds. Miners' varied niches were created by their capacity to take advantage of laboring and small or large investment opportunities made possible by capitalist crisis and opportunity alike.

Besides technologically, how different are these recent gold-mining territories from those of the eighteenth century? Then, simple reproduction depended on miners' abilities to subsist from crops they had to grow themselves, to make alliances with previous and new settlers, and to produce this highly coveted global commodity (Yuan 2000). Chinese pioneers brought capitalist relations, cooperative organization, and direct connections to global markets, moving into the interior of early-nineteenth century Borneo from its coasts and rivers. They brought waterwheels and other technologies (Jackson 1970), often working with indigenous talent to transform iconic, forested Bornean landscapes by creating resource worlds below and above ground (Yuan 2000).

Indonesian gold miners in West Kalimantan today are of more varied ethnic heritages and have initiated their own resource frontiers and territories. "Small-scale mining" has produced larger mines and varied labor and extraction practices through the deployment of local and imported technologies and mobility. Large crews using heavy equipment work alongside smaller crews and independent miners and gleaners. Not only have new and competing land uses emerged in these extraction frontiers and mining spaces, but also miners have created new modes of claiming land and resources as property. Notwithstanding the presence of state actors in the regulation of formal claims or the cloaking of informal ones, a myriad of mechanisms for negotiating informal access to these subterranean sites of enrichment and tragedy abound.

For agrarian subjects in Indonesia today, participation in gold production is one livelihood option alongside other agrarian commodity productions and entrepreneurial activities open to people with high-valued resources in their subterranean lands. Some miners—many of whom are also farmers—turn their profits from mining into agrarian capital, financing production of staple foods that are also commodities, such as rice and corn. With their mining returns, miner-farmers of many ethnicities buy land, rubber, fruit tree, or oil palm seedlings, reproducing at a broader scale not only vast mines and their tailings, but also extending and changing garden and agroforested landscapes. At the same time, when commodity growers suffer from wavering world market prices, they supplement or replace tapping with mining, construction labor, or other wage work abroad (Peluso 2009, 2017, 2018). Flexibility, mobility, and luck—three characteristics that are not usually considered in "classic" agrarian and resource analyses—have helped many smallholder-worker families maintain diverse livelihood portfolios. Though not always capable of preventing household economic crises, these diversifications help them exploit the booms, weather the busts, and weave together flexible territories in the process.

More valuable than gold: rosewood in a contested twenty-first century

As new actors emerge in the global arena, new and often conflicting resource values emerge as well. China's unprecedented growth, for example, has introduced new resource demands—some of which directly conflict with well-established demands from the Global North. Rosewood provides a clear illustration. Chinese demand for many endangered species such as rosewood directly conflicts with demands deriving from the United States and Europe to conserve these species. This is not simply a question of divvying up resources so that new actors get a share of global supply, but rather it reveals how resources can be valued—and even defined—differently by different actors across the globe in sometimes irreconcilable ways.

Much like gold, rosewood has been valued, extracted, traded, and produced around the world for centuries. China has the longest history of rosewood production, with demand for rosewood

Figure 29.2 A Ming-Dynasty styled rosewood bed on display at the offices of a rosewood furniture factory in Shenzhen, China, March 2018. Photo by Annah Zhu.

dating back to the Ming and Qing Dynasties (1368–1911), when the wood was used to craft an elaborate style of furniture made for emperors and the Chinese social elite (Figure 29.2). Since 2000, this furniture style—deeply associated with the wood with which it is made—has been revived as a cultural icon (Zhu 2017, 2020). Having exhausted supplies from Asia, China has been soliciting new imports from across the tropics.

Europe and the United States have a long history of rosewood consumption as well for furniture, instruments, and decorations. Since the late 1990s, however, in the face of dwindling species, Europe and the United States have promulgated domestic laws prohibiting imports of endangered rosewood. In 2013, international trade in many rosewood species became restricted under the Convention on International Trade in Endangered Species of Flora and Fauna, and in 2016, all rosewood in the genus *Dalbergia* was officially restricted.

At the same time that the United States and Europe began restricting the trade in precious hardwoods, demand in China began to pick up. Rosewood has since become the most trafficked group of endangered species in the world (UNODC [United National Office on Drugs and Crime] 2016). As China embraces capitalism and seeks to redefine its modernity with strong reference to its cultural past, classical rosewood furniture is experiencing a commercial renaissance. The same increase in consumer demand has occurred for ivory, rhino horn, and products of other iconic endangered species that have deep cultural associations in China.

Within the context of a late capitalist economy, however, Chinese investors have begun to embrace these markets not only for their cultural value but also for their potential for financial appreciation (Zhu 2020). Far from a simple marketplace for consumers, rosewood has become a type of stock exchange—"a playground for investors" (China Daily 2011). Since around 2005, rosewood in China has transformed into a speculative cultural currency. Similar to gold, rosewood is now purchased for pure speculation—in both its raw form as logs and finished as furniture. With speculative investment driving the price of this endangered timber to unprecedented highs, conservationists find it difficult to keep up. In moments of intensive speculation, the rarest raw timbers can be worth their weight in gold (Sina Collection 2010). As finished antique furniture, these woods can sell for more than their weight in gold (SICA 2011).

Speculative dynamics in the world's rising economies are felt deeply across the agrarian South. China's speculative demand for rosewood has triggered logging booms in no less than 30 tropical countries, with one of the most controversial sites being Madagascar (see Remy 2017; Zhu 2017; Anonymous 2018). Rosewood from Madagascar is the most highly trafficked species according to United Nations seizure data, accounting for over 60 percent of all global rosewood seizures by volume (UNODC [United National Office on Drugs and Crime] 2016). Limited primarily to the spaces now enclosed as national parks, Madagascar's rosewood grows only in the northeastern corner of the country. Since 2009, thousands of loggers have entered these parks in search of rosewood. Legions of Chinese ships have sailed off the island nation's coasts to pick up logs brought there directly from the forest. Meanwhile, the World Bank and United Nations have funded a special rosewood "task force" to fight against this trade. Over 1 billion US dollars have been channeled into Madagascar to finance either rosewood conservation or its logging, in a place where most people make less than a dollar a day.[8]

Finding themselves amidst these contradictory world-making processes, Malagasy villagers alternate opportunistically between logging and conservation work depending on the global demands of the moment. Guides for the national parks often engage in rosewood logging when the market is booming, while forest officials try their hand at trading the wood—some even becoming high-level traders. Cash crops provide another volatile opportunity for Malagasy rural dwellers as a consequence of late capitalist financial speculation. As the region's rosewood boom subsided toward the end of 2014, a vanilla boom replaced it. Global prices for vanilla—the region's other main export—spiked from lows of around $5 per kg to more than $600. Shifting from rosewood to vanilla, loggers returned from the forest to tend their abandoned fields, injecting these gardens with "green" gold (Zhu 2018).

Now, along with the current vanilla boom, a renewed trend for "healing" crystals has swept the United States and Europe. Within the last five years, global crystal prices have surged, sending some former rosewood loggers and vanilla farmers back to mining, often in the same gold mines they abandoned years earlier (McClure 2019). These dynamics make clear the deep connection between late capitalist financial flows and ephemeral consumer desires in the Global North and the flexible rural subjects in the Global South which find themselves at the receiving end of these erratic ups and downs.

Global demands for rosewood, vanilla, and other world-transforming boom and bust commodities make little sense from the perspectives of the rural Malagasy people who experience them most dramatically. In terms of rosewood, villagers marvel as to why Chinese buyers pay so much to take this tree from their shores and why, in contrast, Western donors want so badly to conserve it in place. They are equally confused by the demand for vanilla, laughing in disbelief at the explanation that this commodity that can be worth as much as silver by weight is used for nothing more than flavor and fragrance. Most avoid examining too closely the global demands that find their way to Madagascar—the mysterious cultural economies and peculiar conservation campaigns—instead hoping only to take advantage of the opportunities of the moment.

Conclusion

Using the examples of gold and rosewood, this chapter has shown how some resources acquire widespread social, economic, and political value and how these values can transform worlds (see also Carton and Edstedt, Chapter 34 in this volume). Gold provides a classic example. Managing to become a common—and as close as we have come to a "universal"—standard of value for more than a century, gold and the circulation of gold currencies have in many ways *made* the world what it is today. It enabled a globally unified monetary system and initiated the unprecedented exchange of goods, services, resources, and people. This is largely attributable not to any intrinsic value of the metal but rather to a set of biophysical properties that make gold both appealing to the eye and practical for a time as a global currency. Gold's transformation into a speculative commodity on every global stock exchange, and its volatility and daily calculated prices visible on cell-phone applications around the world, have constituted the metal's postindustrial legacy as a new type of world-maker. Although fetishized as an "already made" commodity, gold in the contemporary era is still dependent on resource makers small and large embedded in contested material extraction processes.

Rosewood, too, has been enrolled in world-making processes for centuries. But more recently, the value of rosewood has been subject to sharp contestation. Since 2005, rosewood furniture crafted in styles from the Ming and Qing Dynasty has been the target of intensive financial speculation. As with speculative demands for gold—arguably even more so—demand for rosewood varies drastically, shifting by orders of magnitude within months, days, even moments. Yet, China's speculative quest for rosewood, sending timber importers to fetch it from the furthest corners of the world, has been condemned by Western governmental and nongovernmental organizations that have identified the species as an international conservation priority.

As historic commodities, gold and rosewood have been "making" worlds for centuries; we have shown here that they have also made new, flexible subjects out of rural producers, who embrace their production and experience the effects of speculation through booms and busts. Although the work of those engaged in mining, logging, and other extractive or agrarian endeavors remains highly manual, the global demands to which they must respond, in contrast, have become increasingly volatile and fragmented. These speculative dynamics have compounded the need for even more flexible agrarian subjects that can navigate between conflicting global desires and local livelihood demands (see also Bustos-Gallardo, Chapter 31 in this volume).

This is the survival strategy of flexible frontier-makers in the rural Global South. Amidst the contested world-making of the twenty-first century, they alternate between different extractive, subsistence, and wage labor opportunities through ever more flexible arrangements, poised—and hoping—to respond to the fleeting global demands of the moment. Juggling the volatility of late capitalist financial speculation, fleeting consumer trends, and rapid technological development or stagnation, the composite worker-farmer-extractors of Indonesia and Madagascar whom we have described in this chapter provide a window into the type of flexible frontier-making that we believe will define the rural twenty-first century. They provide an agrarian parallel to the flexible and fragmented subjects so commonly evoked in the Global North.

Notes

1   When rosewood prices are at these highs and the government is permitting exports, logging provides better financial returns than gold mining or cash cropping.
2   They also differ from the worker-peasants described by Stoler (1995) in Sumatra's early twentieth-century capitalist plantations. Javanese laborers were initially contracted by Dutch-, English-, and American-held private plantations, traveling long distances from their homes on the island of Java to work in Sumatra. Stoler shows how many struggled with the plantation, hoping to get small plots of plantation land to augment their households' subsistence with gardens. This changed, although minimally, their dependence on the plantations for wages. Stoler shows that having the land was critical to their identities. Not happy to be workers alone, they wanted to be peasants, or at least worker-peasants. In contrast, in our research, we examine how rural smallholders—not subsistence-oriented peasants—maintain a smallholder agrarian identity through other kinds of work and resource extraction as laborers and entrepreneurs. Moreover, these smallholders largely embrace commodity production, rather than eschewing it.
3   This chapter is based on ethnographic fieldwork conducted in West Kalimantan, Indonesia; northeastern Madagascar; and Shanghai and Guangzhou, China. The section on gold mining in Indonesia is based on in-depth interviews and observations with over 200 small-scale miners, financiers, landholders, village heads, swidden agroforesters, rubber growers, and other smallholders in Sambas, Bengkayang, Singkawang, Pontianak, Ketapang, and Sekadau, West Kalimantan province over seven months of fieldwork conducted by Nancy Peluso between 2013 and 2017. The section on rosewood logging and new forms of Chinese demand is based on interviews with rosewood loggers and traders conducted in northeastern Madagascar by Annah Zhu during the summers of 2014 and 2015 as well as interviews with timber wholesalers and furniture factory owners in Shanghai and Guangzhou, China, during the winters of 2014, 2015, 2017, and the spring of 2018.
4   Silver, in this sense, was the first "gold" (Graulau 2011). Because of its greater availability, silver served as the dominant world-making currency between the fifteenth and seventeenth centuries.
5   Yet, consumers of "small" amounts of gold in the form of wedding rings and other jewelry continue to drive demand and extraction. More than 50 percent of the gold "above ground" in the world has been made into jewelry (Statista 2019).
6   At first, they came at the behest of a sultan, but later they came through emergent routes independently.
7   This happened well before Chinese and Australian models invaded the Indonesian mining scene.
8   Most of this money, however, has been spent on rosewood logging, with that figure alone allegedly reaching the equivalent of over a billion US dollars (Randriamalala 2013). Conservation finance, especially that specifically mentioning rosewood or intended to curb rosewood logging, remains in the hundreds of millions. For example, among many other conservation finance packages, the World Bank offered a $52-million conservation grant with the explicit condition that rosewood logging legislation be enforced.

References

Anonymous. 2018. "Rosewood Democracy in the Political Forests of Madagascar." *Political Geography* 62: 170–183. doi: 10.1016/j.polgeo.2017.06.014.

Barral, Stephanie. 2018. "In Between Agro-Industries and Family Business Farms: Ephemeral Smallholder Family Plantations in Indonesia." In *Diversity of Family Farming Around the World*, edited by Pierre-Marie Bosc, Jean-Michel Sourisseau, Philippe Bonnal, Pierre Gasselin, Élodie Vallette, and Jean-François Bélières, 137–148. Dordrecht: Springer.

Bernstein, Henry, and Terrence J. Byres. 2001. "From Peasant Studies to Agrarian Change." *Journal of Agrarian Change* 1 (1): 1–56. doi: 10.1080/03066150902820297.

Clark, Grahame. 1986. *Symbols of Excellence: Precious Materials as Expressions of Status*. Cambridge, UK: Cambridge University Press.

China Daily. 2011. "Rosy Times Ahead for Rosewood Furniture." China Daily, January 17.

de Groot, J.J.M. 1885. *Het Kongsiwezen van Borneo. Eene Verhandeling over den Grondslag en den Aard der Chineesche Politieke Vereenigingen in de Koloniën, met eene Chinesche Geschiedenis van de Kongsi Lanfang.* 's Gravengage.

Froncek, Thomas. 1969. *The Horizon Book of the Arts of China.* New York, NY: American Heritage.

Graulau, Jeannette. 2011. "Ownership of Mines and Taxation in Castilian Laws, from the Middle Ages to the Early Modern Period." *Continuity and Change* 26 (1): 13–44. doi: 10.1017/S0268416011000051.

Harvey, David. 1989. *The Condition of Postmodernity.* Oxford: Blackwell.

Heidhues, Mary Somers. 2003. *Golddiggers, Farmers, and Traders in the "Chinese Districts" of West Kalimantan, Indonesia.* Ithaca, NY: Cornell University Press.

Jackson, James C. 1970. *Chinese in the West Borneo Goldfields: A Study in Cultural Geography.* Hull: University of Hull.

Jameson, Frederic. 1991. *Postmodernism, or, the Cultural Logic of Late Capitalism.* Durham, NC: Duke University Press.

Johnson, Heather. 2004. "Subsistence and Control: The Persistence of the Peasantry in the Developing World." *Undercurrent* 1 (1): 55–65.

Lahiri-Dutt, Kuntala. 2018. "Extractive Peasants: Reframing Informal Artisanal and Small-Scale Mining Debates." *Third World Quarterly* 39 (8): 1561–1582. doi: 10.1080/01436597.2018.1458300.

Mandel, Ernst. 1975. "The Industrial Cycle in Late Capitalism." *New Left Review* I 90 (Mar./Apr.): 3–25.

McClure, Tess. 2019. "Dark Crystals: The Brutal Reality Behind a Booming Wellness Craze." *The Guardian*, September 17. https://www.theguardian.com/lifeandstyle/2019/sep/17/healing-crystals-wellness-mining-madagascar.

Middleton, Townsend. 2019. "Frontier 2.0: The Recursive Likes and Death of Cinchona in Darjeeling." In *Frontier Assemblages*, edited by Jason Cons, and Michael Eilenberg, 195–212. Oxford: Wiley.

Miksic, John N. 1990. *Old Javanese Gold.* Honolulu, HI: University of Hawaii Press.

Miksic, John N. 2011. "Gold and Cultural Evolution in Early Southeast Asia. The Ayala Museum's Gold Collection in Regional and Historical Context." In *Philippine Ancestral Gold*, edited by Florina H Capistrano-Baker, John N. Miksic, and John Guy. Makati City: Ayala Foundation; Singapore: NUS Press.

Murra, John V. 1962. "Cloth and Its Functions in the Inca State." *American Anthropologist* 64 (4): 710–728.

Nash, June. 1979. *We Eat the Mines and The Mines Eat Us: Dependency and Exploitation in Bolivian Tin Mines.* New York, NY: Columbia University Press.

Ozinga, Jacob. 1940. *De Economische Ontwikkeling der Westerafdeeling van Borneo en de Bevolkingsrubbercultuur.* Wageningen: Zomer en Keuning.

Peluso, Nancy Lee. 2009. "Rubber Erasures, Rubber Producing Rights: Making Racialized Territories in West Kalimantan, Indonesia." *Development and Change* 40 (1): 47–80. doi: 10.1111/j.1467-7660.2009.01505.x.10.1111/j.1467-7660.2009.01505.x.

Peluso, Nancy Lee. 2017. "Plantations and Mines: Resource Frontiers and the Politics of the Smallholder Slot." *Journal of Peasant Studies* 44 (4): 834–869. doi: 10.1080/03066150.2017.1339692.

Peluso, Nancy Lee. 2018. "Entangled Territories in Small-Scale Gold Mining Frontiers: Labor Practices, Property, and Secrets in Indonesian Gold Country." *World Development* 101: 400–416. doi: 10.1016/j.worlddev.2016.11.003.

Randriamalala, Hery. 2013. "Study of the Sociology of Malagasy Rosewood Operators." *Madagascar Conservation & Development* 8 (1): 39–44.

Remy, Oliver. 2017. "Rosewood Democracy." In *Corruption, Natural Resources and Development: From Resource Curse to Political Ecology*, edited by Aled Williams, and Philippe Le Billon, 142–153. London: Edward Elgar.

Roseberry, William. 1976. "Rent, Differentiation, and the Development of Capitalism Among Peasants." *American Anthropologist* 78 (1): 45–58.

Ross, Laurie Margot. 2016. *The Encoded Cirebon Mask: Materiality, Flow, and Meaning Along Java's Islamic Northwest Coast.* Leiden: Brill.

Schoenberger, Erica. 2011. "Why Is Gold Valuable? Nature, Social Power and the Value of Things." *Cultural Geographies* 18 (1): 3–24. doi: 10.1177/1474474010377549.

Scoones, Ian. 2009. "Livelihoods Perspectives and Rural Development." *The Journal of Peasant Studies* 36 (1): 171–196. doi: 10.1080/03066150902820503.

Scott, James. 1976. *The Moral Economy of the Peasant: Rebellion and Subsistence in Southeast Asia.* New Haven, CT: Yale University Press.

Scott, James. 1985. *Weapons of the Weak: Everyday Forms of Peasant Resistance.* New Haven, CT: Yale University Press.

Sina Collection. 2010. "Hainan Huanghuali is Now More Expensive than Gold" (海南黄花梨现在贵比黄金). November 2. http://finance.sina.com.cn/money/collection/jjmq/20101102/14208886674.shtml.

SICA (Shanghai International Commodity Auction Co.). 2011. "Huanghuali Furniture Set a Record Auction Price of 70 Million." 木比金贵 黄花梨家具拍出7000万刷新拍卖记录. January 26. http://www.alltobid.com/contents/40/738.html.

Statista. 2019. "Demand for Gold Worldwide from 1st Quarter of 2016 to 3rd quarter of 2019, by Purpose." https://www.statista.com/statistics/274684/global-demand-for-gold-by-purpose-quarterly-figures/.

Stoler, Ann Laura. 1995. *Capitalism and Confrontation in Sumatra's Plantation Belt, 1870–1979.* Ann Arbor, MI: University of Michigan Press.

Taussig, Michael. 1980. *The Devil and Commodity Fetishism in South America.* Chapel Hill, NC: University of North Carolina Press.

Thompson, E. P. 1963. *The Making of the English Working Class.* New York, NY: Vintage.

Thompson, E. P. 1975. *Whigs and Hunters: The Politics of the Black Act.* New York, NY: Vintage.

Tsing, Anna. 2004. "Inside the Economy of Appearances." In *The Blackwell Cultural Economy Reader,* edited by Ash Amin, and Nigel Thrift, 83–100. Oxford: Blackwell.

UNODC (United National Office on Drugs and Crime). 2016. *World Wildlife Crime Report: Trafficking in Protected Species.* Vienna: United Nations Office on Drugs and Crime.

Williams, Raymond. 1973. *The Country and the City.* New York, NY: Oxford University Press.

Wolf, Diane L. 1994. *Factory Daughters: Gender, Household Dynamics, and Rural Industrialization in Java.* Berkeley, CA: University of California Press.

Yuan, Bingling. 2000. *Chinese Democracies: A Study of the Kongsis of West Borneo (1776-1884).* Leiden: Research School of Asian, African, and Amerindian Studies, Universiteit Leiden.

Zhu, Annah. 2017. "Rosewood Occidentalism and Orientalism in Madagascar." *Geoforum* 86: 1–12. doi: 10.1016/j.geoforum.2017.08.010.

Zhu, Annah. 2018. "Hot Money, Cold Beer: Navigating the Vanilla and Rosewood Export Economies in Northeastern Madagascar." *American Ethnologist* 45 (2): 253–267. doi: 10.1111/amet.12636.

Zhu, Annah. 2020. "China's Rosewood Boom: A Cultural Fix to Capital Overaccumulation." *Annals of the American Association of Geographers* 110 (1): 277–296. doi: 10.1080/24694452.2019.1613955.

30

Conservation and the production of wildlife as resource

Elizabeth Lunstrum and Francis Massé

Introduction

Descending into Toronto's bustling Bloor and Bay subway station, its drab walls come to life with an image of 13 elephants lumbering upon the shore of a serene lake, their image reflected in the glasslike water (Figure 30.1). The land between the elephants and a distant sunset is vast and empty with bush morphing into mountain. This is land shaped by nature, land untouched by people. Joined by an image of the South African liqueur Amarula and the tagline "Amarula. The Spirit of Africa," this advertising campaign embodies a familiar narrative: one in which African wildlife captures the essence of an entire continent and more deeply brings us back to a time before the arrival of modernity and even people. This feel of a timeless, untouched Africa abounds in nature specials on African wilderness and is used to sell African commodities like Amarula and more substantially African tourism experiences, both nonconsumptive safaris and big-game hunting expeditions.

Disrupting this well-worn storyline of Africa's apparent premodern nature—at once its wildlife and apparent defining spirit—is the reality that African wildlife, as with wildlife in general, is far from timeless. In fact, like other natural resources, wildlife has been actively created as such by human intervention. In this chapter, we chart wildlife's ongoing production as resource through the practice of wildlife conservation and in particular its legal, managerial, and discursive techniques. We begin with the early days of conservation and move through the contemporary commodification and militarization of wildlife as resource. Through this movement, we highlight both how the production of wildlife reflects processes of resource-making highlighted by the critical resource geography literature as well as how wildlife is distinct. We also show how conservation's production of wildlife has led to additional forms of world-making that produce not just animals but new categories of the human, namely "the poacher." We close by examining the possible danger of exposing wildlife as a human-produced resource especially in the age of antienvironmental activism and suggest how we might begin to address this. We ground our observations in the North American and sub-Saharan African conservation experience and highlight the experience of Southern Africa, which is the primary site of our empirical work.

Figure 30.1 Wildlife as timeless: Amarula advertisement at the Bay and Bloor subway station, Toronto, Canada. Photo by Elizabeth Lunstrum.

Wildlife as natural resource: a critical resource geography approach

Like other natural resources, to say that wildlife is a natural resource is to say, first, that it is a type of nonhuman nature in which humans have invested value and extract use. This begins with wildlife's economic value to nature-based tourism economies, including safari-based ecotourism and big-game hunting (Igoe and Brockington 2007; Snijders 2012). Next is the symbolic value of wildlife, which includes how various animals—typically large mammals and birds—embody the nation or a sense of national belonging and pride (Carruthers 1995; Lunstrum 2014). Wildlife also holds value for the health of ecological systems, particularly keeping them diverse, in balance, and/or resilient. When these systems benefit people, wildlife plays a key role in protecting ecosystem services, or the human-derived benefits from ecosystems (Alcamo et al. 2005). This includes the fact that wildlife is a vital source of protein for many communities around the world and has additional economic value through its commercial use, both legal and illegal, as food, medicine, and decorative objects. Wildlife is also often recognized as having inherent value, or value that exists independently of humans (Rolston 1994). While this takes us outside the realm of wildlife as a natural resource, it nonetheless articulates with many of the other values invested in wildlife to provide rationales to protect or conserve species and their habitats. This is precisely the goal of wildlife conservation science and practice. Taken together, these values underscore a key respect in which wildlife differs from many other natural resources like oil, timber, and commercial fish stocks: namely the value of wildlife rests largely, although not entirely, in its nonexploitation or at least nonremoval from its habitat.

Taking a critical resource geography approach, we also begin to see that, like other natural resources, wildlife as resource does not "naturally" exist but has been created as resource through human desire, ingenuity, and intervention. Modifying the words of Zimmermann's (1933) well-worn observation on resources, "Wildlife is not: wildlife becomes." We take this as our point of departure to denaturalize wildlife as resource to show it is dynamic, relational, and always becoming through socioeconomic-political-environmental processes (Bridge 2009; Banoub 2017). We illustrate later how this production of wildlife as resource has happened largely through the practice of wildlife conservation.

How conservation produced wildlife as resource—the early years

Let us turn to the creation of protected areas and specifically national parks as these are the exemplary spaces through which wildlife conservation has produced wildlife as resource. The modern conservation movement and early conservation science began in the late 1800s largely in the United States with roots stretching back to the earlier colonial period (Grove 1995; Jacoby 2001). Fundamental to the movement was the national park ideal, a model of environmental management that advocated the cordoning off of apparently wild, untouched spaces and saving them from human destruction. This was sparked by concern over unbridled urban industrial expansion spreading westward and related fears that the loss of these wild spaces and animals would mean the loss of what presumably defined the American character, i.e., rugged masculine individualism. This model soon took material form in the early national parks including Yellowstone in 1872, often recognized as the world's first national park (Cronon 1996; Jacoby 2001). Reflecting a disquieting and often overlooked aspect of this history, the creation of early parks also routinely translated into the expulsion of Indigenous peoples, which proved central to a larger settler colonial project of dispossession, confinement, and genocide (Spence 1999; Jacoby 2001; Carroll 2014).

The national park ideal soon began to spread and underpin the development of protected areas elsewhere, most famously across sub-Saharan Africa. This gave rise to early game reserves in South Africa in the late 1800s and Congo's Virunga National Park, the continent's first national park, in 1926. These early African protected areas were created to protect dwindling wildlife, which was threatened by disease and unregulated colonial hunting. In terms of the latter, during the early years of Africa's colonization, Europeans saw animals like elephants, rhinos, and lions as a natural resource mostly to be exploited for their skins, ivory, and other trophies (Beinart and Hughes 2007). In fact, the desire to protect these animals so they could be hunted sparked the first global movements and organizations to conserve Africa's megafauna (Adams 2004). This movement was only later followed by the development of nonconsumptive "safari" economies (Neumann 1998). Parks like South Africa's iconic Kruger National Park, moreover, were also developed to help consolidate a (white, European) national identity (Carruthers 1995). This reminds us that even early parks and their resident megafauna have played a role in nation-building, with wildlife today symbolizing many African nations as seen in the animals splashed across their currency.

With the rationales behind protected areas in place, governments and conservation authorities began instituting early conservation measures aimed at regulating, controlling, managing, and—through these—producing wildlife as resource. First was new legislation and legal designations that gave official recognition to protected areas including parks. Land could then be surveyed, mapped, delimited, demarcated as protected, and at times fenced. This typically meant that everything encompassed within the reserve boundaries would be subject to new restrictions on hunting but also human habitation, agriculture, and entry. Early parks were also policed, albeit sparsely, by game wardens and rangers tasked with everything from dealing with problem animals and monitoring environmental health to policing humans, both tourists and trespassers (Carruthers 1995; Neumann 1998; Brockington 2002; Ramutsindela 2004; see also Spence 1999).

On a more troubling front, restrictions underpinning conservation practice and hence core to the production of wildlife often translated into the removal of Indigenous peoples. This was motivated by fear that Indigenous presence would disrupt tourist experiences and that Indigenous hunting, agriculture, and livestock rearing were environmentally destructive, as well as a deep devaluing of Indigenous relations to the land and nonhuman natures (Carruthers 1995; Spence 1999; Neumann 1998; Ramutsindela 2004; see also Lunstrum and Ybarra 2018). Indigenous eviction made space not only for wildlife but more symbolically "wilderness," or peopleless spaces teeming with wild animals that seem to harken back to a time before modernity or even the arrival of humans. Here, visitors could presumably escape the ills of urbanization and industrial capitalism, in effect returning temporarily to simpler times. This feel of timelessness was harnessed and inflated by both nascent tourism industries and early conservation practice. Such wilderness, to be sure, did not preexist its symbolic and literal construction; rather it was a myth that was imagined and actively imposed upon certain landscapes through the practices outlined previously, including human eviction and active wildlife management (Adams and McShane 1992; Neumann 1998; Brockington 2002).

What is particularly interesting about wildlife as opposed to other resources in this context is that part of its value—especially to tourism economies—rests on the assumption that it is timeless and exists outside of human intervention. Wildlife's value thus relies on the active obfuscation or erasure of the conditions of its own production (Igoe 2017). The irony, of course, is that it is human intervention that produced this timeless understanding and feel of wildlife and related notions of wilderness. In this sense, the production of wildlife as timeless masks its own contentious production as a resource.

Resource-making: the science of conservation management

The science of wildlife and conservation management has played a central role in producing wildlife as natural resource. While its roots stretch back to colonial science and early park development, wildlife and conservation management as we know it today became a coherent discipline with Aldo Leopold's ([1933] 1986) *Game Management*. Central to Leopold's notion of conservation and the management of "game" was his disagreement with the common assumption that wildlife was "something that must eventually disappear, [as opposed to] a resource that could be produced at will through prescribed management" (Leopold [1933] 1986, xvii). Turning this assumption on its head, Leopold explored how to manage and actively produce a supply of wild animals in the face of expanding human population and land use change. *Game Management* hence set out the basic theory, techniques, and administrative tools to achieve these ends. These were based on the premise that wildlife managers could control or influence certain variables to increase the productivity (i.e., birth rate) of wild animals above what natural conditions would permit. These interventions included hunting restrictions, predator controls, designation of protected lands, artificially stocking and farming game, and control over environmental factors like diseases and water supplies. Wildlife management was no longer a mere issue of enacting laws to prolong the survival of a species by limiting their killing; humans could now more actively and intensively produce wildlife through a set of prescribed techniques.

Much like the national park model, active wildlife management has travelled widely to underpin mainstream conservation efforts and more contemporary conservation sciences.

The history of South Africa's Kruger National Park, for example, is one of active human intervention to control animal diseases and invasive species, create artificial watering holes, and implement population control measures such as culling certain animals and increasing the birth rates of others (Whyte, van Aarde, and Pimm 2003), although these have ebbed and flowed over time. More broadly, the techniques, science, and practices of how to effectively manage wildlife have evolved. The 1980s saw the development of the new field of conservation biology, which focused on the application of biological science to confront human-induced threats to nonhuman natures (Soulé 1985). The broader framework of conservation science emerged in the 2000s in recognition of the intertwined relationship between human and ecosystem well-being (Kareiva and Marvier 2012). What we see in the evolution from *Game Management* to conservation science is a persistent interest in how humans can intervene to protect and ultimately produce nonhuman natures as wildlife.

Following the material turn in critical resource geographies (Bakker and Bridge 2006, Chapter 4 this volume), it is important to remember that, as conservation science and management produces wildlife, they do so in light of the materiality of the resource itself. For instance, the white rhino adapts rather well to confined breeding and raising much like domestic livestock, unlike many other large wild mammals. On the other hand, parks like Kruger have historically culled elephant populations to keep their numbers in check given their voracious appetite and penchant for knocking down trees to the detriment of both other species and broader ecosystem health (Whyte et al. 2003). In addition, wildlife reserves are often fenced as a key management strategy aimed at containing animals both for their own protection (from being killed) and to prevent crop damage and other dangers to surrounding communities as animals wander where they are not wanted. This underscores that the materiality of wildlife, here its mobility, shapes how the resource is produced and that such materiality can destabilize the values invested in wildlife as it becomes "out of place." Hence, the value of wildlife rests in their being "in place" and managed/produced as such. In this sense, as an "unruly" mobile resource, wildlife "frustrate[s] human desire" (Banoub 2017, 7; see also Valdivia, Wolford, and Lu 2014).

As conservation science and management have taken seriously the values invested in wildlife along with their biophysical properties, the field has been responsible for many conservation successes over the last century. Notable in their own right, these illustrate once more how wildlife is actively produced. Perhaps no success is more impressive than Operation Rhino that brought the white rhino back from the brink of extinction in the 1960s. This entailed the translocation of breeding-aged white rhinos from South Africa's Hluhluwe-Umfolozi Reserve to protected areas and breeding sanctuaries throughout the world over a 12-year period (Player 1973). The approximately 6000 white rhinos in Kruger National Park today can largely be traced to the 345 that were shipped there during this time. As the technologies available to conservation management advance, so too does the ability to intervene and control nature and threats to it. We see this, for example, with technologies that further facilitate translocation and breeding, as well as digital technologies, drones, and satellites that provide ever-increasing monitoring of wildlife and surveillance of threats (van der Wal and Arts 2015; Massé 2019).

Despite these successes, conservation science and practice today face a dire challenge in their mission to protect and produce wildlife. What many call the sixth mass extinction refers to the loss of over 60 percent of vertebrate species globally between 1970 and 2014 (WWF 2018). The major causes of biodiversity decline are the overexploitation of species, caused in part by illegal hunting, along with habitat fragmentation and loss due largely to agriculture. This is exacerbated by the effects of climate change (WWF 2018). This reality of biodiversity-under-threat sets the stage for more intensive conservation efforts, both noteworthy and concerning, a point we return to later.

Producing an economic resource

Arguably the most substantial value shaping the production of wildlife as resource is economic value. Economic values have always shaped the production of resources, and this has only intensified over the last century with resources in general (Bridge 2009; Banoub 2017) and wildlife in particular. The current production of wildlife is deeply embedded in market-based logics that value wildlife and their habitats as economic resources upon which to build tourism enterprises. Market-based conservation rests on the principle that if wildlife is to be conserved, it must be given an economic value so that the incentives to protect it are higher than not. This has led to the commercialization of wildlife, or managing it for commercial gain, and often its privatization, or its transformation from a publicly held entity into a privately held commodity to be bought and sold. These practices are constitutive of a broader neoliberalization of conservation (Igoe and Brockington 2007; Kelly 2011; Mbaria and Ogada 2017; Matusse 2019). In fact, today the commercialization and profit-generating potential of wildlife are *drivers* of protected area development rather than mere consequences of their creation (Kelly 2011). This economic value, however, is not inherent but rather must be produced through legal and administrative means (see Huber, Chapter 14 this volume). And as this value is produced, so too is wildlife as (economic) resource.

Central to producing the economic value of wildlife are the concepts of ownership and control. For much of history, wildlife around the world was a public good or managed as part of a commons (Thompson [1975] 2013; Mbaria and Ogada 2017). The last century has seen an erosion of these arrangements through the "capture" of wildlife and its habitat by state and private elites. With the former, state elites, often working in conjunction with private actors, have enacted legislation to take land and animals that were originally part of Indigenous commons and transform them into state-owned and state-controlled resources. With this legal change in place, wildlife could then be actively managed and produced through the mechanisms outlined previously. While the land and wildlife are not privatized per se, the state seizure and underlying erasure of Indigenous tenure and use have enabled private actors to generate profit through tourism development (Kelly 2011; Mbaria and Ogada 2017). This happens through privately owned projects and concessions adjacent to and within protected areas. While less frequent, there have been efforts to extract economic value from communal or Indigenous-owned wildlife and protected areas through community-based conservation projects. One common critique, however, is that these projects may keep Indigenous tenure or ownership intact but in practice erode Indigenous *control* of wildlife and other resources, including who profits from them (Matusse 2019; Mbaria and Ogada 2017; Ramutsindela 2004). Similar to ownership, control here refers to "practices that fix or consolidate forms of access, claiming, and exclusion for some time" (Peluso and Lund 2011, 668).

Changing patterns of ownership also enable the production of wildlife on private lands where wildlife itself has been privatized and commodified. South Africa, for example, passed the Game Theft Act in 1991 to provide economic incentives for private landowners to conserve wildlife inhabiting their land. The law made wildlife a resource that could be owned, commodified, and used by private landowners as the basis for wildlife businesses if owners followed certain game management guidelines (Snijders 2012). The value of wild animals on private land in South Africa transformed overnight into a profitable economic resource. This, in turn, has led to incentives to enclose large tracts of land, often dispossessing human residents, for the purposes of conservation and wildlife-based business (Snijders 2012). This has come to be known as green grabbing, or "the appropriation of land and resources for environmental ends" (Fairhead, Leach, and Scoones 2012, 38). Moreover, these legal-economic changes have meant that those who do not own large tracts of land and cannot meet certain technical specifications, such as is often the case with local communities, are unable to own and in turn produce wildlife. Whether through defining, granting, and ultimately changing patterns of ownership and control over land and wildlife, wildlife as resource is produced for some often at the expense of others.

The latest round of producing wildlife as resource: threatened wildlife, militarized wildlife

Conservation practice and its production of wildlife are both dynamic and have changed in response to broader political-economic-environmental trends over the last century. We see this most clearly in the recent militarization of conservation practice and space, which reflects a broader militarization of resources (Mehta, Huff, and Allouche 2019). The last decade has seen a precipitous increase in demand for wildlife largely through the illegal wildlife trade (IWT). In sub-Saharan Africa, this includes megafauna like elephants killed for their ivory and rhinos for their horn along with the pangolin, the only mammal wholly covered in scales. Depending on the animal in question, these are consumed for their presumed medicinal properties, displayed to convey wealth and prestige (reflecting their exorbitant costs), eaten as exotic food, and exploited as an investment opportunity especially after the 2007–2009 global financial crisis. While this trade is global in scope, it has largely been driven by growing affluence and related demand in parts of Asia, particularly China and Vietnam. Sub-Saharan Africa has been particularly hard hit. For instance, South Africa, which is home to the majority of the world's remaining 28,000 rhinos, has lost over 1,000 animals yearly between 2013 and 2017 to illegal international trade (Ferreira et al. 2018).

The primary response throughout sub-Saharan Africa has been to ramp up enforcement, which is becoming increasingly militarized. This is captured in the concept of green militarization, or the "use of military and paramilitary (military-like) actors, techniques, technologies, and partnerships in the pursuit of conservation" (Lunstrum 2014, 817) throughout Africa and parts of Asia (Mabele 2016; Barbora 2017; Duffy et al. 2019). South Africa, for instance, has seen the deployment of the Army into protected areas, increased ranger paramilitarization, use of military surveillance and pursuit aircraft, broader partnerships with defense corporations, development of heavily fortified intensive protection zones (IPZs), and hiring of former military leaders to spearhead anti-poaching efforts (Lunstrum 2014, 2018). While conservation has long had ties to military activity (Ellis 1994), the difference today is the intensity and sophistication of different layers of green militarization, making it increasingly difficult to distinguish where conservation ends and military activity begins. These changes have translated into the death and displacement of individuals and communities involved in the trade (Lunstrum 2014; Lunstrum and Ybarra 2018). Such measures are equally producing wildlife as a progressively *militarized* resource, one that is increasingly protected and produced through militarized conservation practices. This further highlights the tripartite value of wildlife as resource: first as a lucrative resource largely for illicit sale; second (and reflecting the conservation-related value of wildlife) as a resource important for national tourism economies, biodiversity, and natural heritage; and third as a means of income for conservation protection officers and profit for suppliers of military and surveillance products (Lunstrum 2014, 2018; Mbaria and Ogada 2017). Grasping the multiple ways in which wildlife is valued in the current context explains why people on different sides of the conservation/IWT equation are willing to kill, and risk being killed, to protect this value or exploit it.

Interestingly this IWT-green militarization dynamic is transforming the "safari" encounter of experiencing wildlife as resource. Today the loud humming of anti-poaching surveillance and pursuit helicopters routinely join the cries of hyenas and roars of lions filling the safari soundscape (Lunstrum 2018; Massé 2018). In many ways, this does not disrupt but rather adds to the feel of wildness and excitement. In addition, tourists now actively seek out opportunities to observe anti-poaching in action and witness the rotting carcass of a rhino killed for its horn (Massé 2019). In some sense, this latter desire is not surprising given that some of the most sought-after safari experiences have been the gory spectacle of animals attacking one another. This new trend offers a novel spin on this classic. Here the detritus of wildlife as resource in one respect (black market commodity) becomes a lucrative resource in another (sought-after tourism spectacle).

World-making: wildlife's production and the creation of new categories of the human

Who gets to manage and produce wildlife, for whom, and for what ends are dynamics that weave through conservation's often tense history. Moreover, these dynamics extend beyond the production of wildlife to include additional forms of world-making, including the production of new categories of the human, from the presumably sneaky environmental trespasser to the hypermasculine conservation hero (e.g., ranger) and big-game hunter (Neumann 2004). One of the most consequential of these categories is "the poacher." We saw previously that the criminalization of largely customary practices, including habitation, hunting, gathering, and farming has been central to the establishment of protected areas. Through these legislative changes, these customary practices *became* criminal, illegal activities, while the people behind these practices *became* criminals. In short, those seeking passageway through these lands and use of these resources became trespassers, and in particular those hunting animals were transformed from hunters into poachers.

In both North America and sub-Saharan Africa, the legitimate hunter-illegitimate poacher dyad took on deeply classed and especially racialized connotations. Here "legitimate" (i.e., state- and business-sanctioned) hunters were largely wealthy, white men looking for leisure and adventure through sport. Those designated as poachers on the other hand were largely poor, often Indigenous, and commonly had been displaced from their lands and livelihoods to make way for conservation (Neumann 1998; Spence 1999; Mavhunga 2014). In the African context, this rich-white/poor-native-black dichotomy still largely shapes Western understandings of who constitutes a legitimate hunter (i.e., sportsmen) versus an illegitimate poacher, often seen as someone lurking within the bush cunningly waiting to take what is not his (Neumann 2004; Mavhunga 2014; Lunstrum and Ybarra 2018; Mbaria and Ogada 2017). The dichotomy is equally responsible for authorizing some of the brutal tactics directed at the latter, from shoot-on-sight policies to killings and forced displacements sanctioned within a logic of green militarization. Hence, conservation practice has produced not only wildlife as a natural resource but also novel forms of world-making that include new and consequential *human* categories like the poacher.

Conclusion: the dangers and promises of exposing the production of wildlife as resource

Let us return to the wildlife-themed advertisement donning the walls of the Toronto subway station, one selling African liquor and more subtly tapping into and reproducing notions of Africa's timeless wilderness. As we have seen, the animals that we have come to embrace and protect as wildlife, such as the elephants in the image, are in no way timeless despite how they have been packaged. They have been actively produced as a natural resource, or an aspect of non-human nature in which various groups have invested value: ecological, symbolic, and economic.

And it is precisely through conservation—including its legal, managerial, and discursive techniques—that this production of wildlife as resource unfolds. In unpacking this process and in turn denaturalizing wildlife, we have engaged in perhaps the defining practice of critical resource geography. In doing so, we can begin to disrupt a taken-for-granted world in which wildlife is timeless and exists outside of and separate from the human realm. This is key to grasping the ways in which conservation practice and its production of wildlife as resource, despite their notable successes, have routinely disrupted already-existing ways of living with wildlife and dispossessed Indigenous groups of their lands and animals, both as a source of livelihood and integral facet of their identity and worldviews (Mavhunga 2014; Mbaria and Ogada 2017).

This exposing of wildlife as a produced resource, however, also poses a *very real danger* at this particular political moment, that of playing into the hands of a growing antienvironmental populist movement. Championed by influential politicians like Donald Trump and Jair Bolsonaro, there is an active global movement to undermine environmental protections, largely in the name of unbridled economic growth and a politics of individualism and anti-regulation. If wildlife's apparent timelessness, existence apart from humans, and so forth generates support for conservation, then revealing these as myth is dangerous territory.

Treading carefully, we close by suggesting that recognizing—and indeed embracing—wildlife as a (partially) human-produced resource imbued with human-defined value can help us protect/produce wildlife but do so in less exploitative ways. The question thus becomes how best to produce wildlife and what values to invest in it, not to return to some fantasy that wildlife should just be left alone to be "wild." Given that conservation has largely excluded Indigenous communities and in many cases been made possible by their exclusion, a more just and effective production of wildlife must include opening up the value of wildlife to include its importance to Indigenous peoples, including their livelihoods, worldviews, and alternative world-making processes. It is also here where we may find less exclusionary reasons to protect and produce wildlife and more effective models, both ecologically and socially, for doing so.

References

Adams, Jonathan S., and Thomas O. McShane. 1992. *The Myth of Wild Africa: Conservation Without Illusion.* New York, NY: W.W. Norton.

Adams, William Mark. 2004. *Against Extinction: The Story of Conservation.* London: Earthscan.

Alcamo, Joseph, Neville J. Ash, Colin D. Butler, J. Baird Callicott, Doris Capistrano, Stephen R. Carpenter, and Juan Carols Castilla, et al. 2005. *Ecosystems and Human Well-Being: A Framework for Assessment.* Washington, DC: Island Press.

Bakker, Karen, and Gavin Bridge. 2006. "Material Worlds? Resource Geographies and the 'Matter of Nature." *Progress in Human Geography* 30 (1): 5–27. doi: 10.1191/0309132506ph588oa.

Banoub, Daniel. 2017. "Natural Resources." In *The International Encyclopedia of Geography: People, the Earth, Environment and Technology,* edited by Douglas Richardson, Noel Castree, Michael F. Goodchild, Audrey Kobayashi, Weidong Liu, and Richard A. Marston. doi:10.1002/9781118786352.wbieg0496.

Barbora, Sanjay. 2017. "Riding the Rhino: Conservation, Conflicts, and Militarisation of Kaziranga National Park in Assam." *Antipode* 49 (5): 1145–1163. doi: 10.1111/anti.12329.

Beinart, William, and Lotte Hughes. 2007. *Environment and Empire.* Oxford, UK: Oxford University Press.

Bridge, Gavin. 2009. "Material Worlds: Natural Resources, Resource Geography, and the Material Economy." *Geography Compass* 3 (3): 1217–1244. doi: 10.1111/j.1749-8198.2009.00233.x.

Brockington, Dan. 2002. *Fortress Conservation: The Preservation of the Mkomazi Game Reserve, Tanzania.* Bloomington, IN: Indiana University Press.

Carroll, Clint. 2014. "Native Enclosures: Tribal National Parks and the Progressive Politics of Environmental Stewardship in Indian Country." *Geoforum* 53: 31–40. doi:10.1016/j.geoforum.2014.02.003.

Carruthers, Jane. 1995. *The Kruger National Park: A Social and Political History*. Pietermaritzburg, South Africa: University of Natal Press.

Cronon, William. 1996. "The Trouble With Wilderness: Or, Getting Back to the Wrong Nature." *Environmental History* 1 (1): 7–28. doi: 10.2307/3985059.

Duffy, Rosaleen, Francis Massé, Emile Smidt, Esther Marijnen, Bram Büscher, Judith Verweijen, and Maano Ramutsindela, et al. 2019. "Why We Must Question the Militarisation of Conservation." *Biological Conservation* 232: 66–73. doi:10.1016/j.biocon.2019.01.013.

Ellis, Stephen. 1994. "Of Elephants and Men: Politics and Nature Conservation in South Africa." *Journal of Southern African Studies* 20 (1): 53–69.

Fairhead, James, Melissa Leach, and Ian Scoones. 2012. "Green Grabbing: A New Appropriation of Nature?" *Journal of Peasant Studies* 39 (2): 237–261. doi: 10.1080/03066150.2012.671770.

Ferreira, Sam M., Cathy Greaver, Zoliswa Nhleko, and Chenay Simms. 2018. "Realization of Poaching Effects on Rhinoceroses in Kruger National Park, South Africa." *South African Journal of Wildlife Research* 48 (1): 1–7. doi: 10.3957/056.048.013001.

Grove, Richard. 1995. *Green Imperialism: Colonial Expansion, Tropical Island Edens, and the Origins of Environmentalism, 1600–1860*. Cambridge, UK: Cambridge University Press.

Igoe, Jim. 2017. *The Nature of Spectacle: On Images, Money, and Conserving Capitalism*. Tucson, AZ: University of Arizona Press.

Igoe, Jim, and Dan Brockington. 2007. "Neoliberal Conservation: A Brief Introduction." *Conservation and Society* 5 (4): 432–449.

Jacoby, Karl. 2001. *Crimes Against Nature: Squatters, Poachers, Thieves, and the Hidden History of American Conservation*. Berkeley, CA: University of California Press.

Kareiva, Peter, and Michelle Marvier.. 2012. "What Is Conservation Science?" *BioScience* 62 (11): 962–969. doi: 10.1525/bio.2012.62.11.5.

Kelly, Alice B. 2011. "Conservation Practice as Primitive Accumulation." *The Journal of Peasant Studies* 38 (4): 683–701. doi: 10.1080/03066150.2011.607695.

Leopold, Aldo. [1933] 1986. *Game Management*. Madison, WI: University of Wisconsin Press.

Lunstrum, Elizabeth. 2014. "Green Militarization: Anti-Poaching Efforts and the Spatial Contours of Kruger National Park." *Annals of the Association of American Geographer* 104 (4): 816–832. doi: 10.1080/00045608.2014.912545.

Lunstrum, Elizabeth. 2018. "Capitalism, Wealth, and Conservation in the Age of Security: The Vitalization of the State." *Annals of the American Association of Geographers* 108 (4): 1022–1037. doi: 10.1080/24694452.2017.1407629.

Lunstrum, Elizabeth, and Megan Ybarra. 2018. "Deploying Difference: Security Threat Narratives and State Displacement from Protected Areas." *Conservation and Society* 16 (2): 114–124. doi: 10.4103/cs.cs_16_119.

Mabele, Mathew Bukhi. 2016. "Beyond Forceful Measures: Tanzania's 'War on Poaching' Needs Diversified Strategies More Than Militarised Tactics." *Review of African Political Economy* 44 (153): 487–498. doi: 10.1080/03056244.2016.1271316.

Massé, Francis. 2018. "Topographies of Security and the Multiple Spatialities of (Conservation) Power: Verticality, Surveillance, and Space-Time Compression in the Bush." *Political Geography* 67: 56–64. doi:10.1016/j.polgeo.2018.10.001.

Massé, Francis. 2019. "Anti-Poaching's Politics of (in)Visibility: Representing Nature and Conservation Amidst a Poaching Crisis." *Geoforum* 98: 1–14. doi:10.1016/j.geoforum.2018.09.011.

Matusse, A. 2019. "Laws, Parks, Reserves, and Local Peoples: A Brief Historical Analysis of Conservation Legislation in Mozambique." *Conservation & Society* 17 (1): 15–25. doi: 10.4103/cs.cs_17_40.

Mavhunga, Clapperton Chakenetsa. 2014. *Transient Workspaces: Technologies of Everyday Innovation in Zimbabwe*. Cambridge, MA: MIT Press.

Mbaria, John, and Mordecai Ogada. 2017. *The Big Conservation Lie*. Auburn, WA: Lens & Pens Publishing.

Mehta, Lyla, Amber Huff, and Jeremy Allouche. 2019. "The New Politics and Geographies of Scarcity." *Geoforum* 101: 222–230. doi:10.1016/j.geoforum.2018.10.027.

Neumann, Roderick P. 1998. *Imposing Wilderness: Struggles Over Livelihood and Nature Preservation in Africa*. Berkeley, CA: University of California Press.

Neumann, Roderick P. 2004. "Moral and Discursive Geographies in the War for Biodiversity in Africa." *Political Geography* 23 (7): 813–837. doi: 10.1016/j.polgeo.2004.05.011.

Peluso, Nancy Lee, and Christian Lund. 2011. "New Frontiers of Land Control: Introduction." *Journal of Peasant Studies* 38 (4): 667–681. doi: 10.1080/03066150.2011.607692.

Player, Ian. 1973. *The White Rhino Saga*. New York, NY: Stein and Day.

Ramutsindela, Maano. 2004. *Parks and People in Postcolonial Societies: Experiences in Southern Africa*. Dordrecht, The Netherlands: Kluwer Academic Publishers.

Rolston, Holmes. 1994. "Value in Nature and the Nature of Value." In *Philosophy and the Natural Environment*, edited by Robin Attfield, and Andrew Belsey, 13–30. Cambridge, UK: Cambridge University Press.

Snijders, Dhoya. 2012. "Wild Property and Its Boundaries—On Wildlife Policy and Rural Consequences in South Africa." *Journal of Peasant Studies* 39 (2): 503–520. doi: 10.1080/03066150.2012.667406.

Soulé, Michael E. 1985. "What Is Conservation Biology?" *BioScience* 35 (11): 72–734. doi: 10.2307/1310054.

Spence, Mark David. 1999. *Dispossessing the Wilderness: Indian Removal and the Making of the National Parks*. Oxford, UK: Oxford University Press.

Thompson, Edward Palmer. (1975) 2013. *Whigs and Hunters: The Origin of the Black Act*. London, UK: Breviary Stuff Publications.

Valdivia, Gabriela, Wendy Wolford, and Flora Lu. 2014. "Border Crossings: New Geographies of Protection and Production in the Galapagos Islands." *Annals of the Association of American Geographers* 104 (3): 686–701. doi: 10.1080/00045608.2014.892390.

van der Wal, René, and Koen Arts. 2015. "Digital Conservation: An Introduction." *Ambio* 44: 517–521. doi:10.1007/s13280-015-0701-5.

Whyte, Ian J., Rudi J. van Aarde, and Stuart L. Pimm. 2003. "Kruger's Elephant Population: Its Size and Consequences for Ecosystem Heterogeneity." In *The Kruger Experience: Ecology and Management of Savanna Heterogeneity*, edited by Johan T. Du Toit, Kevin H. Rogers, and Harry C. Biggs, 332–348. Washington, DC: Island Press.

WWF. 2018. *Living Planet Report, 2018: Aiming Higher*. Gland, Switzerland: World Wildlife Fund.

Zimmermann, Erich W. 1933. *World Resources and Industries*. New York, NY: Harper & Brothers Publishers.

31

Anadromous frontiers

Reframing citizenship in extractive regions. The salmon industry in Los Lagos, Chile

Beatriz Bustos-Gallardo

Introduction

This chapter discusses the role of the salmon farming industry in shaping the configuration of citizens as political, ecological, and economic subjects in southern Chile. Over the last 30 years, southern Chile has undergone a radical transformation associated with the inception and expansion of the farmed salmon industry (Barton and Fløysand 2010; Barton and Murray 2009; Fløysand, Barton, and Román 2010; Katz, Iizuka, and Muñoz 2011; Phyne and Mansilla 2003). Salmon is not the first commodity to transform the spaces and ecologies of the region, which was a commodity frontier long before its introduction (Romero 2018). However, while previous state attempts to secure territorial control over the region (European colonization in the 1880s and industrialization policies in the 1950s–1970s) succeeded in enforcing privatization, enclosure, and control over resources, the salmon farming era, like other similar transformations, "profoundly rework[ed] patterns of authority and institutional architectures" (Rasmussen and Lund 2018, 388), and in doing so, produced new forms of state-nature-subject interaction.

As I elaborate in this chapter, environmental change is at the core of these new forms of state-nature-subject interactions. Atlantic salmon, the primary species produced in Chile, is not native to local ecosystems. To make the investment viable, firms needed to adapt fish biology to the temperatures and microbial conditions found in Chilean rivers, lakes, estuaries, and marine waters. This adaptation involved technological innovations, from genetic experiments to accelerate fish growth and improve virus resistance, to the adjustment of the stomach to digest vegetables instead of animal protein as replicating predation on fish as in the wild would be costly and ecologically intensive. Ecosystems, both terrestrial and aquatic, have become recipients of animal residues leading to eutrophication in lakes and marine channels. Salmon farming has introduced land-use changes and greater demand for water to meet the increased demand for feed. As the industry seizes natural resources for salmon production, other economic activities, such as artisanal fisheries and tourism, have changed and in many cases, been harmed.

In the rest of the chapter, I explain how the nature of salmon production in Chile configures new ecological landscapes and thus new forms of rural living. The first section covers the institutional and ecological work involved in facilitating salmon production in the Los Lagos region, which includes the state's role in creating concessions, the construction of port and road infrastructure, and firms' investments in efficiency-enhancing production models. Next, I expand on the notion of what I call "extractive citizenship," the political practices rural subjects undertake as they navigate streams of political and economic action set by the state and industry. These practices align subject movements along with rhythms and metabolic transformations of the salmon monoculture mode of production based on state evaluation of opportunities and the environmental realities that sustain rural life in salmon-producing regions. The next section focuses on the responses of local populations, and on the effect of these transformations on how people exercise their citizenship. I conclude by discussing the relevance of this case to Critical Resource Geography.

Producing salmon in southern Chile

The introduction of Atlantic salmon into Chilean waters required a combination of intense technological, legal, economic, and political processes. The Chilean state laid the foundations by creating legal forms of private control of marine spaces—in the form of aquaculture concessions[1]—and financial incentives designed to attract capital. The private sector responded. The number of aquaculture firms multiplied throughout the 1990s, however, by 2007, a mere 25 accounted for 90 percent of total exports, half of which were produced by only five companies (Katz, Iizuka, and Muñoz 2011). Universities also played a key role, training cadres for the industry and developing technological solutions to production-related problems.

The salmon farms experienced periods of boom and bust as the industry integrated into the global market and confronted the localized ecological effects of large-scale, intensive, exotic-species farming. In 1988, the industry faced its first mass mortality event caused by algal blooming, and over subsequent years, multiple reports raised concerns about salmon escapes, increased threat of pathogens, and massive mortalities. Disaster struck in 2008 with an outbreak of the ISA (infectious salmon anemia) virus, an infectious disease that brought the Chilean industry to its knees and forced the state to tighten regulations. The resulting increase in production costs drove industry players to merge (Bustos-Gallardo 2013). Boom periods (exponential growth in production and profit) shortened following the ISA disaster; by 2016, the industry faced another major algal bloom event that once again raised the question of ecological sustainability (Bustos-Gallardo and Román 2019; Mascareño et al. 2018).

Salmon farming takes place in three stages: breeding, growing, and processing. The first involves the selection of the best genetic traits and management of the transition of the animals from fresh- to saltwater in controlled pools. During the growing stage, salmon are seeded into sea cages and then fed until reaching commercial weight, at which point they are shipped to processing plants and prepared for market. It is during the growing stage that salmon production is most exposed to ecological unpredictability and, consequently, that its effect on the landscape is most severe: water pollution and eutrophication affects the availability of other marine species for artisanal fishing.

Over time, the use of technology has increased industry capacity to control key factors that drive salmon production. Particularly during the breeding stage, genetic advances—such as DNA sequencing—facilitated the improvement of egg quality, and the move to closed pools and hatcheries increased control over the "when" and the "where" of production. By contrast, the marine stage continues to be the period during which producers have the least control over the fish. This stage is when the flexibility of salmon biology becomes key to its successful production in Chile, transforming the landscape through three interconnected elements: location, time frame, and growing, which I treat in turn.

In terms of location, while the biological flexibility of salmon facilitated its initial reproduction in Chilean waters, the evolution of the industry demanded adaptations and improvements

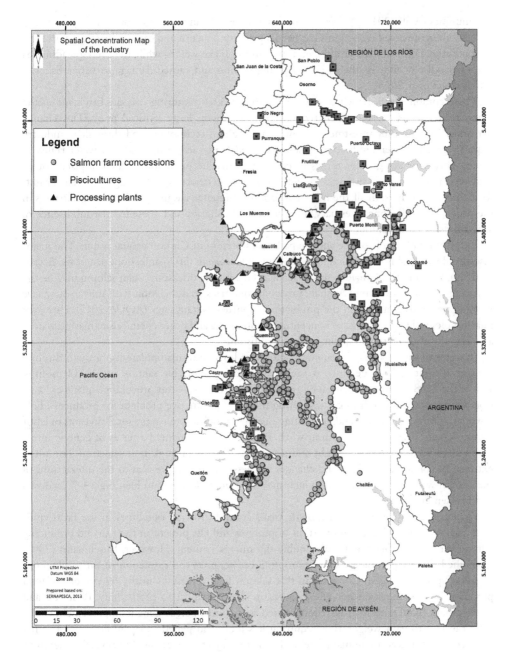

Figure 31.1 Salmon production in Los Lagos. Map by Gino Sandoval.

to local landscapes, infrastructure, and technology. The Los Lagos region (see Figure 31.1) contains 11 lakes, 4 major hydrographic basins, the waters of the Reloncaví estuary, and 300 km of marine coastline. It is the *interconnection* of these bodies of water that make salmon production possible at global scale there. Building this interconnection involved state-funded construction of ports and roads, investments that had the effect of reducing isolation of the region's rural communities (Avilés 2015). Moreover, the environmental damage caused by salmon production in these interconnected ecosystems has been a primary concern and cause of friction with local communities. Fish hatcheries caused eutrophication of lake waters, while the ever-spreading sea cages periodically allowed the escape of young salmon and caused damage to native fisheries (Niklitschek et al. 2013).

These combined effects led to new legislation in 2010 requiring that the first stage of the lifecycle be moved to land-based recirculation systems in large artificial pools. This change demanded a significant investment in infrastructure and technology and through it, industry improved control over production. It also involved firms' acquiring land and water rights in nearby ports and along other routes to connect to marine operations, which displaced other activities, such as artisanal fishing, and removed and/or generated overlapping water usage rights for farmers and others. In all, the fish farms' 30-year presence has increased levels of nutrients in the waters of the Reloncaví estuary, displaced native fisheries, and disrupted the subsistence of artisanal fisher communities.

Industry is vulnerable to the aquatic pathogens present in Chilean coastal waters. The major diseases affecting salmon production are *Piscirickettsia salmonis* (SRS), infectious pancreatic necrosis (IPN), ISA, and the *Caligus* genus of sea lice. Genetically disease-resistant salmon may be the key to production continuity. As such, private laboratories, in association with universities, have invested nearly $17 million in the production of virus-resistant eggs (AQUA 2017). However, pharmacological treatments have concentrated antibiotics and other chemicals in marine waters, affecting the availability and quality of artisanal fishing.

In terms of time frame, salmon companies have learned to manipulate fish growth through control of light and temperature at each stage, rearing tradeable salmon in just 14 months. Before the 2008 ISA crisis, rotation of growing cycles was non-stop, provoking aquatic exhaustion, with anaerobic conditions constituting the main productive problem for producers. Following the crisis, the state mandated resting periods, which also changed local rhythms of labor by introducing seasonality to salmon work. Firms also implemented a series of genetic interventions to increase harvest weight and accelerate growth through trait selection. The more extreme the biological and genetic changes, the less certainty there is as to the interactions of the salmon with the ecosystems in which they are placed (Einum and Fleming 1997; Jacobsen and Hansen 2001).

Finally, in terms of biological growth, firms control growing outcomes in the farm environment by determining the proportion of proteins and fats present in the fish and producing salmon three times fatter and with double the omega content. However, Buschmann (2001) established that each kilogram of salmon produced requires the extraction of three kilograms of native species for feed. Considering that salmon feed represents nearly 60 percent of total production costs, the food-per-kilo conversion rate is critical to profitability. Research and development (R&D) efforts have concentrated on two main areas: genetic improvement of the fish's capacity to gain weight, thus requiring less food, and genetic adaptation of fish to process vegetables instead of animal proteins. The latter objective is essential to several dimensions of resource-making: vegetable proteins are cheaper to produce than animal proteins, making salmon production more economically viable; they are more environmentally friendly, enhancing production viability and rebutting a central critique of the sector by reducing demand for native fishmeal; and their production creates demand among farmers, creating an economic link between aquaculture and rural farming communities.

The salmon industry's successful insertion into an exotic ecosystem and its increasing levels of production have depended upon a combination of technology, environmental conditions, and capital, fostered by state policies. However, the blending of these elements over time has degraded the local environment, exposing ecological contradictions inherent to salmon-production (Bustos-Gallardo and Irarrázaval 2016). During the ISA (2008) and algal bloom (2016) crises, the state and firms introduced spatial and ecological fixes to these contradictions, including (see also, Irarrázaval and Bustos-Gallardo 2018):

a. New government regulations required the introduction of "Agrupaciones de Concesión Salmonera (ACS)," or "salmon neighborhoods," the restriction of cage density to 15–18 kg/m³, the imposition of mandatory sanitary rest periods between harvest and seed-ing, and a freeze on new concessions in the Los Lagos region. These measures addressed time frame and location, expanding and organizing the geographical production area by promoting a southward shift of salmon farms and defining the permissible level of stress to the salmon (i.e., farm density).

b. Movement of farms to off-shore locations, reduction of ecological impact in coastal areas, genetic advancements to improve resistance to diseases and other ecological threats, and transfer of the growth stage to land-based facilities all required significant technology and capital input. The most competitive firms are actively exploring a future scenario in which global demand continues to increase in parallel with calls for more conscious production practices at the local level. Firms are exploring if this is a situation in which higher produc-tion costs would be justifiable and acceptable from a business point of view.

The two crises demonstrated to local people the fragility of the industry, as well as the com-mitment of the state and the private sector to maintaining it. At the same time, the state response directly affected local peoples' modes of living. New sanitary regulations also applied to artisanal fishing and the harvesting of algae, increasing the costs associated with these economies and worsening labor precarity for those involved in them. This postcrisis scenario prompted ques-tioning of how and to whom resource access was granted, and the socio-natural contract upon which salmon production was based drove the emergence of new notions of rights, preroga-tives, and state membership. I will now address these notions in greater detail and their effects on citizenship.

The emergence of extractive citizenship in the Los Lagos region

Citizenship in modern societies is recognized through international, national, and local institu-tions that grant rights, entitlements, and membership in a community, meaning that citizenship is "formed through scalar configuration and engagement with place" (Desforges, Jones, and Woods 2005, 440). Individuals, in turn, cultivate varieties of citizenship through engagement with place and community, and in relation to globalized economies. In this section, I look more closely at these practices, to show how the production of what I call "extractive citizenship" has emerged in relation to the making of salmon aquaculture in southern Chile.

"Extractive citizenship" describes how a rural subject becomes responsible for his/her eco-nomic choices and logics through an entrepreneurial relationship with nature, and by adopting more flexible and mobile economic practices, such as switching among multiple economic practices (e.g., agriculture and seaweed harvesting) and/or migrating or commuting for eco-nomic opportunity. Extractive citizenship is closely related to Ong's (2006) formulation of neo-liberal citizenship. For Ong, neoliberalism is a malleable technology of governing the social through capitalist innovation, adoptable by diverse regimes, and operating through strategic exceptions in order to outcompete others. Simply put, neoliberalism "is a mode of political optimization" (Ong 2006, 3) where interactive modes of citizenship emerge from capitalist entrepreneurship—from economic zones to flexible labor. These new, interactive modes of citizenship reorganize people according not to their membership in national communities but according to the marketability of their skills and knowledge.

In Chile, a neoliberal citizen is an individual who secures the satisfaction of his or her rights (welfare, work, etc.) through the market, in ways that suggest that his or her political rights are conditioned by the dynamism, adaptability, and flexibility of their economic practices (Ortiz Gómez 2014; Gaudichaud 2015; Gómez Leyton 2010). Meanwhile, the state shares decision-making with the private sector, and legal frameworks foster private investment and access to resources, according to the prevailing logic of the market. In the case of salmon farming, this creates a landscape of inclusion and exclusion (Budds 2013; Bustos-Gallardo et al. 2019; Schurman 1996; Taylor 2002), which individuals in rural areas are forced to adapt to by "choosing" to follow practices from among the possibilities presented by the market and state agency programs that make them more competitive, e.g., labor flexibilization, mobility, and exploitation.

While neoliberal citizenship describes forms of subjectification related to an economic logic, extractive citizenship starts with the relationship with the resource itself, in order to showcase how citizen political practices and actions are also shaped by the dynamics and logics of resource extraction (see also, Perreault, Chapter 11 of this volume). Salmon farming, and the spatial and ecological fixes that the state and industry have implemented to ensure its survival, have contributed to the emergence of a particular kind of citizenship practices. Extractive citizenship takes into consideration the material and discursive conditions of resource extraction practices, and how these depend as much on firms and states as on the technologies and ecologies of the resource. Spatially, they rely on community spaces created to arrange access to resources (such as neighborhood associations or water committees) as a point of intermediation between individual and collective interests. It is in these spaces that subjects politically navigate and negotiate the distinctions that the new legal and institutional regimes create within communities. For example, technical assistance to fisherman pushed them into production logics of salmon production: their shift to mussels farming, supported by the state, responded to their expulsion from the sea, lack of native fisheries to hunt, and excessive nutrients in the coastal waters due to salmon farming which provided feed for mussel growing. Technical assistance pushed fisherman to harvest mussels, seizing those "opportunities." It also moved them to organize in new ways (fisherman unions) and created us/them dynamics in rural communities between those who gained access and those who could not.

In other words, the materiality of the resource and the actions that states and firms take to support industrial aquaculture affect the individual's identification with the commodity and expectations about reciprocity, power, and possibilities. In practical terms, the ability to access nature has effects on the expectations about the pact between the state-subjects-companies, which in turn fluctuate depending on the rhythms of production (boom and crisis cycles, but also production cycles) (see also, Zhu and Peluso, Chapter 29 of this volume). Later, I offer an example of how these relations work.

Table 31.1 illuminates contours of extractive citizenship by providing an overview of the dynamics of rural subjects with salmon production in the Los Lagos region. It illustrates the evolving relationship among salmon farming, state devices, and the everyday practices of three, at times overlapping, groups of rural subjects—salmon workers, agricultural farmers, and artisanal fishers—in the period between the ISA and algae bloom crisis. These two crises drove changes in legislation (see previous section) that affected local people and raised questions about the real effect of the salmon industry on the region and its future. Plainly, during the ISA crisis of 2008 and the algal bloom event of 2016, the ecological contradictions that caused them also

Table 31.1 Contours of extractive citizenship in Los Lagos

	Subject		
	Salmon workers	*Artisanal fishermen*	*Subsistence farmers*
Effect of salmon farming	Access to work Consumer power	Dispossession and expulsion from marine areas Access to work Entrepreneurship and mobility	Access to work Abandonment of fields Entrepreneurship
State devices	Labor code	Fishing Law no. 18.892 Sanitary regulations for aquaculture DS no. 320-2001	Indigenous Law no. 19253/Law for indigenous groups access to marine and coastal areas no. 20.249 Farm Subsidies
ISA crisis effect/ reaction Algal bloom effect/ reaction	Large-scale layoffs/ Union mobilization Large-scale layoffs/none	Large-scale flow of redundant salmon labor into illegal fishing/Conflict over access to resources Prohibition to work/ Large-scale mobilizations	Some flows of redundant salmon labor to the fields/ Increased insecurity of family incomes Prohibition to work/Large-scale mobilizations
Effects of institutional solutions to both crisis	Forced labor mobility Labor insecurity (layoffs during mandatory rest periods) Job reduction	Quotas and forced mobility Dealing with degraded ecosystems	Institutionalization of access to coastal areas Dealing with coastal pollution
Grievance	Labor exploitation	Ecological degradation of means of subsistence	Distribution of costs and benefits/ changes in cultural identity
Resulting citizenship Resource identity strategies	Conditioning Assimilation	Fragmentation Resistance	Disenfranchisement Adaptation

triggered different reactions among subjects and broader questioning of the hegemonic presence of the salmon industry in the region.

Notice that rural subjects flow between economic roles (workers, farmers, fishermen), first, because it is part of their traditional livelihood practices, but recently, as a strategy to coinhabit the resulting landscape of salmon production. As the state creates and legitimizes subjects through legal and institutional devices that define the tools and venues (including to expose grievances), firms' mode of production conditions the emergence of particular economic subjects, either through the replication of practices or the assimilation of them. Together, state devices and firms' mode of production configure a mix that pushes rural subjects to new practices of citizenship.

Over several years of fieldwork and nearly 100 interviews,[2] I have identified three main categories of practices that citizens undertake in relation to the expansion and crises of the salmon industry: conditioning (acceptance of and adherence to the rules and mechanisms imposed to access resources), fragmentation (individuation of strategies and interactions with the state as a way to maximize individual benefits), and disenfranchisement (conscious disconnection from state mechanisms and decision-making spaces perceived as formal and meaningless).

To understand conditioning as a mode of extractive citizenship, we need to look at salmon workers. They provide the labor required to transform fish into a commodity. According to Dresdner et al. (2016), in 2014, the industry employed nearly 30,000 direct workers and provided 14,500 indirect jobs, most of them in processing plants. These numbers highlight the contradictions that salmon production represents in terms of citizenship practices. On the one hand, as an employee of the salmon farms, an individual gains access to formal work and consumer choice, achieving a position from which to negotiate their material living conditions with the salmon companies; on the other, as salmon aquaculture expands and dominates the landscape and infrastructure in the region, they lose access to resources from which to make an autonomous living. Furthermore, intermittency of labor availability and anti-union behavior limit workers' negotiating power, as seen during both the ISA and bloom crises. Large-scale layoffs were the first effect, but in both cases, the instinct of workers was to secure their jobs and industry continuity in the region, while rejecting labor destabilization and anti-union practices.

Precarity lies at the core of their citizenship practices because rural subjects become dependent on state subsidies to satisfy their livelihood.[3] In turn, they are pushed into adopting the industry's logic of accumulation, resulting in their political subordination to the prevailing mode of production. The following quotation from a Chilote (a resident of the largely rural island of Chiloé in the region), reflecting on the aftermath of the ISA crisis layoffs, outlines the political effect of accepting state subsidies:

> So Chilotes—a self-sufficient people with a culture of bartering, of dignity, of working as a community—have become accustomed … what's it called? Protectionist; someone who expects things to be given to him, who is dependent on the government's gifts, on project X that gives them this, on project Y that gives them fences … they become welfare workers
> (Chilote local leader, 2018)

Artisanal fishers' extractive citizenship coheres through fragmentation. Their lives and livelihoods are connected with the sea and directly affected by environmental degradation caused by the salmon industry. They also reflect the neoliberal citizenship ideals of entrepreneurship and mobility through their constant search for opportunities and business mentality. This puts them in a contradictory situation: while they are the most vocal group in denouncing the ecological degradation that salmon farming causes, their history of overharvesting and overfishing are also part of ecological problems in the region.

Artisanal fishers are free to move between labor spaces and to exercise their entrepreneurial spirit to pursue social reproduction. However, fishing quotas and other constraints imposed by the Fisheries Law (No. 18.892/1991)—such as the requirement to register for fishing rights and the geographical delimitation of fishing areas according to regional administrative boundaries— have removed their access to marine resources and forced them to choose between joining the salmon workforce or illegally overexploiting marine resources.

Meanwhile, the state has, through different agencies, split the concerns and interests of artisanal fishers into pieces, fragmenting their political action across multiple locations:

> That's what I'm telling you: they split you up. The same thing happened with the PDTI programs [Indigenous Territorial Development Program]; that is, I don't know … They

split you up. You have an orchard, a greenhouse, and three cows, and you're segment two … you have an orchard and less than half a hectare, and you're segment one … You have more than three cows, and you're segment three or who knows what. So, you can apply to investment project XYZ and ask for many millions. And they go off and split us up … And then the comparisons begin: "segment one who knows, segment two has these options, segment three is going to be given … I don't know, compensation of I don't know how much

(Chilote woman, 2018)

This quote shows that the fragmentation of help and opportunities from the state requires that rural subjects adopt a logic of flexibility to access them: they need to become whatever the subsidy or program privileges, and thus, like the salmon themselves, subjects are left to navigate the various streams of political and economic action through which state and industry control access to resources. Thus, during the ISA and bloom crises, artisanal fishers' demands extended no further than compensation for the environmental damage suffered.

Finally, to understand disenfranchisement as a form of extractive citizenship, let's review agricultural subsistence farmers. Farmers are the labor force in times of salmon boom: farmers migrate to urban areas to work in the processing plants. They also offer refuge for workers and artisanal fishers in times of crisis by providing work, food, and housing. Salmon production represents a paradox for farmers. Demand for produce represents stability for rural livelihoods, but industry expansion has led to some farmers abandoning agricultural land, due in part to the loss of labor to the salmon processing plants, and in part to increases in farming costs driven by tighter regulation (higher production standards, sanitary regulations, taxes, etc.). These challenges put farmers in a highly precarious position (Caniguan Verlarde 2016; Valenzuela Silva 2016). Only recently, the Indigenous Law (No. 19,253/1993) and related legislation (No. 20,249/2009) have guaranteed access to marine areas for indigenous people and provided farmers with the tools necessary to regain access to and control of both terrestrial and marine spaces. State subsidies financially support farming, but production is insufficient to provide a proper living (Barton et al. 2013; Saliéres et al. 2005), leading to a vicious circle of informality: eligibility for subsidies requires farmers to prove their vulnerability, in turn driving them to maintain their informal and subsistence status. This leads farmers to claim that the state has abandoned them, that their prosperity is secondary to that of the salmon industry, and that the state is oblivious of Chilotes. Grievances voiced by subsistence farmers during the two crises concerned the loss of cultural identity and autonomy. This sentiment is shared regionally: a survey applied to 567 homes in the Los Lagos region found that 80 percent of respondents believed that the state favored the salmon industry. Furthermore, 69 percent claimed that the state had not protected the community from industry abuses (FR1151215). An older farmer reflecting on the evolution of the industry in Chiloé and its relationship with local communities expressed:

The state played a role … [negative] in that, it has applied an "anything goes" policy in Chiloé… . They accepted this because the industry gave work. To put it simply and rather bluntly, the state pulled down its pants to allow this industry to exist

(Chilote farmer, 2018)

As the quotes reveal, there is a shared perception that it does not matter what locals think, and thus a sense that it is useless to engage in debates with authorities. This disenfranchisement has had two outcomes. Some people have adapted by navigating the available subsidies to sustain their rural existence. They also have advocated for the implementation of the indigenous and marine access laws that secure exclusive access to natural resources, accepting with resignation, the privatization logics that these concessions entail. Others have used their identities as farmers and indigenous people as a political device for interaction on more equal terms with the state and the salmon industry. Applications for ECMPO (Coastal Areas for Indigenous People) rights have sky-rocketed over the last decade because they allow indigenous communities to dispute salmon firms' control of the sea. By June 2018, 46 percent of ECMPO applications involved overlap with salmon concessions in the Los Lagos region (Escobar Mendoza 2018); if granted to the subsistence farmers who have applied for them, they will gain rights and the ability to exclude salmon farms.

Conclusion

The case presented here invites Critical Resource Geography researchers to seek a deeper understanding of how commodity making affects and is affected by citizenship practices, how local communities identify with the commodity in question and the impact of its production upon territorial development paths, and the individual capacities of citizens to access and use resources. To guide these explorations, I have introduced the concept of extractive citizenship. Extractive citizenship is shaped by a combination of institutional constraints, economic opportunities, and ecological configurations that emerge from the prevailing mode of production: neoliberal economics. I have explored some of the contours of extractive citizenship for people working in the aquaculture industry, artisanal fishers, and subsistence farmers. The analysis reveals that in the case of salmon farming in southern Chile, individuals strategically assume multiple roles in attempts to maximize their access to resources, and an individual might embody multiple roles at a given time. Thus, it is up to each individual to negotiate the balance of incentives and penalties of each capacity. This is a key feature of citizenship in commodity frontiers, while the state offers multiple venues for involvement, each one requiring individuals to assess whether collective action is preferable to individual solutions.

While the ecological and spatial fixes implemented by the state and firms may have (temporarily) solved the problem of accumulation by the industry, they also created new challenges for companies and the state in the form of citizenship practices. The relationship between the salmon production process and its environment, based on flexibility and control, has seeped into social relations, evolving into the phenomenon of extractive citizenship conducted through processes of disenfranchisement, fragmentation, and conditioning. The situation, in turn, offers opportunities to build new spaces of political action and contestation, empowering subjects to either align themselves with the predominant mode of accumulation or challenge it. The more stress in the system, the greater the likelihood of crisis; however, the solution does not guarantee a transformation in modes of production or democratic participation.

Notes

1   Regulation of aquaculture concessions began with the 1991 General Fishing Law (18,892). Previously, firms obtained administrative authorization from the Undersecretariat of Fishing on a case-by-case basis, the first being awarded in 1976.
2   Research for this chapter was made possible by two grants (FI11121451 and FR1160848), 2013–2019. Fieldwork took place in 13 municipalities within the Los Lagos region. A team of researchers conducted interviews (from which the quotations used here were extracted), participatory mapping, and ethnographic work, mostly with rural dwellers in salmon-producing areas.
3   It is beyond the scope of this chapter to explain the wide range of subsidies in detail, though subsidies to farmers and fishermen come through multiple different agencies and government levels and are a main site of interaction between people and the government.

References

AQUA. 2017. "En Chile: La millonaria apuesta en salud de peces." AQUA: Acuicultura y Pesca. https://www.aqua.cl/reportajes/chile-la-millonaria-apuesta-salud-peces/.

Avilés, D. 2015. "Construcción de una economía política híbrida: Análisis comparativo de las inversiones públicas y privadas desde una óptica neoestructural." In *Revolución salmonera: paradojas y transformaciones territoriales en Chiloé*, edited by Álvaro Román, Jonathan R. Barton, Beatriz Bustos-Gallardo, and Alejandro Salazar, 79–122. Santiago de Chile: RIL Editores; Instituto de Estudios Urbanos y Territoriales UC.

Barton, Jonathan R., and Arnt Fløysand. 2010. "The Political Ecology of Chilean Salmon Aquaculture, 1982–2010: A Trajectory from Economic Development to Global Sustainability." *Global Environmental Change* 20 (4): 739–752. doi: 10.1016/j.gloenvcha.2010.04.001.

Barton, Jonathan R., and Warwick E. Murray. 2009. "Grounding Geographies of Economic Globalisation: Globalised Spaces in Chile's Non-Traditional Export Sector, 1980–2005." *Tijdschrift Voor Economische En Sociale Geografie* 100 (1): 81–100.

Barton, Jonathan, Álvaro Román, Alejandro Salazar, and Bernardita McPhee. 2013. "¿Son nuevas las ruralidades de Chiloé? Transformaciones territoriales y la modernización de los modos de vida rurales." In *Anales de la Sociedad Chilena de Ciencias Geográficas*, 197–203. Santiago de Chile: Sociedad Chilena de Ciencias Geográficas.

Budds, Jessica. 2013. "Water, Power, and the Production of Neoliberalism in Chile, 1973–2005." *Environment and Planning D: Society and Space* 31 (2): 301–318.

Buschmann, Alejandro H. 2001. *Impacto ambiental de la acuicultura: el estado de la investigacion en Chile y el mundo; Un analisis bibliografico de los avances y restricciones para una produccion sustentable en los sistemas acuaticos*, edited by. Santiago: Fundacion TERRAM.

Bustos-Gallardo, Beatriz, and Álvaro Román. 2019. "A Sea Uprooted: Islandness and Political Identity on Chiloé Island, Chile." *Island Studies Journal*. Published ahead of print, doi: 10.24043/isj.91.

Bustos-Gallardo, Beatriz. 2013. "The ISA Crisis in Los Lagos Chile: A Failure of Neoliberal Environmental Governance?" *Geoforum* 48: 196–206. doi: 10.1016/j.geoforum.2013.04.025.

Bustos-Gallardo, Beatriz, Michael Lukas, Caroline Stamm, and André Torre. 2019. "Neoliberalismo y gobernanza territorial: propuestas y reflexiones a partir del caso de Chile." *Revista de Geografía Norte Grande* 73: 161–183.

Bustos-Gallardo, Beatriz, and Felipe Irarrázaval. 2016. "'Throwing Money into the Sea': Capitalism as a World-Ecological System; Evidence from the Chilean Salmon Industry Crisis, 2008." *Capitalism Nature Socialism*. doi:10.1080/10455752.2016.1162822.

Caniguan Verlarde, Francisca. 2016. "Desafíos y transformaciones en el desarrollo rural en la región de Los Lagos post crisis del virus ISA." Geografía, Facultad de Arquitectura y Urbanismo, Universidad de Chile.

Desforges, Luke, Rhys Jones, and Mike Woods. 2005. "New Geographies of Citizenship." *Citizenship Studies* 9 (5): 439–451.

Dresdner, Cid, C. Chávez, M. Estay, N. González, C. Salazar, O. Santis, Y. Figueroa, A. Lafon, C. Luengo, and F. Quezada. 2016. "Evaluación socioeconómica del sector salmonicultor, en base a las nuevas exigencias de la Ley General de Pesca y Acuicultura." SUBPESCA.

Einum, S., and I. A. Fleming. 1997. "Genetic Divergence and Interactions in the Wild Among Native, Farmed and Hybrid Atlantic Salmon." *Journal of Fish Biology* 50 (3): 634–651.

Escobar Mendoza, Laura Magali. 2018. "Conflictos por uso de territorio entre espacios costeros marinos de pueblos originarios y concesiones de acuicultura, a la luz de lo dispuesto en la Ley No. 20.479." B.A. Thesis, Universidad de Chile.

Fløysand, Arnt, Jonathan R. Barton, and Álvaro Román. 2010. "La doble jerarquía del desarrollo económico y gobierno local en Chile: El caso de la salmonicultura y los municipios chilotes." *EURE (Santiago)* 36 (108): 123–148.

FR1151215_grant. 2017.

Gaudichaud, Franck. 2015. *Las fisuras del neoliberalismo chileno: Trabajo, crisis de la "democracia tutelada" y conflictos de clases*. Buenos Aires: CLACSO.

Gómez Leyton, Juan Carlos. 2010. *Política, democracia y ciudadanía en una sociedad neoliberal (Chile: 1990–2010)*. Santiago de Chile: Editorial Arcis.

Irarrázaval, Felipe, and Beatriz Bustos-Gallardo. 2018. "Global Salmon Networks: Unpacking Ecological Contradictions at the Production Stage." *Economic Geography* 1–20. doi: 10.1080/00130095.2018.1506700.

Jacobsen, Jan Arge, and Lars Petter Hansen. 2001. "Feeding Habits of Wild and Escaped Farmed Atlantic Salmon, Salmo Salar L., in the Northeast Atlantic." *ICES Journal of Marine Science* 58 (4): 916–933.

Katz, Jorge, Michiko Iizuka, and Samuel Muñoz. 2011. "Creciendo en base a los recursos naturales, 'tragedias de los comunes' y el futuro de la industria salmonera chilena." *Serie Desarrollo Productivo No. 191*. Santiago de Chile: CEPAL.

Mascareño, Aldo, Rodrigo Cordero, Gabriela Azócar, Marco Billi, Pablo A. Henríquez, and Gonzalo A. Ruz. 2018. "Controversies in Social-Ecological Systems: Lessons from a Major Red Tide Crisis on Chiloé Island, Chile." *Ecology and Society* 23 (4): 15.

Niklitschek, Edwin J., Doris Soto, Alejandra Lafon, and Pamela Toledo. 2013. "Southward Expansion of the Chilean Salmon Industry in the Patagonian Fjords: Main Environmental Challenges." *Reviews in Aquaculture* 5: 172–195.

Ong, Aihwa. 2006. *Neoliberalism as Exception: Mutations in Citizenship and Sovereignty*. Durham, NC: Duke University Press.

Ortiz Gómez, María Guadalupe. 2014. "El perfil del ciudadano neoliberal: la ciudadanía de la autogestión neoliberal." *Sociológica (México)* 29: 165–200.

Phyne, John, and Jorge Mansilla. 2003. "Forging Linkages in the Commodity Chain: The Case of the Chilean Salmon Farming Industry, 1987-2001." *Sociologia Ruralis* 43 (2): 108–127.

Rasmussen, Mattias Borg, and Christian Lund. 2018. "Reconfiguring Frontier Spaces: The Territorialization of Resource Control." *World Development* 101: 388–399.

Romero, Diego. 2018. "El rol del Estado en la producción de territorios commodity: una revisión histórico-geográfica del proceso de comoditización de la región de Los Lagos (1845–2017)." Magister en Geografía, Geografía, Universidad de Chile.

Saliéres, Magali, Mathieu Le Grix, Waldo Vera, and René Billaz. 2005. "La agricultura familiar chilota en perspectiva." *Lider: revista labor interdisciplinaria de desarrollo regional* 13: 79–104.

Schurman, Rachel A. 1996. "Snails, Southern Hake and Sustainability: Neoliberalism and Natural Resource Exports in Chile." *World Development* 24 (11): 1695–1709.

Taylor, Marcus. 2002. "Success for Whom? An Historical-Materialist Critique of Neoliberalism in Chile." *Historical Materialism* 10 (2): 45–75.

Valenzuela Silva, Rosario. 2016. "La naturaleza no dá. Un análisis de la trayectoria de apropiación de la naturaleza en el contexto post crisis del ISA en Cochamó." B.A. Thesis, Universidad de Chile.

32

Extracting fish

Elspeth Probyn

Introduction

Fishing is one of the oldest forms of extraction.[1] It is a very particular form of extraction that brings humans and nonhumans together in very different ways than, say, mining. For millennia, fishing was sustainable. Fish sustain humans, who in turn practice different techniques to ensure that the fish and the marine habitat could continue to reproduce. At its simplest, fishing is about humans getting fish out of water. Its forms are many, and much of it takes place far from human habitation where it is "out of sight and out of mind" (Rayfuse 2020). Traditional forms of fishing, however, emphasized a different material relationship with fish than forms of extraction that are dominant today.

To begin to flesh out the more-than-human worlds of fishing, I turn briefly to Australia—home of the oldest continuing culture in the world—where Aboriginal people have fished for tens of thousands of years. From ancient artwork, we see how fish featured in daily and spiritual life. This history and philosophy continue to inform current practices. As Tom Butler, a Murramarang Elder and retired commercial fisherman, states, "We've been doing it [fishing] in this country thousands of years more than white men put together have been doing it; we know how fish react, we know how they travel, we know where they travel, we know what time of year they travel. Our parents taught us, we were brought up living off the sea" (as quoted in AIASTSIS 2020). Jonathan Yalandhu, a Yolŋu man of the Gupapuyŋu clan, says "Tides will tell us [when to go out fishing] ... the weather will tell us. Those balanda [white people], they've got fish radar but we don't use that one. We know the places to get what we want, it's there. What we don't want, we just leave it" (as quoted in AIASTSIS 2020).

This is a more-than-human engagement whereby there is little distinction between human, environment, and fish. It is a way of knowing that comes with ways of living *with* sea country. It is not surprising that many, if not most contemporary industrial modes of fishing extraction, are at odds with Indigenous ways of thinking and living. Leanne Betasamosake Simpson, a First Nations writer and activist, notably articulates how for Indigenous people, dominant forms of extraction "remove all of the relationships that have given whatever is being extracted meaning" (Simpson 2015; as quoted in Szeman 2017, 440).[2]

In this chapter, I focus on how fishing as a mode of extraction operates within a web of inter-relationships of fish, ocean, human, and history. I hope to demonstrate that considering extraction as always more-than-human—as always involving other ways of knowing and being—might add to critical studies of resources. I begin by examining what a more-than-human framework and attendant notions of materiality mean for an analysis of fishing as extraction. I then relate how my fieldwork with bluefin tuna fishers reveals the more-than-human connections, and the shifting material relationships of people who live with and from the seas. I examine how forms of fisheries management have changed the relationship between fish and humans as they created different subjectivities imbricated with different extractive practices.

More-than-human materiality

Across a wide range of disciplines, the question of materiality—what it is and why it matters—has been much debated; these debates have much relevance to conceptualizing a more-than-human notion of extraction. For instance, Emma-Jayne Abbots argues in relation to food studies that "the recent turn to 'new materialisms' (cf. Coole and Frost 2010) in the social sciences and humanities has opened up new avenues for the critical study of food by emphasizing the factic-ity of food's material components" (2016, 1–2). As she continues, "it destabilizes human-centric thinking and offers an approach in which the agentic capacity to cause affect is located in the relations between human and nonhuman bodies, discourse, and the material world … As a result, the materiality of food—its very stuff—comes to the fore" (3). Abbots here engages with the ways in which food has often been symbolically analyzed, following Lévi-Strauss' (1962) famous dictum that "food is good to think with." While it obviously is, this focus on human meaning-making often obscures the "facticity" of the material and structural factors that its production relies upon. For instance, thinking *with* food demands attention to the intertwining of human and nonhuman, animate and non-animate. As recent research on food and environmental epigenetics demonstrates, human bodies are porous and open to more-than-human factors in ways that blur traditional boundaries (Mansfield 2012, 2018, 2019; Mansfield and Guthman 2015; Probyn 2020). What this research means for extraction studies is that it demonstrates that human bodies are deeply connected to and with the environment: Human activities of extraction—be they mining or fishing—affect the environment, just as those human bodies are affected in return.

Thinking *with* the more-than-human is aligned with and extends Sarah Whatmore's influential argument that "cultural geographers have found their way (back) to the material in very different ways that variously resonate with what I take to be amongst the most enduring of geographical concerns – the vital connections between the geo (earth) and the bio (life)" (2006, 601). For Whatmore, "this return to the livingness of the world shifts the register of materiality from the indifferent stuff of a world 'out there,' articulated through notions of 'land,' 'nature,' or 'environment,' to the intimate fabric of corporeality that includes and redistributes the 'in here' of human being" (609, see also Naslund and McKeithen, Chapter 6 this volume). As Beth Greenhough notes, "For Whatmore (1999), the current division of geographical labour into the study of nature (earth surface processes and landforms) and culture (society and space) has seemingly failed to capture the liveliness and agency of non-human living beings" (2016, 37).

As Greenhough raises, Whatmore's initial argument was directly addressed to geographers for whom "geo" is pivotal to the discipline. However, the "geo" is often solely understood as terrestrial. As Christopher Bear puts it, "[G]eographers in the main continue to treat the sea as a feature to be crossed, or as a space of the unknown, ignoring the variety of life and activity that exists within it" (2013, 22). Drawing on Lambert, Martins, and Ogborn (2006, 482), Bear points to the need for "investigations that engage with the sea as something with a lively and energetic materiality of its own" (22). This understanding of materiality needs to include questions about fish stocks and legal regulation that as Bear says, have tended to be "the territory of economics and politics" (22). As Lambert, Martins, and Ogborn put it, "the relationships between different elements and materials—water, wind, wood, salt, cloth, metal, coal, rope, plastics—and the cultures of nature that combine them within different practices and technologies" (2006, 486) are core material realities integral to considerations of fishing as extraction.

However, Tanya Richardson and Gisa Weszkalnys warn that forms of "materiality [have] come to be treated as something that can be conceptually separated out" (2014, 18). There is a tendency within some of the more-than-human literature to seize on one aspect as "lively materiality"—to take Jane Bennett's (2010) influential term—often disregarding other essential but perhaps more mundane but important material aspects. To take a point that I have discussed elsewhere (Probyn 2016), in Bennett's argument about the vitality of matter of fish oil, she pays scant attention to where fish oil comes from and what its production does to the ecologies of humans and fish. I suggest we not fetishize "matter" or "materiality" but use a more-than-human orientation to focus in on the forms of materiality that can better inform our understanding of fishing or other modes of extraction. As Abrahamsson et al. (2015, 6) write, it "is not matter itself all by itself, but rather matter in context."

Fishing as an extractive activity is marked by "matter in context." Each element (of gear, of regulation, of marine environment, of species) has its own forms of materiality, which then produces more when brought together. In his "Afterword" to an important volume that seeks to extend and trouble cultural studies of extraction, Imre Szeman points out how, in general, "we are doubly distanced from the spaces and material processes of extraction … first by physical distance and second by techno-utopian fantasies of a quotidian reality shaped by immaterial forces" (2017, 443). Szeman's understanding of extraction is firmly terrestrial, but his point is pertinent to oceanic extraction. The materiality of the ocean comes to the fore as it is legally circumscribed and delineated. In terms of fishing as extraction, as we will see, different forms of codification and fishing management tools attempt to overcome the spatial and temporal distance of fishers, fish, and ocean.

Fishers, fish, and Individual Transferable Quota (ITQ)

The problematic nature of how to regulate and manage human-fish relationships has long been recognized. As I learned from a lengthy interview in June 2010 with a senior Australian fisheries manager, the enterprise of managing fishing relies on balancing extraction versus reproduction. And of course, as he put it "fish have tails and they don't recognize jurisdictional lines." He saw his role as a custodian of fishing resources, with "fish as the principle clients." It is an understatement to say that there is, at times, antipathy between managers and fishers—he received death threats when he closed a fishery for several years to allow it to recover. As Andrea Nightingale writes, management and fishers at times inhabit different spheres even when in the same room: "The meeting room [of fishery managers] shifts fishers' subjectivities from one of economic anchor of small coastal villages to one wherein they are blamed for overexploitation in the fishery" (2013, 2372).

The main levers available to fishery managers are control of input (e.g., number of boats or nets or fishing hours) or output (number of fish extracted from the sea). Both aim to curtail what is called "the race to fish" by setting limits in extractive practices that allow for reproduction of species. According to my interviewee, there was no effective management until the science was able to (roughly) estimate stock size, population dynamics, and, in turn, use these to advise a species-specific "sustainable" Total Allowable Catch (TAC). But this science is far from perfect, not least because of the difficulties of counting moving fish in the sea: For instance, as Daniel Pauly and his team have demonstrated, the "science" relies heavily on human reporting, which can be fallible for many reasons (Pauly et al. 1998; Pauly and Palomares 2005).

In the late 1970s, several countries began to allocate TAC (total quota) to fishers in the form of the Individual Transferable Quota (ITQ): Portions of the total quota allocated to individual users, who are then free to sell (the transferable dimension of ITQs) their allocation. This, it was hoped, would put a brake on the rush to fish and undermine competition among fishers. Economic theory suggested that ITQs would enhance efficiency by creating incentives for less profitable fishers to sell their quotas to more profitable fishers. However, the system has been heavily critiqued for the social havoc that it can wreak, for instance, by driving consolidation and forcing smaller producers out of business—as we will see in my discussion of the Southern Bluefin tuna fishery.

When I did research in Port Lincoln in 2009–2011, some of the financial outcomes of the ITQ system could be seen in the magnificent mansions that dot this fairly small town at the bottom of the Eyre Peninsula in South Australia. Port Lincoln has the distinction of being home to the most multimillionaires per capita in Australia for quite some time. They are the "tuna barons," who went from being immigrant peasant fishers to wealthy farmers of the seas. During my fieldwork, I became aware of the deep affective ties that many fishers have with the more-than-human marine. As the wife of a fisher spontaneously said in an interview: "I just love the fish."

The Southern Bluefin tuna is a particularly amazing fish. They spawn in the Java Sea, which is only 180 mi from the Australia coastline. They follow a route that takes them off the coast of Western Australia toward the southernmost tip of the state. Then they either veer toward Japan or turn left toward the Great Australian Bight. This huge bay stretches to South Australia's Eyre Peninsula where seasonal upwellings provide plentiful food for the juvenile tuna. This is also where they are caught by the bluefin tuna fishery located in Port Lincoln. As Brian Jeffries, the CEO of The Australian Tuna Association, puts it, "This fish is a gift" (in McGlashan 2019). The gift seemed to endlessly keep on giving. Thanks to the Japanese market, bluefin tuna was the most expensive fish in the sea. But by the 1990s, "Japan was catching more than 40,000 tonnes of SBT and Australia about 21,000 tonnes, but the population was estimated to be at barely five per cent of the 1960s levels" (McGlashan 2019).

With their mansions, high-tech boats and spotter planes, and toys—such as racehorses—the tuna fishermen could be said to epitomize the widespread image of commercial fishers as plunderers of the sea. However, the backstory of the tuna fishers is a fascinating tale of the migration of humans lured by the desire to extract fish. As Andrea Nightingale has argued across her studies of Scottish fishers, it is important to consider "how individuals come to inhabit multiple relationships to the fishery; relationships that are material, embodied, spatial, and political" (2013, 2364; see also, Bustos-Gallardo, Chapter 31 in this volume). While fishermen often are depicted and self-depict as rugged individualistic workers devoid of emotion, what Nightingale alerts us to—and what emerged in my own research—is that "there is a strong link between [their] attachments to the sea, [and] the emotional relationships produced" (2013, 2364). In turn, "the kind of emotions which people feel and display are deeply socionatural – products of relational, embodied interactions with both human and nonhuman others" (2369). These are the sorts of materialities that one can find in many more-than-human situations, but the point I will pursue is that the interaction of fisher, fish, and fisheries management produces yet more material relationships.

Historically, waves of immigration from the northern hemisphere were definitive in forming the Australian tuna fishery. In the 1920s, Italian fishers began to arrive from the south of Italy to settle on the southern coast of New South Wales. The Puglisi family was one of the earliest and most influential. They form an intricate network of kin. Tory Puglisi, called "the old man of the sea," often said that "fishing was in his family's blood … His father, grandfather and great grandfather were all fishermen in the Aeolian Islands off the coast of Sicily" (Bevitt 2019). They suffered the double stigmatization of being Italian in "white" Australia and being fishers. One of the couples I interviewed told me of how the wife's pastoral family was against her to-be-husband because he was a fisherman as much as because he was Italian. There was also governmental racism: "During WW2 the authorities requisitioned all the Italian owned fishing vessels, reselling them back after the war. Italians from Ulladulla during the war were forced to report to the local police once a week" (Dunn 2017).

While the Puglisis were important in the development of the prawn fishery in South Australia, from the 1960s Croatians became dominant in the tuna fishery based in Port Lincoln. The early families had escaped Josip Broz Tito's regime in Yugoslavia. Several of my interviewees told me tales of their families fleeing from the little fishing port of Kali on Ugljan Island, illegally crossing by boat into Italy. Rick told me his father's story: Stowed away on a boat with his pregnant wife, brother, and fifty other people, they managed to get to Italy in 1958. They then walked to Paris, and from Paris somehow got to Sydney where they were put in immigration camps. They had one precious contact in Port Lincoln, and they finally made it there. The Croatian fishers worked hard and by the early 1960s, they were catching 8000 tonnes of bluefin using the line and pole method, now promoted by canned tuna companies as "sustainable" extraction. The line and pole method—hugely strenuous work heaving large tuna out of the sea—became the basis for the World Championship Tuna Toss held every year at the Tunarama Festival in Port Lincoln. (With the price of tuna, rubber ones have replaced real tuna in the contest now.)

From my interviews, it is clear that the 1960s and 1970s were heady times of fast money, hard work, and play in Port Lincoln. However, by 1980, it was apparent that the party was over. Scientists from Australia, New Zealand, and Japan concurred that bluefin biomass was severely diminished, and a TAC was decided upon by all three countries. In 1984, ITQs were allocated by the Australian Government—individual fishers were allocated ITQs based on a formula incorporating the highest annual catch in the three years from October 1, 1980, to September 30, 1983.

This meant that the big boat owners received the major share, which was part of the government's motivation hoping to reduce the number of licenses. At the time, Australia had racing inflation and the Central Bank drove interest rates as high as 18.45 percent. Boat owners were faced with a dilemma. High prices for quotas and heavily in debt, many sold out. ITQ combined with high interest rates reduced the number of boats to about 20 in a relatively short amount of time.

Later, as the overexploitation of stock became ever more apparent, fishers had to find new ways to get the most out of their quota. But how to extend the value of dwindling stock? What about if you caught small ones and put them into a feedlot like cattle to grow? This plan would turn fishers into farmers and radically change the extraction of bluefin tuna. While the younger fish would still be caught using the mixture of technologies and gear—purse seine nets and sonar spotter planes—once rounded up they would become less "wild" and certainly less extracted from the wild. Once in pens the fish would become somewhat domesticated and usher in fundamental changes to the more-than-human relationships that are at play out in the mobile and ever-changing milieu of the oceans. Hunters of the last wild food resource would become industrial operators.

The (in)famous tuna baron, Dinko Lukin, one of my interviewees, told me that he came up with the idea of catching juvenile bluefin in the Great Australian Bight. He knew that he could fatten them up with sardines, which he said "could never run out." In 1993, he began:

Tuna were carefully moved into cages and then slowly towed back to Port Lincoln. They are then transferred—again very carefully—into pens in the open sea and fed large quantities of frozen sardines and pilchards. In this way, they could be grown out and fattened up before sale to Japanese buyers. The Japanese buyers are known to come over to inspect the fish in their cages, and I have even heard that they slice slivers from near the tail to judge the texture and the taste, which the fisher/farmers then remedy if not up to their standards. I have deliberately empha-sized the care with which these operations were done. Understanding the Japanese market and their concern with beautiful unblemished tuna, Dinko knew that the rough days of handling tuna were over. The fish were bled and gutted and either sent by air fresh to Tokyo or snap frozen at −60°C and sent by boat.

It was an ingenious idea, and Dinko maintained that "everyone copied me." Both the Tuna Boat Owners Association of Australia and the Fisheries Research and Development Corpora-tion take credit for the plan (ASBTIA 2019). When I talked to Dinko, still feisty in his 70s, he vehemently restated his ownership of the concept. Whoever came up with it, the system was to fundamentally change the extraction of bluefin tuna. Soon everyone was into it. The Croatians exported the idea back home to Kali, and it spread to Spain, Malta, South Korea, Italy, Mexico, and Libya. It is hard to underestimate how farming tuna changed fishing as we know it. While they made money, for the "old guys of the sea" the turn to fish farming must have been a wrench. During one of our sessions, Dinko spoke to me of his "life in tuna." Blue eyes flashing he regaled me with how "crew did not feel any fear" even in "the strongest elements that God created." "The crew believed that the boat was magic." In the same breath, he boasted that "he'd never give up one inch to the weather … I only go out in bad weather." Although Dinko saw himself as the creator of the ranching system, it was very clear that while it was financially lucra-tive, it was not the same as being a true fisherman in the elements.

As Nightingale argues, there are definite shifts in "subjectivity that occur when individu-als move between the different spaces and places of fisheries, subjectivities that are inherently embodied and emerge relationally in space" (2013, 2366). Elsewhere, she calls for a framework that better understands "the practices and interactions that are required to be considered a 'fish-erman'" (Nightingale 2011, 121). The more-than-human relationships of sea, of different gear and species of fish, the topography of fishing as extraction produce very different effects and affects. The introduction of ITQs by fisheries management produced, or was the condition of possibility, for bluefin tuna fish farming. It also produced a very different "fisherman," perhaps one that even is "not a fisherman … just a businessman," as one of Nightingale's interviewees puts it (2013, 2372). Some of my interviewees complained about how much time they spent with accountants.

Conclusion: rethinking the materialities of extracting the oceans

We are a long way from the Indigenous fishing practices I described at the outset of this chap-ter. For Indigenous fishers, the more-than-human relationship of fishing operates within a cosmology that insists on the interconnectedness of all life. In the case of the bluefin tuna fish-ery, we see very different subjectivities emerge from and through cultures and histories, regu-lation, and management. I want to reiterate the need to frame a more-than-human account of the extraction of fish that is attuned to the materialities of extraction. These materialities take many different forms. As I have argued, the tacit knowledge and love of some fishers for fish takes the form of affective relationalities. The value of markets drives a more technologi-cal connection where fishers rely on spotter planes and sonars rather than the generations-old passing of knowing through doing. Then legal formulations produce yet other forms of material relationality between human, fish, and ocean. As I have argued, forms of regulation produce or alter relations among humans, fish, and the ocean.

In this chapter, I have tried to go beyond using the more-than-human as an abstracted framing of fishing as extraction to ask, what are the material bases that allow or disallow human connection with fish and oceans? What types of material relationality do these bases encourage or disavow? In the case of the bluefin tuna fishers and farmers, ITQ regulation made some fishers very wealthy but it also reformulated the materialities of extraction whereby fish end up as private livestock in pens. This produces different mentalities and regimes of care for fish and transforms fishers' identities and relations to and with the sea. It also changes the status of fish—from wild to somewhat domesticated (Lien 2015).

While I have focused on fishing as a particular exemplar, the necessities of bringing the material into the more-than-human relationship extends to the analysis of all forms of extraction. Extraction forcefully brings together how the more-than-human is valued—or not. In theorizing extraction, we therefore must foreground how forms of the human are produced in the process of extracting the ocean. Using a framework of the more-than-human alerts us to the multidimensionality of extraction: it is never as straightforward as the previously dominant model of the mastery of man extracting goods from nature that then become abstracted commodities. The age-old lessons of Australian Indigenous practices of care for sea country instruct us that land and sea are not separate and that forms of human and nonhuman life are necessarily intertwined. Highlighting the different ways in which bodies are brought together in fishing reveals the deep human-marine relationality of ocean extraction as well as the ways in which dominant extractive logics focused on rationalization and efficiency compromise this relation.

Notes

1 My thanks to the people who gave of their time to talk to me, and to the Australian Research Council for funding. I also thank Bradon Ellem for his comments and the editors for their astute direction.

2 There are, of course, several non-Indigenous geographers who acknowledge the deep disconnect between western scholarship and Indigenous knowledge. From her position as a Pākehā (non-Maori) New Zealander, Amanda Thomas makes clear that it is up to non-Indigenous researchers to engage in "decoloniz[ing] more-than-humanisms" (2015, 978). This must be a material engagement with the more-than-human. As Ruth Panelli argues in the context of New Zealand, non-Indigenous people need to engage with "forms of co-relation with place via whakapapa (genealogy and cultural identity), whenua (land), moana (sea) and mahinga kai (traditional food resources and practices)" (2010, 6). See also De Alessi (2012) for a discussion of the political economy of Maori fisheries.

References

Abbots, Emma-Jayne. 2016. "Introducing a Special Issue on Food Stuffs: Materialities, Meanings, and Embodied Encounters." *Gastronomica: The Journal of Critical Food Studies* 16 (3): 1–4.

Abrahamsson, Sebastian, Filippo Bertoni, Annemarie Mol, and Rebecca Ibáñez Martín. 2015. "Living With Omega-3: New Materialism and Enduring Concerns." *Environment and Planning D: Society and Space* 33 (1): 4–19.

AIASTSIS. 2020 "A Brief History of Indigenous Fishing." The Australian Institute of Aboriginal and Torres Strait Islander Studies. Accessed June 4, 2020. https://aiatsis.gov.au/exhibitions/brief-history-indigenous-fishing.

ASBTIA. 2019 "Industry Background." Tuna Boat Owners Association of Australia. Accessed January 26, 2019. https://asbtia.com.au/industry/tuna-industry-background.

Bear, Christopher. 2013. "Assembling the Sea: Materiality, Movement and Regulatory Practices in the Cardigan Bay Scallop Fishery." *Cultural Geographies* 20 (1): 21–41.

Bennett, Jane. 2010. *Vibrant Matter: A Political Ecology of Things*. Durham, NC: Duke University Press.

Bevitt, Kate. 2019. "'He Just Loved Catching Fish': Frank Puglisi's Lifelong Passion." Tuna Australia. https://www.tunaaustralia.org.au/news/he-just-loved-catching-fish-frank-puglisis-lifelong-passion/.

Coole, Diana, and Samantha Frost, eds. 2011. *New Materialisms: Ontology, Agency, and Politics*. Durham, NC: Duke University Press.

De Alessi, Michael. 2012. "The Political Economy of Fishing Rights and Claims: The Maori Experience in New Zealand." *Journal of Agrarian Change* 12 (2–3): 390–412.

Dunn, Cathy. 2017. "History of Ulladulla's Fishing Industry." Ulladulla Info. Accessed January 1, 2017. http://www.ulladulla.info/wp-content/uploads/2017/12/fishing.pdf.

Greenhough, Beth. 2016. "Vitalist Geographies: Life and the More-Than-Human." In *Taking-Place: Non-Representational Theories and Geography*, edited by Ben Anderson, 51–68. London: Routledge.

Lambert, David, Luciana Martins, and Miles Ogborn. 2006. "Currents, Visions and Voyages: Historical Geographies of the Sea." *Journal of Historical Geography* 32 (3): 479–493.

Lévi-Strauss, Claude. 1962. *La pensée sauvage*. Paris: Plon.

Lien, Marianne. 2015. *Becoming Salmon: Aquaculture and the Domestication of a Fish*. Oakland, CA: University of California Press.

Mansfield, Becky. 2012. "Race and the New Epigenetic Biopolitics of Environmental Health." *BioSocieties* 7 (4): 352–372.

Mansfield, Becky. 2018. "A New Biopolitics of Environmental Health: Permeable Bodies and the Anthropocene." In *The Sage Handbook of Nature*, edited by Terry Marsden, 216–234. London: Sage.

Mansfield, Becky. 2019. "Mercury." In *Keywords in Radical Geography: Antipode at 50*, edited by the Antipode Editorial Collective, 181–183. Hoboken, NJ: Wiley Blackwell.

Mansfield, Becky, and Julie Guthman. 2015. "Epigenetic Life: Biological Plasticity, Abnormality, and New Configurations of Race and Reproduction." *Cultural Geographies* 22 (1): 3–20.

McGlashan, Al, dir. 2019. "Life on the Line: The Amazing True Story of the Southern Bluefin Tuna." McGlashan Media and Klinik. Accessed November 20, 2020. https://www.youtube.com/watch?v=hjdb1AVUnVI.

Nightingale, Andrea. 2011. "Beyond Design Principles: Subjectivity, Emotion, and the (Ir)Rational Commons." *Society and Natural Resources* 24 (2): 119–132.

Nightingale, Andrea. 2013. "Fishing for Nature: The Politics of Subjectivity and Emotion in Scottish Inshore Fisheries Management." *Environment and Planning A: Economy and Space* 45 (10): 2362–2378.

Panelli, Ruth. 2010. "More-Than-Human Social Geographies: Posthuman and Other Possibilities." *Progress in Human Geography* 34 (1): 79–87.

Pauly, Daniel, Villy Christensen, Joahanne Dalsgaard, Rainer Froese, and Francisco Torres. 1998. "Fishing Down Marine Food Webs." *Science* 279 (5352): 860–863.

Pauly, Daniel, and Maria-Lourdes Palomares. 2005. "Fishing Down Marine Food Web: It Is Far More Pervasive Than We Thought." *Bulletin of Marine Science* 76 (2): 197–212.

Probyn, Elspeth. 2016. *Eating the Ocean*. Durham, NC: Duke University Press.

Probyn, Elspeth. 2020. "Wasting Seas: Oceanic Time and Temporalities." In *The Temporalities of Waste*, edited by Fiona Allon, Ruth Barcan, and Karma Eddison-Cogan. London: Routledge.

Rayfuse, Rosemary. 2020. "'Out of Sight, Out of Mind': The Challenge of Regulating the High Seas." In *Sustaining Seas: Oceanic Space and the Politics of Care*, edited by Elspeth Probyn, Kate Johnston, and Nancy Lee. London: Rowman & Littlefield.

Richardson, Tanya, and Gisa Weszkalnys. 2014. "Introduction: Resource Materialities." *Anthropological Quarterly* 87 (1): 5–30.

Simpson, Leanne Betasamosake. 2015. "I Am Not a Nation-State." Accessed January 1, 2015. https://www.leannesimpson.ca/writings/i-am-not-a-nation-state.

Szeman, Imre. 2017. "On the Politics of Extraction." *Cultural Studies* 31 (2–3): 440–448.

Whatmore, Sarah. 1999. "Culture-Nature." In *Introducing Human Geographies*, edited by Paul Cloke, Philip Crang, and Mark Goodwin, 4–11. London: Arnold.

Whatmore, Sarah. 2006. "Materialist Returns: Practising Cultural Geography in and for a More-Than-Human World." *Cultural Geographies* 13 (4): 600–609.

33

Human tissue economies
Making biological resources

Maria Fannin

Introduction

Resource geography is intimately concerned with the relation between nature and society, between the human and physical worlds. Resources such as oil, coal, and minerals or even materials like food (including living beings consumed as food) are the key entities through which geographical concerns over environmental exploitation and extraction, social justice, ecological risk, and sustainability are staged. However, much of the geographical work on natural resources tends to presume that these resources are nonhuman if they are living, or part of the world beyond or outside the human if they are geological or mineral. This enduring division between the "natural" and the "social" underpins the call by Karen Bakker and Gavin Bridge in 2006 to suggest that studies of embodiment and the materiality of the body offer lessons to resource geographers for how to move beyond the familiar analytical frames of the "production of nature" and the "social construction of nature." One of the ways, they argue, to achieve this is through drawing from how research on bodily geographies "provides a way to acknowledge physicality/corporeality—and the sociopolitically productive nature of physical variation—without surrendering the social to the biological" (Bakker and Bridge 2006, 15). In searching for a renewed orientation for resource geography, Bakker and Bridge suggest that work on corporeality and the materiality of the lived body offers a generative conceptual model for how to navigate the ongoing tensions between the "social" and the "natural" in new ways.

Responding to this call, work on the biopolitics of environmental health by feminist political ecologists challenges the presumption that the environment begins outside the body. For example, Becky Mansfield (2012) demonstrates how dietary recommendations for pregnant women in the United States disrupt presumed boundaries between the pregnant body and the environment. Dietary recommendations connect the bodily spaces of pregnancy to the traces of chemicals found in water and food and in turn place new burdens on the pregnant person to protect the fetus from environmental toxins. Julie Guthman (2012) also focuses on environmental toxins and their relationship to the body's materiality in her research on obesity, questioning the assumptions of mainstream obesity researchers and arguing for a closer engagement between political ecology and health geography. Their collaborative research (see Guthman and Mansfield 2013) suggests that scholarship on the openness of the human body to toxic chemicals would enhance understandings of the body as an ecology itself (rather than the body conceived primarily as a single part of a wider ecology). This work, in Guthman's (2012) words, seeks to open up the "black box of the human body" to greater scrutiny, an effort echoed by the return within feminist theory to bodily materiality and biology (Dixon 2015; England, Fannin, and Hazen 2019; Frost 2016).

Yet the distinction between the human and the nonhuman (and thus body/environment) continues to underpin how a "natural resource" is defined despite, as Mansfield (2012, 2018) writes, the widespread elaboration of non-dualist understandings of the environment as socionatural. Indeed, socionatures of the environment—of the making and remaking of the worlds of nonhuman beings, living and nonliving, and biotic and geologic—have been more thoroughly explored than those of the body. This suggests that the distinction between the human and the nonhuman continues to endure, despite efforts to disrupt or erase it in many subfields of the discipline (Panelli 2010; Sundberg 2014; Williams et al. 2019; see also Barua 2017).

This chapter brings the corporeal and material into resource geography by considering how the human body itself—its biophysical elements of cells, tissues, and organs—becomes a resource. I argue that research on bodily geographies, following Bakker and Bridge's provocation, offers not only a useful conceptual model for resource geographers but also evidence of how bodies have become, and are increasingly conceived as, resource geographies in themselves. Rather than shifting geographical reflection to the nonhuman, animal, mineral, or other material entities in order to challenge geographers' persistent anthropocentrism, this chapter approaches the problem of the human/nonhuman distinction and the effort to overcome this dualism in a different way. I highlight how presuppositions about what constitutes "the human" obscure the important ways in which human bodies and body parts in medicine and the life sciences are (and indeed have long been in many ways) treated as resources: detached, fragmented, extracted, and transformed in order to circulate through multi-scalar flows and networks in the fields of medicine and the life sciences, much as animal, mineral, and other material entities more familiar to resource geographers circulate in other sectors of the economy. This intervention is important because geographical literatures on natural resources continue to view their typical objects of study as "outside" or beyond the realm of the human body.

By analyzing how human bodies and their parts are constituted as *biological resources*, this chapter examines what Catherine Waldby and Robert Mitchell (2006) call the "tissue economies" of bodily materials that circulate around the globe. Research on the tissue economies of healthcare and the life sciences draws attention to the work that transforms bodies and body parts into scientific objects, research tools, or new forms of medical therapy. This process reconstitutes parts of bodies as scientific objects with potential economic value that may in other contexts be viewed as sacred or profane, as objects of adornment or reverence or disgust, or as entities that bear witness to relations of kinship or connections to land and place. Research on tissue economies thus offers a means to analyze how capitalist processes of valorization, extraction, and exchange act directly on and through the human body.

Of course, labor within capitalist economies has always meant the extraction of value from the embodied labor of the worker. But the ability to separate and exchange parts (cells, tissues, organs) from bodies is central to the value-making practices of the life sciences, the pharmaceutical industry, healthcare, and medicine in what has been termed a new era of capitalist economy, that of biocapital (Rajan 2006; see also Naslund and McKeithen, Chapter 6 in this volume). These sites of biocapital encompass both human and animal bodies transformed into biotechnologies designed to further human ends, laboring as experimental subjects or disaggregated into research tools at a cellular or molecular level.

In what follows, I synthesize work focused on human biological resources to argue for closer dialogue between the scholarship on resource geographies and the tissue economies of human bodily materials. There is now an extensive literature outside the field of geography—and influencing ongoing geographical work—on the circulation of breast milk, eggs, sperm, and other materials into scientific and therapeutic settings (for example, see Boyer 2010; Fannin and Kent 2015; Parry 2015). The social science research on tissue economies discussed here seeks to address key questions about the processes that make human bodies and body parts into resources. How are biological materials extracted and separated from human bodies? How are these materials transformed into scientific or therapeutic resources? What are the political, social, and economic dimensions of this process?

In line with the provocation by the editors of this book regarding the "world-making" capacities of human biological resources, this chapter also asks how are such resources—circulating as material entities with shifting ontological status as waste, tool, or medical therapy—co-constitutive of new forms of biological capital? What are the contours of the new forms of value attached to bodies and body parts in the contemporary moment? And how do these approaches to the body as a biological resource prefigure the increasing ways in which one's biology (including not only tissues and cells but also the seemingly immaterial entities such as cognition and behavior) is subject to new forms of control and relations of power? This chapter's contention is that the transformation of bodies and body parts into resources is inseparable from these wider incursions into the defining of life as a form of capital, or "lively capital" (Haraway 2007).

Ultimately, this chapter argues for the value of considering bodily materials through a resource-production lens, signaling how the "political, economic, and cultural work" that makes natural resources, as Bridge (2011, 820) writes, also operates in healthcare, medicine, the life sciences, and related fields to make biological resources out of human bodies and their parts. Processes familiar to scholars of resource extraction such as commodification, circulation, and exchange are at work on human bodies and body parts in ways that echo other forms of material extraction and exploitation, from "discovery" to "mining" to notions of scarcity and surplus.

The discussion mentioned later moves from an overview of the bodily materials routinely defined as resources, in particular blood and gametes (sperm and eggs) to the institutional sites of biobanks, or repositories of human tissue, that act as key nodes in the making of the bodily resource economy. Alongside biobanks, where biological materials are collected and stored, the interlinked domains of biotech research laboratories and clinical settings also condition how biological resources are made and valued. I close with the possibilities opened up by expanding resource geographies to include human biological materials and the complex regulatory, social, and political dimensions of these tissue economies that invite more sustained attention by resource geographers.

Human tissue economies

Biological resources, in the most expansive sense, include anything defined as living, from viruses and other microbial forms of life to larger organisms and their organs, tissues, and cells. This section will focus on the human body and its identification and exploitation as a biological resource, defined as the living parts (cells, tissues, organs) of bodies and the information that can be derived from them (see Parry and Gere 2006). This latter element is crucial: bodies and body parts are viewed as untapped resources that can be probed and analyzed to reveal the "secrets" of their generation, growth, and development.

The collection of human biological materials as scientific resources has its origins in the early work of anatomists who sought to open up the body and make its interior visible for observation and study (Park 2006). Like the extraction of biophysical resources, these processes have intensified dramatically over the last 300 years. The demands for human biological material have increased since the development of tissue culture techniques, transfusion, and transplantation technologies; assisted in vitro reproductive technologies; and, in the wake of the mapping of the human genome, the ability to modify or alter the body at the level of the gene. Defining human-derived materials as biological resources now encompasses sites and practices of organ and blood donation, biobanks, anatomical collections, the production of new therapeutics, experimental materials used to test and produce new drugs, and assisted reproduction in in vitro fertilization (IVF) clinics.

Bodies and body parts may be repurposed and utilized in many different medical and scientific research fields. Indeed, all bodies are potentially biological resources—living and dead. The collection of bodily materials for research use includes everything human bodies shed: teeth, hair, nail clippings, urine, fecal matter, umbilical cords, and placentas. Bodily materials used for research and therapy also include materials that are extracted from the body such as organs (including blood), fat, muscle, and other tissues such as breast milk, eggs, and sperm. Parts of bodies are routinely collected and stored in hospital pathology departments to enable diagnosis and treatment of disease. In a forensic context, bodily materials may be valued for their capacity to reveal something of the identity of the body from which they originated and become "evidence" in a judicial setting. Bodily materials may be cultured as living entities to further develop and grow in a laboratory (Landecker 2007). Parts of bodies may be minimally manipulated (in the case of whole-organ transplantation) or extensively engineered and transformed to become a new therapeutic transplant tissue in what is now designated the field of "regenerative" medicine. Technologies of preservation shape the potential value of bodily materials and their intended uses, with innovations in chemical fixing, low-temperature freezing (Radin and Kowal 2017), and tissue culture greatly expanding the ability to arrest and then restart cellular development in order to preserve, study, and manipulate forms of life.

Distinctions between the use of bodily materials for research, for therapy, or in the case of breast milk, for feeding, are reflected in the regulatory structures and ethical considerations surrounding the "mobility" of materials outside the body (Kent, Fannin, and Dowling 2019). Bodily materials may be attributed "individuality" through genetic analysis, HLA (human leukocyte antigen)-typed for transplant compatibility, or assessed for their cell "stem-ness" to ascertain their ability to grow and develop new cells. Research on the social categories and modes of classification informing scientific discourses on human tissues also reveal how biological materials may be aged, sexed, racialized, ethnicized, and gendered (Kent and Farrell 2015; Kierans and Cooper 2011; TallBear 2013). Bodily materials may also be valued for their multiplicity, in which parts are viewed not as one but many, for example, as hosts to microbes, or accumulations of chemicals, viral infections, parasites, and small amounts of another organism's genetic material, as in fetal-maternal microchimerism, in which fetal genetic material circulates in the maternal bloodstream during and beyond pregnancy (Mansfield 2012; Martin 2010).

The alienation of bodily materials from the purview of a subject is one of the dynamics that drives the production of biological resources *as* resources (Andrews and Nelkin 1998; Sharp 2000). Whether body parts are wholly or incompletely separated from the identity of the subject from whom they originated is also central to the moral and ethical debates over the commodification and objectification of biological materials in medicine, science, and technology (Parry 2008; Parry and Gere 2006). Depending on context, biological materials can be valorized precisely *because* they circulate as anonymized samples stored in a biobank or, conversely, gain greater value for their "identity" and connection to a specific community or population. They can also embody what Donna Haraway calls the "encounter value" of coming into proximity and contact with an/other life, a value implicated in the political economies of assisted reproduction described by Charis Thompson (2005) in which what is produced is not a "product" but a child. Alienated from the subject from whom they originated, body parts not only can become commodities exchanged on markets or gifts circulated through personal or public networks, but they can also become potent carriers and embodiments of much more.

Given the diversity and extent to which bodily materials are repurposed in scientific and clinical settings, the following discussion focuses specifically on the economies of blood and gametes (eggs and sperm) in order to examine the dynamics of each more closely. Many other kinds of bodily tissues and cells circulate through networks of tissue banks, research laboratories, and reproductive technology clinics.[1] This discussion of human biological resource-making is inspired by Haraway, Thompson, and other feminist theorists in techno-science studies and by the early-twenty-first-century turn to new feminist materialisms of the body and technology. Human bodily resources are implicated in biocapital as resources and "microbiotechnologies," but they are also lively materials, objects of care as well as commodification. This section closes with a discussion of the value generated by biobanks, the institutional sites where bodily materials are stored, and which are central to the making of bodily materials into resources.

Gifts and goods: the human blood economy

The blood economy is often viewed as the paradigmatic example of a human tissue economy, and blood could be viewed as an exemplary human tissue extracted and valued as a renewable resource. Blood is widely used in therapeutic settings to treat a range of chronic diseases such as hemophilia and thalassemia as well as acute conditions requiring transfusion during surgery or after childbirth. The vital role of blood as a resource in therapeutic settings has led to the development of complex technical and organizational infrastructures, including state-run blood services, networks of local and regional blood banks, and Red Cross and Red Crescent managed systems (Healy 2006). Indeed, blood's central role in the clinic means it is defined by the World Health Organization (WHO) as a "vital health care resource" (WHO [World Health Organization] 2017) and an "essential health technology" (Busby, Kent, and Farrell 2014, citing WHO [World Health Organization] 2004).

While giving underpins the public health orthodoxy that nonremunerated donation is safer, blood selling (or "paid donation" in the blood donation literature) and family "replacement donation," in which a family member donates blood to replace what is withdrawn from a blood bank, continue to supply over half of all blood donated in 71 countries around the globe (WHO [World Health Organization] 2017). Blood is thus unevenly commodified: where public blood donation systems do not operate, often in resource-poor contexts, family or voluntary "replacement" donors, as well as paid "donation," provide access to blood products.

Richard Titmuss's (1970) landmark text, *The Gift Relationship: From Human Blood to Social Policy*, compared the blood economies of the United States and the United Kingdom to argue for the broader public good of giving—whether blood or money in the form of wealth redistribution—to strangers. Despite the discourses surrounding the blood economy as predominantly organized around unremunerated voluntary "gifts," close analysis of blood and transfusion policy and practice reveals "moral hierarchies and scales of exchange and use value which have excluded MSM [men who have sex with men] and migrants as donors, and segregated blood along racial lines in some countries" (Kent and Farrell 2015, 37). Although Titmuss proposed that blood donation should be regarded as an example of a fundamental "right to give," this right is by no means evenly extended to all. The extraction of blood as a resource is shaped by social and cultural dynamics that bely universalist claims about donation.

Blood is also routinely described by blood donation policymakers as a precious "national resource" and the self-sufficiency of national blood services a policy objective. Like other resources, blood may be imported or exported to meet demand; however, the direct commercialization of whole blood for transfusion is viewed by many states and the WHO as antithetical to the principles protecting the safety of the blood supply. However, markets for blood components, such as plasma, exist in many jurisdictions, including the United States. For example, in the UK blood economy, public health concerns about the risks associated with donor exposure to variant Creutzfeldt-Jakob Disease (vCJD) mean that the National Health Service (NHS), the UK's public health-care system, purchases selected blood products from overseas, often from paid donors (Kent and Farrell 2015).

The "resourcification" of blood may also take less obvious forms than exchange on an outright market. For example, scholarship on late-nineteenth-century European efforts to understand the transmission of blood-borne disease reveals how the blood of colonized people was viewed as a vital resource for the study of public health, illustrating the "entanglement" of imperial and experimental logics of colonialism (Tilley 2014). Today, alongside public blood collection and transfusion services, blood is subject to commercialization as a resource to be banked and stored for an individual's future personal use or to provide research communities with cell lines and other research tools (Fannin 2011). Commercial firms in the private biobanking sector may also seek to enroll individuals as biobank account holders, for example, umbilical cord banks aim to enroll parents who will store their child's material for future, uncertain use (Brown and Kraft 2006; Fannin 2011; Martin, Brown, and Turner 2008). These cord blood biobanks have received the most scholarly attention to date in part because of their early rhetorical promises of providing therapeutic treatment for a host of diseases. The call by commercial cord blood banks to make "wise use" of precious bodily resources, such as cord blood, involves appeals to the perceived therapeutic value of the resource in question and the construction of cord blood as a scarce resource that cannot be easily renewed. Similar to the wise use movement surrounding state-managed resources of water and forests in resource geography (see McCarthy 2002), political and ethical debates over the appropriate use of blood resources stem from competing constituencies and debates over the multivalent value of the resource in question. The construction of blood as a scarce resource also shapes the efforts to extract blood for research purposes from individuals and groups perceived as having "rare" blood types, including indigenous peoples around the globe, and to whom such projects often bring little material benefit (TallBear 2013).

Although umbilical cord blood, as discussed earlier, has been successfully used to treat some rare blood disorders and is an effective replacement for bone marrow, cord blood banking in the private sector remains a speculative practice (Brown, Machin, and McLeod 2011). The late-twentieth-century emergence of private cord blood banks, alongside their public counterparts modeled on bone marrow banks, suggests that the boundaries between market and gift in blood economies are messy and complex (Brown, Machin, and McLeod 2011; Waldby and Mitchell 2006). A resource geography of human blood would thus not only take into account the moral, social, and economic dimensions of blood deeply embedded in the politics of citizenship, identity, and belonging but also subject to resource flows between states, commercial entities, and individual donors/sellers/bankers and recipients of blood.

Gamete economies: extracting and exchanging sperm and eggs

There is now a considerable body of work in the social sciences on the extraction and circulation of gametes (sperm and eggs) in the reproductive technology industry. Eggs and sperm circulate in some jurisdictions, such as the United States, in a mostly unregulated market

(Almeling 2011). The extraction and exchange of gametes operates at the global scale through cross-border movements that take advantage of changing regulatory environments and shifts in client demand. Ethnographic work on the movement of eggs and sperm across borders for fertility treatments reveals how national imaginaries of race and ethnicity also shape the gamete economy (Nahman 2013). This is also starkly visible in US egg and sperm vending where sellers are profiled and gametes often selected based on the seller's physical traits and educational background (Parry 2015).

Similar to blood, human eggs and sperm are increasingly viewed as national resources, their value enabled by freezing technologies that permit their preservation and transport both within and beyond national borders. The two largest global suppliers of sperm are located in the United States and Denmark, jurisdictions that continue to permit anonymous donation and that advertise their services to an international clientele. Denmark's Cryos International touts the high-quality of its donor sperm, while California Cryobank advertises "the industry's largest and most diverse donor selection recruited from world class universities including UCLA, USC, Stanford, Harvard and MIT" (California Cryobank 2020; see also Wahlberg 2008). These and other instances illustrate how reproductive technologies trade on notions of national identity shaped by hierarchies of race, education, and other social differences.

Commercial banking services have also extended to other bodily tissues. Most recently, innovations in methods of material preservation have resulted in the growth of commercial biobanks targeting consumers seeking to bank their eggs (Martin 2010; Waldby 2014). The improvement of egg-freezing techniques has made long-term storage of eggs a feature of US corporate culture in the tech sector, with the highly publicized support of Google, Amazon, and Facebook for employee egg freezing (Ikemoto 2015) enabled through policies that extend insurance coverage to personal, nonmedical storage. The extension of insurance coverage of egg freezing to employees at leading tech firms demonstrates how resource-making of bodily materials, such as gametes, supports both the individualization of control over one's reproductive body (one becomes one's own egg donor) and the extraction of workers' labor through their prime reproductive years in the interest of economic growth. Employers are not the only beneficiaries; such developments also situate workers' future reproductive trajectories within the lucrative fertility industry.

The use of eggs and sperm for research relies significantly on the extraction of these materials in fertility clinics; once no longer intended for fertilization, such materials may be donated and recirculated for research use. Eggs and sperms are also used as research tools, put to work to support national ambitions for scientific innovation. Egg extraction for direct research use, although a less common route for procuring eggs, received global media attention in the mid-2000s when the scandal surrounding researcher Woo-Suk Hwang's embryonic stem cell research revealed that "donors" were paid to undergo egg extraction but were not given details of how their eggs would be used and, in some cases, were students or employees of Hwang's laboratory, thus raising serious concerns about consent and the ethics of paying for eggs for research use (Baylis 2009; Ikemoto 2009). The gamete economy is thus governed by a heterogeneous regulatory landscape of laws governing human tissue or organ use and exchange, on which there is no international consensus.

Geographies of the tissue economy: biotech, biomedicine, and biobanks

Biological materials are stored in what are described as "biobanks," a term originating in early-twentieth-century efforts to distribute blood for medical use. Gottweis and Peterson (2008, 5) define biobanks as "collections of human biological material, often combined with personal medical, genealogical, environmental, and lifestyle information, within healthcare systems and the medical sciences. Clinical settings, research projects and the judiciary field are typical sites for biobank collections." The growth of large-scale biobanks is evident in the regional and international efforts to standardize biobanking protocols for sample collection, processing, and use. In Europe, large biobanks have been established as part of national-scale research studies but also to provide research resources for the scientific communities. Biobanks in this sector typically fall into two groups: biobanks that collect material related to a specific disease condition or from individuals with a specific condition and "population-based" biobanks that are viewed as representative of an entire population.

Biobanks can include very local collections of biological resources—for example, hospital-based biobanks that store material collected for testing, diagnosis, or future study. Biobanks can also include collections established and maintained by universities and their staff as part of scholarly research projects. At the regional and national scale, biobanks may be established by governments as part of efforts to provide researchers with resources for further study. Depending on their organization around various diseases or populations, biobanks may serve as national or global "reserves" of biological materials. Biobanking also carries with it the notion of the body as a fungible or standardized commodity like money (Fannin 2013; Swanson 2014). Efforts to create larger and larger biobanks have increased as preservation technologies improve and the ability to analyze genetic information gleaned from banked materials becomes less costly and time-consuming.

Making resources out of bodies and body parts also generates financial value. The value of the market for the storage and use of human tissues in biobanks in 2016 was estimated at $198.2 billion and projected to increase to $240.2 billion by 2021 (BCC Research 2016). Biotechnology firms in the life sciences and health sectors are reliant on a regular supply of tissues (human and nonhuman) as research tools and as resources for genetic databases, a practice increasingly implicating willing consumers whose genetic data derived from their bodily material is transformed into a product to be aggregated and monetized. 23andMe and other consumer genomics services are estimated to now have access to the genotypes of around 16 million consumers, an often under-recognized resource created through the amassing of millions of saliva samples (Lussier and Keinan 2018; see also Nash 2013).

While there is considerable variability in the regulatory frameworks governing the access and use of human tissue, initiatives to create local, regional, and national repositories of biological materials, and importantly, accompanying medical, behavioral, and other data have multiplied as the costs of genomic and other analytical tools have decreased (Chalmers et al. 2016). Biobanks aim to render stored materials more valuable through standardized procedures and protocols as well as by linking banked materials to additional data, including not only medical records but also lifestyle data, physical activity, and occasionally for forensic use by the police (see Dranseika, Piasecki, and Waligora 2016). Efforts to enhance the value of banked biological samples are intensifying as both public and commercially driven organizations seek to combine information derived from biological materials with "lifestyle" data collected from social media, health apps, and fitness tracking devices (see Ajana 2017).

Conclusion

As resource geographers have long reiterated, water, minerals, and other resource entities have no intrinsic or essential identity as substances but are made into resources through the processes of commodification, fragmentation, and objectification entailed in their extraction and use. This chapter focuses on efforts to transform and put to work parts of human bodies, a process that also encompasses the "mining" and "mapping" of genetic materials in research endeavors that draw on the tropes of imperial discovery and exploitation.

Just like other resources defined as "natural," human biological resources are not given but made. The "resourcification" of human biological materials in the twentieth and twenty-first centuries is characterized by extractive dynamics, discourses of scarcity, and exhortations of "wise use" as well as by powerful rhetorical calls to view the data derived from biological materials as "unexplored territory" (Khan and Mittelman 2018; on scarcity in resource geography, see Mehta [2005], Yapa [1996], and Huber [2011]). These themes echo many of the signal concerns of resource geographers.

A reconceptualization of the political economies of resource materialities to encompass human biological resources would begin by recognizing that human bodies are made into what Donna Haraway terms "lively capital" in at least three ways: as workers, tools, and products. For research on human biological resources, a pluralist approach to what bodies and body parts *are* (their ontology) and also to how they are valued across different times and spaces offers the potential for deeper collaboration with resource geographers interested in the ontological pluralism of resources, in which land, seas, and other forms of life may indeed *be* different things in different cosmologies (Bawaka Country 2016; Charpleix 2018; Howitt and Suchet-Pearson 2006; see also Ahlborg and Nightingale, Chapter 2 in this volume). Value, in the richest conception of the term, does not simply encompass economic value or the value of a human biological resource as a commodity or asset but also the value attached to an object that promises to improve health (as in a transplanted tissue with therapeutic value) or that enables the generation of novel insights into a developmental process (as in tissue cultured as a laboratory research tool) or that is shaped by deeply held social and cultural norms.

The making of biological resources differs in many ways from the processes of constituting parts of the biophysical world as natural resources. To take but one example, ethical deliberation over the constitution of human biological resources is often underpinned by notions of the sanctity and exceptionalism of human life in ways that complicate any simple analogy between the extraction of human and nonhuman resources. However, there are insights to be drawn from the constitution of the human body as a resource that can inform the wider study of natural resource geographies and that requires rethinking bodily biology as the underlying and undertheorized support for world-making.

The underlying presumption of a fundamental division between human biological and natural biophysical worlds might explain why studies of human biological resources have tended to occupy scholars of health, medicine, and the life sciences, the research of which could be brought into more generative conversation with scholars of water, minerals, forests, or food. To not interrogate the overlapping and co-constituting processes by which human bodies become exploitable resources—just as the land, the seas, and other living beings become resources—is to miss out on a whole host of connections that draw together the human and the nonhuman in concentric and overlapping processes of valorization, extraction, use, and exchange.

By expanding the purview of resource geographers' critical perspectives onto the sociopolitical processes at work in medicine and the life sciences, insights into the production of resources; the discursive and material framing of bodily resources in terms of scarcity and surplus; and the dynamics of commodification, extraction, and exchange can be brought to bear on human bodies and body parts. In addition, attention to how bodies have historically been constituted as resources opens up resource geography to new objects of study, to worlds within and around the body's materiality, and to the "new natures" of the body's biology. It also offers resource geographers' new avenues of closer conversation with scholars of bodily resources in feminist, reproductive, and health geographies.

Note

1   An organ's capacities for growth and development do not necessarily need to be separated from the body to become valorized as "resources." The literature on surrogacy and clinical trial participation emphasizes that these practices are precisely about gaining access to the in vivo biology—the reproductive or metabolic system—of the surrogate or trial participant. Because these practices involve the in situ exploitation of the body's capacity for living and developing, they are imprecisely described as "donation" and more aptly termed forms of "clinical labor" (Abadie 2010; Cooper and Waldby 2014).

References

Abadie, Robert. 2010. *The Professional Guinea Pig: Big Pharma and the Risky World of Human Subjects*. Durham, NC: Duke University Press.

Ajana, Btihaj. 2017. "Digital Health and the Biopolitics of the Quantified Self." *Digital Health* 3: 1–18. doi: 10.1177/2055207616689509.

Almeling, Rene. 2011. *Sex Cells: The Medical Market for Eggs and Sperm*. Berkeley, CA: University of California Press.

Andrews, Lori, and Dorothy Nelkin. 1998. "Whose Body Is It Anyway? Disputes Over Body Tissues in a Biotechnology Age." *The Lancet* 351 (9095): 53–57.

Bakker, Karen, and Gavin Bridge. 2006. "Material Worlds? Resource Geographies and the 'Matter of Nature." *Progress in Human Geography* 30 (1): 5–27.

Barua, Maan. 2017. "Nonhuman Labour, Encounter Value, Spectacular Accumulation: The Geographies of a Lively Commodity." *Transactions of the Institute of British Geographers* 42 (2): 274–288.

Bawaka Country including Wright, Sarah, Sandie Suchet-Pearson, Kate Lloyd, Laklak Burarrwanga, Ritjilili Ganambarr, Merrkiyawuy Ganambarr-Stubbs, Banbapuy Ganambarr, and Djawundil Maymuru. 2016. "The Politics of Ontology and Ontological Politics." *Dialogues in Human Geography* 6 (1): 23–27.

Baylis, Françoise. 2009. "For Love or Money? The Saga of Korean Women Who Provided Eggs for Embryonic Stem Cell Research." *Theoretical Medicine and Bioethics* 30 (5): 385–396.

BCC Research. 2016. "Biobank Industry a Ripe Market for Opportunities." *Press Releases*. Accessed 4 June 2020. https://www.bccresearch.com/pressroom/bio/biobank-industry-a-ripe-market-for-opportunities.

Boyer, Kate. 2010. "Of Care and Commodities: Breast Milk and the New Politics of Mobile Biosubstances." *Progress in Human Geography* 34 (5): 5–20.

Bridge, Gavin. 2011. "Resource Geographies I: Making Carbon Economies, Old and New." *Progress in Human Geography* 35 (6): 820–834.

Brown, Nik, and Alison Kraft. 2006. "Blood Ties: Banking the Stem Cell Promise." *Technology Analysis and Strategic Management* 18 (3/4): 313–327.

Brown, Nik, Laura L. Machin, and Danae McLeod. 2011. "Immunitary Bioeconomy: The Economisation of Life in the International Cord Blood Market." *Social Science & Medicine* 72 (7): 1115–1122.

Busby, Helen, Julie Kent, and Anne-Maree Farrell. 2014. "Revaluing Donor and Recipient Bodies in the Globalised Blood Economy: Transitions in Public Policy on Blood Safety in the United Kingdom." *Health* 18 (1): 79–94.

California Cryobank. 2020. "Experience the CCB Difference: Donor Selection." Accessed 4 June 2020. https://www.cryobank.com/why-use-us/.

Chalmers, Don, Dianne Nicol, Jane Kaye, Jessica Bell, Alastair V. Campbell, Calvin W. L. Ho, Kazuto Kato, et al. 2016. "Has the Biobank Bubble Burst? Withstanding the Challenges for Sustainable Biobanking in the Digital Era." *BMC Medical Ethics* 17 (1): 39. doi: 10.1186/s12910-016-0124-2.

Charpleix, Liz. 2018. "The Whanganui River as Te Awa Tupua: Place-Based Law in a Legally Pluralistic Society." *The Geographical Journal* 184 (1): 19–30.

Cooper, Melinda and Catherine Wadlby. 2014. *Clinical Labour: Tissue Donors and Research Subjects in the Global Bioeconomy*. Durham, NC: Duke University Press.

Dixon, Deborah P. 2015. *Feminist Geopolitics: Material States*. Farnham, UK: Ashgate.

Dranseika, Vilius, Jan Piasecki, and Marcin Waligora. 2016. "Forensic Uses of Research Biobanks: Should Donors Be Informed?." *Medicine, Health Care and Philosophy* 19 (1): 141–146.

England, Marcia R., Maria Fannin, and Helen Hazen, eds. 2019. *Reproductive Geographies: Bodies, Places and Politics.* Abingdon, UK: Routledge.

Fannin, Maria. 2011. "Personal Stem Cell Banking and the Problem with Property." *Social & Cultural Geography* 12 (4): 339–356.

Fannin, Maria. 2013. "The Hoarding Economy of Endometrial Stem Cell Storage." *Body & Society* 19 (4): 32–60.

Fannin, Maria, and Julie Kent. 2015. "Origin Stories from a Regional Placenta Tissue Collection." *New Genetics and Society* 34 (1): 25–51.

Frost, Samantha. 2016. *Biocultural Creatures: Toward a New Theory of the Human.* Durham, NC: Duke University Press.

Gottweis, Herbert, and Alan Peterson, eds. 2008. *Biobanks: Governance in Comparative Perspective.* London: Routledge.

Guthman, Julie. 2012. "Opening Up the Black Box of the Body in Geographical Obesity Research: Toward a Critical Political Ecology of Fat." *Annals of the Association of American Geographers* 102 (5): 951–957.

Guthman, Julie, and Becky Mansfield. 2013. "The Implications of Environmental Epigenetics: A New Direction for Geographic Inquiry on Health, Space, and Nature-Society Relations." *Progress in Human Geography* 37 (4): 486–504.

Haraway, Donna. 2007. *When Species Meet.* Minneapolis, MN: University of Minnesota Press.

Healy, Kieran. 2006. *Last Best Gifts: Altruism and the Market for Human Blood and Organs.* Chicago, IL: University of Chicago Press.

Howitt, Richard, and Sandra Suchet-Pearson. 2006. "Rethinking the Building Blocks: Ontological Pluralism and the Idea of 'Management." *Geografiska Annaler: Series B, Human Geography* 88 (3): 323–335.

Huber, Matthew T. 2011. "Enforcing Scarcity: Oil, Violence, and the Making of the Market." *Annals of the Association of American Geographers* 101 (4): 816–826.

Ikemoto, Lisa C. 2009. "Eggs as Capital: Human Egg Procurement in the Fertility Industry and the Stem Cell Research Enterprise." *Signs: Journal of Women in Culture and Society* 34 (4): 763–781.

Ikemoto, Lisa C. 2015. "Egg Freezing, Stratified Reproduction and the Logic of Not." *Journal of Law and the Biosciences* 2 (1): 112–117.

Kent, Julie, Maria Fannin, and Sally Dowling. 2019. "Gender Dynamics in the Donation Field: Human Tissue Donation for Research, Therapy and Feeding." *Sociology of Health and Illness* 41 (3): 567–584.

Kent, Julie, and Anne-Maree Farrell. 2015. "Risky Bodies in the Plasma Bioeconomy: A Feminist Analysis." *Body & Society* 21 (1): 29–57.

Khan, Razib, and David Mittelman. 2018. "Consumer Genomics Will Change Your Life, Whether You Get Tested or Not." *Genome Biology* 19: 120. doi: 0.1186/s13059-018-1506-1.

Kierans, Ciara, and Jessie Cooper. 2011. "Organ Donation, Genetics, Race and Culture: The Making of a Medical Problem." *Anthropology Today* 27 (6): 11–14.

Landecker, Hannah. 2007. *Culturing Life: How Cells Became Technologies.* Cambridge, MA: Harvard University Press.

Lussier, Alexandre A., and Alon Keinan. 2018. "Crowdsourced Genealogies and Genomes." *Science* 360 (6385): 153–154. doi: 10.1126/science.aat2634.

Mansfield, Becky. 2012. "Environmental Health as Biosecurity: 'Seafood Choices,' Risk and the Pregnant Woman as Threshold." *Annals of the Association of American Geographers* 102 (5): 969–976.

Mansfield, Becky. 2018. "A New Biopolitics of Environmental Health: Permeable Bodies and the Anthropocene." In *The SAGE Handbook of Nature, Volume 1,* edited by Terry Marsden, 216–234. London: Sage.

Martin, Aryn. 2010. "Microchimerism in the Mother(land): Blurring the Borders of Body and Nation." *Body & Society* 16 (3): 23–50.

Martin, Lauren Jade. 2010. "Anticipating Infertility: Egg Freezing, Genetic Preservation and Risk." *Gender & Society* 24 (4): 526–545.

Martin, Paul, Nik Brown, and Andrew Turner. 2008. "Capitalizing Hope: The Commercial Development of Umbilical Cord Blood Banking." *New Genetics and Society* 27 (2): 127–143.

McCarthy, James. 2002. "First World Political Ecology: Lessons from the Wise Use Movement." *Environment and Planning A* 34 (7): 1281–1302.

Mehta, Lyla. 2005. *The Politics and Poetics of Water: The Naturalisation of Scarcity in Western India*. New Delhi: Orient Blackswan.

Nahman, Michal. 2013. *Extractions: An Ethnography of Reproductive Tourism*. London: Palgrave Macmillan.

Nash, Catherine. 2013. "Genome Geographies: Mapping National Ancestry and Diversity in Human Population Genetics." *Transactions of the Institute of British Geographers* 38 (2): 193–206.

Panelli, Ruth. 2010. "More-Than-Human Social Geographies: Posthuman and Other Possibilities." *Progress in Human Geography* 34 (1): 79–87.

Park, Katharine. 2006. *Secrets of Women: Gender, Generation, and the Origins of Human Dissection*. New York, NY: Zone Books.

Parry, Bronwyn. 2008. "Entangled Exchange: Reconceptualising the Characterisation and Practice of Bodily Commodification." *Geoforum* 39 (3): 1133–1144.

Parry, Bronwyn. 2015. "Narratives of Neoliberalism: 'Clinical Labour' in Context." *Medical Humanities* 41 (1): 32–37.

Parry, Bronwyn, and Cathy Gere. 2006. "Contested Bodies: Property Models and the Commodification of Human Biological Artefacts." *Science as Culture* 15 (2): 139–158.

Radin, Joanna, and Emma Kowal, eds. 2017. *Cryopolitics: Frozen Life in a Melting World*. Cambridge, MA: The MIT Press.

Rajan, Kaushik Sunder. 2006. *Biocapital: The Constitution of Postgenomic Life*. Durham, NC: Duke University Press.

Sharp, Lesley A. 2000. "The Commodification of the Body and Its Parts." *Annual Review of Anthropology* 29: 287–328.

Sundberg, Juanita. 2014. "Decolonizing Posthumanist Geographies." *Cultural Geographies* 21 (1): 33–47.

Swanson, Kara W. 2014. *Banking on the Body: The Market in Blood, Milk and Sperm in Modern America*. Cambridge, MA: Harvard University Press.

TallBear, Kimberly. 2013. *Native American DNA: Tribal Belonging and the False Promise of Genetic Science*. Minneapolis, MN: Minnesota University Press.

Thompson, Charis. 2005. *Making Parents: The Ontological Choreography of Reproductive Technologies*. Cambridge, MA: The MIT Press.

Tilley, Helen. 2014. "Experimentation in Colonial East Africa and Beyond." *International Journal of African Historical Studies* 47 (3): 495–505.

Titmuss, Richard. 1970. *The Gift Relationship: From Human Blood to Social Policy*. London: Allen & Unwin.

Wahlberg, Ayo. 2008. "Reproductive Medicine and the Concept of 'Quality'." *Clinical Ethics* 3 (4): 189–193.

Waldby, Catherine. 2014. "'Banking Time': Egg Freezing and the Negotiation of Future Fertility." *Culture, Health & Sexuality* 17 (4): 470–482.

Waldby, Catherine, and Robert Mitchell. 2006. *Tissue Economies: Blood, Organs, and Cell Lines in Late Capitalism*. Durham, NC: Duke University Press.

Williams, Nina, Merle Patchett, Andrew Lapworth, Tom Roberts, and Thomas Keating. 2019. "Practising Post-Humanism in Geographical Research." *Transactions of the Institute of British Geographers* 44 (4): 637–643.

WHO (World Health Organization). 2004. *Global Database on Blood Safety. Report 2001–2002*. Accessed 5 June 2020. http://www.who.int/bloodsafety/GDBS_Report_2001-2002.pdf.

WHO (World Health Organization). 2017. *Global Status Report on Blood Safety and Availability 2016*. Accessed 5 June 2020. https://apps.who.int/iris/bitstream/handle/10665/254987/9789241565431-eng.pdf;jsessionid=B756E9EB06D0F2B6A8BC59C7D03A45B2?sequence=1.

Yapa, Lakshman. 1996. "Improved Seeds and Constructed Scarcity." In *Liberation Ecologies: Environment, Development, Social Movements*, edited by Richard Peet, and Michael Watts, 69–85. New York, NY: Routledge.

34

Making, and remaking, a world of carbon

Uneven geographies of carbon sequestration

Wim Carton and Karin Edstedt

Introduction

At the COP24 climate negotiations in Katowice, Poland, in December 2018, the Polish presidency tabled a much-discussed "Declaration on Forests for the Climate." Signatories to this declaration pledged to step up their efforts "to ensure that the global contribution of forests and forest products [to climate change mitigation] is maintained and further supported and enhanced by 2050" (UNFCCC 2018, 2). This two-page document, approved by over 60 countries, is just the latest in a long line of initiatives, declarations, and international agreements that seek to promote carbon sequestration in forests as a key resource in the fight against climate change. This reframing of forests as a supposedly inexpensive and convenient way to bridge the current mitigation gap raises pertinent questions about the implications of this strategy, particularly in terms of trade-offs with the way land and forest resources are currently utilized.

In this brief chapter, we explore the kinds of world-making that are occurring on the back of this climate change strategy and that risk becoming more prevalent if ongoing efforts to increase sequestration fail to take on board past lessons and experiences. At stake are the environments, landscapes, and socio-ecological relations that are sidelined when constructing forest carbon as a resource as well as the wider ramifications this has for both rural communities and nonhuman natures. Our argument proceeds by way of, first, a background discussion on the emerging political economy of carbon removal, its relationship to continued fossil fuel combustion, and the specific concerns this raises for critical resource geographers. We then bring some of these concerns to life by detailing two examples of climate-justified a/reforestation in Uganda, where carbon sequestration is being used to offset greenhouse gas emission from the Global North. While our cases are meant to be illustrative rather than representative, they show how justice, sociopolitical, and ecological dimensions across multiple scales demand careful consideration when promoting forest-based sequestration as a climate mitigation pathway.

Not seeing the forest for the carbon

The Katowice forest declaration continues longstanding global efforts to frame forests as a key focus area to help achieve ambitious climate targets. This is a discussion best known, perhaps, for the ongoing controversies and political battles waged over REDD+, an international scheme meant to help finance reductions in deforestation and forest degradation in the Global South (Corbera and Schroeder 2017; see also Shapiro-Garza et al., Chapter 20 this volume). And surely, forests are a necessary priority area in any discussion on climate change mitigation. An estimated 12 percent of greenhouse gas emissions are attributable to deforestation and general land use change (Smith and Bustamante 2014), while intact forests function as an important carbon sink. Forests also provide an impressive range of other ecosystem functions and are a vital source of human subsistence, not in the least for indigenous people and rural communities across the Global South. On the face of it therefore, it seems hard to be critical of the kind of initiatives that the international community is launching, which appear as necessary efforts to help protect vitally important ecosystems.

Closer scrutiny, however, reveals that the ongoing push for carbon sequestration is about more than just countering forest loss. Observers of the COP24 meeting were quick to point out the irony of Poland—a country refusing to confront its deep dependence on coal, the most polluting of all fossil fuels—pushing for enhancing carbon sinks while remaining mostly silent on fossil fuel production (FERN 2018). The Katowice declaration, moreover, leaves "forests" undefined, creating a vague category that includes everything from old-growth forests to monoculture tree plantations. Using "Forests for Climate" could, in this way, just as well allow for the intensified use of forest products as supposedly "carbon neutral" materials (Jong 2019). Instead of protecting natural forests, this risks encouraging a wood product industry whose alleged sustainability has long been questioned while doing little to incentivize the kind of structural transformations of the energy system that are needed.

What appears to be unfolding in much of the policy discussions on forests and climate, then, is a subtle move away from reducing direct land use emissions, toward the promotion of forestry—including in its productive, narrowly economic sense—as a stand-in for other forms of mitigation. This is facilitated by a recent discursive move from "zero" to "net zero" emissions as the international community's ultimate mitigation target and the corresponding inclusion in the Paris Agreement of the idea that countries should seek to achieve "a balance between anthropogenic emissions by sources and removals by sinks" (UNFCCC 2015). Implied here is the belief that climate change mitigation is no longer a question of just reducing emissions but that it involves the enhancement of carbon sinks, such as forests, through large-scale carbon dioxide removal as well.

The problem is that the "net" in net zero turns out to be a rather convenient smokescreen for governments and businesses who have long acted as if drastic emission cuts are too onerous, too politically sensitive, or simply against prevailing economic interests (Carton 2019). All too often it hides staggering amounts of assumed carbon removal, which promise to "balance out" the continuation of fossil fuel emissions long into the second half of the century (Anderson and Peters 2016). Thus, someone like the CEO of Shell can subscribe to the temperature limits of the Paris Agreement not by abandoning his company's oil and gas business but by promoting "massive reforestation." "Think of another Brazil in terms of rainforest," he told a recent gathering of shareholders (Vaughan 2018). With a logic like this, the proposed balance between sequestration and direct emission cuts risks tilting ever more perilously toward so-called natural climate solutions (see Griscom et al. 2017), such as a/reforestation, as the allegedly most convenient and "cost-effective" mitigation option.

Of course, few policymakers are as unwavering in their support for fossil fuels as the oil and gas industry (though some names come to mind). Many are adamant that increased sequestration should not serve as a replacement for emission cuts. In the face of decades of lackluster ambition on the latter and a dominant focus on the prioritization of cost-effective solutions, this assurance seems unconvincing. Even the Intergovernmental Panel on Climate Change (IPCC)

(2018) has incorporated the dominant logic that designates some emissions as simply unavoidable and that articulates progress on mitigation as inevitably gradual and dictated by the prevailing logics of mainstream economics (Carton, 2020). Its scenarios to keep warming within 1.5°C/2°C are replete with so-called negative emissions, or large-scale carbon dioxide removal (including afforestation) in the second half of the century, for the simple reason that models deem short-term emission cuts economically unfeasible (Parson 2017). IPCC scenarios in this way prioritize sequestration "technologies" like afforestation or the speculative bioenergy with carbon capture and storage (BECCS) over more urgent and short-term mitigation strategies. They use increased carbon sinks to make up for decades of failed direct mitigation and put forward future carbon removal as a substitute for reducing emissions at source.

This climate policy trend should be of immediate interest to critical resource geographers. It suggests that a particular resource—tree carbon—is being discursively constructed at least in part to help shore up the economic value of another, supposedly more socially important resource—fossil fuels. In essence, the increasing importance of carbon sinks as an alternative to direct emission reductions puts new aspects of nonhuman nature in the service of the "old" carbon economy. It mobilizes specific landscapes and ecosystem functions for the maintenance and reproduction of fossil capital, in many cases in tandem with the production of more traditional forest commodities, such as timber (for a similar relationship focused at the scale of municipalities, see Purifoy, Chapter 10 this volume). Borrowing from Boyd, Prudham, and Schurman (2001), we could perhaps speak of the subsumption of mitigation to capital, that is, the reworking of climate change policy—as a specific set of socioecological relations—to fit the short-term priorities of dominant, fossil-fuel-invested capitalist interests (and while attempting to create new avenues for capital investment, such as carbon markets). This occurs by way of a wider process of economization in which mitigation everywhere is deemed functionally equal and increases in carbon sinks are rendered equivalent to emission cuts, allowing the most cost-effective option to be prioritized (Lohmann 2011). These also tend to be the mitigation strategies most compatible with the political and socioeconomic status-quo (Carton 2017).

In what sense should scholars—and resource geographers in particular—be critical of this? At the most general level, the substitution of carbon sequestration for direct emission benefits those countries, sectors, or businesses that are most invested in fossil fuels, hence those that bear the largest responsibility for climate change. Much as with the subsumption of nature more broadly, the political construction of forest carbon as resource is a process ultimately oriented toward the assertion of control over the conditions of capitalist production, in this case over the kind of transitions (and most significantly perhaps, their temporality) that fossil capital needs to undergo. There is, therefore, a clear risk that sequestration ends up a justification to delay or altogether avoid structural changes in the way fossil fuel-dependent production processes are organized, hence also in the power relations that operate through them. This could be detrimental from a justice *and* a mitigation perspective.

At the same time, there is an unavoidable geographical, uneven dimension to the notion that "we" will be able to sequester ourselves out of the climate crisis. Indeed, carbon sequestration vividly illustrates the dynamics of what Fairhead, Leach, and Scoones (2012) call the global "economy of repair" or the idea that "unsustainable use 'here' can be repaired by sustainable practices 'there', with one nature subordinated to the other" (242). The unavoidable local dimensions of this "there" is where a second set of key concerns arise. Despite Shell's vision of a vastly increased forest cover, few places comes to mind where one could plant "another Brazil in terms of rainforest" that do not already fulfill some social, cultural, or economic function for someone somewhere or indeed provide ecosystem functions that fall outside of human valuation. There are, in other words, inevitable trade-offs and conflicts involved with promoting increased carbon sequestration. The nature of these trade-offs as well as how they are (not) resolved will determine to what extent "natural climate solutions" constitute real and sustainable alternatives to cutting emissions at source. While most people would probably agree that protecting and promoting forest growth is beneficial and therefore desirable, questions over what exactly constitutes a forest, who gets to decide and manage it, who carries the costs and reaps the benefits, and so on are inevitably political questions and a common source of disagreement (Cavanagh and Benjaminsen 2014; Leach and Scoones 2015).

Ultimately then, the making of tree carbon into a resource is of global importance; the mobilization of forest ecosystems and rural landscapes to "repair" and sustain a world made of fossil carbon is necessarily also the mobilization of some person or community's land and resources. It is the expression of specific social relations and the assertion of the power to decide over resources and direct the way in which these are to be utilized. The global logics of sequestration and the local dynamics of tree planting are inseparable and therefore need to be considered together. What form these connections take and whether there are substantial conflicts and injustices involved are empirical questions the answers to which will differ from one context to the other.

For now, much remains unknown about the scale at which proposed carbon removal scenarios will unfold. However, already-existing carbon sequestration projects provide an important window into the kind of outcomes that are possible. To get a sense of the "world-making" that could occur if the international community's increasing fondness for "natural climate solutions" is carried through, it is thus useful to turn to one of the sectors where the "economy of repair" has perhaps been articulated most explicitly thus far: the carbon offset industry.

Carbon offsetting as world-making

The mechanisms by which carbon offset markets operate will be familiar to many readers. Individuals, companies, and even governments that want to shrink their carbon footprint but feel it is unfeasible to do so by reducing, or foregoing high-carbon activities, can instead "compensate" for their continued emissions by buying carbon offsets. There is a veritable mess of different offsetting schemes and project types through which this can occur, the details of which it is not possible to go into here (for overviews see, e.g., Hamrick and Gallant 2017; World Bank 2018).

In this chapter, we are concerned with projects that claim offsets through a/reforestation. We offer two examples of such projects from our own work in Uganda, each representing a different approach to carbon forestry. This allows us to say something about the different ways in which tree carbon—and the communities on whose land offsetting occurs—enters into carbon markets, and in turn, how landscapes and everyday lives are restructured in the process. Despite the differences, there are also important commonalities between the projects, most notably in how both double as sites of timber production. The two projects in this sense illustrate how forest carbon as resource is closely entangled with (and justifies) existing political economies and "more-than-carbon" (Asiyanbi 2016) processes of resource creation/extraction.

While we certainly do not want to claim that sequestration inevitably takes the forms that they do in these projects or results in the same kinds of problems, our discussion illustrates the importance of looking beyond globally dominant narratives of tree planting as somehow easier or more straightforward than other forms of mitigation. A global carbon sequestration agenda, like the one being pushed by the international climate community, needs to carefully consider the intertwined social and ecological dimensions of forestry and their implications for "sustainability." Our cases illustrate two broad kinds of "world-making," which, loosely inspired by Leach and Scoones (2015), we identify as the creation of carbon sequestration "sacrifice zones" and the recruitment of poor communities as "carbon stewards" in the emerging economy of repair.

The Kachung Forest Project—a CDM sacrifice zone

Our first example is the Kachung Forest Project (KFP), an afforestation project in central Uganda that operates through the Clean Development Mechanism (CDM; see Box 34.1). KFP is one of the very few a/reforestation projects in the CDM and represents an industry-focused take on the promotion of carbon sequestration. The project is operated by the Norwegian-based forest company Green Resources and financed by carbon offset sales to the Swedish Energy Agency. All financial benefits of KFP accrue directly to the company, but the project claims to provide a number of sustainable development benefits to surrounding communities. Mostly, these take the form of training in the construction of energy-efficient cooking stoves, infrastructure investments, the provision of boreholes and wells, the handing out of pine tree seedlings to communities, and labor opportunities in the company's tree nursery (Green Resources 2012).

These sustainable development claims notwithstanding, the project has been highly controversial. It sits on one of Uganda's Central Forest Reserves—which, though owned by the government, has long been used by local communities for a range of subsistence purposes: from food cultivation to firewood collection, livestock grazing, and access to water, among others. A number of households were also directly living within the area. The implementation of the project therefore required the forceful eviction of local people from the reserve, and their exclusion from most of the activities that they had carried out there before with the notable exception of firewood collection. Studies, including our own, have elucidated extensive negative effects on local livelihoods because of this (Lyons and Ssemwogerere 2017; Edstedt and Carton 2018). While the project has created (predominantly seasonal and low-paid) employment opportunities

---

Box 34.1   Key concepts

Carbon sequestration	The process by which CO_2 is taken up and stored by plants. For sequestration in trees, we here interchangeably use the term **carbon forestry**. The term **carbon removal** encompasses sequestration but also includes nonbiological technologies.
A/reforestation	Depending on whether an area did or did not previously have tree cover, tree planting is referred to as either **reforestation** or **afforestation**.
Voluntary Carbon Market (VCM)	A voluntary mechanism through which individuals or companies can offset their emissions.
Clean Development Mechanism (CDM)	The regularly offsetting mechanism established under the United Nations Framework Convention on Climate Change (UNFCCC). The CDM can be used by governments and businesses to meet international emission reduction obligations. It will be succeeded by a new market mechanism under the Paris Agreement.

for some 250–400 people in the district, it also reduced surrounding communities' access to agricultural and grazing lands with corresponding effects on opportunities to sell crops or livestock on the market and therefore on incomes. For some community members, this has directly led to increased poverty and food insecurity. The need to cultivate the remaining land more intensely has in turn raised concerns over land degradation and observations of declining yields and poorer crop quality (Edstedt and Carton 2018).

It is worth highlighting the *kind* of forestry that is practiced here. Behind the discourse of afforestation and sustainable development sits what is essentially an industrial tree plantation, utilizing fast-growing exotic pine and eucalyptus species planted in monoculture set-ups. This forestry model is not accidental of course: it corresponds to the objective of Green Resources to maximize timber production while simultaneously generating carbon offsets. There are, however, clear environmental trade-offs with sequestering carbon like this. Fast-growing tree plantations, and particularly eucalyptus, have high water demands and generally do not support a lot of other vegetation (Scott 2005). Moreover, Green Resources keeps competing plants at bay through a combination of manual weeding and the application of herbicides. Effects on biodiversity are likely to be negative given that native grasses, shrubs, and some trees had to be removed for the project. Several of the affected smallholders mentioned through interviews how, from their perspective, the forest reserve had been degraded as a consequence of the plantation—which is interesting given that forest degradation by local communities is an explicit justification for the implementation of KFP in the first place (Hajdu, Penje, and Fischer 2016). Interviewees, for example, mentioned how people used to hunt in the reserve but that animals had now disappeared, how they lost access to good-quality firewood when the native trees and shrubs were removed, or how the remaining grasslands in the plantation were of low quality. In one community bordering the reserve, people argued that wells had dried out as a consequence of the eucalyptus that was planted nearby. The plantation's ecological characteristics thus reinforce its impacts on neighboring communities that have long relied on resources in the "national forest."

This makes KFP a clear example of what some have called "fortress conservation" (Brockington 2002): i.e., the forced exclusion of local communities from lands and resources they previously had access to in order to obtain conservation goals demanded and implemented from afar—in this case for the creation of internationally traded carbon offsets. If the relationship between local communities and the ecosystem in question is recognized at all in these kinds of projects, it is a predominantly negative one characterized by "encroachment," resource extraction, or degradation. This characterization is not only often incorrect (e.g., Fairhead and Leach 1996; Hajdu, Penje, and Fischer 2016) but also ignores complex and dynamic relations to local ecosystems, structural drivers of poverty and marginalization, and long-contested rights to land. With the exception of some native forest patches that were left intact and a wetland area that the company agreed to maintain, KFP's variant of fortress conservation also extends to its treatment of local "natures." Native shrubs and trees are "unwanted" and kept out to guarantee space for even-aged blocks of fast-growing pine and eucalyptus.

In sum, the KFP afforestation project created a global space that is managed by an international forestry company, financed by the Swedish government, made up of species adapted to other continents, used to "solve" a climate crisis that is largely the responsibility of the Global North, and whose solutions are devised in boardrooms and negotiating spaces where Kachung communities have no say. This globalized "forest" becomes a sacrifice zone; a "sacrificing" possible because the sequestration strategy is abstracted from local ecologies, livelihood needs, and subsistence strategies. Token development interventions and the creation of employment opportunities in the plantation offer little of substance to change this.

The Trees for Global Benefits Project—a VCM zone of recruitment

A very different form of world-making occurs in our second example, the Trees for Global Benefits (TFGB) project in the west of Uganda. TFGB is a carbon offsetting project operating on the voluntary carbon market (VCM) where forestry projects are popular. Essentially, TFGB recruits smallholder farmers to plant trees on their land and keep them there for at least 25 years, after which farmers can harvest the trees and keep the resulting income for themselves. During the first 10 years and conditional upon meeting specified intermediate targets, farmers receive a number of incentive payments, the sum of which is based on the carbon price specified in their contracts and the total area that is planted. These payments are financed through the sale of carbon offsets to various companies in the Global North, primarily but not exclusively in Sweden. As with other offset projects, TFGB claims a number of benefits on top of the carbon sequestered. The project uses indigenous trees species and thus aims to conserve local biodiversity; it creates a number of new income sources for participants; it provides a source of firewood and is therefore meant to relieve deforestation pressures on nearby natural forests; it helps conserve soils; etc. (Ecotrust 2016).

TFGB is an offsetting project that has been far less controversial than KFP. Indeed, in 2013 it won the UN SEED award for "entrepreneurship in sustainable development," and it has been hailed both nationally and internationally as a best practice example of participatory carbon offsetting, through which local communities directly benefit. Based on interviews, participants generally seem pleased with their trees and those we talked to commonly argued that the project had improved the local environment. Indeed, the project has grown substantially to include upwards of 6000 farmers by 2017, giving an indication of its popularity. This is not to say that the project is without problems. Research has shown substantial challenges in at least some project areas to do with stalled or missing payments, widespread failure to meet interim targets, dissatisfaction about the level of payments, confusion about spacing and thinning requirements, miscommunication between participants and project organizer, and maladaptation and dieback of some tree species giving rise to dissatisfaction about the choice of trees (Carton and Andersson 2017; Fisher et al. 2018). As with many other payments for ecosystem services (PES) initiatives, benefits from the project—in terms of who can and does participate—tend to be unevenly distributed within communities, and accrue mostly to more well-off farmers (Schreckenberg, Mwayafu, and Nyamutale 2013; Fisher 2013). Many farmers also appear narrowly motivated to join the project because of the incentive payments. Since the project depends on farmers maintaining their trees for at least 15 years after the payments have stopped, this poses questions for the long-term sustainability of the project (Fisher 2012). In addition, farmers generally tend to have a poor understanding of the project's rationale, which raises important justice concerns (Carton, 2020; Fisher 2013; Fisher et al. 2018).

Despite TFGB being framed as a "cooperative community-based" or even "community-led" project (Plan Vivo 2016), community members have practically no say in how the project operates as a whole, the kind of trees they can plant (i.e., outside the technical specification they choose), or how they choose to grow them. They also have no say over the price they are assigned in their carbon contracts. A number of farmers argued that they had no way to voice concerns or communicate problems to the project organizers. Though TFGB does have mechanisms in place for participants to give feedback, mainly through community meetings, it appears that these are poorly attended (Ecotrust 2017). Despite the "community" narrative then, this is in many ways a project focused on developing the entrepreneurial capacities of individual smallholders. This occurs mostly by way of the conditionality of payments, through which farmers are required to manage their trees in line with the specific planting, thinning, and pruning guidelines prescribed by carbon market accounting (Carton and Andersson 2017). They are also asked to arrange for their own seedlings; they largely carry the risk for failure individually; and they are encouraged to establish "tree-based enterprises" (Ecotrust 2017, 22) peripheral to their plantations.

In terms of world-making, TFGB offers an entirely different sequestration model than KFP. The project is certainly no less "global" in its reach, but rather than establishing a "sacrifice zone," it establishes what we could call a "zone of recruitment," in which smallholder farmers in the Global South are enrolled as what the project organizers referred to as "proud carbon farmers." Participants thus become "green custodians" (Fairhead, Leach, and Scoones 2012), willing stewards of the "new" carbon economy. This recruitment takes the form of environmental disciplining: i.e., the promotion and enforcement—through contract—of a particular environmental behavior, anchored in the local environmental and livelihood benefits of tree planting, and with the explicit agreement—in many cases enthusiasm—of participants. Despite this consensual partnership and the apparent ability of the project to align itself, at least in part, with the interests of (some) smallholders, TFGB is similar to KFP in that it is designed, framed, and structured in a way that serves a global environmental agenda. The "participation" of smallholders in the reorganization of their livelihoods and landscapes, in line with environmental objectives, is itself a commodity to be consumed by businesses and consumers in the Global North. Meanwhile, through the enrichment of their own environments, TFGB participants are *also* implicitly enrolled in the perpetuation of the "old" carbon economy: their activities mobilized as justifications for continued emissions in the Global North.

Conclusion

Many of the dynamics described in the two cases here are not new. The landscapes and livelihoods of people in Uganda and elsewhere have long been part of a wider global political economy, serving the interests and resource needs of governments and businesses in the Global North. What existing experiences with carbon forestry—and indeed with wider conservation efforts in the Global South—show is how easily these uneven dynamics resurface in what seems like a necessary and entirely well-intended effort to promote sustainable development, create "win-win" solutions, and bring about what some are framing as a "good Anthropocene" (Asafu-Adjaye et al. 2015): a socio-nature created in line with what is deemed ecologically desirable. The two cases illustrate the processes of "sacrifice" and "recruitment" through which this inclusion within the globalizing logics of climate change mitigation can operate. It would be imprudent to try and generalize from this necessarily concise sketch, yet these cases make evident that the creation of forest carbon as a global resource is taking a range of forms, leading to radically different ways in which local livelihoods and ecologies (or "worlds") are being remade. Pertinently, the two cases illustrate different extents to which communities, or at least certain groups within these, manage to derive benefits and pursue their own interests within the conditions that carbon sequestration projects create. As such, they underscore the need for resource geographers to be mindful of how global processes of resource-making "act not simply against, but also through existing livelihoods" (Holmes and Cavanagh 2016, 203) and, indeed, how the reworking of local livelihoods and ecologies becomes a precondition for new (or renewed) forms of resource extraction.

Experiences with carbon forestry thus illustrate the importance of seeing local and global processes of world-making as intertwined, as mutually constitutive and legitimating of each other. We could perhaps refer to this as a multi-scalar form of world-making in which what is being remade is necessarily also the geographical unevenness of world-making itself. In both cases, the transformation of local socio-ecological realities takes on a meaning only when considering the global purposes it serves, which is to help justify—and/or reproduce—the fossil fuel–dependent economic landscapes of the Global North. Critical geographers' examinations of carbon sequestration as an increasingly important global resource need to pay close attention to these interactions between local and global processes of world-making. They raise important questions as to who gains from climate change solutions, and on whose conditions specific mitigation strategies are pursued while others are not.

References

Anderson, Kevin, and Glen Peters. 2016. "The Trouble with Negative Emissions." *Science* 354 (6309): 182–183. doi: 10.1126/science.aah4567.

Asafu-Adjaye, John, Linus Blomqvist, Stewart Brand, Barry Brook, Ruth Defries, Erle Ellis, Christopher Foreman, et al. 2015. "An Ecomodernist Manifesto." April. Accessed July 23, 2020. https://static1.squarespace.com/static/5515d9f9e4b04d5c3198b7bb/t/552d37bbe4b07a7dd69fcdbb/1429026747046/An+Ecomodernist+Manifesto.pdf.

Asiyanbi, Adeniyi P. 2016. "A Political Ecology of REDD+: Property Rights, Militarised Protectionism, and Carbonised Exclusion in Cross River." *Geoforum* 77: 146–156. doi: 10.1016/j.geoforum.2016.10.016.

Boyd, William, W. Scott Prudham, and Rachel A. Schurman. 2001. "Industrial Dynamics and the Problem of Nature." *Society & Natural Resources* 14 (7): 555–570. doi: 10.1080/08941920120686.

Brockington, Dan. 2002. *Fortress Conservation: The Preservation of the Mkomazi Game Reserve, Tanzania.* Bloomington, IN: Indiana University Press.

Carton, Wim. 2017. "Dancing to the Rhythms of the Fossil Fuel Landscape: Landscape Inertia and the Temporal Limits to Market-Based Climate Policy." *Antipode* 49 (1): 43–61. doi: 10.1111/anti.12262.

Carton, Wim. 2019. "'Fixing' Climate Change by Mortgaging the Future: Negative Emissions, Spatiotemporal Fixes, and the Political Economy of Delay." *Antipode* 51 (3): 750–769. doi: 10.1111/anti.12532.

Carton, Wim, and Elina Andersson. 2017. "Where Forest Carbon Meets Its Maker: Forestry-Based Offsetting as the Subsumption of Nature." *Society & Natural Resources* 30 (7): 829–843. doi: 10.1080/08941920.2017.1284291.

Carton, W. (2020). Carbon unicorns and fossil futures. Whose emission reduction pathways is the IPCC performing? In: Sapinski JP., Buck H., Malm A. (eds) *Has it Come to This? The Promises and Perils of Geoengineering on the Brink.* Rutgers University Press, Chapter 3, pp. 34–49.

Carton, Wim. 2020. "Rendering Local: The Politics of Differential Knowledge in Carbon Offset Governance." *Annals of the American Association of Geographers* 110 (5): 1353–68. https://doi.org/10.1080/24694452.2019.1707642.

Cavanagh, Connor, and Tor A. Benjaminsen. 2014. "Virtual Nature, Violent Accumulation: The 'Spectacular Failure' of Carbon Offsetting at a Ugandan National Park." *Geoforum* 56: 55–65. doi: 10.1016/j.geoforum.2014.06.013.

Corbera, Esteve, and Heike Schroeder. 2017. "REDD+ Crossroads Post Paris: Politics, Lessons and Interplays." *Forests* 8 (12): 1–11. doi: 10.3390/f8120508.

Ecotrust. 2016. *Plan Vivo Project Design Document (PDD): Trees for Global Benefits, Version 2.0.* Kampala, Uganda: Enironmental Conservation Trust of Uganda. http://www.planvivo.org/project-network/trees-for-global-benefits-uganda/.

Ecotrust. 2017. *Trees for Global Benefits Annual Report 2016.* Kampala, Uganda: Enironmental Conservation Trust of Uganda. http://www.planvivo.org/docs/TGB-2016-AR_public.pdf.

Edstedt, Karin, and Wim Carton. 2018. "The Benefits That (Only) Capital Can See? Resource Access and Degradation in Industrial Carbon Forestry, Lessons from the CDM in Uganda." *Geoforum* 97: 315–323. doi: 10.1016/j.geoforum.2018.09.030.

Fairhead, James, and Melissa Leach. 1996. *Misreading the African Landscape: Society and Ecology in the Forest-Savanna Mosaic.* Cambridge: Cambridge University Press.

Fairhead, James, Melissa Leach, and Ian Scoones. 2012. "Green Grabbing: A New Appropriation of Nature?" *The Journal of Peasant Studies* 39 (2): 237–261. doi: 10.1080/03066150.2012.671770.

FERN. 2018. "Katowice Forest Declaration: It's in the Way That You Use It." *ForestWatch News*, 18 December. https://fern.org/node/1515.

Fisher, Janet. 2012. "No Pay, No Care? A Case Study Exploring Motivations for Participation in Payments for Ecosystem Services in Uganda." *Oryx* 46 (1): 45–54. doi: 10.1017/S0030605311001384.

Fisher, Janet. 2013. "Justice Implications of Conditionality in Payments for Ecoystem Services: A Case Study from Uganda." In *The Justices and Injustices of Ecosystem Services*, edited by Thomas Sikor, 21–45. Oxon, UK: Routledge.

Fisher, Janet, Connor J. Cavanagh, Thomas Sikor, and David Mwayafu. 2018. "Linking Notions of Justice and Project Outcomes in Carbon Offset Forestry Projects: Insights from a Comparative Study in Uganda." *Land Use Policy* 73: 259–268. doi: 10.1016/j.landusepol.2017.12.055.

Green Resources. 2012. "Project Design Document – Kachung Forest Project: Afforestation on Degraded Lands." January 28, 2019. https://cdm.unfccc.int/Projects/DB/TUEV-SUED1301918616.32/view.

Griscom, Bronson W., Justin Adams, Peter W. Ellis, Richard A. Houghton, Guy Lomax, Daniela A. Miteva, William H. Schlesinger, et al. 2017. "Natural Climate Solutions." *Proceedings of the National Academy of Sciences* 114 (44): 11645–11650. doi: 10.1073/pnas.1710465114.

Hajdu, Flora, Oskar Penje, and Klara Fischer. 2016. "Questioning the Use of 'Degradation' in Climate Mitigation: A Case Study of a Forest Carbon CDM Project in Uganda." *Land Use Policy* 59: 412–422. doi: 10.1016/j.landusepol.2016.09.016.

Hamrick, Kelley, and Melissa Gallant. 2017. *Unlocking Potential: State of the Voluntary Carbon Markets 2017.* Washington, DC: Forest Trends' Ecosystem Marketplace. https://www.forest-trends.org/wp-content/uploads/2017/07/doc_5591.pdf.

Holmes, George, and Connor J. Cavanagh. 2016. "A Review of the Social Impacts of Neoliberal Conservation: Formations, Inequalities, Contestations." *Geoforum* 75: 199–209. doi: 10.1016/j.geoforum.2016.07.014.

Intergovernmental Panel on Climate Change (IPCC). 2018. *Global Warming of 1.5°C: An IPCC Special Report on the Impacts of Global Warming of 1.5°C above Pre-Industrial Levels and Related Global Greenhouse Gas Emission Pathways, in the Context of Strengthening the Global Response to the Threat of Climate Change.* Switzerland: IPCC. http://report.ipcc.ch/sr15/pdf/sr15_spm_final.pdf.

Jong, Hans Nicholas. 2019. "COP24: Green Groups Warn of Pitfalls in 'Forests for Climate' Deal." Mongabay. 3 January. Accessed July 22, 2020. https://news.mongabay.com/2019/01/cop24-green-groups-warn-of-pitfalls-in-forests-for-climate-deal/.

Leach, Melissa, and Ian Scoones, eds. 2015. *Carbon Conflicts and Forest Landscapes in Africa.* New York, NY: Routledge.

Lohmann, Larry. 2011. "The Endless Algebra of Climate Markets." *Capitalism Nature Socialism* 22 (4): 93–116. doi: 10.1080/10455752.2011.617507.

Lyons, Kristen, and David Ssemwogerere. 2017. *Carbon Colonialism: Failure of Green Resources' Carbon Offset Project in Uganda.* Oakland, CA: Oakland Institute. https://www.oaklandinstitute.org/carbon-colonialism-failure-green-resources-carbon-offset-project-uganda.

Parson, Edward A. 2017. "Climate Policymakers and Assessments Must Get Serious about Climate Engineering." *Proceedings of the National Academy of Sciences* 114 (35): 9227–9230. doi: 10.1073/pnas.1713456114.

Plan Vivo. 2016. "Trees for Global Benefits: Uganda." Plan Vivo Foundation. 2016. Accessed July 23, 2020. http://www.planvivo.org/project-network/trees-for-global-benefits-uganda/.

Schreckenberg, Kate, David M Mwayafu, and Roselline Nyamutale. 2013. *Finding Equity in Carbon Sequestration: A Case Study of the Trees for Global.* Kampala: Uganda Coalition for Sustainable Development. http://www.espa.ac.uk/files/espa/Case%20Study%20Trees%20for%20Global%20Benefits%20Project%2C%20Uganda.pdf.

Scott, David F. 2005. "On the Hydrology of Industrial Timber Plantations." *Hydrological Processes* 19 (20): 4203–4206. doi: 10.1002/hyp.6104.

Smith, Pete, and Mercedes Bustamante. 2014. "Agriculture, Forestry and Other Land Use." *Climate Change 2014: Mitigation of Climate Change.* Cambridge University Press, Cambridge, United Kingdom. https://doi.org/10.1104/pp.900074.

UNFCCC. 2015. *Report of the Conference of the Parties on Its Twenty-First Session, Held in Paris from 30 November to 13 December 2015.* United Nations Framework Convention on Climate Change. 29 January 2016. Accessed July 23, 2020. https://undocs.org/pdf?symbol=en/FCCC/CP/2015/10.

UNFCCC. 2018. "The Ministerial Katowice Declaration on Forests for the Climate." COP 24. 14 December. https://cop24.gov.pl/fileadmin/user_upload/Ministerial_Katowice_Declaration_on_Forests_for_Climate_OFFICIAL_ENG.pdf.

Vaughan, Adam. 2018. "Shell Boss Says Mass Reforestation Needed to Limit Temperature Rises to 1.5C." *The Guardian*, 9 October 2018. Accessed July 23, 2020. https://www.theguardian.com/business/2018/oct/09/shell-ben-van-beurden-mass-reforestation-un-climate-change-target.

World Bank. 2018. *State and Trends of Carbon Pricing 2018.* May 2018. Washington, DC: World Bank Group. https://openknowledge.worldbank.org/bitstream/handle/10986/29687/9781464812927.pdf.

35

World-making and the deep seabed

Mining the *Area* beyond national jurisdiction

Anna Zalik

Introduction

Mineral and hydrocarbon extraction in the seabed beyond state jurisdiction is in its physical infancy but global debates concerning property rights in, as well as access and claims to, the high seas have been central to the history of international law (Grotius [1609] 2009).[1] In the late twentieth century, the long-standing debate concerning the "freedom of the seas" became intertwined with post-WWII discussions of global equity and common property in the international seabed, the zone beyond national jurisdiction referred to as "the *Area*." The *Area* is regulated through the United Nations Convention on the Law of the Sea (UNCLOS), the negotiation of which extended over various decades in the post-WWII period. Attention to the *Area* is once again ramping up as technological advancement and demands for rare earth minerals spur resource extraction on and below the ocean floor.

Under UNCLOS, regulatory power for deep-sea mineral extraction beyond national jurisdiction is the responsibility of the International Seabed Authority (ISA), a UN agency established in 1994 and headquartered in Jamaica. While its regulatory purview includes all nonliving resources attached to or below the ocean floor, the ISA primarily regulates exploration for three minerals—polymetallic nodules, polymetallic sulfides, and ferromanganese (cobalt rich) crusts—in specific locations beyond national jurisdiction. The most important of these is the Clarion Clipperton Zone (CCZ) in the Pacific between Hawaii and Mexico. And while there are 29 contracts for exploration under ISA auspices, an exploitation regime has yet to be finalized for any seabed zone under ISA control.

The ISA's location in Jamaica reveals the important role held by the G-77/Global South in the twentieth-century UNCLOS negotiations. The ISA remains the only UN agency that holds an annual assembly meeting in the Global South. Through the work of the 77 formerly colonized Global South countries referred to as the G-77, the principle of the "common heritage of (hu)mankind" (CHM) was established as a guideline for redistributive policy for resource development in the seabed beyond state jurisdiction. CHM is today considered a key global equity principle in international law (Okereke 2008). However, with the rise of global neoliberalism, including in the mining sector (Himley 2010), the CHM principle in the UNCLOS regime for the *Area* became overshadowed by market demands. While CHM was central to Part XI of

UNCLOS concerning the *Area* and the establishment of the ISA, the 1994 Implementation Agreement (IA) on Part XI inserts the need to consider intellectual property rights and market principles. It also notes that in cases of inconsistency between UNCLOS and the IA, the IA's provisions will prevail (UN 1994), embedding neoliberal logics at the expense of the CHM.

As industrial interest in minerals and hydrocarbon extraction from the deep seabed have increased with the twenty-first-century commodity boom, distributional, ecological, and geopolitical concerns are influencing the emerging ISA extraction regime (Sammler 2016). On the point of distribution, UNCLOS indicates that the ISA may collect royalties on minerals, including hydrocarbons, extracted beyond 200 nmi from a state's coastline even if extraction takes place within a country's extended continental shelf, a zone defined by the UNCLOS as the "submerged prolongation of the land territory of the coastal state."[2] This provision, however, has yet to be implemented. The potential impact of extraction in the poorly understood ecologies of these depths is gaining further attention from global environmental NGOs and advocacy organizations as moves toward establishing a deep-seabed mining code have ratcheted up. The rare earth mineral resources in the *Area* are promoted as a potential contributor to post-carbon energy technologies. As various analysts emphasize, however, many of these metals could be collected via recycling, including of e-waste (Golev et al. 2014), undermining claims that call for seabed mining as a component of global transition to renewable energy. Finally, questions of geopolitics and jurisdiction in the *Area* are apparent in the contemporary ISA proceedings. Despite the fact that the United States has not ratified UNCLOS, its persistent influence—alongside that of Brazil, Russia, India, China, and South Africa (the BRICS; see Zalik 2015, 2018)—is notable. In the midst of these dynamics, US philanthropic foundations, research institutes, government agencies, and industry are active at the ISA. Private and state firms, including global defense and security firm Lockheed Martin, are also significant figures. Legal and scientific consultants and ISA contracting firm staff make up a broader "epistemic community" shaping the ISA seabed extractive regime and the emerging, but contested, mining code.

Drawing on these dynamics, this chapter examines historically situated and intersecting processes mobilized by states and firms in the international arena to transform seabed minerals into commercially available resources (i.e., resource-making, see also, Oliveira, Chapter 28 in this volume, Fannin, Chapter 33 in this volume). I do this in three parts, including the conclusion. Each section also demonstrates an attribute of an explicitly critical resource geography. First, I examine the ISA as well as state legislation established by nonparties to UNCLOS (namely, the United States), which give weight to private capital in shaping the emerging extraction regime for the *Area*. I analyze how information generated via corporate consortia in the post-WWII period has shaped epistemological control over deep-sea spatial data and technology required for industrial extraction of seabed minerals (Squire 2016; Zalik 2018). Second and informed by the first section, I offer a geo-juridical analysis (Hernández Cervantes 2014) of two contemporary moves toward mining the *Area*. A geo-juridical analysis attends to contesting and overlapping legal codes, here of the ISA and US legal regimes in two sites. These sites are (a) transboundary hydrocarbons in the Western Gap of the Gulf of Mexico between Mexico and the United States—a zone not formally under ISA authority but to which UNCLOS indicates it holds royalty rights—and (b) overlapping US and ISA exploration contracts held by US firm Lockheed Martin and its British subsidiary in the CCZ. To contextualize these two moves, I discuss (c) how extraction from the seabed is co-constituted with the extension of finance capital necessary to harness "high risk" deep ocean resources.

Taken together, features that constitute an extractive regime for the international seabed via the UNCLOS/ISA express a world-making process enabling or creating different kinds of socio-environmental organization. The chapter also points to key methodological attributes of an explicitly critical geography of resources: (a) attention to the institutions and firms that control <u>knowledge</u> on resources-in-the-making (see also Kama, Chapter 5 this volume); (b) a geo-juridical analysis concerning how overlapping and competing multilateral, bilateral, and state <u>legislation</u> shape exploitation and regulation of emerging resources; and (c) a consideration of how financial practices and venture <u>capital</u> influence ongoing and future development of such resources in the contemporary period. As suggested previously in each methodological attribute, these run parallel to what may be understood as the epistemologies, jurisdictions, and commodifications shaping the development of contemporary ocean frontiers, elements which are prerequisite to world-making in the deep seabed and the transformation of its minerals into resources (Havice and Zalik 2018).

Technologies and control: UNCLOS and the International Seabed Authority

Contemporary dynamics at the ISA, including the use of seabed mining data and technology, have emerged from centuries long debates over access to, and sovereignty over, marine space. These debates collectively shaped early international law, most notably through Grotius' work *Mare Liberum—The Free Sea* (Grotius [1609] 2009). In the twentieth century, a number of key tensions, in particular between (a) Global North and South and (b) the United States and USSR and their allies during the Cold War, influenced the emerging UNCLOS deep-seabed mining regime.[3]

In the immediate postwar period, the UNCLOS I (1958) and UNCLOS II (1960) emerged out of the Geneva Conventions. UNCLOS I produced certain agreements but was unable to settle whether a coastal state's jurisdiction should be allowed to extend past the 12 nmi formally recognized as "territorial waters." The UNCLOS III negotiations (1973–1982) occurred within Cold War maneuvers and thus were influenced by tensions between the United States and the Soviet Union. In this period, the international seabed was subject to activity by various state navies alongside large deep-sea mining consortia, the members of which incorporated the oil industry and branches of the military (Hayashi 1989).

A key dividing line in the UNCLOS III negotiations was that between the G-77 under the umbrella of the New International Economic Order (NIEO) and former colonizing and global powers. The G-77/NIEO emerged from the 1955 Bandung Conference that built Third World solidarity to demand compensation for resources taken under colonialism. The CHM principle was articulated in 1967 by the late Arvin Pardo, UN representative for Malta (Okereke 2008; Tladi 2014), to advance the objectives of G-77. In addition to siting the ISA in Jamaica, Global South countries pursued mandated technology transfer (TT) as a mechanism to avoid the privatization of knowledge and technology on seabed minerals and build greater global equity (Yarn 1984). This remains a point of tension between formerly colonized countries and the Global North at today's ISA. TT provisions continue to weigh significantly in US non-ratification of UNCLOS.

UNCLOS also proposes the formation of "the Enterprise," a kind of global parastatal entity under ISA auspices, to implement the redistributionist intentions of the CHM. The purpose of the Enterprise as firm would be to facilitate fair involvement in extracting and processing seabed resources and to compensate the Global South, including its landlocked countries, for systematic underdevelopment arising from colonial extraction (Girvan [1976] 2017). However, as this chapter discusses, neoliberal governance logics reinforced through US non-ratification, geo-juridical arrangements, and finance capital have trumped the implementation of UNCLOS' more emancipatory components.

In the lead up to President Reagan's rejection of UNCLOS in 1982, the United States passed the Deep Seabed Hard Mineral Resources Act. This Act, which remains in place today, permits multilateral or bilateral agreements that allow seabed mining in the oceans beyond state jurisdiction. Within months of the US rejection of UNCLOS, the "Agreement Concerning Interim Arrangements Relating to Polymetallic Nodules of the Deep Seabed" was passed by the United States and signed by France, West Germany, and the United Kingdom. Made up of bilateral notes, the agreement entailed parallel laws being passed by each member state to form a "reciprocating states regime"; the Soviet Union passed similar national legislation (Churchill and Lowe 1999). Such agreements subsequently impelled the protection, under UNCLOS, of "pioneer investors" which had conducted research and activities in the *Area* in advance of subsequent LOS ratification. These "pioneer investors" comprised both parastatal and private firms. Among them was the Ocean Minerals Company (OMCO), the members of which included Royal Dutch Shell and the Lockheed Martin Corporation (Larson 1986). Lockheed Martin's wholly owned British subsidiary, United Kingdom Seabed Resources (UKSRL), continues to hold rights to the activities and interests of OMCO. Despite the fact that the USSR and the G-77 criticized this bilateral and multilateral regime among Western powers, India and the USSR were ultimately among the first to register as "pioneer investors" under UNCLOS in 1987, so as to protect their own pre-ratification research and activities (Hayashi 1989).

By the time UNCLOS was ultimately ratified in 1994, geopolitical conditions had shifted in favor of the capitalist West, epitomized in the fall of the Berlin Wall, and the apparent triumph of global neoliberalism via the World Trade Organization inaugurated the same year.

The IA that accompanied UNCLOS' ratification qualified certain redistributive aspects of the convention and tipped the balance in favor of corporations. Notably, "market principles" were emphasized in key sections of the IA, including the preamble. In contrast to the goal of economic equity and decommodification of knowledge that undergirded the principle of TT to the Global South, the IA states that TT would be made on "fair and reasonable commercial terms and conditions, consistent with the effective protection of intellectual property rights" (IA, Section b). The IA also holds that no subsidies, tariffs, or nontariff barriers may apply to resources produced from the *Area* on international markets and that "in accordance with sound commercial principles, the General Agreement on Tariffs and Trade and successor or superseding agreements shall apply with respect to activities in the *Area*" (IA, Section 6).

Despite these neoliberal modifications, the proposal for the Enterprise—as a kind of global parastatal promoting the redistributive policies neoliberalism sought to dismantle—remained in place, although trumped through the IA by "commercial principles." Today, at annual ISA sessions, the potential establishment of the Enterprise is regularly mentioned, notably by state representatives from the Global South and Eastern Europe, as a means to implement equitable sharing of data, technology, and outputs (personal observation ISA Annual Session 2013, 2017). However, the current secretariat of the ISA rarely refers to the Enterprise in their public discussions of the emerging exploitation regime. Indeed, two recent articles concerning financial and regulatory mechanisms for mining the *Area* coauthored by the ISA Secretary General—British attorney Michael Lodge—and US-based academics and government researchers do not mention the Enterprise (Lodge, Segerson, and Squires 2017; Lodge and Verlaan 2018).[4] Thus, despite the provisions for redistribution and technology-sharing endorsed by UNCLOS, the tensions arising from neoliberalism have weakened the common heritage approach to resource-making in the international seabed. In the next section, we consider two examples of how potentially conflicting international and state jurisdictional moves seek to enclose seabed resources, and their relationship with risk financing which determines the viability of costly seabed extraction.

The enclosure of resources in the *Area*

A geo-juridical analysis of resource exploitation examines various and contesting forms of law—in this case laws whose enactment constitutes a world-making process shaping seabed exploration, extraction, and regulation. UNCLOS negotiations, on the one hand, and competing state legislation and multilateral negotiations, on the other, pursue an extractive regime for deep-sea minerals and hydrocarbons with divergent goals in mind. In the two cases discussed in this section, the role of US and global capital in "world-making" is significant. Both cases pertain to regions under UNCLOS/ISA jurisdiction, yet current developments in each indicate the significant geopolitical weight that the United States holds in the *Area* despite, and to some extent because of, its non-ratification of the Convention. The first case concerns hydrocarbon development in the Western Gap of the Gulf of Mexico—a zone where the United States and Mexico hold bordering and potentially conflicting offshore claims. The second case, introduced previously, concerns the role of Lockheed Martin/UKSRL in the CCZ where the firm holds rights to an exploratory concession under both the ISA and the US DSHMRA. This latter example underlines the power that Lockheed/OMCO amassed historically through geopolitical relations and control over data. Both instances reveal overlapping regulatory authority under UNCLOS and state legislation (see also Havice 2018).

The Western Gap in the Gulf of Mexico

Oil and gas exploration and production in the Gulf of Mexico's Western Gap, one of two gaps in the Gulf of Mexico's *Area* that occur in "donut holes"—spaces otherwise enclosed by the Exclusive Economic Zones of the United States, Mexico, and other Caribbean/Central American states—has been governed via US-Mexico accords. Offshore drilling in the Western Gap was facilitated by a bilateral agreement signed in 2000 dividing the space beyond 200 nmi of each country's coastline (Exclusive Economic Zone) that is enclosed by their respective maritime boundaries. This space is potentially within both of their extended continental shelves. As discussed in this chapter, even in cases where a country's extended continental shelf has been proven to extend 200 nmi beyond its coastline, UNCLOS mandates a royalty payment to the ISA for extraction beyond 200 nmi. The Gap's 5092 nmi^2 (17,467 km^2) space, slightly smaller than the state of New Jersey, holds an estimated 172 million barrels of oil equivalent (BOE) and 304 billion ft^3 of natural gas (Hagerty and Uzel 2013).

The subsequent 2012 Transboundary Agreement on Hydrocarbons lifted a moratorium on extraction along a 135-mi portion of the border (covering almost 158,584 ac) of the Western Gap that had been off-limits under the 2000 agreement. For some decades, critical scholars in Mexico have pointed to the possibility that US deepwater drilling in the Gulf of Mexico siphoned resources out of deposits, including this gap, that cross into Mexico's territorial seabed (Barbosa Cano 2003).[5] The 2012 agreement permits the United States and Mexico to drill in the zone under moratorium through the process of "unitization" that allows bordering or overlapping assets to be shared among operators and then divided. It is a common practice among oil and gas operators in the United States but less so internationally, and for reserves that cut across national jurisdictions. Formal unitization would offer the certainty/securitization required for investment in deep-sea resources in this zone, thus meeting the venture financing requirements for ongoing development (Hagerty 2014).[6]

Mexican critical media portrayed the agreement as a surrender of sovereignty (Arroyo and Zalik 2016) in part because Mexico's twentieth-century constitutional provisions prohibited foreign entities from exploiting the country's hydrocarbons. The US-Mexico agreement was co-constituted with denationalization and privatization of the Mexican oil and energy sector, in particular reforms passed in 2008 and subsequently in 2013. Mexico's 2013 energy reform reopened the country's petroleum sector to foreign investment, unprecedented since the 1938 nationalization of the oil industry that made the parastatal Pemex the country's sole operator for decades thereafter. La Huasteca and El Aguila, the key firms Mexico expropriated as a result of the 1938 nationalization, were subsidiaries of what would today be Shell, Exxon, and Chevron. These firms are actors in the US Gulf of Mexico and have invested in the Mexican deepwater in the auctions that followed upon Mexico's 2013 reform. Indeed, as part of its current emphasis on deepwater investment (Raval 2018), Shell has strategically positioned itself in the Gulf of Mexico through both US and Mexican deepwater auctions. The 2012 Transboundary Agreement and Mexico's 2013 energy reform thus signal further integration of the North American oil and gas market (Hernández Cervantes and Zalik 2018) via private firms that previously did not extract in Mexico.

The 2012 Transboundary Agreement makes no reference to UNCLOS, although it is pertinent. In fact, activities in the Gulf of Mexico's Western Gap manifest the limited weight of the ISA and UNCLOS provisions in the international seabed. According to US proponents of UNCLOS, jurisdictional certainty in the international seabed through US ratification would assist firms to secure financial support for deep-sea exploration, essential for costly offshore extraction. But US opponents of UNCLOS ratification use the example of the Gulf of Mexico to argue the opposite. A piece published by the now defunct conservative US think tank National Center for Policy Analysis (NCPA) stresses that "US companies are already successfully investing in an area of the extended continental shelf—the 'western gap' in the Gulf of Mexico" (Murray 2013, 6). That is, the Western Gap activities demonstrate that the United States need not ratify UNCLOS in order to extract with confidence beyond 200 nmi.

There is weight to NCPA's empirical point. Under UNCLOS, key investors in the US and Mexican oil industries arguably achieve collective advantages through US *non-ratification* even though Mexico has ratified. In the 1970s, during the early UNCLOS negotiations, Mexico held that the Western Gap reserves were the CHM but ultimately agreed that the region could be exploited by the United States and Mexico through their extended continental shelves[7] (Vargas 1996; Heaton 2013). While Mexico's extended continental shelf claims have been submitted to the UN Commission on the Limits of the Continental Shelf, the United States as a non-ratifier has not submitted such claims for international review (Magnússon 2017). Geopolitically, Mexico's decision to submit its claims reflects a response to both the realpolitik of US non-ratification of UNCLOS and existing US drilling in the Gap.

US non-ratification provides advantages to both the United States and its industry. Under Article 82 of UNCLOS, 1 percent of revenues per annum generated from extraction beyond 200 nmi (even if on an internationally recognized extended continental shelf) must be paid to the ISA after the first five years of production (ISA 2009; Harrison 2017). The United States has not ratified and is thus exempt, as was US drilling in the Western Gap in advance of the 2012 agreement.[8] Mexico also *appears* exempt as a result of the denationalization and privatization of the Mexican petroleum sector through which it has become dependent on imported US natural gas (Hernandez-Cervantes and Zalik 2018). UNCLOS Article 82 makes an exception to the 1-percent rule for a developing state which is a "net importer of the mineral resource produced from its continental shelf," with mineral resources compromising hydrocarbons (Neff 2014, 61). After 1938, Mexico was self-sufficient in petroleum products for many decades. However, this position reversed due to the 2013 Mexican energy reform and continental energy restructuring from the US shale boom (S&P Global/Platts 2017), which made the United States an energy superpower. As of 2015, Mexico became a *net importer* and thus external to mandated redistribution embedded in UNCLOS Article 82.

The case of the US-Mexico Gulf suggests how UNCLOS measures may be employed to facilitate enclosure of seabed areas beyond national jurisdiction designated as CHM. UNCLOS provides both states a channel to avoid revenue sharing with the international community by unitizing US-Mexico oil fields in areas beyond national jurisdiction that are mined by private operators. The dynamics in the Western Gap exemplify how geo-juridical processes that structure oil extraction on the extended continental shelf unfold around issues of revenue distribution and enclosure.

Lockheed Martin/UKSRL in the Clarion Clipperton Zone

Our second example of geo-juridical dynamics shaping seabed mining practices arises from Lockheed Martin's presence in the CCZ. It exemplifies how the ISA's current exploration contracting process in the *Area*, overlapping jurisdiction with US legislation, and the relative weight carried by US interests shape geo-juridical agreements in seabed mining. Lockheed's unique position at the ISA, as a US firm operating in a venue governed by a UN convention that the United States has not ratified, parallels the growing role of US civil society at ISA meetings. As discussed previously, Lockheed is the progenitor of OMCO, a major Western mining consortium of that period (Larson 1986; Hayashi 1989). OMCO's activities in the 1970s and 1980s are protected under pioneer investor clauses of UNCLOS (article 305) and the IA (Annex Section 1; 5b) as well as US domestic legislation under the DSHMRA. In the current ISA allocation process, the Lockheed subsidiary UKSRL is the only firm to hold more than one exploration contract for polymetallic nodules in the CCZ and, importantly, the only firm to hold rights under *both* US legislation and the ISA to mining concessions in the CCZ. Observers note that the existence of these overlapping contracts may impel the ISA to develop an extractive regime quickly, to avert the possibility that Lockheed would seek to mine under US domestic legislation. It is also holder of important proprietary data employed by many contractors and consultants (Larson 1986).

As I have documented elsewhere (Zalik 2018), competitor firms identify OMCO's data as fundamental to their contractual and financial arrangements on concessions in the CCZ (Nimmo, M./TOML GOLDER (2013). In addition, Lockheed has partnered with countries such as Singapore; the firm Ocean Mineral Singapore (OMS) explicitly recognized UKSRL/Lockheed as a joint owner in its 2014 application to the ISA (International Seabed Authority [ISA] 2014).

NGOs in the Pacific and the United States have criticized Lockheed/UKSRL's activities for undue influence and the evasion of environmental protection. In 2013, Lockheed's relations with the Fijian government and the Pacific Islands Applied Geoscience Commission (SOPAC) were under scrutiny for undue influence on the country's seabed-mining legislation, in part because of work a British attorney was conducting for both SOPAC and Lockheed (Saiki 2013). In 2015, the US-based NGO Center for Biological Diversity (CBD) unsuccessfully sued the US National Oceanic and Atmospheric Administration on the grounds that Lockheed had not carried out requirements for environmental impact assessment; consequently, the CBD argued that Lockheed's US DHSMRA permit for exploration in the CCZ should not be renewed (Center for Biological Diversity 2015, 2016).

ISA documents and competitor seabed mineral firms refer specifically to the Lockheed proprietary database concerning mineral resources in the *Area* (International Seabed Authority [ISA] 2010, 6). The data OMCO collected in the 1970s and 1980s remains the basis for many competitor firms' applications for financing; the possibility of monopolization is raised by various actors in ISA processes given the private contracts that competitors may need to secure with Lockheed in order to access Lockheed's data. If the United States ratifies UNCLOS, Lockheed/UKSRL data would in principle be incorporated into the TT provision associated with the

CHM. US ratification of UNCLOS would thus undermine the financial and epistemological weight that Lockheed, and the United States to which it is a key military contractor, carries at the ISA through its control over data.

Contending ISA/US jurisdiction and authority also extends to less formal aspects of regulation. US environmental NGOs, consultants, and universities are active as formal observers in the ongoing negotiation of the ISA's exploitation regime and the environmental debate surrounding it. Among these, the US Pew Charitable Trusts and the Robert Kaplan Fund in New York supported the Clarion Clipperton Regional Environmental Management Plan adopted in 2012 (Lodge and Verlaan 2018). Consultants associated with the Massachusetts Institute of Technology have presented models for fiscal arrangements for mining the *Area* that emphasize investor risk but not the risk of environmental harm (Roth and Munoz Royo 2018; Zalik 2018, Earth Negotiations Bulletin 2018).

These cases of current developments in the *Area* reflect the geo-juridical component of overlapping and/or contested jurisdiction shaping potential seabed mining. Each exemplifies the practical implementation of the UNCLOS international regime for seabed extraction, including how that regime is co-produced with broader geopolitical developments. Each reveals contradictions in the nominal UNCLOS objective of promoting technology, data sharing, and equitable revenue distribution from transnational space and underlines the political economic conditions shaping extraction therein. While nominally US-based industrial actors—notably the oil and gas industry—present themselves as globalists supportive of the US ratification of UNCLOS, US non-ratification also carries relative competitive advantages for global firms and states, as per the example of possible protection from the ISA royalty. US firms and state institutions are also "free" from certain regulatory requirements required of UNCLOS ratifiers, while they access opportunities open to UNCLOS parties through subsidiaries or partners. In particular, the possession of commercializable data and technology protected from UNCLOS' TT requirements allow US-based firms, such as Lockheed, and the global capital blocs with which they are associated, competitive advantages in financial markets for high risk, offshore activity.

Securing deep-sea resources through finance

In recent years, geographic scholarship has paid increasing attention to the role of finance in shaping spatial outcomes, including in carbon trading, climate risk insurance, and oil markets (Pike and Pollard 2010; Zalik 2010; Johnson 2013). Understood by scholars in Marxist traditions as fictitious capital, the possibility of future profits attracts financial investment to activities in the present. In the absence of financing, endeavors such as offshore oil extraction and deep-seabed mining would be impossible. Deep offshore extraction in heavily traded hydrocarbons, or in resources not yet commercialized such as seabed minerals, are typically categorized as relatively high risk by financial institutions.

In the case of both the deep-offshore Gulf of Mexico and seabed minerals in the ISA-governed *Area*, the availability of finance is key to enacting deep-sea mining via UNCLOS (world-making), without which contemporary commercialization of seabed minerals (resource-making) is improbable or impossible. Powerful industrial actors in industries such as oil, gas, and mining help make "high-risk" projects possible by projecting models of *future* demand for the commodities under production (Zalik 2010). In the deep offshore, petroleum engineers are important agents in constituting a "subterranean geopolitics" of volume (Kama 2014; Valdivia 2015; Lehman 2016). Via their representation and measurement of hydrocarbon reserves, the work of engineers and geologists is central to the reserve-based lending structure often employed to finance offshore oil and gas production.

The emergence of so-called Reserve-Based Lending (RBL) or Reserve-Based Finance (RBF) in hydrocarbon exploration and production was shaped by the oil shocks and co-constituted with global demand for energy commodities that surrounded the UNCLOS III negotiations. The RBL approach was developed in Texas and advanced in the UK North Sea Market in the 1970s and 1980s (Fox, Gonsoulin, and Price 2014). RBL proffers capital to develop high-cost producing and nonproducing reserves. To reduce risk to creditors, it requires reassessment of borrowing conditions—including fluctuations in transnational oil and gas prices. Generally speaking, this is based on a *projection of future demand* in that finance provided for a given asset is assessed on the net present value of a borrower's *future* income from a given resource (to be produced at that reserve). Financing for deep-seabed minerals exploration is similarly reliant on socially produced forecasts of demand. The socially constructed (Mackenzie 2008; Zalik 2010) nature of lending rates and projection of future value is clear in that the process is guided by financial analysts and is *dependent on the work of energy experts*. In consonance with Valdivia's (2015) analysis concerning engineers, borrowers must submit regular reserve reports authored by independent petroleum engineers.

The conditions shaping the fiscal conditions for the ISA exploitation regime also require balancing assessment of profitability and investor risk. Firms currently exploring in ISA managed areas such as the CCZ must negotiate investment financing in advance of baseline environmental sensitivity studies. Various marine biologists consider such baseline studies prerequisite to determining whether mining should be permitted in the first place (DOSI [Deep-Ocean Steward Initiative] 2018). Extractive projects that rely on financing in advance of production impede application of the precautionary principle because the finance is extended on the basis of future production and profits. As described explicitly in a presentation on a potential ISA fiscal regime by Massachusetts Institute of Technology consultants, "Investors will only take on [a] project if discounted future revenues are large enough to provide a return on their investment that is competitive with other investment opportunities" (Roth and Munoz Royo 2018). Similarly, and also pertinent to the Gulf of Mexico case discussed previously, Council on Foreign Relations analyst Stewart Patrick points out that while "it is true that the ISA collects royalties for deep sea mining... these remain extremely modest—as one would expect from an arrangement that was effectively negotiated by US oil companies" (Patrick 2012). Given extremely low royalties, the prospects for a substantively redistributive fiscal regime from profits generated in the *Area* are low.

The 2017 articles by the ISA Secretary General and US colleagues referred to the aformentioned outline of possible fiscal models for an ISA exploitation regime (Lodge, Segerson, and Squires 2017). They depict a context of uncertainty, potentially meager revenues, and the need to provide returns to investors. Overlooking *the Enterprise*, their proposed fiscal regime does not entail direct redistributive payments to developing states, but rather proposes that the ISA would fund "projects," such as development programs to benefit said states, or a fund for investment in human and physical capital. To address the environmental risk of seabed mining and reduce insurance premia, Lodge, Segerson, and Squires (2017, 454) suggest that the ISA could employ a mix of "regulation, environmental fees (put into an environmental fund), and negligence-based liability." It should be noted that negligence-based liability is far weaker than strict liability for firms, a variation which Lodge and Verlaan (2018) discuss openly. In contrast, Norway employs strict liability and is considered one of the most effectively regulated offshore oil and gas sectors from an environmental perspective. It is through considering such proposed fiscal arrangements, alongside juridical and epistemological components of mining the *Area* that this chapter offers a critical approach to the constitution of seabed resources.

Toward a critical resource geography of seabed extraction

The previous discussion of seabed mining as world-making seeks to employ characteristics of what I have described as an explicitly *critical geography* of deep ocean resources. *First, the opening section **not only describes current knowledge and technology** to constitute seabed minerals-as-resources, but also **probes the institutions and firms who control** the dissemination of, and/or restriction of access to, this knowledge;* in turn this requires considering the historical context in which they amassed it. *Second, a critical resource geography **requires geo-juridical analysis** of* world-/resource-making. In this case, I examine the conditions under which UNCLOS was constituted and negotiated, modified in the 1994 IA, as well as the current geopolitical conditions shaping proposed contractor terms described in draft exploitation regulations.[9] Geo-juridical study also entails attention to how overlapping jurisdiction asserted under US domestic legislation (DSHMRA) and UNCLOS, as per Lockheed concessions in the CCZ, may impel the ISA to develop a mining code quickly in order to prevent Lockheed seeking exploitation rights under US legislation. Even in the absence of US ratification of UNCLOS, US government agencies, NGOs, and private capital are playing significant roles in shaping the developments at the ISA, in part via control over proprietary knowledge and technology. **These two attributes—a critical interrogation of epistemology/knowledge and of law/jurisdiction**—inform the *third: an examination of how venture finance for resource development* is **accessed, harnessed, and ultimately accumulated** *via financial instruments and markets,* in this case arising from forecasts on seafloor or subsea minerals and the technologies to extract them.

These three methodological attributes share parallels with categories we have used elsewhere to conceptualize the emergence of ocean frontiers (Havice and Zalik 2018): the *epistemologies, jurisdictions, and commodifications* constituting the *Area* are crucial elements in the creation of a market in seabed minerals as intersecting world- and resource-making processes. Knowledge, jurisdictions, and commodification combine to transform the seabed from remote, empty, and inaccessible to exploitable resource (see also Campbell et al., Chapter 36 in this volume). A critical geography of the materials and ecologies uncovered in this under-researched, deep-open space necessitates analysis at the intersection of all three. The constitution of knowledge, jurisdiction, and finance capital may co-create conditions that permit enclosure of nature nominally part of our global "common heritage", thus making these minerals salable as "natural resources." But there is an additional attribute of critical geographical work largely missing from the analysis in this chapter: an assessment of sociopolitical resistance to the ecological consequences of resource exploitation. Accordingly, the growing international movement for a moratorium on deep-sea mining is an important direction for future work (Deep Sea Mining Campaign, London Mining Network, Mining Watch Canada, 2019). On that point, the reader is urged to consult the work of the transnational Deep Sea Mining Campaign at http://www.deepseaminingoutofourdepth.org/ and a scholars' letter available at www.ourcommonheritage.org.

Notes

1 The author thanks the handbook editors and Christopher Adams for careful reading and commentary on drafts of this chapter, and the Social Sciences and Humanities Research Council of Canada and York University for research funding. All errors are hers.

2 For a more fulsome definition of the term as applied under the UNCLOS, see https://www.un.org/Depts/los/clcs_new/continental_shelf_description.htm#definition.

3 In addition to the ISA, the UNCLOS established the International Tribunal of the Law of the Sea, which includes a Seabed Disputes Chamber to oversee tensions over coastal jurisdiction beyond territorial waters. Materially, the tension between Global North and South has manifest in the

ability of wealthier and more powerful states to map their extended continental shelves, prolonging their territory seaward in advance of states with less technological or financial capacity. The United States, as a nonparty to UNCLOS, has not sought UN recognition for extended shelf claims (Magnússon 2017).

4   Beyond the scope of this paper, the election of Lodge as Secretary General in 2015 was contested by the Global South, notably the African caucus. Interviews indicate a commonly held view among ISA observers that Lodge's position reflects the interest of the Global North, notably the United Kingdom and the United States.

5   For a summary of US-Mexico treaties pertinent to the region, see https://www.boem.gov/Treaties/.

6   Beyond the scope of this chapter, the 2012 agreement establishes a geopolitical baseline for the broader Gulf of Mexico and Caribbean basin, which prompted a subsequent (2013) agreement between Mexico and Guatemala and has fueled ongoing discussion on ocean territorial boundaries with Cuba in the so-called Eastern Gap of the Gulf of Mexico.

7   The delineation of the extended continental shelf under the UNCLOS regime and its relationship to current EEZ sovereignty disputes is outside this paper's scope, given this paper's focus on the *Area* under international jurisdiction.

8   An ISA Technical Study on the implementation of Article 82 points out that the US Minerals and Management Service has alerted contractors that royalty to the ISA could apply were it to ratify UNCLOS (International Seabed Authority [ISA] 2009, 6).

9   Beyond the scope of this analysis, ISA contracts are confidential and discussed only in the ISA Finance Committee, whose lack of transparency has been critiqued by official observers.

References

Arroyo, Michelle, and Anna Zalik. 2016. "Displacement and Denationalisation: The Mexican Gulf 75 Years after the Expropriation." *Area* 48 (2): 134–141.

Barbosa Cano, Fabio. 2003. *El Petróleo en las "Hoyas de Dona" y Otras Áreas Desconocidas del Golfo de México.* México: Instituto de Investigaciones Económicas, UNAM/Editorial Miguel Angel Porrua.

Center for Biological Diversity. 2015. *Complaint for Declaratory and Other Relief to US Secretary of Commerce and National Oceanic and Atmospheric Administration.* Washington, DC. https://www.biologicaldiversity. org/campaigns/deep-sea_mining/pdfs/Deep-seabedMiningComplaint_05-12-2015.pdf.

Center for Biological Diversity. 2016. *Settlement: Trump Administration Must Study Environmental Risks Before Approving Deep-Sea Mining.* Press Release. Washington DC, November 30. https://www.biologicaldiversity.org/news/press_releases/2016/deep-sea-mining-11-30-2016.html.

Churchill, Robin R., and Alan V. Lowe. 1999. *The Law of the Sea.* Manchester: Manchester University Press.

Deep Sea Mining Campaign, London Mining Network, Mining Watch Canada. 2019. *Why the Rush? Seabed Mining in the Pacific Ocean.* July. http://www.deepseaminingoutofourdepth.org/wp-content/uploads/Why-the-Rush.pdf.

DOSI (Deep-Ocean Steward Initiative). 2018. *Commentary on the "Draft Regulations on Exploitation of Mineral Resources in the Area."* http://dosi-project.org/wp-content/uploads/2018/10/DOSI-Comment-on-ISA-Draft-Exploitation-Regulations-September-2018.pdf.

Earth Negotiations Bulletin. 2018. ISA 24, Part 2 Highlights 25 (160), July 17. https://enb.iisd.org/vol25/enb25160e.html.

Fox, Jason, Dewey Gonsoulin, and Kevin Price. 2014. "Reserve Based Finance: A Tale of Two Markets Part 1." *Oil and Gas Financial Journal* 11: 26–31.

Girvan, Norman. [1976] 2017. *Corporate Imperialism: Conflict and Expropriation.* London: Routledge.

Golev, Artem, Margaretha Scott, Peter D. Erskine, Saleem H. Ali, and Grant R. Ballantyne. 2014. "Rare Earths Supply Chains: Current Status, Constraints and Opportunities." *Resources Policy* 41: 52–59.

Grotius, Hugo. [1609] 2009. *Mare Liberum, 1609-2009.* Leiden, The Netherlands: Brill.

Hagerty, Curry. 2014. *Legislation to Approve the Transboundary Hydrocarbons Agreement.* Congressional Research Service. https://fas.org/sgp/crs/row/R43610.pdf

Hagerty, Curry, and James Uzel. 2013. *Proposed US-Mexico Transboundary Hydrocarbons Agreement: Background and Issues for Congress.* Congressional Research Service. https://fas.org/sgp/crs/row/R43204.pdf

Harrison, Rowland. 2017. "Article 82 of UNCLOS: The Day of Reckoning Approaches." *The Journal of World Energy Law & Business* 10 (6): 488–504.

Havice, Elizabeth. 2018. "Unsettled Sovereignty and the Sea: Mobilities and More-than-Territorial Configurations of State Power." *Annals of the American Association of Geographers* 108 (5): 1280–1297.

Havice, Elizabeth, and Anna Zalik. 2018. "Ocean Frontiers: Epistemologies, Jurisdictions, Commodifications." *International Social Science Journal* 68: 219–235.

Hayashi, Moritaka. 1989. "Registration of the First Group of Pioneer Investors by the Preparatory Commission for the International Sea-Bed Authority and for the International Tribunal for the Law of the Sea." *Ocean Development & International Law* 20 (1): 1–33.

Heaton, S. Warren. Jr.. 2013. "Mexico's Attempt to Extend Its Continental Shelf Beyond 200 Nautical Miles Serves as a Model for the International Community." *Mexican Law Review* 5 (2): 433–450.

Hernández Cervantes, Aleida. 2014. *La Producción Jurídica de la Globalización Económica. Notas de una Pluralidad Jurídica Transnacional.* México: Universidad Autónoma de San Luis Potosí-CEIICH-UNAM.

Hernández Cervantes, Aleida, and Anna Zalik. 2018. "Canadian Capital and the Denationalization of the Mexican Energy Sector." *Journal of Latin American Geography* 17 (3): 42–72.

Himley, Matthew. 2010. "Global Mining and the Uneasy Neoliberalization of Sustainable Development." *Sustainability* 2 (10): 3270–3290.

International Seabed Authority (ISA). 2009. *Issues Associated with the Implementation of Article 82 of the United Nations Convention on the Law of the Sea. ISA Technical Study 4.* Jamaica: Kingston.

International Seabed Authority (ISA). 2010. *A Geological Model of Polymetallic Nodule Deposits in the Clarion Clipperton Fracture Zone. ISA Technical Study 6.* Jamaica: Kingston.

International Seabed Authority (ISA). 2014. "Report and Recommendations to the Council of the International Seabed Authority Relating to an Application for the Approval of a Plan of Work for Exploration by Ocean Mineral Singapore Pte Ltd." ISBA/20/C/7. https://www.isa.org.jm/documents/isba20c7.

Johnson, Leigh. 2013. "Index Insurance and the Articulation of Risk-Bearing Subjects." *Environment and Planning A* 45 (11): 2663–2681.

Kama, Kärg. 2014. "On the Borders of the Market: EU Emissions Trading, Energy Security, and the Technopolitics of 'Carbon Leakage.'" *Geoforum* 51: 202–212.

Larson, David L. 1986. "Deep Seabed Mining: A Definition of the Problem." *Ocean Development & International Law* 17 (4): 271–308.

Lehman, Jessica. 2016. "A Sea of Potential: The Politics of Global Ocean Observations." *Political Geography* 55: 113–123.

Lodge, Michael, Kathleen Segerson, and Dale Squires. 2017. "Sharing and Preserving the Resources in the Deep Sea: Challenges for the International Seabed Authority." *International Journal of Marine and Coastal Law* 32 (3): 427–457.

Lodge, Michael, and Philomene A. Verlaan. 2018. "Deep-Sea Mining: International Regulatory Challenges and Responses." *Elements* 14 (5): 331–336.

MacKenzie, Donald. 2008. *An Engine, Not a Camera: How Financial Models Shape Markets.* Boston, MA: MIT Press.

Magnússon, Bjarni Már. 2017. "Can the United States Establish the Outer Limits of Its Extended Continental Shelf Under International Law?" *Ocean Development & International Law* 48 (1): 1–16.

Murray, Iain. 2013. *Lost at Sea.* National Center for Policy Analysis. 25 March 2013. http://www.ncpathinktank.org/pub/bg167.

Neff, Erik A. 2014. "Deepwater Transboundary Hydrocarbons: Considerations for Exploitation at the Edge of Continental Margins under the United Nations Convention on the Law of the Sea (1982) between Coastal States and the International Seabed Authority." *University of Miami International & Comparative Law Review* 22 (1): 45–76.

Nimmo, M./TOML GOLDER. 2013. *Technical Report Updated NI 43–101 Clarion Clipperton Zone Project-Pacific Ocean.* Submitted to TML, March. Nautilus Minerals. Available at http://www.nautilusminerals.com/irm/PDF/1313_0/UpdatedTechnicalReportCCZProjectPacificOcean.

Okereke, Charles. 2008. "Equity Norms in Global Environmental Governance." *Global Environmental Politics* 8 (3): 25–50.

Patrick, Stewart M. 2012. *Everyone Agrees: Ratify the Law of the Sea.* Council on Foreign Relations. 8 June 2012. https://www.cfr.org/blog/everyone-agrees-ratify-law-sea?cid=oth_partner_site-atlantic.

Pike, Andy, and Jane Pollard. 2010. "Economic Geographies of Financialization." *Economic Geography* 86 (1): 29–51.

Raval, Anjli. 2018. "Shell Taps Its Deepwater Legacy to Fund Its Future." *Financial Times*, October 30. https://www.ft.com/content/66110162-dae2-11e8-8f50-cbae5495d92b.

Roth, Richard, and Carlos Munoz Royo. 2018. *Update on Financial Payment Systems: Seabed Mining for Polymetallic Nodules.* International Seabed Authority Council Meetings, July 16. https://www.isa.org.jm/document/mit-presentation-council-july.

S&P Global/Platts. 2017. *Mexico's Energy Transformation Takes Hold: Platts Special Report*, August. https://s3-ap-southeast-1.amazonaws.com/sp-platts/Mexico-Energy-Liberalization-Special-Report.pdf.

Saiki, Arnie. 2013. "SOPAC Expedites New Seabed Mining Legislation for Lockheed Martin." *Foreign Policy in Focus*, March 20. https://fpif.org/sopac_expedites_new_seabed_mining_legislation_for_lockheed_martin/.

Sammler, Katherine Genevieve. 2016. "The Deep Pacific: Island Governance and Seabed Mineral Development." In *Island Geographies*, edited by Elaine Stratford, 24–45. Abingdon: Routledge.

Squire, Rachel. 2016. "Immersive Terrain: The US Navy, Sealab and Cold War Undersea Geopolitics." *Area* 48 (3): 332–338.

Tladi, Dire. 2014. "The Common Heritage of Mankind and the Proposed Treaty on Biodiversity in Areas beyond National Jurisdiction: The Choice between Pragmatism and Sustainability." *Yearbook of International Environmental Law* 25 (1): 113–132.

UN. 1994. *Agreement Relating to the Implementation of Part XI of the United Nations Convention on the Law of the Sea.* United Nations. https://www.un.org/Depts/los/convention_agreements/convention_overview_part_xi.htm.

Valdivia, Gabriela. 2015. "Oil Frictions and the Subterranean Geopolitics of Energy Regionalisms." *Environment and Planning A: Economy and Space* 47 (7): 1422–1439.

Vargas, Jorge A. 1996. "The Gulf of Mexico: A Binational Lake Shared by the United States and Mexico." *The Transnational Lawyer* 9 (2): 459–488.

Yarn, Douglas. 1984. "The Transfer of Technology and UNCLOS III." *Georgia Journal of International and Comparative Law* 14 (1): 121–153.

Zalik, Anna. 2010. "Oil 'Futures': Shell's Scenarios and the Social Constitution of the Global Oil Market." *Geoforum* 41 (4): 553–564.

Zalik, Anna. 2015. "Resource Sterilization: Reserve Replacement, Financial Risk, and Environmental Review in Canada's Tar Sands." *Environment and Planning A* 47 (12): 2446–2464.

Zalik, Anna. 2018. "Mining the Seabed, Enclosing the Area: Proprietary Knowledge and the Geopolitics of the Extractive Frontier Beyond National Jurisdiction." *International Social Science Journal* 68 (229/230): 343–360.

World-making through mapping

Large-scale marine protected areas and the transformation of global oceans

Lisa M. Campbell, Noella J. Gray, Sarah Bess Jones Zigler, Leslie Acton, and Rebecca Gruby

---

World-making through mapping: large-scale marine protected areas and the transformation of global oceans

Ocean conservation is a global concern and marine protected areas (MPAs) are prominent on the ocean conservation agenda (Campbell et al. 2016). Although terrestrial protected areas (PAs) date to the late 1800s, MPAs have a more recent history. From fewer than 500 MPAs prior to 1985 (Kelleher 1999), MPAs number around 18,000 in 2020, covering almost 8 percent of global oceans and 18 percent of oceans within the 200-mi exclusive economic zone (EEZ) of coastal states (as of April 13, 2021, https://www.protectedplanet.net/marine, Protected Areas Coverage in 2021).

In this chapter, we theorize MPA expansion as a world-making project through which global oceans are transformed. World-making concerns the relationship between "representing" and "being in" the world (Goodman 1978; Hollinshead, Ateljevic, and Ali 2009; Escobar 2018) and can occur at a variety of social and geographic scales; it is additionally meaningful for our analysis given the global scale of ocean transformation (see also, Zalik, Chapter 35 in this volume). Transformation can take many forms, including materially when MPAs restrict or eliminate resource use. From a critical resource geography perspective, this represents an "unmaking" of resources with value in extraction (e.g., as fish, oil, or other commodities) and a "remaking" of resources with value in non-extraction (e.g., as carbon sinks, recreation opportunity, or political capital). We are interested in how MPA expansion—and in particular, global maps documenting that expansion—transforms "imaginative geographies" (Gregory 1995) of the oceans, shaping the way oceans and ocean resources are seen and, in turn, possibilities for governing them.

Drawing on critical cartography, we understand maps of MPA expansion as both processual (in a state of "becoming" rather than settled) and performative (both reflecting and shaping the world). Mapping is key to the world-making we describe, as global maps of MPAs, incrementally and repeatedly added to, perform progress toward "global oceans conserved." To support our argument, we organize the chapter as follows. First, we review imaginative geographies of oceans and situate our work in critical cartography. Second, we draw on our research on the global expansion of MPAs generally and emergence of large scale MPAs (LSMPAs) specifically to illustrate the processual and performative nature of mapping. We analyze the work done by maps of progress toward "global oceans conserved" and we look for points of "slippage" (Gibson-Graham 2008), where alternative ocean worlds are possible, by making our own maps.[1]

Imagining oceans, governing oceans

In *The Social Construction of the Ocean*, Steinberg (2001) explores the link between how humans imagine and govern oceans. A dominant western construction of oceans as "unpeopled" spaces of nature supports prioritizing the free, frictionless, and unfettered movement (of military, capital, and communications) across a smooth surface. Oceans have also been constructed as an inexhaustible source of resources and sink for pollution, an imaginative geography that suggests governance is unnecessary (Gray 2018). These long-standing constructions increasingly compete with alternatives that "people" oceans in various ways. For example, the "middle passage"—the transatlantic route followed by ships carrying enslaved Africans to the Americas—invokes a geography of the Atlantic Ocean that is "only thinkable through … traces of extreme violence and loss" (Lehman 2018, 298) and that, in turn, directs efforts to commemorate it. Contemporary discourses of Blue Economy position oceans as a frontier for capitalist expansion (Havice and Zalik 2018) and also as spaces where small-scale fishers make their living and where Blue Justice is needed (Silver et al. 2015;Voyer et al. 2018). In contrast, MPA proponents concerned about biodiversity invoke an "overuse narrative" (Steinberg 2008) that describes oceans as in crisis and need of intervention. This narrative has recently been extended to the high seas, as conservationists concerned with "governance gaps" construct an imaginative geography that recognizes remote open water spaces as amenable to enclosure for conservation (Gray 2018).

Since many people are "underexposed" to oceans (Steinberg 2008), with limited or no material interactions, imaginative geographies are particularly important to how people understand ocean worlds. Although discourses (e.g., Blue Economy), narratives (e.g., of overuse), and metaphors (e.g., sources and sinks) all inform and reflect imaginative geographies of oceans and ocean resources, in this chapter, we focus on the world-making accomplished through mapping. From a starting point in 2006, maps of global oceans have changed incrementally with dramatic results. What was once mapped as a vast, undifferentiated "blank" space has been disrupted by more, and especially larger, LSMPA polygons (see Figures 36.1 and 36.2). Maps "stimulate the imagination of their audiences" (Caquard 2011, 136) and inform how society conceptualizes human-ocean relations (St Martin 2001; Steinberg 2009; Boucquey et al. 2019). Again, "underexposure" to oceans means that maps and visualizations are central to how oceans are seen and known and geographic imaginaries formed.

Seeing and knowing are always partial, and mapmakers choose what to represent on their maps (Bryan, Chapter 37 in this volume). Prior to the critical turn in cartography, mapmakers grappled with the implications of choice but treated it as a technical necessity that could be made objectively. Beginning in the 1980s, geographers began to question the "truth value" of maps as objects (Kitchin and Dodge 2007) and to treat choice as reflecting the values and biases of mapmakers. Thus, critical cartography sought to reveal the "hidden, and sometimes hideous, narratives and agendas embedded in maps" (Caquard 2011, 136), or their "second text" (Kitchin and Dodge 2007). The point in identifying hidden agendas was to circumvent or oppose them and counter-mapping emerged as one way of doing so. In short, cartographers recognized maps as political.

The critical approach broadened beyond revealing the politics of maps to consider maps as both performative and processual (Caquard and Cartwright 2014). A performative understanding of maps attends to the "work that maps do, how they act and shape our understanding of the world" (Pickles 2004, 12). In representing, maps simultaneously "provide the very conditions of possibility for the worlds we inhabit and the subjects we become" (Pickles 2004, 5). A processual understanding of maps shifts the focus from "ontology (what things are) to ontogenetic (how things become)" (Caquard 2015, 229). Maps become through data, knowledge, and practices that combine to coproduce reality (Crampton 2009), such that maps are not so much copies of the world but achievements (compare to Hollinshead, Ateljevic, and Ali 2009). If early critical work deconstructed and exposed the "second text" of maps, this phase of critical cartography situated maps as a "compelling form of storytelling" (Caquard 2011, 136). It is in the use of the map, the repetition of the map's story in a "particular setting for a particular purpose" (Caquard 2015, 229) that world-making can occur. Maps are made not just by the mapmaker but every time someone engages with them (Rossetto 2012).

Drawing on data, knowledge, and practices arising from our research on the human dimensions of LSMPAs (Gruby et al. 2016, 2017), we engage with maps of progress toward "global oceans conserved" in two ways. First, we engage as map-readers, identifying the dominant story of progress toward "global oceans conserved" and contextualizing it in the politics of global marine conservation. Second, we engage as mapmakers. We look for points of slippage and ontogenetic possibility by remapping LSMPAs to tell alternative stories about the transformation of global oceans and ocean resources. Our goal is to think through the role of mapping in the LSMPA phenomenon and, in doing so, bring critical cartographic theory to bear on questions of the making and unmaking of ocean spaces and resources.

Performing progress toward "global oceans conserved"

MPAs are central to global conservation efforts, as reflected in global targets for biodiversity conservation. The Convention on Biological Diversity's (CBD) Aichi Target 11 encourages expansion of PAs, and its marine component envisions that, by 2020, 10 percent of "coastal and marine areas, especially areas of particular importance for biodiversity and ecosystem services, are conserved through … protected areas and other effective area based conservation measures…" (CBD 2010). Among the CBD's 20 Aichi Targets, the MPA component of Target 11, specifically its area coverage component, is one of the few likely to be met on schedule (UNEP-WCMC, IUCN, and NGS 2018; Campbell and Gray 2019).

Progress toward Target 11 in part reflects opportunity. When the CBD adopted its first MPA target in 2006, MPAs covered less than one percent of oceans, leaving ample room for expansion. Although MPAs have increased in number, most MPAs are coastal and small (almost two-thirds are less than 5 km^2); we exclude them from our global maps as most cannot be seen at that scale. In contrast, LSMPAs are defined as larger than 100,000 km^2. Three LSMPAs were established prior to the 2006 target (Figure 36.1), but LSPMAs emerged as a phenomenon beginning in 2006 when the United States declared the Papahānaumokuākea US Marine National Monument and Kiribati announced its intention to create an LSMPA in 10 percent of its EEZ, the 12th largest EEZ in the world. LSMPAs have proliferated since and as of this writing, there are 37 LSMPAs (Figure 36.2), the majority in EEZs of small islands. More than half of LSMPAs are larger than 250,000 km^2 and five are larger than 1,000,000 km^2. Whether or not individual LSMPAs are established because of CBD Aichi Target 11, the LSMPA phenomenon will underwrite any success in reaching it. The 20 largest LSMPAs account for more than 65 percent of global MPA coverage (as of April 17, 2020, https://www.protectedplanet.net/marine, Size Distribution)

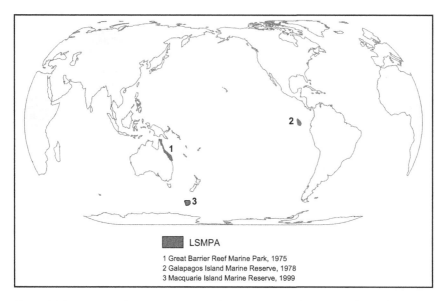

Figure 36.1 Large-scale marine protected areas prior to 2006. Map by authors.

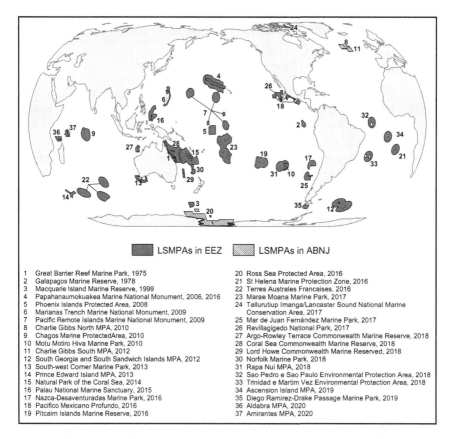

Figure 36.2 Large-scale marine protected areas as of March 2020. Map by authors.

The global expansion of LSMPAs is a world-making project, supported by and illustrative of processual and performative mapping. In moving from Figures 36.1 to 36.2 over an 14-year period, the map of global oceans has been transformed from an undifferentiated blank space to one marked with LSMPA polygons, each representing a space of conservation in which resources once valued for extraction are reimagined and remade as resources valued for non-extraction (e.g., coral reefs supporting fisheries are transformed into tourist attractions). While this transformation occurs with MPAs generally, the scope of transformation via LSMPAs is unprecedented with entire EEZs and their resources remade. Although individual LSMPAs have their own stories, we argue that the power of the world-making project is in the collection, the iterative expansion of LSMPAs and maps of LSMPAs through repetition. As each additional LSMPA moves Parties to the CBD closer to achieving CBD Aichi Target 11, global maps of LSMPAs inspire the creation of more LSMPAs. Figure 36.2 reflects incremental, cumulative progress by countries toward shared goals as negotiated in multilateral forums like the CBD, which in turn reinforces the relevance of those forums and targets-based governance. In performing progress toward "global oceans conserved," the map does work for global environmental governance.

Our maps are made with data from the World Database on Protected Areas (WDPA), the official dataset the CBD uses to measure progress toward Aichi Target 11. Figure 36.2 is our map of progress toward "global oceans conserved," but there are others. The Atlas of Marine Protection (hereafter MPA Atlas) starts with WDPA data and replaces or updates MPA records using national or regional databases. It also distinguishes among LSMPAs according to level of protection (ranging from strictly protected to multiple-use areas) and legal status (whether the site is declared, designated, or implemented) (as of April 13, 2021, http://www.mpatlas.org/. The WDPA produces its own map that includes all MPAs, large and small (as of April 21, 2020, https://www.protectedplanet.net/MPA_Map.pdf).

The differences among these maps arise from choices made by mapmakers. For example, we include the full extent of the Marae Moana Marine Park (Cook Islands) as registered in the WDPA; the MPA Atlas does not. We exclude the announced by undesignated Kermadec Islands Ocean Sanctuary (New Zealand): the MPA Atlas also excludes it, but the WDPA includes it. The implications of our choices will be discussed further, but in spite of the differences among Figure 36.2, the MPA Atlas map, and the WDPA map, all three perform progress toward "global oceans conserved."

As LSMPAs have expanded so too has debate about their ecological, economic, political, and social costs and benefits. Debate is captured in numerous scientific articles and the popular press. While we review and contribute to this debate elsewhere (Artis et al. 2020, Gruby et al. 2017), here we are interested in two features of the relationship between the debate about LSMPAs and LSMPA expansion. First, regardless of the merits of different points in the debate, the debate's existence, extent, and nature reflect the significance of LSMPAs as a world-making project. LSMPAs have expanded at a fast rate and over a large extent, with each successive "largest" LSMPA announcement garnering mainstream press coverage. This attention draws a response, both from those supportive of LSMPAs and concerned that they are being oversold as the solution for ocean conservation. Second, maps performing progress toward "global oceans conserved" (Figure 36.2) motivate and simultaneously obscure points in the debate. For example, motivated by concerns that many LSMPAs allow resource extraction (Sala et al. 2018), the MPA Atlas map differentiates levels of protection and legal status in order to accurately describe protection "on the water." The map can be read as a "check" on the story of progress. However, the map simultaneously reinforces the sense of progress: distinguishing among different levels of protection implies differentiated conservation action and suggests that further progress is possible as LSMPA status can be updated when new implementation measures are enacted. The map changes and evolves as progress continues.

With the dominant performance of progress toward "global oceans conserved" in mind, we turn to four alternative maps. We made two of these: a map of LSMPAs in small island developing states (SIDS) (Figure 36.3) and a map of countries with the most LSMPA coverage (Figure 36.4). Two maps were made by others: a map of LSMPAs included in the Pew Bertarelli Ocean Legacy project (Figure 36.5, replicated by us) and a map of LSMPAs that are members of Big Ocean Network (Figure 36.6, reproduced with permission). Each alternative counters or modifies the dominant performance of progress toward "global oceans conserved."

LSMPAs in Small Island Developing States (SIDS)

We are interested in how SIDS, motivated by the inequitable distribution of costs and benefits of global conservation initiatives, have responded to increased interest in ocean conservation. SIDS (the group designation of 38 island nations within the United Nations [UN]) have emerged as important actors in global forums with ocean interests, positioned as "ripe for developing, testing, and implementing new or expanded governance arrangements for ocean conservation" (Silver and Campbell 2018, 241). Relative to their position in the international political economy, SIDS play an outsized role in climate change diplomacy in part by exercising "moral leadership" (Betzold 2010). For marine biodiversity, Pacific SIDS have positioned themselves as leaders within the CBD and leverage this position to access resources and exert influence (Gruby and Campbell 2013). Establishing LSMPAs is well aligned with these efforts, and with work by several SIDS to rebrand SIDS as "Large Ocean States," a challenge to institutionalized discourses that emphasize their small size and vulnerability (Chan 2018; Mawyer and Jacka 2018; Silver and Campbell 2018).

Maps of LSMPAs make islands visible collectively by directing attention to oceans and individually by making islands too small to see on a global map visible with their large EEZs filled in to highlight an LSMPA. LSMPA maps thus reinforce the Large Ocean State discourse, particularly when based on a "realist conception of power in international politics" as linked to territorial size (Chan 2018, 540). LSMPA maps project sovereignty over state space, and some states may realize increased sovereignty by using the LSMPA to protect biodiversity and/or claim benefits from resource use, e.g., in fisheries, tourism, or seabed mining. Although some LSMPAs are EEZ-wide "no-take" zones where resource extraction is prohibited, others include no-take LSMPAs within part of the EEZ or have designed an EEZ-wide LSMPA as multi-use. Palau, for example, has divided its EEZ into three zones, including the no-take Palau National Marine Sanctuary and an area for domestic fisheries intended to reclaim sovereignty over tuna resources (Gruby et al. 2017). Sovereignty is further strengthened when visibility is enhanced via remote monitoring and surveillance through, for example, the Project Eyes on the Seas program, a partnership between The Pew Charitable Trusts and Satellite Applications Catapult. Made available to help monitor and enforce LSMPAs specifically, these technologies can help SIDS overcome long-standing challenges associated with exercising sovereignty more generally (Chan 2018).

The global LSMPA map can support a story of increased island visibility and sovereignty and SIDS as leaders in ocean conservation. However, only three countries in the UN SIDS grouping—Kiribati, Palau, and Seychelles—have implemented LSMPAs (Figure 36.3). Most LSMPAs have been implemented by former colonial states in what are now territories, dependencies, and protectorates of those states. We mapped Figure 36.3 to think through the relationship between LSMPAs and sovereignty as well as potential sovereignty gains that LSMPAs present for non-sovereign territories. For example, we include Cook Islands in Figure 36.3 even though it is not in the UN SIDS group due to its "free association" with New Zealand. However, some UN

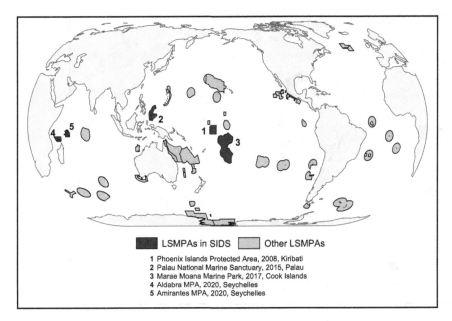

Figure 36.3 Large-scale marine protected areas in Small Island Developing States. Map by authors.

agencies recognize it as a SIDS. Further, the country is self-governed, a Cook Islander was the original LSMPA proponent, and a notion of Cook Islands' nationalism was mobilized through the LSMPA project (Durbin 2018). There may be islands not fully sovereign that see LSMPAs as a means to bring attention to or facilitate governance ambitions. However, the global map of LSMPAs obscures the complex and heterogeneous sovereign status of many islands and the historical context from which sovereign status has emerged. States are not named in many global LSMPA maps (including Figure 36.2). In mapping Figure 36.3, we attempt to balance a story that acknowledges SIDS and island agency while revealing that most LSMPAs are not in SIDS. The question is not just whether and how LSMPAs enhance sovereignty, but sovereignty for whom? We explore this question in relation to contested state sovereignty in the Map 36.4.

LMPAs as an extension of neocolonial conservation

Figure 36.4 maps LSMPAs according to countries that have the most MPAs coverage: the United States, the United Kingdom (UK), France, Cook Islands, Australia, and Chile. With the exception of Australia and Cook Islands, these countries have achieved this coverage in large part via LSM-PAs established in former colonies now governed as territories, dependencies, and protectorates, and often thousands of miles distant from the administrative center of sovereign power. This global LSMPA map visualizes the ongoing influence of a handful of wealthy and powerful nations that "rule the waves" militarily and economically. Figure 36.4 tells a story of a neocolonial ocean conservation project in which wealthy countries that have benefited from extractive activities in global oceans meet international conservation targets by establishing LSMPAs around distant oceanic islands. In doing so, these countries strengthen their own moral leadership in international forums like the CBD, where they can leverage MPA accomplishments in pursuit of other goals.

In revealing a neocolonial ocean conservation project, we draw attention to the role of LSM-PAs in asserting sovereignty in places where it may be contested. The most egregious example is

431

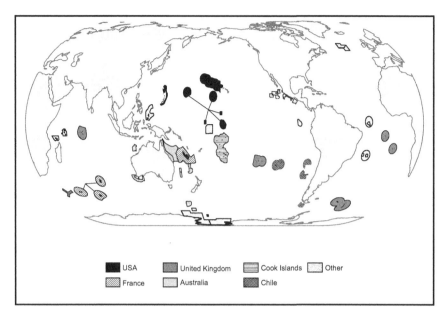

Figure 36.4 Nations with most large-scale marine protected areas coverage. Map by authors.

the UK government's establishment of a no-take LSMPA in the British Indian Ocean Territory (BIOT), or Chagos Archipelago, in 2010. The Chagos Archipelago was excised in 1965 from the former British colony of Mauritius. Mauritius and displaced native Chagossians dispute UK sovereignty (Sand 2012) and the United Kingdom was widely criticized for using the LSMPA to reinforce territorial claims (De Santo 2020). Although the UN High Court ruled in favor of Mauritius' territorial claim in February 2019 and the UN General Assembly supported the ruling in May 2019, the United Kingdom has refused to abide by the decision. Other LSMPA designations in UK overseas territories have been less politically fraught; e.g., Pitcairn Islands Marine Reserve reportedly enjoys unanimous support among inhabitants (Alger and Dauvergne 2017). Regardless, it is clear that LSMPAs are a priority of and initiated by the UK government; the UK government's Blue Belt Program aims to establish MPAs in 4 million km² of oceans in UK overseas territories. As of this writing, it has reached 3.6 million km² (as of April 17, 2020, https://www.protectedplanet.net/marine, Coverage of National Waters).

The United Kingdom has also pursued marine conservation in domestic waters but via a decentralized and participatory approach to siting, establishing, and regulating marine conservation zones (MCZs). As of this writing, 211,000 km² were protected or 29 percent of the domestic EEZ. However, less than 0.001 percent of MCZs are strongly or fully protected versus at least 50 percent of LSMPAs in overseas territories. Similar trends are seen in other countries with LSMPAs in island dependencies, with more and more fully protected MPAs than in domestic waters. Although the United States is an exception with almost equal percentage protected in domestic waters and overseas, this is only because Papahānaumokuākea is in Hawai'i, a state within the United States. Without Papahānaumokuākea, the United States has less than one percent of its domestic waters in strong or full protection where most forms of resource extraction are prohibited (O'Leary et al. 2016).

However, a map that reveals LSMPAs as a neocolonial conservation project can miss the story of how Indigenous peoples in particular places have mobilized in response to LSMPAs. For example, native Hawaiians have been involved with the establishment and ongoing management of Papahānaumokuākea US Marine National Monument, which, according to Kikiloi et al. (2017), increasingly reflects native Hawaiians' priorities and values, enhances cultural connection, and facilitates cultural practice. In Rapa Nui (Easter Island), Chile, a contested and initially top-down LSMPA process provided an opening to negotiate the relationship between the Chilean state and Rapa Nui as a "special territory" of Chile (Gruby et al. 2017). Through resistance by Rapanui people and subsequent consultations, an initial vision for a no-take marine park transitioned to a multiuse MPA that will be coadministered by a Sea Council with majority Indigenous Rapanui members (Zigler 2020). These examples illustrate that LSMPAs can provide opportunities for Indigenous peoples to strengthen connections or claims to ocean spaces. They temper the story of LSMPAs as neocolonial conservation and highlight sovereignty as a concern of Indigenous peoples within states. But the possibilities should not be overstated. In both Hawai'i and Rapa Nui, Indigenous responses to LSMPAs are varied. For example, some Rapanui assert that an LSMPA established without Chilean recognition of Rapanui sovereignty over land and sea is an extension of neocolonial power (Zigler 2020). A global map revealing LSMPAs either as neocolonial control or Indigenous empowerment fails to capture place-based politics at particular sites.

LSMPAs as NGO accomplishment

Environmental nongovernment organizations (NGOs) have played a central role in LSMPA expansion and "the emergence of a large MPA norm" (Alger and Dauvergne 2017, 31). National Geographic has led scientific expeditions to support the establishment of LSMPAs in remote areas through its Pristine Seas project. Conservation International supported the establishment of Kiribati's Phoenix Islands Protected Area. The Nature Conservancy innovated a debt for "blue" nature swap to support marine spatial planning throughout Seychelles' EEZ, which resulted in the recent designation of two LSMPAs. The Pew Charitable Trusts' Pew Bertarelli Ocean Legacy project (hereafter Pew) aims to establish 15 fully protected marine parks of 200,000 km^2 or larger by 2022. We focus on Pew in this alternative map because it was the main NGO involved in four of five LSMPA sites we have studied (Gruby et al. 2017), and because it updates its own map of progress (as of April 21, 2020, https://www.pewtrusts.org/en/projects/pew-bertarelli-ocean-legacy, Where We Work).

We replicate Pew's map as at April 21, 2020 (Figure 36.5). Although the scale is global, the map omits LSMPAs where Pew has not campaigned and sites where Pew has campaigned but LSMPAs were not established. Sites where LSMPAs are established and Pew campaigns complete appear as polygons. Dots denote active campaigns, even if LSMPAs are already established; these are sites where Pew is campaigning for enhanced protection. Overall, the map performs progress toward "global oceans conserved," but this time emphasizing Pew's work. More generally, the map performs NGOs as legitimate and effective actors in global ocean conservation.

Again, this map obscures as much as it reveals. The role of NGOs in establishing particular LSMPAs can be contentious. Opponents of LSMPAs mobilized concerns about Pew as an "outside" and "foreign" organization to successfully resist an LSMPA in Bermuda (Gruby et al. 2017) and unsuccessfully resist one in CNMI (Richmond et al. 2019). In Kiribati, views on Conservation International's role in the Phoenix Islands Protected Area are mixed (Mitchell 2017). Alternatively, Alger and Dauvergne (2017) credit Pew with its investment in Pitcairn Islands, where the head of the campaign knocked on "every door" to garner unanimous support for the LSMPA. However, concerns about NGO involvement are also broader than individual campaigns. Even among LSMPA advocates, the role of NGOs and NGO claims about that role are sometimes contested. For example, Pew lists on its website LSMPAs "that governments have already designated as a result of ***our work*** [emphasis added]" (https://www.pewtrusts.org/nb/projects/pew-bertarelli-ocean-legacy/where-we-work, as of April 17, 2020). Pew, like all

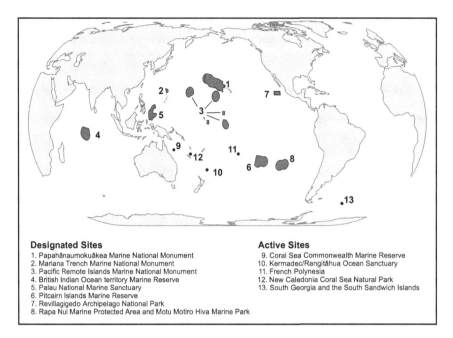

Designated Sites
1. Papahānaumokuākea Marine National Monument
2. Mariana Trench Marine National Monument
3. Pacific Remote Islands Marine National Monument
4. British Indian Ocean territory Marine Reserve
5. Palau National Marine Sanctuary
6. Pitcairn Islands Marine Reserve
7. Revillagigedo Archipelago National Park
8. Rapa Nui Marine Protected Area and Motu Motiro Hiva Marine Park

Active Sites
9. Coral Sea Commonwealth Marine Reserve
10. Kermadec/Rangitāhua Ocean Sanctuary
11. French Polynesia
12. New Caledonia Coral Sea Natural Park
13. South Georgia and the South Sandwich Islands

Figure 36.5 Pew Bertarelli Ocean Legacy campaign sites. Map by authors.

NGOs, needs to demonstrate effectiveness to retain support from funders; this is their reality (Jepson 2005). However, in our research, some national and local actors complained that their own efforts to secure LSMPAs are overshadowed or undermined by NGO claims.

These complaints reflect more than interpersonal politics or resentment about who gets credit for what. They speak to concerns about sovereignty. Occurring at a time when many governments are in retreat from large projects to govern public resources, LSMPA expansion has only been possible through support by mostly US-based NGOs and philanthropies. NGOs and their funders are negotiating "hybrid forms of governance" (Campbell et al. 2016, 531), private-public ventures often labeled as partnerships (Chan 2018), which bring funding and technology to support the drive to designate, implement, and monitor LSMPAs. In return, NGOs are able to demonstrate their effectiveness in securing ocean conservation, but they also sometimes gain authority over ocean space. In Kiribati, Conservation International and the New England Aquarium have seats on the board of directors for the Phoenix Islands Protected Area Conservation Trust (Mawyer and Jacka 2018). In Seychelles, The Nature Conservancy has a seat on the Seychelles Conservation and Climate Adaptation Trust (SeyCCAT) (Silver and Campbell 2018). These groups decide how finances will be allocated and to what kinds of activities, and thus influence state decision-making. Maps detailing NGO accomplishment only hint at the complexities in this kind "articulated" sovereignty (Lunstrum 2013).

LSMPAs as spaces of management

The final global LSMPA map (Figure 36.6) is made by Big Ocean, a peer learning network "created by managers for managers," to support information exchange and communicate the value of LSMPAs to the broader conservation community (https://bigoceanmanagers.org/, as of April 17, 2020). Big Ocean has published a research agenda for LSMPAs (Wagner 2013) and cohosted a workshop on the human dimensions of LSMPAs (Christie et al. 2017). In partnership with the

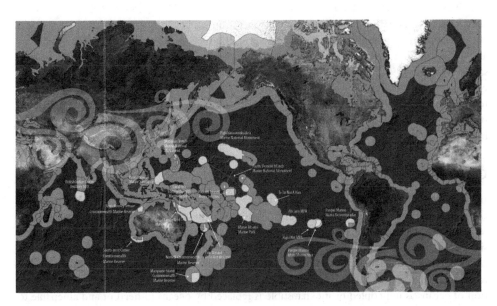

Figure 36.6 Big Ocean network of large-scale marine protected areas.

Source: Big Ocean.

IUCN World Commission on Protected Areas Large-Scale MPA Task Force, it published a guide for design and management of LSMPAs (Lewis et al. 2017). In emphasizing LSMPAs as spaces of management, Big Ocean pushes back on critiques that LSMPAs are "paper parks." Big Ocean describes LSMPAs as the "greatest hope for achieving marine conservation goals such as the Convention on Biodiversity's Aichi Target 11" and the leverage this affords, particularly when LSMPA managers act together: "The size of LSMPAs accentuates their inter-governmental and global significance; they can often affect international marine policies in ways that smaller scale MPAs cannot" (https://bigoceanmangers.org/impact, as of April 17, 2020).

Figure 36.6 maps Big Ocean members. Members are marked as belonging to the network with infill (bright yellow in the color version, https://bigoceanmanagers.org/big-ocean-site-map, as of June 30, 2020) but also as distinct, with LSMPA names in the language of the site included on the map. The work of Big Ocean "enlivens" the LSMPA seascape as a social space, where managers cooperate, collaborate, and learn together. Relations among managers are simultaneously relations among states (LSMPA managers are representatives of national government agencies) and among island people linked through shared ocean values and sometimes histories. Although not exclusive to the Pacific Ocean and not all LSMPA managers are Pacific Islanders, the map with its artwork suggestive of wind or waves invokes a sense of Pacific Islands' identity.

We engage Figure 36.6 informed by our broader understanding of LSMPAs and of Big Ocean. We co-organized the human dimensions' workshop referenced earlier and have presented our research results to Big Ocean. But our engagement is also informed by anthropologist Epeli Hau'ofa's (1994, 1998) essays on the need for a regional Oceanic identity that recognizes mobility and connection and is centered on the sea. He preferred the term Oceania to the more common label of Pacific Islands, a label that for him was connected "to an official world of states and nationalities" (Hau'ofa 1998, 402) and their interests. In contrast, "Oceania is a world of social networks that crisscross the ocean…. It is a world that we have created largely through our own efforts" (Hau'ofa 1998, 392). As we engage with Figure 36.6, we envision

LSMPA managers (rather than states) moving and connecting across space, focused on a shared commitment to safeguarding the oceans. However, Big Ocean also reflects the challenges Oceania faces within a political economic order based on "neocolonial dependency" and "patron states" (Hau'ofa 1998, 399). Many Big Ocean member sites are in island dependencies of former colonial powers, and Big Ocean has largely been funded by the US government. We read these tensions in the map, one that both reinforces territories of nation states while invoking an alternative space of oceans as connected, mobile, and dynamic, a "sea of islands" rather than "islands in the sea" (Hau'ofa 1994).

Conclusions

In writing about US President George W. Bush's decision in 2006 to establish what was then the largest MPA on the planet, Steinberg (2008, 2083) suggested there was "ample evidence to question the long-range social significance" of that action. With hindsight, we see the social significance in the repeated performance (37 times and counting) of LSMPAs. The proliferation of LSMPAs is a world-making project and mapping is key to that project. The map of progress toward global oceans conserved supports a new imaginative geography in which the construction of oceans as unpeopled or inexhaustible is replaced by oceans in need of and amenable to area-based conservation (see also Gray 2018). The map reinforces existing ocean governance by highlighting nation states and their EEZs. It also promotes the role of multilateral cooperation generally, and the CBD specifically; the value of targets-based governance generally, and Aichi Target 11 specifically. It centers islands, and particularly Pacific islands, as key to these efforts and makes them newly visible, both as Large Ocean States and as important players in global marine conservation efforts. Performance and process interact; each LSMPA inspires the next, as the global cache increases and achieving Aichi Target 11 comes closer to reality. Maps document LSMPA expansion and motivate LSMPA expansion; they are products of ocean conservation and productive of it.

Whether or not CBD Aichi Target 11 is reached by the 2020 deadline, the progress visualized in the map will be celebrated. But it will also be challenged, and an important challenge will come from conservationists committed to MPA expansion. They will draw on the MPA Atlas map that differentiates LSMPAs according to level of resource protection and legal status. Their goal is for more, more strictly protected MPAs, not in 10, but 30–50 percent of the ocean (Campbell and Gray 2019). Although they acknowledge the importance of LSMPAs in reaching Aichi Target 11 and in bringing attention to marine conservation, some conservationists are concerned that without remaking resources such that value accrues from non-extractive use, LSMPAs may be "paper parks" and their collection "just" a mapping project. We suggest these concerns reflect an understanding that a mapping project is never "just" that. Maps of progress toward "global oceans conserved" underwrite a new imaginative geography of oceans as increasingly conserved, and the power of that imaginative geography may undermine claims that further conservation is needed.

A processual understanding of mapping in critical cartography attends to the ontogenetic, i.e., how things become. The dominant global LSMPA map *becomes* though individual countries taking actions that, collectively, are counted toward a shared goal. The incremental addition of each new (and often larger) LSMPA to the map contributes to the performance of progress and inspires further action. We have also engaged in a map-making process to look for slippage in the dominant story of LSMPAs and propose alternative or modified stories. Our maps emerge (or become) from our research on the human dimensions of LSMPAs and our interests in the politics of global marine conservation. These interests guided some technical choices in the mapping process, for example, our decision to categorize Cook Islands as a SIDS rather than a nation in "free association" with New Zealand. We wanted to recognize Cook Islands agency and sovereignty in Figure 36.3 more than we wanted to strengthen our critique of neocolonial oceans in Figure 36.4. By engaging as mapmakers, we gained new insights into research we have been conducting for over a decade; the maps we made sometimes challenged our assumptions about the LSMPA phenomenon, giving us direct experience with the power of mapping. However, we do not imply that our maps have power equal to those produced by NGOs, multilateral organizations, or management agencies. The world-making effects of our alternatives, published in a book for an academic audience of resource geographers, will be limited if they go no further.

Critical cartography directed us to reflect on our own mapping process. We mapped alternatives and then interrogated them by, for example, pointing out the ways our map of neocolonial oceans (Figure 36.4) obscures relations among particular LSMPAs, island governments, NGOs, and people, including Indigenous people. Doing so revealed our central concern about the LSMPAs world-making project: its scale. Although "world-making" does not imply a particular geographic or social scale, it is apt here given the size of LSMPAs and the global scope of the effort to conserve oceans. "The size of LSMPAs accentuates their inter-governmental and global significance" (https://bigoceanmanagers.org/impact, as of April 17, 2020), and part of that significance is that they can be mapped and made visible at a global scale. The map of progress toward "global oceans conserved" also performs "the global" as a relevant scale where conservation can be done and outcomes and impacts assessed. But just as a map of progress toward "global oceans conserved" obscures the various points of debate about LSMPAs, all global maps of LSMPAs—including those we produced—obscure the outcomes and experiences of individual LSMPAs, nationally and locally. Individual LSMPAs are the "outcomes of particular political processes and struggles" (Gray, Gruby, and Campbell 2014, 64) and succeed or fail in particular places. Their impacts on people, resources, and the marine environment will be multiple, complex, uneven, interacting and evolving, and place and context specific. A global map of LSMPAs cannot be the only way we understand LSMPAs as a world-making project. The world enacted in the global map has consequences in particular places.

Note

1   Our maps were made by cartographer Marie Puddister, Dept. of Geography, Environment & Geomatics, University of Guelph.

References

Alger, Justin, and Peter Dauvergne. 2017. "The Politics of Pacific Ocean Conservation: Lessons from the Pitcairn Islands Marine Reserve." *Pacific Affairs* 90 (1): 29–50. doi: 10.5509/201790129.

Artis, Evan, Noella J. Gray, Lisa M. Campbell, Rebecca L. Gruby, Leslie Acton, Sarah Bess Zigler, and Lillian Mitchell. 2020. "Stakeholder Perspectives on Large-Scale Marine Protected Areas." *PLoS ONE* 15 (9): e0238574. doi: 10.1371/journal.pone.0238574.

Betzold, Carola. 2010. "'Borrowing' Power to Influence International Negotiations: AOSIS in the Climate Change Regime, 1990–1997." *Politics* 30 (3): 131–148.

Boucquey, Noëlle, Kevin St Martin, Luke Fairbanks, Lisa M. Campbell, and Sarah Wise. 2019. "Ocean Data Portals: Performing a New Infrastructure for Ocean Governance." *Environment and Planning D: Society and Space* 37 (3): 484–503. doi: 10.1177/0263775818822829.

Campbell, Lisa M., Noella J. Gray, Luke Fairbanks, Jennifer J. Silver, Rebecca L. Gruby, Bradford A. Dubik, and Xavier Basurto. 2016. "Global Oceans Governance: New and Emerging Issues." *Annual Review of Environment and Resources* 41: 517–543. doi: 10.1146/annurev-environ-102014-021121.

Campbell, Lisa M., and Noella J. Gray. 2019. "Area Expansion Versus Effective and Equitable Management in International Marine Protected Areas Goals and Targets." *Marine Policy* 100: 192–199. doi: 10.1016/j.marpol.2018.11.030.

Caquard, Sébastien. 2011. "Cartography I: Mapping Narrative Cartography." *Progress in Human Geography* 37 (1): 135–144. doi: 10.1177/0309132511423796.

Caquard, Sébastien. 2015. "Cartography III: A Post-Representational Perspective on Cognitive Cartography." *Progress in Human Geography* 39 (2): 225–235. doi: 10.1177/0309132514527039.

Caquard, Sébastien, and William Cartwright. 2014. "Narrative Cartography: From Mapping Stories to the Narrative of Maps and Mapping." *The Cartographic Journal: Cartography and Narratives* 51 (2): 101–106. doi: 10.1179/0008704114Z.000000000130.

CBD. 2010. *Decisions Adopted by the Conference of the Parties to the Convention on Biological Diversity at Its Tenth Meeting (Decision X/2, Annex IV).* Nagoya: Japan: Convention on Biological Diversity.

Chan, Nicholas. 2018. "Large Ocean States': Sovereignty, Small Islands, and Marine Protected Areas in Global Oceans Governance." *Global Governance* 24 (4): 537–555. doi: 10.1163/19426720-02404005.

Christie, Patrick, Nathan J. Bennett, Noella J. Gray, T. 'Aulani Wilhelm, Nai'a Lewis, John Parks, Natalie C. Ban, Rebecca L. Gruby, Lindsay Gordon, Jon Day, Sue Taei, and Alan M. Friedlander. 2017. "Why People Matter in Ocean Governance: Incorporating Human Dimensions into Large-Scale Marine Protected Areas." *Marine Policy* 84: 273–284. doi: 10.1016/j.marpol.2017.08.002.

Crampton, Jeremy W. 2009. "Cartography: Performative, Participatory, Political." *Progress in Human Geography* 33 (6): 840–848. doi: 10.1177/0309132508105000.

De Santo, Elizabeth M. 2020. "Militarized Marine Protected Areas in Overseas Territories: Conserving Biodiversity, Geopolitical Positioning, and Securing Resources in the 21st Century." *Ocean and Coastal Management* 184. doi: 10.1016/j.ocecoaman.2019.105006.

Durbin, Trevor J. 2018. "'What Now, Fishgate?': Scandal, Marae Moana, and Nation Making in the Cook Islands." *Contemporary Pacific* 30 (1): 1–31. doi: 10.1353/cp.2018.0002.

Escobar, Arturo. 2018. *Designs for the Pluriverse: Radical Interdependence, Autonomy, and the Making of Worlds.* Durham, NC: Duke University Press.

Gibson-Graham, J. K. 2008. "Diverse Economies: Performative Practices for 'Other Worlds'." *Progress in Human Geography* 32 (5): 613–632. doi: 10.1177/0309132508090821.

Goodman, Nelson. 1978. *Ways of Worldmaking.* Hassocks, Sussex: Harvester.

Gray, Noella J. 2018. "Charted Waters? Tracking the Production of Conservation Territories on the High Seas." *International Social Science Journal* 68 (229–230): 257–272. doi: 10.1111/issj.12158.

Gray, Noella J., Rebecca L. Gruby, and Lisa M. Campbell. 2014. "Boundary Objects and Global Consensus: Scalar Narratives of Marine Conservation in the Convention on Biological Diversity." *Global Environmental Politics* 14 (3): 64–83.

Gregory, Derek. 1995. "Imaginative Geographies." *Progress in Human Geography* 19 (4): 447–485.

Gruby, Rebecca L., and Lisa. M Campbell. 2013. "Scalar Politics and the Region: Strategies for Transcending Pacific Island Smallness on a Global Environmental Governance Stage." *Environment and Planning A* 45 (9): 2046–2063.

Gruby, Rebecca L., Luke Fairbanks, Leslie Acton, Evan Artis, Lisa M. Campbell, Noella J. Gray, Lillian Mitchell, Sarah Bess Jones Zigler, and Katie Wilson.. 2017. "Conceptualizing Social Outcomes of Large Marine Protected Areas." *Coastal Management* 45 (6): 416–435. doi: 10.1080/08920753.2017.1373449.

Gruby, Rebecca L., Noella J. Gray, Lisa M. Campbell, and Leslie Acton. 2016. "Toward a Social Science Research Agenda for Large Marine Protected Areas." *Conservation Letters* 9 (3): 153–163. doi: 10.1111/conl.12194.

Hau'ofa, Epeli. 1994. "Our Sea of Islands." *The Contemporary Pacific* 6 (1): 147–161.

Hau'ofa, Epeli. 1998. "The Ocean in Us." *The Contemporary Pacific* 10 (2): 392–410.

Havice, Elizabeth, and Anna Zalik. 2018. "Ocean Frontiers: Epistemologies, Jurisdictions, Commodifications." *International Social Science Journal* 68 (229–230): 219–235. doi: 10.1111/issj.12198.

Hollinshead, Keith, Irena Ateljevic, and Nazia Ali. 2009. "Worldmaking Agency—Worldmaking Authority: The Sovereign Constitutive Role of Tourism." *Tourism Geographies* 11 (4): 427–443. doi: 10.1080/14616680903262562.

Jepson, Paul. 2005. "Governance and Accountability of Environmental NGOs." *Environmental Science & Policy* 8 (5): 515–524.

Kelleher, Graeme. 1999. *Guidelines for Marine Protected Areas*. Gland, CH: IUCN.

Kikiloi, Kekuewa, Alan M. Friedlander, 'Aulani Wilhelm, Nai'a Lewis, Kalani Quiocho, William 'Āila Jr, and Sol Kaho'ohalahala. 2017. "Papahānaumokuākea: Integrating Culture in the Design and Management of One of the World's Largest Marine Protected Areas." *Coastal Management* 45 (6): 436–451. doi: 10.1080/08920753.2017.1373450.

Kitchin, Rob, and Martin Dodge. 2007. "Rethinking Maps." *Progress in Human Geography* 31 (3): 331–344. doi: 10.1177/0309132507077082.

Lehman, Jessica. 2018. "Marine Cultural Heritage: Frontier or Centre?" *International Social Science Journal* 68 (229–230): 291–301. doi: 10.1111/issj.12155.

Lewis, Nai'a, Jon C. Day, Daniel Wagner, Carlos Gaymer, Alan Friedlander, John Parks, and 'Aulani Wilhelm, et al. 2017. *Large-Scale Marine Protected Areas: Guidelines for Design and Management. Vol. 26 of Best Practice Protected Areas Guidelines Series*. Gland, CH: IUCN.

Lunstrum, Elizabeth. 2013. "Articulated Sovereignty: Extending Mozambican State Power Through the Great Limpopo Transfrontier Park." *Political Geography* 36: 1–11. doi: 10.1016/j.polgeo.2013.04.003.

Mawyer, Alexander, and Jerry K. Jacka. 2018. "Sovereignty, Conservation and Island Ecological Futures." *Environmental Conservation* 45 (3): 238–251. doi: 10.1017/S037689291800019X.

Mitchell, Lillian. 2017. "Governing Large Marine Protected Areas: Insights from the Remote Phoenix Islands Protected Area." Master's Thesis, University of Guelph. http://hdl.handle.net/10214/11420.

O'Leary, Bethan C., Marit Winther-Janson, John M. Bainbridge, Jemma Aitken, Julie P. Hawkins, and Callum M. Roberts. 2016. "Effective Coverage Targets for Ocean Protection." *Conservation Letters* 9 (6): 398–404. doi: 10.1111/conl.12247.

Pickles, John. 2004. *A History of Spaces: Cartographic Reason, Mapping and the Geo-Coded World*. Oxon, UK: Routledge.

Richmond, Laurie, Rebecca L. Gruby, Dawn Kotowicz, and Robert Dumouchel. 2019. "Local Participation and Large Marine Protected Areas: Lessons from a U.S. Marine National Monument." *Journal of Environmental Management* 252: 109624. doi: 10.1016/j.jenvman.2019.109624.

Rossetto, Tania. 2012. "Embodying the Map: Tourism Practices in Berlin." *Tourist Studies* 12 (1): 28–51.

Sala, Enric, Jane Lubchenco, Kirsten Grorud-Colvert, Catherine Novelli, Callum Roberts, and U. Rashid Sumaila. 2018. "Assessing Real Progress Towards Effective Ocean Protection." *Marine Policy* 91: 11–13. doi: 10.1016/j.marpol.2018.02.004.

Sand, Peter H.. 2012. "Fortress Conservation Trumps Human Rights? The "Marine Protected Area" in the Chagos Archipelago." *Journal of Environment & Development* 21 (1): 36–39. doi: 10.1177/1070496511435666.

Silver, Jennifer J., Noella J. Gray, Lisa M. Campbell, Luke W. Fairbanks, and Rebecca L. Gruby. 2015. "Blue Economy and Competing Discourses in International Oceans Governance." *Journal of Environment & Development* 24 (2): 135–160. doi: 10.1177/1070496515580797.

Silver, Jennifer J., and Lisa M. Campbell. 2018. "Conservation, Development and the Blue Frontier: the Republic of Seychelles' Debt Restructuring for Marine Conservation and Climate Adaptation Program: Conservation, Development and the Blue Frontier." *International Social Science Journal* 68 (229–230): 241–256. doi: 10.1111/issj.12156.

St Martin, Kevin. 2001. "Making Space for Community Resource Management in Fisheries." *Annals of the Association of American Geographers* 91 (1): 122–142.

Steinberg, Philip E. 2001. *The Social Construction of the Ocean*. New York, NY: Cambridge University Press.

Steinberg, Philip E. 2008. "It's So Easy Being Green: Overuse, Underexposure, and the Marine Environmental Consensus." *Geography Compass* 2 (6): 2080–2096.

Steinberg, Philip E. 2009. "Sovereignty, Territory, and the Mapping of Mobility: A View from the Outside." *Annals of the Association of American Geographers* 99 (3): 467–495. doi: 10.1080/00045600902931702.

UNEP-WCMC, IUCN, and NGS. 2018. *Protected Planet Report 2018*. Cambridge, UK: UNEP-WCMC, IUCN and NGS.

Voyer, Michelle, Genevieve Quirk, Alistair McIlgorm, and Kamal Azmi.. 2018. "Shades of Blue: What Do Competing Interpretations of the Blue Economy Mean for Oceans Governance?" *Journal of Environmental Policy & Planning* 20 (5): 595–616. doi: 10.1080/1523908X.2018.1473153.

Wagner, Daniel, ed. 2013. "Big Ocean: A Shared Research Agenda for Large-Scale Marine Protected Areas." In *Big Ocean: Papahanaumokuakea Marine National Monument & UNESCO World Heritage Site, and the NOAA Office of National Marine Sanctuaries*. Auckland, NZ: 25th International Congress for Conservation Biology.

Zigler, Sarah Bess Jones. 2020. "MPA: Marine Protected Area or Marine Pluriverse Area? A Political Ontology of Large Marine Conservation in Rapa Nui (Eastern Island Chile)." PhD diss., Duke University.

37

Mapping resources

Mapping as method for critical resource geographies

Joe Bryan

Introduction

Mapping is often an assumed, taken-for-granted activity when it comes to understanding resources. Maps help enumerate the qualities of resources, making resources into observable facts, the distributions of which can be calculated and managed. So it is that we have maps of property rights, mining claims, oil concessions, and cultural resources, each underpinned by powerful notions of land, geology, and nature (Braun 2002). None of these things simply appears on maps, however. Each requires practices that involve the production of maps and their use. Pressed further, mapping never simply identifies resources. Mapping helps make them (Wood and Fels 2008). The production, use, and circulation of maps therefore offer a practical means to grasping critical resource geography's core claim that resources are "irreducibly social" (Bridge 2009), grasping resources in terms of how they come to be, the possibilities they generate, and the worlds they make.

This chapter takes up mapping as a method for understanding how resources are produced relationally in ways that are both historically contingent and politically contested (Braun 2002). That approach underscores this volume's approach to doing critical resource geographies, moving beyond critiques of maps as representations or instruments and pushing against the temptation to see all knowledge and experience as mappable. That assumption reinforces the idea that there is something universal about space, to say nothing of maps themselves. This chapter argues for an approach to mapping to be understood in terms of practices of making, using, and reading maps. An emphasis on practice helps separate mapping from cartographic ideals about the universality of space (Edney 2019), offering an important, first methodological step to grasping how what counts as an object and its spatial relations is specific to a particular worldview, episteme, or ontology (Turnbull 1989). In turn, that claim situates mapping as a set of practices that occurs within a broader set of social relations that determines "who speaks for whom, about what, and with what authority" (Pickles 2004, 91). Such concerns call for a methodological approach to mapping that is experimental and exploratory. While this may sound like a call to return to mapping's colonial origins, it is anything but. A renewed empiricism will not lead to better, more accurate maps. An experimental approach begins instead by taking stock of the limitations of previous approaches to mapping, critiquing the discursive foundations that claims to empiricism draw on. That effort opens up a space for critique as opposed to resolution, exploring the possibilities for transformation.

Two further points flow from that claim. The first concerns the importance of seeing mapping beyond the map and its production, examining the range of situations in which people read, use, and otherwise engage with maps. That approach highlights how dominant forms of knowledge circulate through mapping, trafficking in ideals about space and knowledge masked by cartography's claim to scientific objectivity (Bryan 2009). Turning to a broader range of practices associated with mapping not only brings questions of power into view. It also recenters attention to the relational production of people, land, and resources (Sundberg 2014). Methodologically, this approach approximates a kind of grounded deconstruction of dominant forms of knowledge (Coulthard 2014). By treating mapping as a practice, it foregrounds how it is used to bring new forms of existence to life (Blaser 2014). Who one maps with and where that mapping occurs are therefore fundamental questions. My discussion of mapping as method turns to a trio of projects that I have worked on directly in collaboration with Indigenous Mapuche, Maidu, and Zapotec communities nominally located in Chile, the United States, and Mexico. Each of these projects engages mapping as practice organized by Indigenous concepts of space in ways that decenter the idea of "resources" and even "land" (Coombes, Johnson, and Howitt 2012). Those projects all require an additional methodological commitment to mapping *with* Indigenous peoples. That effort moves beyond notions of maps as objects or representations to consider mapping as means of spatializing power relations. Mapping thus becomes a methodology for diagnosing and altering relations of power as they operate through maps and resources (see also, Campbell et al., Chapter 36 in this volume).

"Territory is not a checklist": mediating difference through resources

"What do you see?" Rubén stopped the truck abruptly, skidding the tires on the gravel road. For the last hour or so, we'd been driving toward the Mapuche community where Rubén's grew up. Our goal was to test a method that he'd designed for mapping Mapuche territory, demonstrating their existence to a perennially skeptical Chilean state. For the test, Rubén had flipped the script of his methods, making me the researcher and him the guide. I took a minute to look at our surroundings. I'd already missed the entry to the community's territory, preoccupied by taking in the cornfields and the occasional forest plantation. "I see trees. Eucalyptus, mainly. Rocks, a ravine, fallen branches…" My list trailed off as I turned back to look at Rubén. He was smiling. "I don't see any of that," he replied. "I see stories." He continued with a round of stories, each of which deftly placed the objects I had named into nodes in a much wider web of relations. "You don't see any of that," he said, "but I can, even if it can't fit on a map." He started up the truck again, and we continued on down the road toward lunch with his family. "Territory is not a *checklist*," he added, saying the last word in English for emphasis.

Rubén's last line captured a key challenge for the mapping project and for Mapuche people in general. The conflict between Mapuche communities and the Chilean state is as seemingly intractable and complex as they come. Part of the problem is that neither side sees the same thing, even as their visions collide and tangle in a given locale (Blomley 1998). Since militarily occupying the region in the mid-nineteenth century, the Chilean state has seen the region in terms of resources. At first, they saw agricultural land ripe for settlement by Chileans and European immigrants as property (Klubock 2014). Property continues to dominate how state officials see this landscape, now immeasurably changed by settlement, industrial agriculture, plantation forests, and water resources for hydroelectricity. Mapuche people see all of these things as impositions overlaid on *Wallmapu*, the territory or surroundings in which Mapuche life is situated (Cayuqueo 2017; Melin, Mansilla, and Royo 2017). Mapuche people and Chilean officials ostensibly see the same trees, hills, and fields, even as they understood them in radically different terms as captured by their radically different understandings of space. Throughout, a distinctly Chilean resource imaginary dominated, forcing the struggle time and again back to the very resources created by their dispossession of Mapuche people—property, plantation forests, land, *nature* (Di Giminiani 2018). None of these concepts had any inherent ontological stability to them. Rather they were all effects of a particular set of social forces broadly in line with what is elsewhere termed "settler colonialism" (Coulthard 2014; Simpson 2017).

The emphasis on resources sets the terms of the debate and historical understandings of Mapuche struggle. Soon after the military occupation of Mapuche territory and its reframing as the Araucania, Chilean officials oversaw the displacement of Mapuche communities from native forests and communal lands by issuing land grants, or *títulos de merced*. The titles dramatically reduced Mapuche access to land and resources, reorganizing them into settled agrarian communities that structured the terms of their existence under Chilean rule. Even those titles were not respected, and Mapuche communities continued to lose land on up through the 1960s. The tide of losses was briefly reversed during the administration of Salvador Allende from 1970 to 1973. Following the US-backed coup against Allende, General Augusto Pinochet led an agrarian counterreform that further stripped Mapuche communities of land previously titled to them (Mallon 2005). Pinochet's counterreform allowed for a new wave of land consolidation led by several of his close allies who, in turn, used the newly acquired land to expand the region's burgeoning plantation forestry industry.

Mapuche communities resisted the expansion during the Pinochet dictatorship, despite widespread government suppression. After the end of the dictatorship in 1990, Mapuche communities returned their claims to land to the public realm, citing land restitution as essential to democratization. The Chilean government refused, insisting instead on the sanctity of private property rights. The conflict quickly escalated with Mapuche communities using a full suite of legal and political tactics to demand restoration of their land rights. Chilean officials responded by prosecuting Mapuche activists for crimes against property, using anti-terrorism laws passed by the Pinochet regime. Throughout their campaign, Mapuche actions have targeted plantation forests occupying community lands by way of demanding their rights to territory.

The Chilean state has only dealt with Mapuche demands for land in a piecemeal fashion, focusing almost entirely on property (Bauer 2016). One of the state's leading efforts to date involved transferring property from willing sellers to Mapuche communities. Communities that accepted the reforms often received land that was remote and marginal for agriculture. In several instances, state officials granted communities property located at a considerable distance away from where people lived. At the same time, there has never been a single coordinated Mapuche land claim that would allow the state to deal with the issue comprehensively, despite the fact that many Mapuche communities frame their calls for land in terms of regaining control over all of *Wallmapu*, the Mapuche world.

Throughout this contentious history, resources such as land, water, and soil have been used to mediate relationships between the Mapuche, settlers, and the Chilean state. In keeping with the state's emphasis on property, discussion has focused almost entirely on ownership of land and resources. That emphasis has repeatedly worked against Mapuche conceptions of space as expansive and all-encompassing (Melin, Mansilla, and Royo 2017). In response, Mapuche activists have made the point that each of these resources refers to categories imposed by Chilean occupation (Hirt 2012; Bauer 2016). From this vantage point, property rights are at best a limited facet of a space that Mapuche people grasp in relational terms that draw together soil, water, trees, and people, among other elements. That expansive understanding provides the reference against which the violence of Chilean occupation is measured. It is also notoriously difficult to map.

That was the point that Rubén was trying to convey to me as we drove along the backroads in his community that day. Where I saw resources as bounded objects that could be located, he saw connections and histories better narrated than mapped cartographically. If the latter emphasized location and categorization, as cartography often requires, the former concerned itself with relations and experience. The distinction is more than epistemological. It is ontological, generative of relations that elude effort to understand them in terms of the specificity of a location (place) mapped with coordinates or the historical and geographical evolution of a territory through improved technical accuracy. It was a difference that I could apprehend, even if I could not fully understand it (de la Cadena 2015). What made that distinction problematic was the fact that in order to prove Mapuche land claims, we were both compelled to gloss over those differences in sight in order to locate them on a map legible to state officials, certifiers, and forestry companies, among others.

In Rubén's case, the immediate demand for a map stemmed from private forestry companies' efforts to certify the sustainability of their plantations from the Forest Stewardship Council (FSC). One of the criteria involved demonstrating that there were no standing claims to company lands from Mapuche communities. The evidence for those claims was everywhere visible for Mapuche people, running from place-names to observed changes in surface water flow and soil degradation, to visceral memories of forced dispossession in the mid-1970s. As apparent as all that was to Mapuche communities, it was a space that Chilean officials and plantation owners all but refused to see. And when they were able to see it, they insisted on it being on their terms—documented resource damage, land titles, cultural sites, and so on. Their formulation of resource indexed the way this landscape had been brought into being "by reason or by force" of Chilean occupation, to apply the national motto.

Toward that effect, Rubén and his colleagues had been compelled to find a way of mapping that remained legible to state officials and certifiers without sacrificing their ontological commitments to principles on which Wallmapu is founded. It was not an easy task. Eventually, the team had identified the *lof,* the basic socio-spatial category of Mapuche life, as their reference point for analysis (Hirt 2012). Loosely defined in the anthropological literature as a territory occupied by a group of related families, the category worked well enough in principle. To map it, however, required breaking it down into its elements, all of which emphasized culture (Melin Pehuen, Mansilla Quiñones, and Royo Letelier 2019). Rubén was thoroughly frustrated by the approach since it required turning the *lof* into a checklist of bounded objects to locate and map: cemeteries, cultural sites, villages, place-names, and so on. His frustration became a refrain over the course of our work together: "Territory is not a checklist."

Nonetheless, Rubén's team made the maps and submitted them to the FSC as a part of their report on Mapuche rights (Millamán and Hale 2016). The maps showed that nearly all of the land held up for certification was claimed by Mapuche communities, enumerating a continuous wave of violent dispossession stretching from the present back to a least in the mid-nineteenth century. Any discussion of improving resource management had to grapple with the history of dispossession that had produced the resources in question. But could improved resource management, whether done to meet FSC standards or through state recognition, ever establish the relationships necessary for a more spatially just society? Or did it simply reaffirm resources as something separate from society, grounded in nature, and an object of social struggle? Mapping could not solve that problem. It could only draw close to it.

The world-maker is here: public lands and cultural resources in northeastern California

Even if there is nothing inherently natural about resources, the temptation persists to see them in ontological terms. A *lof* simply can't be mapped without doing a great deal of violence to the concept itself, reducing social relations to bounded objects that can be located within a gridded space. Even so, unmappable concepts still shape mapping practices. Concepts of state territory, property, and resources shape mapping all the time. Even as maps help bring those concepts into being, each contains unmappable aspects. There's nothing obvious about looking at a map of a border that says do not cross this line under penalty of death. That information is supplied by the social relations through which maps circulate (Wood 2010). But is it possible to use maps to bring other, nondominant concepts into play?

That was the question that came up, time and again, over the course of a mapping project conducted with members of the Honey Lake Maidu tribe in northeastern California in the United States. For decades, Maidu across the region had been forced to contend with the category of "public lands" owned by the federal and state governments for the benefit of US citizens. As in many other parts of the United States, public lands are strewn with places important to Indigenous peoples. The famed California gold rush in 1849 triggered the displacement and destruction of Indigenous peoples across the region, including the Maidu. Rather than establishing a reservation to move Maidu people to, in this corner of California, the federal government opted for granting private property rights to Maidu in close proximity to where they lived (Manning 2018). By the time this happened in the 1920s, Maidu had already been forced out of places where they had previously lived, flooded out by dams and fenced out by cattle ranchers. Logging companies took care of the rest, "purchasing" land titled to individual Maidu families as part of federal efforts to assimilate Native Americans into US society. The process rendered many groups of Maidu homeless, including the Honey Lake Maidu (Tolley 2006). To this day, they are not recognized as a "Native American tribe" by the US government given that they have no established land base or formally recognized political authority. Instead, they are lumped into the "public." As members of the public, they have further been forced to maintain and protect themselves in part through using public lands.

Much like the public in general, the concept of public lands is structured by hierarchies of race and class. Federal and state "resource management" often prioritizes extractive activities over and against protection of "cultural" activities and resources. In the Maidu case, this often means that they have little to no recourse since their interests and concerns are merely cultural, as opposed to economic or scientific (Manning 2018). It is a problem that persistently shapes Maidu life, and one that was made particularly visible by proposed construction of a four-season resort in the middle of an area steeped in Maidu history. While the resort itself was slated for privately owned timber lands, its impacts stretched across a swath of public land managed by the US Forest Service. For decades, Maidu people had pushed the Forest Service to include protection for a number of key "cultural sites" in the area with important, if limited, success. The proposed resort renewed the threat, and Maidu people joined with other non-Indigenous peoples in the area to oppose the permit. In the Maidu case, their effort focused on demonstrating the threat posed by the resort to what US land management agencies legally refer to as "cultural resources." Toward that end, lawyers representing the Honey Lake Maidu requested help in producing a map of cultural resources that could be used to substantiate the tribe's claim.

Under US law, "cultural resources" encompasses a broad range of objects, sites, buildings, and other tangible remains of past human activity. Federal agencies like the National Park Service

445

have played an important role in further defining the category through including it within the scope of their resource management activities. For Native American tribes, the category has opened up a new avenue for protecting everything from graves and artifacts, to cultural sites and even landscapes. Tribes across the United States have since used this opening to establish greater recognition of their relationships to "resources" found off of reservation lands and managed by a patchwork of federal and state agencies. Much like Federal Indian Law in general, formal strategies for identifying and managing cultural resources have helped locate Native Americans spatially and politically within US society (Carpenter, Katyal, and Riley 2008; Coombe 2009). Despite that legal advance, the fit has been far from perfect. Part of this difficulty involves the reliance on archaeological and anthropological criteria for formally classifying resources. This is particularly true of cultural sites that are formally recognized in terms of property, with all the connotations of individual ownership and entitlement that entails (Harding 1997; Riley 2000). Maps are often used to gloss over these differences, focusing on the location, extent, and distribution of cultural resources in ways that can make the legal definition appear natural. Conversely, a critical approach to those categories shows how the terms of recognition reinforce the hierarchical social and spatial organization of US society (Tuck and Yang 2012; De Leeuw and Hunt 2018).

The terms of recognition present an immediate problem in mapping cultural resources. Some of those problems are practical. A "cultural site" is rarely, if ever, merely a specific bounded location that can be fenced off and protected like property. Instead, it is often a node in web of relations that extends across both time and space, positioned in relationship to ancestry, more-than-human entities, and the route taken to arrive there. Much like the Mapuche *lof*, none of this fits particularly well into the legal-cartographic matrix of Federal Indian Law. In the Honey Lake Maidu case, the matter was all the more complicated by lack of Federal recognition of their existence as a tribe. That difficulty gives way to a larger problem of considering how the very idea of cultural resources conceals its foundation in the theft of Indigenous lands.

The Honey Lake Maidu mapping project routed straight through the core of those difficulties. Recognition hinged on locating Maidu resources within the matrix of public and private land ownership, leaving the question of how ownership was established to history while focusing on resource management in the present. After a good deal of discussion, Honey Lake Maidu families decided that the best way to convey the potential impact of the resort on cultural sites was to locate them within a Maidu account of the creation of their world. Over the course of the next several months, the mapping effort focused on following the Creator's journey that brought the Maidu world into existence.[1] The narration located places that we were able to map, recording coordinates for important sites in the narration. Many of these sites further coincided with identifiable "resources," such as springs, lakes, and rock formations. Others we simply had to approximate due to the fact that they had been flooded by dams, razed by bulldozers, and buried beneath houses. All the same, the Creator's journey had a definite route and sequence that we followed the best we could, locating place-names and important sites *en route* to producing a map of "Kodom Yeponi: The World-Maker's Journey." This information was compiled with more typical resource data, such as hydrography, topography, and land cover. At the same time, we rotated everything to convey Maidu conventions of conceiving space in five directions, oriented by the transit of the sun from east (the bottom of the map) to west (the top). The Creator's path formed a spiral pattern wound tight around the area of the proposed resort, circling out under reservoirs and through the landscape before eventually trailing off to the east. The result was a view that was neither entirely Maidu nor easily understood by non-Indigenous peoples. Instead, it invited engaging in an incomplete or unfinished conversation about land and resources.

Like many Indigenous maps of this sort, the Honey Lake Maidu map hedged its challenge to cultural resources. Despite its tweaks to the orientation and content, it still had to be readable as a map in order to engage federal and state land managers alongside non-Indian residents of the region. With a little prodding, anybody familiar with the terrain could find themselves in it by looking for common points of reference such as mountains, rivers, and towns. With a little care and research, anybody could go out and retrace its route, encountering its twinned history of world-making and dispossession. By the time we were finished, funding for the proposed resort had evaporated in the wake of the 2008 financial crisis. The map itself, however, circulated among Federal land managers in the Forest Service, Bureau of Land Management, and National Park Service. Despite its divergence from a conventional map, all have used it to see new resources of particular importance to Maidu people with a subtle but important shift in management practices. Topping the list has been the restoration of Maidu access and control to a number of sites in the area mapped. The ontological difference between the World-Maker's journey and the public lands have generated sufficient friction to produce new kinds of resources even as those resources leave open the potential to expand state control. Hardly a solution to long-running histories of displacement and dispossession, the map opens the limited possibility of seeing resources differently. Of course, it also exposes the Maidu world to renewed threats from destruction, vandalism, and willful ignorance. One key spot in the World-Maker's path has been almost entirely consumed by a gravel mine. Old hierarchies are hard to shake, even if the possibility is in plain sight.

Territories beyond territory: mapping airspace in Oaxaca, Mexico

If the material practices and effects of mapping matter, so too do the qualities of the terrain that mapping negotiates and helps transform. No matter how hard it may be to map a cultural site, there are still physical locations to experience that work against map's claims to definitive representation. That gap often provides the impetus for making another map under the guise of conserving or refuting what was previously shown. Hence the idea of counter-mapping. The underlying risk is that so much mapping and remapping can serve to harden a notion of space as resource that is primarily understood and conceived of in terms of maps. Put differently, the centrality of maps as a means for understanding space too readily becomes a means of reinforcing a particular "Western" ontology (Turnbull 1989). If the Honey Lake Maidu case opens the door to considering the multiple ontologies that shape how maps are made, used, and read, that occurs because of the "friction" created by different understandings bumping against each other (Tsing 2005). A product of ontological difference, their combination is generative of new understandings that, in turn, have their own ontogenetic potential that embodies the history of its own coming into existence. Despite its apparent difference, the relationship that produces that new world remains structured by inequalities of knowledge whose colonial qualities are unmistakable. "Map your world in ours," they seem to say. What's missing is a sense of how even that question is itself a product of social relations that carry ontological assumptions about what the world is and how to be in it. What then happens when one starts from an altogether different space grasped independent of any maps?

Like many people, Zapotec communities in the Rincón of the Sierra Juárez have a complex way of thinking space and conveying spatial knowledge that functions perfectly well *without* maps. Their approach describes territory as everything, the sum of all the things, and their relationships that extend across space and time voiced in the Rincón variant of Zapotec as *tu xhën*. Often translated as territory, *tu xhën* is anathema to mapping as conventionally practiced (Cruz 2010). You cannot map the territory as a defined thing without severing some of the relationships that give it meaning in the first place. Accordingly, mapping fits into a long series of efforts to subtract or take away from the whole that is territory that begins with Spanish colonialism and carries on through the present. The list of subtractions along the way included the reforms carried out under Benito Juárez (the Sierra's most famous son) that turned their communities into small property holders in the 1850s, before being transformed again through agrarian reforms emphasizing collective ownership in the 1930s threatened by current efforts to dissolve communal lands once again under neoliberal reforms enacted in the 1990s. Each change has sought to reduce the totality of relations of which the community is part, both in the name of giving them rights to land and taking away rights to mineral, hydrologic, and forest resources. That succession has, in turn, shifted the grounds of community resistance, a point captured by recent mobilizations to defend territory from so-called megaprojects focused on hydropower and mining as well as mapping projects. With regard to the latter, these same communities in the Sierra Juárez faced down an effort by US-based geographers to map their territories with funds from the US Military (Cruz 2010; Wainwright 2013; Bryan and Wood 2015). In short, the territory that communities across the Sierra speak of as something to defend from state and private encroachment is not the same as the territory they conceive of on their own terms. Instead, that territory is a product of relations whose weave gives structure and form of *tu xhën*. There is a clear resonance with the Mapuche *lof* and the Maidu World-Maker, underscoring the imposition posed by thinking space in terms of maps.

One unique aspect of the Zapotec case involves how they have used their own concept of space to both ground their critique of maps and mapping and forge their own spatial categories. This last point has been developed in the context of efforts to expand and develop communications capable of linking together a community that is increasingly stretched out spatially. Migration to other parts of Mexico and the United States has been a major aspect of community life in the Sierra Juárez for some time now. Migration has not so much diminished the communal structure of life in the Sierra as reworked it (e.g., Mutersbaugh 2002; Robson et al. 2018). It is not at all uncommon for migrants to return home at the request of the community assembly to meet obligations, and even more common for migrants to send money to pay for town fiestas and collective works. As part of efforts to maintain those ties across new distances, communities across the Sierra (and Oaxaca in general) have developed new communication infrastructure including cooperative cell phone networks, communal Internet, and radios. Each is grounded in the importance of maintaining relationships that make up community life even as the distances are transformed (Stephen 2012).

In 2014, Mexican officials unilaterally intervened to regulate community radios in particular as part of a new law regulating telecommunication and radio. At the federal level, the law was intended to subject airspace to increased regulation as a step toward greater security of licensed, private operators (Cultural Survival 2018). Community radios all across Mexico were subsequently put on notice that they either had to secure federal licensing in order to continue operating or risk the consequences of operating illegally. Even prior to the law's approval, federal officials had begun cracking down on "illegal" community radios by destroying their equipment and jailing staff. While the new law offered a clear path toward legalization, it also ran directly counter to community sensibilities. What else was the "air" other than a part of the commons integral to the community? Opposition to the new law coalesced across Oaxaca around the sense of the radio being a part of the commons managed by each community. Air was a part of territory, a point that resonated with Zapotec speakers in particular for whom the verb "to speak," *didza*, literally translates as "to move air." Radio was an extension of this concept, stretching perceptions of territory as defined in communities' own terms (Cruz 2014; Sánchez 2016). Radio as part of territory was further connected to new efforts to confront the other privatizations affecting communities, including mining, hydropower, and wind energy concessions. Spreading the word over the airwaves was therefore a key element of a much more classic struggle for territory that involved land and resources.

This struggle is far from over. For the purposes of thinking about mapping and resources, it poses interesting questions. Radio space is not something that can easily be mapped. The soundwaves are blocked by mountainsides and storms, amplified by antennas and streaming services. At the same time, what knits this space together is neither the reach of the radio waves nor access to the Web. It has everything to do with listeners tuning in and responding to content (Fanon 1965). Newer technologies like WhatsApp and community Wi-Fi make this conversation increasingly interactive. In the ten radio stations in Oaxaca that I have visited since 2016, the broadcast booth is always a flurry of text messages from listeners, music, and commentaries broadcast by DJs. Interviews with radio staff often wind up adding my voice to the mix in programs where I am the one who is interviewed. Seen from a community perspective, all of these things are relations that together make up the commons as a "living space" (Cruz 2010). In this regard, they create a communal resource that is directly counterposed to state efforts to create resources by carving up space into privatized plots of land, resource concessions, jurisdictions, and licensed radios stations. Maps are but a means of carving up space, of a piece with efforts to enclose of a commons rather than define it. At the same time, communities across the Sierra Juárez have come to see maps as essential to understanding state and private efforts to constrain and catalog their use of resources, broadly construed.

Conclusion

The three projects discussed here present approaches to mapping that illuminate the relations through which resources are made. Their work of making, using, and reading maps further demonstrates just how restrictive cartography is, not to mention how intricately linked it is to dominant understandings of resources and, for that matter, the world. Each suggests a much more dynamic and fluid approach to mapping that contends with cartography's ideal of a well-ordered sphere. As much as it was guided by the demands of recognition, mapping the Mapuche *lof* shows just how limiting recognition can be without ignoring its tactical importance. Even when so much of what animates Mapuche struggles for land and resources cannot fit in the Cartesian space of maps, those concepts persist and shape approaches to mapping. The Honey Lake Maidu project attempts to bring more of those differences into the map itself, producing something that is recognizable as a map though largely unreadable without Maidu explanation. The third instance from the Sierra Juárez considers how new resources can be identified and formed beyond maps through experimentation with space and form.

All three projects present mapping as a means of moving beyond the dominant ideals that shape cartography as a means of documenting resources through identifying their location and distribution. None of these projects replaces the dominance of cartography. At the same time, each refuses to be limited by the dominance of that approach. Their objective is never just to locate their lands and resources within existing frameworks, the scope and reach of which are expanded through reform. Instead, each project is more properly attuned to what happens through mapping. Along the way, they poke holes in the tendency to think that the importance of space to ordering knowledge and experience reinforces the authority of maps. Resources are thus never immutable qualities of the world that they make, so much as they are expressions of a particular understanding of the world (see also, Lunstrum and Massé, Chapter 30 in this volume). As a method, mapping provides a means for doing critical resource geographies that challenges the claim that resources are objects to be observed, enclosing them within scientific and institutional frameworks that perpetuate the marginalization of entire peoples and societies.

Many of these insights are often associated with historically marginalized people and places. Mapping offers a practical means of engaging with how they are anything but, despite their apparent remoteness or isolation from "global" factors. The Mapuche and Maidu are connected by their shared displacement by gold rushes in the mid-nineteenth century as well as their shared use of indigeneity as an analytic for articulating their demands. Zapotec communities in the Sierra Juárez are similarly involved in processes of migration that transit Maidu and other Indigenous worlds. All of these factors shape their involvement with mapping. It is a politics in which many others can partake in so long as they are willing to organize their exclusion and displacement from the status quo as the basis for something different. As a method, these kinds of mapping belong to a "vigorous underground experimentalism" (Unger 1998, 4.) aimed squarely at changing the relationships that structure understandings of the world. That is to say they are all concerned with world-making, even as those worlds are entangled with others (Blaser 2014; Sundberg 2014). The ethics of mapping is nothing if not the practice of care for those relations and perspectives that might otherwise be easily excluded from mapping, opening up the space for the relationships that allow them to flourish (Todd 2016). Put differently, mapping is the difference between whatever the world is supposed to be and whatever else it becomes.[2]

Notes

1   The map can be viewed online at http://www.honeylakemaidu.org/photos/maiduFINALlores.pdf (Accessed February 27, 2018).
2   I am paraphrasing a line from a talk given by Fred Moten, "Recess and Nonsense: The end of the poetry world and the ends of the poet," given at Naropa University, Boulder, Colorado, on February 5, 2018. Moten's line is "poetics is the difference between whatever it is that you think you have to say and whatever it is that language becomes; or, poetics is the relation, and the difference, between content and form; or poetics thinks and enacts the differences that constitute the relation between content and form." I take responsibility for any differences of meaning and intent between Moten's words and what is written here.

References

Bauer, Kelly. 2016. "Land Versus Territory: Evaluating Indigenous Land Policy for the Mapuche in Chile." *Journal of Agrarian Change* 16 (4): 627–645. doi: 10.1111/joac.12103.

Blaser, Mario. 2014. "Ontology and Indigeneity: On the Political Ontology of Heterogeneous Assemblages." *Cultural Geographies* 21 (1): 49–58. doi: 10.1177/1474474012462534.

Blomley, Nicholas. 1998. "Landscapes of Property." *Law & Society Review* 32 (3): 567–612.

Braun, Bruce. 2002. *The Intemperate Rainforest: Nature, Culture, and Power on Canada's West Coast.* Minneapolis, MN: University of Minnesota Press.

Bridge, Gavin. 2009. "Material Worlds: Natural Resources, Resource Geography and the Material Economy." *Geography Compass* 3 (3): 1217–1244. doi: 10.1111/j.1749-8198.2009.00233.x.

Bryan, Joe. 2009. "Where Would We Be Without Them? Knowledge, Space and Power in Indigenous Politics." *Futures* 41 (1): 24–32.

Bryan, Joe, and Denis Wood. 2015. *Weaponizing Maps: Indigenous Peoples and Counterinsurgency in the Americas.* New York, NY: The Guilford Press.

Carpenter, Kristen A., Sonia K. Katyal, and Angela R. Riley. 2008. "In Defense of Property." *Yale Law Journal* 118 (6): 1022–1125.

Cayuqueo, Pedro. 2017. *Historia Secreta Mapuche*. Santiago de Chile: Editorial Catalonia.

Coombe, Rosemary J. 2009. "The Expanding Purview of Cultural Properties and Their Politics." *Annual Review of Law and Social Science* 5 (1): 393–412.

Coombes, Brad, Jay T. Johnson, and Richard Howitt. 2012. "Indigenous Geographies I: Mere Resource Conflicts? The Complexities in Indigenous Land and Environmental Claims." *Progress in Human Geography* 36 (6): 810–821.

Coulthard, Glen Sean. 2014. *Red Skin, White Masks: Rejecting the Colonial Politics of Recognition*. Minneapolis, MN: University of Minnesota Press.

Cruz, Kiado. 2014. "El Espectro Radioeléctrico, Un Bien Común Para La Telecomunicación Indígena." *El Topil* 21: 11–13.

Cruz, Melquiades. 2010. "A Living Space: The Relationship Between Land and Property in the Community." *Political Geography* 29 (8): 420–421.

Cultural Survival. 2018. *Situación de la Radiodifusión Indígena En México*. Cambridge, MA: Cultural Survival.

de la Cadena, Marisol. 2015. *Earth Beings: Ecologies of Practice Across Andean Worlds*. Durham, NC: Duke University Press.

De Leeuw, Sarah, and Sarah Hunt. 2018. "Unsettling Decolonizing Geographies." *Geography Compass* 12 (7): e12376. doi: 10.1111/gec3.12376.

Di Giminiani, Piergiorgio. 2018. *Sentient Lands: Indigeneity, Property, and Political Imagination in Neoliberal Chile*. Tucson, AZ: University of Arizona Press.

Edney, Matthew H. 2019. *Cartography: The Ideal and Its History*. Chicago, IL: University of Chicago Press.

Fanon, Frantz. 1965. "This Is the Voice of Algeria." In *A Dying Colonialism*. New York, NY: Monthly Review Press.

Harding, Sarah. 1997. "Justifying Repatriation of Native American Cultural Property." *Indiana Law Journal* 72: 723–774.

Hirt, Irène. 2012. "Mapping Dreams/Dreaming Maps: Bridging Indigenous and Western Geographical Knowledge." *Cartographica: The International Journal for Geographic Information and Geovisualization* 47 (2): 105–120.

Klubock, Thomas Miller. 2014. *La Frontera: Forests and Ecological Conflict in Chile's Frontier Territory*. Durham, NC: Duke University Press.

Mallon, Florencia E. 2005. *Courage Tastes of Blood*. Durham, NC: Duke University Press.

Manning, Beth Rose Middleton. 2018. *Upstream: Trust Lands and Power on The Feather River*. Tucson, AZ: University of Arizona Press.

Melin Pehuen, Miguel, Pablo Mansilla Quiñones, and Manuela Royo Letelier. 2019. *Cartografía Cultural Del Wallmapu: Elementos Para Descolonizar El Mapa Del Territorio Mapuche*. Santiago, Chile: LOM Ediciones.

Melin, Miguel, Pablo Mansilla, and Manuela Royo. 2017. *Mapu Chillkantukun Zugu: Descolonizando El Mapa Del Wallmapu, Construyendo Cartografía Cultural En Territorio Mapuche*. Temuco: Pu Lof Editores.

Millamán, Rosamel, and Charles Hale. 2016. Chile's Forestry Industry, FSC Certification and Mapuche Communities (Final Report). Temuco: Chile. Accessed January 22, 2020. https://ga2017.fsc. org/wp-content/uploads/2017/10/Chiles-Forestry-Industry-FSC-Certification-and-Mapuche-Communities-FINAL.pdf.

Mutersbaugh, Tad. 2002. "Migration, Common Property, and Communal Labor: Cultural Politics and Agency in a Mexican Village." *Political Geography* 21 (4): 473–494.

Pickles, John. 2004. *A History of Spaces; Cartographic Reason, Mapping and the Geo-Coded World*. London: Routledge.

Riley, Angela R. 2000. "Recovering Collectivity: Group Rights to Intellectual Property in Indigenous Communities." *Cardozo Arts and Entertainment Law Review* 18: 175–225.

Robson, James, Daniel Klooster, Holly Worthen, and Jorge Hernández-Díaz. 2018. "Migration and Agrarian Transformation in Indigenous Mexico." *Journal of Agrarian Change* 18 (2): 299–323.

Sánchez, Griselda. 2016. *Aire No Te Vendas: La Lucha Por El Territorio Desde Las Ondas*. Copenhagen: International Work Group on Indigenous Affairs.

Simpson, Leanne Betasamosake. 2017. *As We Have Always Done: Indigenous Freedom Through Radical Resistance*. Minneapolis, MN: University of Minnesota Press.

Stephen, Lynn. 2012. "Community and Indigenous Radio in Oaxaca: Testimony and Participatory Democracy." In *Radio Fields: The Anthropology of Radio in The Twenty-First Century*, edited by Danny Fisher, and Lucas Bessire, 124–142. New York, NY: New York University Press.

Sundberg, Juanita. 2014. "Decolonizing Posthumanist Geographies." *Cultural Geographies* 21 (1): 33–47.

Todd, Zoe. 2016. "An Indigenous Feminist's Take on the Ontological Turn: 'Ontology' Is Just Another Word for Colonialism." *Journal of Historical Sociology* 29 (1): 4–22.

Tolley, Sara-Larus. 2006. *Quest for Tribal Acknowledgment: California's Honey Lake Maidus*. Norman, OK: University of Oklahoma Press.

Tsing, Anna L. 2005. *Friction: An Ethnography of Global Connection*. Princeton, NJ: Princeton University Press.

Tuck, Eve, and K. Wayne Yang. 2012. "Decolonization Is Not a Metaphor." *Decolonization: Indigeneity, Education and Society* 1 (1): 1–40.

Turnbull, David. 1989. *Maps Are Territories: Science Is an Atlas*. Chicago, IL: University of Chicago Press.

Unger, Roberto Mangabeira. 1998. *Democracy Realized: The Progressive Alternative*. New York, NY: Verso.

Wainwright, Joel. 2013. *Geopiracy: Oaxaca, Militant Empiricism, and Geographical Thought*. New York, NY: Palgrave Pivot.

Wood, Denis. 2010. *Rethinking the Power of Maps*. New York, NY: Guilford Press.

Wood, Denis, and John Fels. 2008. *The Natures of Maps: Constructions of the Natural World*. Chicago, IL: The University of Chicago Press.

Index

Note: *Italicized* page numbers refer to figures, **bold** page numbers refer to tables

Printed in the United States
by Baker & Taylor Publisher Services

Printed in the United States
by Baker & Taylor Publisher Services